ABOUT THE AUTHORS

James W. Dally obtained a Bachelor of Science and a Master of Science degree, both in Mechanical Engineering from the Carnegie Institute of Technology. He earned a Doctoral degree in Mechanics from the Illinois Institute of Technology. He has taught at Cornell University, Illinois Institute of Technology, the U. S. Air Force Academy and has served as Dean of Engineering at the University of Rhode Island. Currently he is a Glenn L. Martin Institute Professor of Engineering at the University of Maryland, College Park.

Professor Dally has also held positions at the Mesta Machine Co., IIT Research Institute, and IBM, Federal Systems Division. He is a fellow of the American Society for Mechanical Engineers, Society for Experimental Mechanics, and the American Academy of Mechanics. He was appointed as an honorary member of the Society for Experimental Mechanics in 1983 and elected to the National Academy of Engineering in 1984. Professor Dally was selected by his peers in the College of Engineering to receive the Senior Faculty Outstanding Teaching Award. He was also a member of the University of Maryland team receiving the Boeing Outstanding Educator Award in 1996.

Dr. Dally has co-authored several other textbooks: *Experimental Stress Analysis, Photoelastic Coatings, Instrumentation for Engineering Measurements, Packaging of Electronic Systems, Production Engineering and Manufacturing, Design Analysis of Structural Elements and Introduction to Engineering Design, Books 1, 2, 3, 4, 5, 6 and 7.* He has written over two hundred scientific papers and holds five patents.

Gary A. Pertmer received his B.S. degree in Aerospace Engineering from Iowa State University and his M.S. and Ph.D. degrees in Nuclear Engineering from the University of Missouri – Columbia. As a faculty member at the University of Maryland, Dr. Pertmer has taught courses at all levels, including Introduction to Engineering Design, Reactor Systems Analysis, Reactor Engineering, Nuclear Technology Laboratory, and Mathematical Techniques for Engineering Analysis. In recognition of his teaching abilities, Dr. Pertmer received the Clark School of Engineering Senior Faculty Outstanding Teaching Award. Dr. Pertmer is currently Associate Dean for Education in the Clark School.

Dr. Pertmer previously worked for Stone and Webster Engineering Corporation where he was involved with a number of power plant design projects and also did plant start-up engineering at Beaver Valley Power Station in Pennsylvania. While at Maryland, Dr. Pertmer spent a sabbatical at the Karlsruhe (Germany) Nuclear Research Center doing research on passive heat removal systems. His research interests include reactor system thermal-hydraulics, computer code applications, and fluid mechanics. He has authored or co-authored more than 20 technical papers, and is a member of the American Nuclear Society, the American Society of Mechanical Engineers, and the American Society for Engineering Education.

PREFACE

This text is a new book in a series of eight dealing with Introduction to Engineering Design. Jim Dally, and other authors working with College House Enterprises, prepares a new or revised textbook each academic year. These textbooks are used to teach first year engineering students at the University of Maryland, College Park, at the University of Nevada, Reno and at Auburn University. It is encouraging that several other Colleges of Engineering are also adopting one or more of the textbooks in this series for their first year engineering classes.

PROCESS AND CONTENT

The procedure followed in offering a design experience to first year students is to:

- Teach the class in moderate size sections.
- Divide each section into product development teams with five or six members per team.
- Assign a project entailing the development of a product that will require a significant amount of student activity over the entire semester.
- All student teams develop the same product.
- In the product realization process, the students:
  - Design
  - Manufacture or procure parts for their prototype.
  - Assemble their system.
  - Test and evaluate the product or system.
- In developing the product or system, the students have an opportunity to learn:
  - Communication skills.
  - Team building skills.
  - Engineering graphics.
  - Software applications including CAD.
  - Design methods and procedures.

This textbook is used to support the students during a semester long project. Some of the material may be discussed in class or in a computer laboratory. Other material is covered in reading assignments. In other instances, the students use the text as a reference document for independent study. Exercises, provided at the end of each chapter, should be used for assignments when the demands of the project on the students' time are not excessive.

The book is organized into six parts to present various topics that the students need as they develop a product or a system. Part I introduces students to engineering and provides a student survival guide. Product and or system development is introduced in Part II. The project deals with the development of a model of an anti-icing system for highway bridges. Chapter 4 describes the system in considerable detail and outlines the methods to be used in testing the model, which the student teams develop. Chapters 5, 6 and 7 provide much of the background technical information that student design teams will need to develop their systems. Chapter 5 provides an introduction to basic electrical circuits. Chapter 6 provides introductory information on sensors with additional material on interfacing sensors with the RCX unit. Finally Chapter 7 describes programming in ROBOLAB, an icon based programming language based on

LabVIEW. Design of a system is treated as an opportunity to integrate a spectrum of knowledge about many topics. Experience shows that students serving on a design team developing a product are highly motivated. They learn much more on their own initiative than in a more typical lecture format. Hands-on participation in a design, building and testing a product significantly enhances the learning process.

Part II presents three chapters on engineering graphics. This coverage includes the use of tables and graphs in communicating engineering information, and a relatively complete treatment showing the student how to prepare three-view drawings. Pictorial drawings including isometric, oblique and perspective are also covered. The emphasis in Part II is on manual preparation of the graphics.

Part IV describes two different software programs useful to the first year engineering student. A spreadsheet program, Microsoft Excel and a computer graphics presentation program—Microsoft PowerPoint. The coverage is intended to provide the students with entry-level skills. Students are expected to perform calculations, plot curves and prepare several types of charts using spreadsheets. The description of the capabilities of PowerPoint is related to preparing design briefings for a design project.

Part V treats the very important topic of communication. A chapter on library research is included to aid the student in transitioning from a high school library to a much larger and more complex university library system. Another chapter on technical reports describes many aspects of technical writing. The most important lesson here is that a technical report is different than a paper for the History or English Departments. An effective professional report is written for a predefined audience with specific objectives. The technical writing process is described and many suggestions to facilitate composing, revision, editing and proofreading are given. In the chapter on design briefings, a distinction is drawn among speeches, presentations and group discussions. Emphasis is placed on the technical presentation and the importance of preparing excellent visual aids for a design briefing

Part VI contains four chapters dealing with engineering and society. A historical perspective on the role engineering played in developing civilization and on improving the lives of the masses. In this chapter, we move from the past into the present and indicate the current relationship between business, consumers and society. A chapter is devoted to introducing concepts related to product safety, reliability and performance. The chapter on ethics, character and engineering includes a large number of topics so the instructor can select from among them. A description of the Challenger and Columbia accidents are covered because they provide an excellent case history covering safety related conflicts between management and engineers. In the final chapter, the concepts of sustainable engineering are introduced. This is a relatively new subject that is beginning to receive attention in engineering colleges. The problem of achieving a balance between economic growth and maintaining the quality of the environment is emphasized. The important role technology will play as engineers design and produce products with manufacturability, remanufacturability and recycling as design objectives is described.

ABET AND EDUCATIONAL OUTCOMES

ABET's Criteria 2000 was considered in writing this book. It is believed that a first course in engineering design could include many of the educational outcomes listed below:

Communication Skills

1. Engineering Graphics
   - Understand the role of graphics in engineering design.
   - Understand orthographic projection in producing multi-view drawings.
   - Understand three-dimensional representation with pictorial drawings.
   - Understand dimensioning and section views.
   - Demonstrate capability of preparing drawings using both manual and computer methods.
   - Demonstrate understanding of engineering graphics by incorporating appropriate high-quality drawings in the design documentation.
2. Design Briefings:

- Each team member presents a design review for the class using appropriate visual aids.
- Each team member demonstrates briefing skills.
3. Design Reports:
   - The team's design is documented in a professional report incorporating time schedules, costs, parts list, drawings and an engineering analysis.

Team Experience

- Develop an awareness of the challenges arising in teamwork.
- Demonstrate teamwork in the product realization process through a systematic design concept selection process involving participation of all team members.
- Demonstrate planning from conceptualization to the evaluation of the prototype.
- Understand and demonstrate sharing responsibility among team members.
- Demonstrate teamwork in preparing design reports and presenting design briefings.

Software Applications

- Demonstrate entry-level skills in using spreadsheets for calculations and data analysis.
- Show a capability to prepare graphs and charts with a spreadsheet.
- Show a capability to prepare professional quality visual aids.
- Demonstrate these computer skills in preparing appropriate materials for design briefings and design reports.

Design Project:

- The design project is the overarching theme of the course.
- Utilize all the skills listed above to assist in the product development process.
- Demonstrate competence in defining design objectives.
- Generate design concepts that meet the design objectives.
- Understand the basis for design for manufacturing, assembly and maintenance.
- Manage the team and the project effectively.

ACKNOWLEDGEMENTS

Acknowledgments are always necessary in preparing a textbook because so many people are involved in many helpful ways. First, it is important to recognize the contributions of my colleagues at the University of Maryland at College Park. Many of them have made suggestions for improving the content in this book over the past five years. In particular, thanks are due to my coauthor Dr. Gary A. Pertmer for his commitment in providing the leadership necessary to operate a multi-section offering of this course to about 800 students per year and in providing support for the many instructors involved each semester. Thanks are also due to Dr. Jesse Adams at the University of Reno for his many suggestions in recent years and to Dr. David Lovell for his work in initially introducing the RCX to the ENES 100 students.

In preparing this revision, thanks are in order for three educators who made contributions by authoring portions of this book. Dr. Sheryl Ehrman, from the University of Maryland, authored the interesting and timely case study on the Bridgestone/Firestone tire recall. Margaret M. Cunningham, University of Maryland Libraries, authored the chapter on Library Research Skills. Dr. Stanley J. Reeves, from Auburn University, coauthored the chapter on Basic Electric Circuits. The authors appreciate their willingness to participate and their valuable contributions.

Jim Dally
Gary Pertmer
University of Maryland
College Park

ABOUT THE AUTHORS
PREFACE

CONTENTS

PART I ENGINEERING AND SUCCESS SKILLS

CHAPTER 1 THE ENGINEERING PROFESSION

CHAPTER 2 A STUDENT SURVIVAL GUIDE

PART II PRODUCT AND SYSTEM DEVELOPMENT PROCESSES

CHAPTER 3 DESIGN AND PRODUCT DEVELOPMENT

CHAPTER 4 DESIGNING ANTI-ICING SYSTEMS FOR HIGHWAY BRIDGES

CHAPTER 5 BASIC ELECTRIC CIRCUITS

CHAPTER 6 SENSORS

CHAPTER 7 PROGRAMMING IN ROBOLAB

PART III ENGINEERING GRAPHICS

CHAPTER 8 TABLES AND GRAPHS

CHAPTER 9 THREE-VIEW DRAWINGS

CHAPTER 10 PICTORIAL DRAWING

PART IV SOFTWARE APPLICATIONS

CHAPTER 11 MICROSOFT EXCEL

PART V COMMUNICATION

PART VI ENGINEERING AND SOCIETY

CHAPTER 17 SAFETY, RISK AND PERFORMANCE

CHAPTER 18 ETHICS, CHARACTER AND ENGINEERING
By James W. Dally and Sheryl H. Ehrman

CHAPTER 19 SUSTAINABLE ENGINEERING

APPENDICES

APPENDIX A DEVELOPMENT TEAMS

PART I

ENGINEERING AND SUCCESS SKILLS

CHAPTER 1

THE ENGINEERING PROFESSION

1.1 WHY ENGINEERING IS A PROFESSION

You may consider engineering from two different viewpoints. First, as a course of study pursued in an accredited college of engineering and second, as an occupation. Let's discuss these two different viewpoints because they are both important in establishing engineering as a profession.

As a student in an engineering college, you are presented with a very structured curriculum. You are required to take many credit hours of mathematics, chemistry, physics and hopefully biology in your first two years of study. Mixed with the courses mathematics and science are several engineering science courses. These are essentially applied science courses taught by engineering faculty members. In the last two years of the curriculum, many discipline oriented engineering courses are required. These courses provide the basic analytical methods so necessary for success after graduation when you begin to practice. Also included in the final two years are engineering courses intended to provide realistic design experiences related to product development. This schedule of courses is designed to prepare you to practice engineering upon graduation. Near the conclusion of your studies, you will be presented with an opportunity to take the first of two examinations leading to a professional license to practice engineering—the Engineering Fundamental Examination. It is a good idea to take this examination before graduation because it provides you with an opportunity to test your analytical skills. It is also an attribute that you can add to your resume. Passing the examination is a certification of your fundamental analytical skills.

Your ability to practice engineering immediately upon graduation is the primary reason for such a highly structured curriculum containing so many technical courses. The on-the-job training opportunities available to you in your first position depend upon the policies of the company that is paying your salary. Some large firms provide a transition period where you serve as an engineer trainee with close supervision and rotate from one division to another to provide an overview of the company's operations. However, if you take a position with a smaller corporation, usually there is little formal training, and you are expected to earn your salary beginning with the first day on the job.

In addition to at least three years of mathematics, science, engineering science and engineering courses, you will be required to take several courses in the other colleges on campus. The university has general education requirements, which dictate that all students become familiar, if not proficient, with basic premises in the social sciences, fine arts, and the humanities. The engineering profession endorses this out-of college exposure because it broadens one's perspective and encourages important assessments of contemporary issues.

Let's now consider the second viewpoint—engineering assignments undertaken after graduation. Whether you will be working as an engineering professional depends to a large degree upon the position you accept. Some engineering graduates are offered positions with investment banking and brokerage houses at very attractive salaries. Managers at brokerage houses like the analytical skills developed in engineering programs and find that engineering graduates perform well in assessing investment opportunities. These graduates pursue an interesting and lucrative career, but they are not practicing engineering.

At the other extreme, a graduate of a civil engineering program takes a position with a construction firm that designs and builds highways, bridges, and large public buildings. In this position, public safety is an important consideration and licensing is essential, because State laws require it. The graduate is expected to pass the Engineering Fundamental examination, and then— after several years to gain practical experience—he or she will take the Principles and Practice of Engineering examination to become licensed. During the period between the two examinations, the individual works with licensed engineers who supervise his or her work and certifies it to be accurate and correct.

For those graduates working in industry, licensing is usually less of an issue. Most corporations have a small staff of licensed engineers who certify work when it is required. However, most of the engineers are working on products in which public safety is not an issue. In these cases, professionalism is not a matter of licensing. Professionalism is more a matter of attitude. Professionals are usually salaried and their remuneration is independent of the hours worked. They do not punch a time clock and they are expected to devote the time necessary to finish a task on schedule. Engineers assume responsibility and are committed to the project and to the development team. They cooperate with management, and freely sharing ideas and concepts to advance the welfare of the corporation and their division. They may decide to move from one corporation to another, but loyalty and respect are important considerations even after they leave.

1.2 CHARACTERISTICS OF ENGINEERING STUDENTS

Students entering colleges of engineering today are bright, with average Scholastic Achievement Test (SAT) scores ranging from 1200 to 1300 in most colleges in the U. S. In fact, the higher ranked engineering colleges attract students with average scores of 1300 or above. In addition, engineering students are usually in the top 5 or 10% of their high school graduating class with grade point averages ranging from 3.8 to 4.0. Student applications to the admissions office of the university often include strong letters of recommendation from their teachers, counselors and principals certifying the student's stellar abilities.

It is well recognized that the abilities of entering engineering students in mathematics is exceptional. SAT scores for the mathematics portion of the examination approach or exceed 700 for most students. What is less well recognized is that engineering students also have excellent verbal skills. With average SAT scores for verbal skills of about 600, engineering students usually rank above their peers in the colleges of arts, humanities and social sciences.

Most engineering students are men; however, 21% of the bachelor degrees in engineering were awarded to women in 1999[1]. The interest of women in engineering depends on the discipline as shown in the bar chart in Fig. 1.1. Statistics show that women are attracted to chemical, biomedical and industrial engineering disciplines, but not to mining and petroleum programs.

The geographic representation in the class depends strongly on whether the university is state supported or privately funded. State universities often restrict enrollment of out-of-state students to a small proportion of the class. A tuition differential is also imposed making it more expensive for out-of-state students to attend. Consequently, undergraduate classes in state universities tend to be much more homogeneous with students representing the backgrounds and cultures of the region. Privately funded universities usually do not have such restrictions and often vigorously seek students with diverse backgrounds from different states and countries.

Clearly, engineering students are very well prepared academically to begin their studies. They have proven themselves in high school as scholars and class leaders. They have gained the respect of

---

[1] ASEE Prism, Volume 10, No. 1, September 2000, p. 14.

their teachers, peers and family. However, the retention rate for engineering students is appalling. Probably one half or more of the entering class of well-qualified students either fail-out or withdraw from the program. Most of these students leave in the first two years. While the exact retention rate varies from college to college across the U. S., the problem is widespread. Why then do so many students leave the engineering program without completing the requirements for a B. S. degree in one of the engineering disciplines? Answers to this question will be explored in Chapter 2.

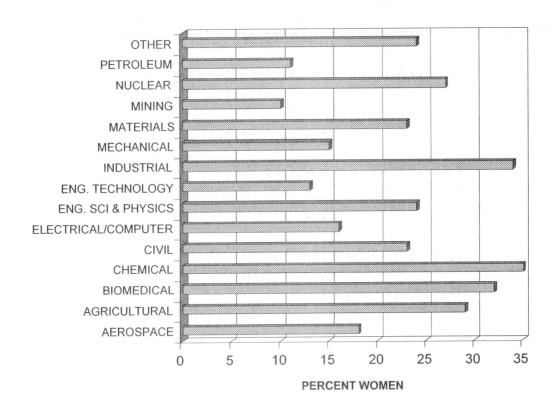

Fig. 1.1 Percent of bachelor degrees awarded to women by discipline.

1.3 SKILL DEVELOPMENT

Pursuing a successful career in engineering is strongly dependent on your ability to develop a number of important skills. Of course, you enter the program with many skills, but it is necessary to improve them and to add new ones to your repertoire. The engineering curriculum is designed to develop additional skills and as you proceed through the program you will find that your ability to perform challenging engineering tasks expands each year.

The Accrediting Board for Engineering and Technology[2] (ABET) has developed a list of requirements for all of the disciplines offered in the accredited colleges of engineering in the U. S. Each department in the college must demonstrate to a team of ABET visitors that their graduates have the skills and knowledge listed in Table 1.1 to maintain the accreditation of each engineering program for which a B. S. degree is offered.

---

[2] The Accreditation Board for Engineering and Technology, Inc., 111 Market Place, Suite 1050, Baltimore, MD 21202, certifies all of the programs (not departments) leading to accredited engineering degrees in the U. S.

Table 1.1
Listing of knowledge and skills that engineering graduates must demonstrate.
ABET Criteria 2000

(a) Ability to apply knowledge of math, engineering, and science
(b1) Ability to design and conduct experiments
(b2) Ability to analyze and interpret data
(c) Ability to design a system, component or process to meet needs
(d) Ability to function on multi-disciplinary teams
(e) Ability to identify, formulate, and solve engineering problems
(f) Understanding of professional and ethical responsibility
(g) Ability to communicate effectively
(h) Broad education necessary to understand the impact of engineering solutions in a global and
   societal context
(i) Recognition of the need for, and the ability to engage in life-long learning
(j) Knowledge of contemporary issues
(k) Ability to use techniques, skills, and tools necessary for engineering practice

Before examining this list, consider a somewhat different set of skills and abilities for engineering graduates prepared by a study committee for the American Society of Mechanical Engineers (ASME). The ASME listing [1] in priority order is presented in Table 1.2.

Table 1.2
Skills considered important for new Mechanical Engineers
With Bachelor Degrees, Priority ranking

1. Teams and Teamwork	11. Sketching and Drawing
2. Communication	12. Design for Cost
3. Design for Manufacture	13. Application of Statistics
4. CAD Systems	14. Reliability
5. Professional Ethics	15. Geometric Tolerancing
6. Creative Thinking	16. Value engineering
7. Design for Performance	17. Design Reviews
8. Design for Reliability	18. Manufacturing Processes
9. Design for Safety	19. Systems Perspective
10. Concurrent engineering	20. Design for Assembly

As you explore both of these lists, the emphasis on skill development in colleges of engineering should become evident. The importance of these skills, as determined by two major engineering societies, should guide your efforts in achieving the proficiencies necessary to insure a successful career.

Design

Most engineering graduates join corporations and participate in some way related to either product or process development. If it is a manufacturing company, the product may be a vacuum cleaner, a lawn mower, an automobile, or some other item. The point is you will design, manufacture, assemble, ship and service an array of products. Design is essential to the success of product development. If the design is flawed, the product or service offered will fail, and the company and your career may suffer. If the design is optimal, the product will be profitable, the company will prosper, and your career will advance. This is a simple viewpoint, but one that has merit.

Design is perhaps the most difficult of the engineering skills to acquire because it is more an art than a science. While there are well-defined design procedures that you will learn (some of them introduced later in this book), one often intuitively senses the good and bad elements of a design. Also, experience is important because design is a complex activity that must incorporate many different aspects that cross discipline lines. Examine Table 1.2 and note the various descriptors ASME used with the word design. For example, engineers design for reliability, for performance, for safety, for cost, for manufacture, and for assembly and one could add "ease of maintenance" and others to the list.

As you proceed through the engineering program, the opportunities you have to practice design will be severely limited. The emphasis in engineering education is on developing analysis methods used to predict reliability and performance of products and systems. Some programs are limited to a single capstone design experience in your final year of study. Other programs have two or three design courses where you will have the opportunity to design a new product or to modify the design of an existing product or process. Engineering programs with more than three design courses (a total of nine credit hours in a program of about 128 credit hours[3]) are rare in the U. S.

Many colleges of engineering participate in design competitions that are sponsored by the professional societies. These competitions involve the design of robots, large model airplanes, automobiles, steel bridges, and human and engine powered vehicles, etc. You are encouraged to become involved in one or more of these projects. The projects give you an opportunity to participate on an interdisciplinary design team and to gain valuable experience in the design process. Usually you can obtain credit towards your gradation requirements for this design competition by registering for a technical elective that has been established by the faculty member in charge of the competitive design projects.

Another method for gaining design experience is by working off campus. Many students work to partially fund their expenses while pursuing their education. If you are able to find a position that involves design, manufacturing or assembly, the experience gained is probably worth more than the salary you earn. Try to obtain a position in a company that produces products rather than with a retail organization. While experience in either type of organization is valuable, the production experience is more applicable to engineering.

Teamwork

In this course, you probably will be required to participate on a development team. There are three reasons for this requirement. First, the project is often too ambitious for an individual to complete in the time available. You need the collective efforts of the entire team to develop a product during the semester. The development teams will be pressed time-wise to complete the project on schedule.

Second, you must begin to learn teamwork skills. Experience has shown educators that most students entering the engineering program are deficient in team skills. From elementary through high school, the educational process has focused on teaching you to work as an individual, often in a setting where you competed against others in your class. However, you should begin functioning as a team member where cooperation, following, and listening are as important as individual effort. Leadership is important in a team setting, but cooperation and following the lead of others are also critical elements for successful team performance.

The final reason is to better prepare you for the real world you will enter upon graduation with a B. S. degree. You will probably be assigned to a development team very early in your career if you take a position in industry. A recent study by ASME, the results of which are shown in Table 1.2, ranked teamwork as the most important skill to develop in an engineering program. Teamwork was also the first skill, in a list of 20, considered important by managers from industry. Hopefully, this

---

[3] The number of credit hours required varies from program to program and college to college. The 128 credit hours cited here is typical.

course will be instrumental in exposing you to team working skills so necessary for a successful career.

A complete discussion on teamwork is presented later in Appendix A. You will be introduced to many concepts that will facilitate your ability to work harmoniously and effectively as a well-regarded team member.

Communication

Communication skills are vitally important in every aspect of your life. On the personal side you must be able to accurately convey your thoughts to your family, friends and peers. Professionally, it is just as important to communicate effectively. What value is a great idea if you cannot express it with sufficient clarity for it to be accepted by management or your associates? With family and friends, most of your communication is by conversation (informal). Occasionally, you write letters or more frequently e-mail messages. You can be casual in communicating with friends and family because they are accommodating and overlook your shortcomings. You must be much more careful in your professional communications. Messages must be clear and unambiguous. The information conveyed must be accurate and timely. The presentation must be a concise and to the point; long rambling prose is not appreciated.

Three different modes of communication are used in engineering—writing, speaking and graphics. All three are important, and engineering versions of all three modes different to some degree from commonly accepted methods.

It is important that you enjoy writing, because engineers often have to prepare several hundred pages of reports, theoretical analyses, memos, technical briefs, and letters during a typical year on the job. Communication, particularly good writing, is extremely important. Advancement in your career will depend on your ability to write well. You will be taking several courses offered by the English Department and by departments in Social Sciences, Arts and the Humanities, which will require writing assignments. These courses will help you with the structure of your composition and the development of good writing skills. Most of the assignments will be to write essays, term papers, or to study selected works of literature. However, there are several differences between writing for an engineering company and writing to satisfy the requirements of courses such as English 101 or History 102. These differences will be described in a Chapter 14 titled **Technical Reports**.

The design briefing is important to both the product development process and to your career. Information about the product must be effectively transmitted to your peers, management, and other people involved with the project. Clear messages that accurately define problems, which the development team will address, are imperative. On the other hand, ambiguous messages are often misunderstood, hinder the definition of the problem, and lead to delays in implementing solutions. The design briefing provides an opportunity to review the status of a specific product development process. It also permits peers to share their ideas with you, and it affords management an open forum for assessing the quality of your work and the progress made by your development team. Because the professional presentation is critically important, Chapter 15 titled **Design Briefings** has been included in this book. This chapter describes some valuable techniques for properly delivering your message to different audiences—strangers, peers, team members and management.

Engineering graphics is the most important method of communication for presenting design concepts and details. When attempting to communicate design ideas, you will find writing and speaking insufficient to express your thoughts. A more visual technique for communicate is required. It is much more effective to present your ideas by means of drawings, sketches, pictures, and graphs of many different types. Visuals aids, such as drawings and photographs, convey your ideas quickly and with remarkable accuracy.

There are two general approaches used in preparing drawings and graphs. The first is manual where drawings and graphs are prepared by hand using a few simple drawing instruments. The second utilizes a computer and suitable software programs that greatly facilitate the preparation of drawings or graphs. In this book, three chapters covering manual methods for preparing drawings and graphs are provided. In Chapter 11, we describe EXCEL—a spreadsheet program used for computation and for preparing several different types graphs and charts. Finally, a graphics presentation program, PowerPoint, employed to prepare slides and overhead transparencies for oral presentations, is explained in Chapter 12. Other textbooks are referenced that discuss computer-aided design (CAD) software programs used in preparing both two and three-dimensional engineering drawings.

Design Analysis

In engineering, you will attempt to predict the performance of a product prior to its final design and construction. Obviously, you would not want to design and construct a bridge to have it fail after a few months or years in service. You must be able to predict with confidence that a bridge will not fail over its entire design life (perhaps a hundred years or more). Constructing analytical models and performing analysis is key to making an accurate prediction regarding performance. In modeling, you reduce your structure into a number of different components or subsystems that are amenable to analysis. You then apply analytical methods that enable you to accurately determine performance parameters for each model. The analytical results permit verification of the adequacy of the design. As such, analysis and design are coupled. You usually begin with a design concept; then you subject the design to an analysis. The results of the analysis are then used to improve the design enhancing its performance.

A significant portion of the engineering curriculum is devoted to developing your analytical skills. About one year, of the four-year curriculum, is devoted to courses in mathematics and science, which provide the foundation for engineering courses. In most colleges of engineering, at least another year is devoted to engineering and engineering science courses where analytical and computational methods are taught. In a typical engineering program, your analytical and computational skills will be honed more than any other skill that is listed in either Table 1.1 and 1.2.

Your performance (read this as grade point average, GPA) will depend on your ability to solve problems posed on hourly examinations or on final examinations. These problems are crafted to test your understanding of a few engineering principles[4]. The mathematics involved is usually not overly complex and the physics entailed is even more straightforward. You will be provided with several useful tips on problem solving methods in Chapter 2.

Experimental Analysis

In predicting the performance of a structure, or machine component, you will often conduct experiments and measure the performance parameters. While it is usually more expensive to perform experimental studies than analytical ones, the costs are warranted when you are not certain that the analytical modeling accurately reflects reality. Sometimes converting the physical reality of a product to an analytical model requires many assumptions. You will often perform carefully designed experiments to verify these assumptions and to insure the successful performance of the product. Extensive tests on almost all products are performed before they are released to the market. The purpose of these tests is to insure safety of the product, and to give management assurance that the product will meet the guarantees specified in the warranty pertaining to performance, life and safety. In this class, you will be required to test a prototype of a product that your team has developed. The test is to determine if the product's performance meets its specification.

---

[4] In practicing engineering, you will find that analyses are based on only a dozen or so engineering principles. The methods developed in engineering courses may appear complex, but they are based on this very limited number of fundamental concepts.

There are four different phases of an experimental program. The first phase is to design the experiment so that it provides all of the answers required to verify the modeling or to insure the adequacy of a design. The second phase is to conduct the experiment using appropriate instrumentation, which will provide accurate and appropriately timed measurements of the controlling performance parameters. The third phase is to analyze the data obtained using suitable statistical methods that predict and bound experimental errors. The final phase is to correctly interpret the data to verify the adequacy of a design or to provide information that may be used to enhance a design.

The laboratory classes offered in conjunction with chemistry and physics courses are not designed to provide you with skills for designing experiments or for making measurements. These are highly structured laboratory experiences intended to reinforce theories and principles introduced during lecture classes. The curriculum of many of the engineering disciplines provides a course on circuits and instrumentation. This course introduces linear circuit theory and gives some information about sensors, transducers and both analog and digital instrumentation. A course of this type is strongly recommended because it provides some of the basic information needed to plan and design experiments.

You are encouraged to enroll in other courses that provide additional exposure to experimental methods. This may be possible with a careful selection of technical electives that are available in the third and fourth year of your program. You might also consider an assignment working in a laboratory directed by one of the professors in your department. Many professors lead significant research programs, and in conducting these studies, they often perform new and novel experiments. The experience gained in such a setting is often more valuable than that obtained in a traditional engineering measurements course.

Understanding Professional and Ethical Responsibilities

When you graduate and begin practicing engineering, you will be expected to follow a professional code of ethics. An example of the ABET code of ethics is presented in Chapter 18 titled Ethics, Character and Engineering. While there are many other codes, each endorsed by a professional society, they all are similar. Basically the codes ask you to uphold and advance the integrity, honor and dignity of the engineering profession. The codes then list fundamental principles and canons that should govern professional behavior. The fundamental principles advanced by ABET are:

1. **To be selective in the use of our knowledge and skills so as to ensure that our work is of benefit to society.**
2. **To be honest, impartial and serve our constituents with fidelity.**
3. **To work hard to improve the profession.**
4. **To support the professional organizations in our engineering discipline.**

Personal behavior out-of-class or away from the work place is also important. The results of several polls indicate that Americans are less ethical today than in previous generations. Sixty-four percent of individuals in a sample of 5000 admit to lying if it does not cause real damage. An even larger percentage of those sampled (74%) will steal providing the person or business that is being ripped off does not miss the item pilfered [3]. Many students (75% high school and 50% college) admit to cheating on an important exam [4]. Many students have not developed a moral code to use as a guide for their behavior. Relatively minor transgressions of a generation ago have been replaced with more serious problems involving mass murder, drug and alcohol abuse, pregnancy, suicide, rape, robbery and assault. Some of the reasons for these changes are explored later in Part VI of this book.

Ethical behavior is important in both your professional and personal lives. Recall the golden rule—do unto others, as you would have them do unto you—to guide your behavior. It is a very simple rule, and it is effective.

Committing to the Need for Life-Long Learning

Most engineering programs require successful completion of about 128 credit hours for graduation with a B. S. degree. This number is higher by about eight credit hours than the requirements for any other college on campus. Engineering educators have always required an extra effort measured in credit hours for their students prior to graduation. This belief is based on the concept that the student should be capable of practicing engineering immediately after graduation.

During the past twenty years the number of credit hours required for graduation has decreased by about 10%, while the amount of important material necessary to practice has increased significantly. Moreover, the scope of the fields in all of the engineering disciplines continues to evolve and to expand. Engineering educators recognize that they cannot increase the number of credit hours to accommodate the evolution and expansion of the knowledge base for each discipline. Instead, it becomes essential that all engineering graduates recognize the need to continue their studies after graduation.

Upon graduation many of you will take a position in industry and be challenged with a new environment and new assignments. You will be able to measure your strengths and weaknesses against this new setting. Perhaps you will conclude that you need advanced courses in your discipline, or basic courses in some other engineering discipline. With more time in practice, you will probably be given an opportunity to manage a small engineering group and become a first level engineering manager. In this case you may decide to enroll in a few courses in the business school or to pursue a masters program in business administration. No one really understands with certainty what knowledge will be important in the future[5]. The idea is to stay flexible and be prepared to devote time each week both to study and professional development.

You are fortunate that many opportunities exist for continuing education. Some companies bring instructors into its facilities to focus on topics of immediate concern to the organization. Professional societies offer short courses on timely topics in their discipline. Well-known universities offer short courses each summer with outstanding professors teaching in their specialty. Local universities offer a wide array of course on a regular basis that lead to M. S. and Ph. D. degrees in a number of engineering disciplines. Professional societies offer many well-written books that cover advances within their discipline. Distance learning opportunities abound, and this mode of education is growing rapidly.

It is essential that you recognize graduation with a B. S. degree as the beginning of your studies and not the end.

1.4 THE ENGINEERING DISCIPLINES

The U. S. Military Academy at West Point, established by an act of Congress in 1802, was the first college of engineering in the America. Military engineering was the first discipline, but in the 1800s several other programs were introduced, and the field of engineering began its division into many different disciplines each with its special emphasis [5]. The number of different disciplines grew slowly in the 1800s and early 1900s and included:

1. Civil engineering
2. Mechanical engineering
3. Electrical engineering
4. Chemical engineering
5. Industrial engineering

---

[5] If someone had told me upon graduation with a B. S. degree that I would spend most of my career in research and education, I would have laughed. At that time, I fully intended to design heavy-duty steel mill equipment.

In more recent years the number of engineering disciplines has continued to grow with more and more technical specialties converted into degree programs. Recently the American Society of Engineering Education (ASEE) [6] published the undergraduate enrollment by engineering discipline during the academic year 2002-2003. These results are presented in Fig. 1.2.

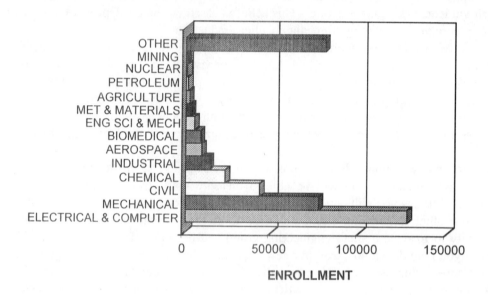

Fig. 1.2 B. S. degrees awarded by engineering discipline (2002-2003 academic year).

The total number of B. S. degrees awarded in engineering in 2002 was slightly more than 67,301. The largest number of degrees was awarded to graduates from electrical and computer engineering programs; however, in recent years many departments of electrical engineering have divided their program and they are now offering two different degrees—an electrical engineering degree and a computer engineering degree. Experience shows students tending to favor the computer-engineering program when they enroll. In addition to the B. S. degrees in engineering, 31,346 M. S. and 5,802 Ph.D. degrees were also awarded in engineering in 2002.

Mechanical and civil engineering programs also have large enrollments. The curriculum in both of these programs is relatively stable in comparison with electrical and computer engineering. However, some civil engineering departments are establishing a separate degree program in environmental engineering, and this separation necessitates significant changes in the curriculum.

A large number of engineering programs are classified as other in Fig. 1.2. This category contains many programs that are offered by only a few accredited colleges of engineering in the U. S. and include:

- Ocean engineering
- Marine engineering
- Fire protection and safety engineering
- Ceramic engineering
- Engineering management
- Systems engineering
- Geological engineering

The enrollment in each discipline reflects perceived demand for the graduates and, to a modest degree, the starting salaries for graduates from each program. Average starting salaries for several of the engineering disciplines with B. S. degrees are listed in Table 1.3.

Table 1.3
Starting salaries of B. S. graduates in Engineering
Data are from the 2003 Salary Surveys National Association of Colleges and Employers

Discipline	Annual Salary
Aerospace Engineering	$48,826
Biomedical Engineering	$51,167
Chemical Engineering	$52,169
Civil Engineering	$41,067
Computer Engineering	$52,665
Electrical Engineering	$50,566
Environmental Engineering	$44,220
Industrial engineering	$47,442
Materials Engineering	$46,276
Mechanical Engineering	$48,659

The difference in starting salaries for engineers in different disciplines is not large. These differences vary somewhat from year to year with the chemical engineers and computer engineers usually leading the other disciplines by 5 to 10% and the civil and environmental engineers usually trailing the other disciplines by 5 to 10%. Starting salaries for engineering graduates are significantly greater than comparable salaries for other majors. For example, offers to business administration majors averaged $37,368 and marketing majors averaged $36,071. Average offers to accounting majors are better at $42,045. Liberal arts majors usually do poorly in starting salaries and in the winter of 2003-2004 their average offer was $30,153. The only exception to this trend was with computer science where strong demand for graduates from this program in the late 1990s produced salaries approaching those offered to starting mechanical engineering graduates. However, since 2001, graduates from computer science and information science and systems have lost substantial ground in starting salaries.

Median income with discipline or the branch of engineering differs considerably from the starting salaries as shown in Table 1.4. The median income is higher than the starting salaries because the engineers' compensation increases time on the job and the experience gained. The median salaries depend on the branch of engineering with higher salaries commanded by those working in aerospace and materials engineering. Those working in civil and general engineering tend to earn lower salaries. However, after about 20 years of service salaries tend to level out for most engineers with small incremental gains from year to year in the later stage of their career.

Table 1.4
Median Annual Salaries for Engineers by Discipline
From Engineers' Salaries: 2003 Special Industry Report,
Engineering Workforce Commission of the
American Association of Engineering Societies

Discipline	Number of Years after B. S.				
	0	5	9-10	13-16	21-25
Aerospace Engineer	—	$65,757	$75,540	$79,923	$89,143
Materials Engineer	—	—	$72,132	$83,865	$96,921
Computer Engineer	—	—	$77,656	$79,225	$84,268
Civil Engineer	$43,800	$56,200	$64,436	$67,665	$74,547
General Engineer	—	$76,160	$82,056	$80,096	$81,176
Mechanical Engineer	$51,323	$63,098	$73,146	$80,955	$88,746

Electrical Engineering

The **electrical engineering** program provides the basic background needed to work in broad fields of electronics, communications, automatic control, computers, materials processing, electromagnetic, signal processing, and power utilization. The program places emphasis on the fundamentals of mathematics and science leading to careers in research, development, design, or operation in such diversified areas as electronics, control systems, computers, communications systems and equipment, biomedical instrumentation, radar and navigation, power generation and distribution, consumer electronics, and industrial devices.

The Institute of Electrical and Electronics Engineers (IEEE), the professional society representing electrical engineering, describes the purpose of electrical engineering as:

> Helping advance global prosperity by promoting the engineering process of creating, developing, integrating, sharing, and applying knowledge about electrical and information technologies and sciences for the benefit of humanity and the profession.

About 1.4 million electrical engineers work to develop a wide array of electronic products ranging from simple household appliances to sophisticated missile intercept systems. The field is so large that the IEEE has divided its organization into 10 regions, 36 technical societies, 4 technical councils, approximately 1,200 individual and joint society chapters, and 300 sections. If this discipline is of interest to you, visit the IEEE web site at http://www.ieee.org/ for a much more complete description[6].

Mechanical Engineering

Mechanical engineers apply the fundamental principles of mechanics, thermo-sciences, and design to fulfill the needs of society. These principles are covered in detail in the subjects of solid and fluid mechanics, thermodynamics, heat transfer, design, systems analysis, control theory, vibrations, and machine design. Mechanical engineers work on a wide variety of different assignments. One of the primary areas is the energy field where mechanical engineers are responsible for generating and distributing electricity, designing heating and air conditioning systems, and developing refrigeration systems. A second significant area of activity is in the design of a wide range of product such as vehicles of all types, appliances, office equipment and machinery. Finally, many mechanical engineers work in production facilities developing manufacturing processes and overseeing production.

The American Society for Mechanical Engineering (ASME International), the professional society representing mechanical engineering, describes its mission as:

> To promote and enhance the technical competency and professional well being of our members, and through quality programs and activities in mechanical engineering, better enable its practitioners to contribute to the well being of humankind.

Mechanical engineering is one of the oldest disciplines with a broad range of industrial activities. Next to electrical engineering, mechanical engineering is the second largest engineering discipline. While membership in the ASME is about 125,000, employment by mechanical engineers in industry probably approaches one million. The organization of the ASME is divided into 39 technical divisions that publish technical papers and sponsor technical conferences.

---

[6] The Web sites for all of the professional societies are listed in Table 1.5 on page 20.

Civil Engineering

Civil engineers are concerned with many societal problems including environmental quality, building and maintaining the nation's infrastructure, and developing safe and rapid transportation systems. Topics of study include structures, soil mechanics, construction materials, environmental engineering, water resources, pavements, and transportation. This program emphasizes fundamentals of mathematics, sciences, engineering principles, and engineering methods leading to careers in engineering practice and/or graduate education. An option in **environmental engineering** or a separate degree in this discipline is often available in civil engineering departments.

The American Society for Civil Engineering (ASCE), the professional society representing civil engineering, describes its function as:

> When new developments occur in research and practice, ASCE's technical divisions and councils and their respective technical committees bring information to the membership through conferences and workshops and by publications such as manuals of practice, pre-standards, journal articles, and policy statements.

The technical activities of the ASCE include more than 14 annual conferences and workshops, and the publication of 10 journals. More than 5,000 ASCE members participate by serving on technical committees. Their work addresses programmatic thrust areas such as sustainable environment, sustainable transportation, extreme environment and information technology, and construction and materials.

Computer Engineering

Computer engineering is a relatively new discipline that is still emerging in the many colleges of engineering in the U. S. Many electrical engineering faculties are dividing the traditional electrical engineering curriculum into two separate programs—one dealing with power, electronic systems and communications and the other concerned with computers and information technology. Computer engineers work in the computer industry designing hardware and writing software. The discipline combines technical knowledge from digital electronics, signal transmission and processing with a programming emphasis of computer science.

The IEEE Computer Society, an element within the IEEE organization, has emerged as the professional society representing computer engineers. The mission statement for the IEEE Computer Society is given below:

> The IEEE Computer Society is dedicated to advancing the theory, practice, and application of computer and information processing technology. Through its conferences and tutorials, applications and research journals, local and student branch chapters, technical committees, and standards working groups, the society promotes an active exchange of information, ideas, and technological innovation among its members. In addition, it accredits collegiate programs of computer science and engineering in the United States.

With nearly 100,000 members, the IEEE Computer Society is the world's leading organization of computer professionals. Founded in 1946, it is the largest of the 36 societies organized under the umbrella of the Institute of Electrical and Electronics Engineers (IEEE).

Chemical Engineering

Chemical engineering is concerned with the processing of materials and the production or utilization of energy through molecular or sub-molecular changes. Chemical or atomic reactions and physical changes are included in this field. Chemical engineers begin their process designs with well-known chemical reactions previously discovered by chemists. They use this information to design large-scale chemical plants to produce large quantities of raw materials, petroleum products, pharmaceuticals, synthetics, etc.

The American Institute of Chemical Engineers (AIChE) is a nonprofit organization providing leadership to the chemical engineering profession. Representing 57,000 members in industry, academia, and government, AIChE provides forums to advance the theory and practice of the profession, upholds high professional standards and ethics, and support excellence in education. Institute members range from undergraduate students, to entry-level engineers, to chief executive officers of major corporations. The AIChE mission statement is given in the bullet listing below:

- Promote excellence in chemical engineering education and global practice;
- Advance the development and exchange of relevant knowledge;
- Uphold and advance the profession's standards, ethics and diversity;
- Enhance the lifelong career development and financial security of chemical engineers through products, services, networking, and advocacy;
- Stimulate collaborative efforts among industry, universities, government, and professional societies;
- Encourage other engineering and scientific professionals to participate in AIChE activities;
- Advocate public policy that embraces sound technical and economic information and that represents the interest of chemical engineers;
- Facilitate public understanding of technical issues;
- Achieve excellence in operations of the Institute.

Industrial Engineering

Industrial engineers design the systems that organizations use to produce goods and services. In addition to working in manufacturing industries, industrial engineers serve to insure quality and productivity in places such as medical centers, communication companies, food service, education systems, government, transportation companies, banks, urban planning departments and an array of consulting firms. Industrial engineers educate and direct these groups in the implementation of Total Quality Management (TQM) principles. Today, especially popular functions are manufacturing, health care, occupational safety and environmental management. Industrial engineering is the most people-focused discipline in engineering. Those who pursue careers in this discipline usually have strong leadership skills and a commitment to working with teams of managers, scientists and other personnel to solve important problems. They enjoy helping organizations serve human needs and accommodate many different concerns

The Institute of Industrial Engineers (IIE) is the society that serves the professional needs of industrial engineers and others involved with improving quality and productivity. Its 24,000 members stay current with new developments in their profession through Institute's life-long-learning approach, as reflected in the educational opportunities, publications, and networking opportunities offered. Members also gain valuable leadership experience and enjoy peer recognition through numerous volunteer opportunities.

Aerospace Engineering

Aerospace engineers are concerned with the physical understanding, related analyses, and creative processes required to design aerospace vehicles operating within and beyond planetary atmospheres. Such vehicles range from helicopters and other vertical takeoff aircraft at the low speed end of the flight spectrum to spacecraft operating at thousands of miles per hour during entry into the atmospheres of the earth and other planets. In between are general aviation and commercial transports flying at speeds below and close to the speed of sound, and supersonic transports, fighters, and missiles which cruise at speeds greater than the speed of sound. Among the subjects studied are aerodynamics, flight dynamics, flight structures, flight propulsion, and the synthesis of all these principles into one system with a specific application such as a complete transport aircraft, a missile, or a space vehicle.

The nonprofit American Institute of Aeronautics and Astronautics (AIAA) is the principal society serving the aerospace profession. Its primary purpose is to advance the arts, sciences, and technology of aeronautics and astronautics and to foster and promote the professionalism of those engaged in these pursuits. Although founded and based in the United States, AIAA is a global organization with nearly 30,000 individual professional members, over 50 corporate members, thousands of customers worldwide, and an active international membership.

Materials Engineering

Materials engineering is the study of mechanical, physical, and chemical properties of engineering materials, such as metals, ceramics, polymers, and composites. The objective of a materials engineer is to predict and control material properties through an understanding of atomic, molecular, crystalline, and microscopic structures of engineering materials. A materials engineer is an essential member of an engineering team responsible for synthesis and processing of advanced materials for manufacturing. A graduate's work may vary from automobile or aerospace applications to microelectronics manufacturing. Opportunities are available through these industries in areas of research, quality control, product development, design, synthesis, and processing operations.

The American Society for Materials (ASM International) serves as one of the technical societies representing materials engineering. For many years it has provided a means for exchanging information and professional interaction. Benefits to members include:

- Access to information for extending and maintaining your professional skills.
- Opportunities for professional development through continuing education.
- Network locally and worldwide.
- Achieving professional recognition.

Biomedical Engineering

Biomedical engineering is the newest engineering discipline, integrating the basic principles of biology with the fundamentals of engineering. With the rapid advances in biomedical research, and the economic pressures to reduce the cost of health care, biomedical engineering will play an important role in the medical environment of the 21st century. Over the last decade, biomedical engineering has evolved into a separate discipline bringing the quantitative concepts of design and optimization to problems in biomedicine. The opportunities for biomedical engineers are wide ranging. The medical device and drug industries are increasingly investing in biomedical engineers. As gene therapies become more sophisticated, biomedical engineers will play an important role in bringing these ideas into real clinical practice. Finally, as technology plays an ever-increasing role in medicine, there will be a larger need for physicians with a solid engineering background. From biotechnology to tissue engineering, from medical imaging to microelectronic prosthesis, from biopolymers to rehabilitation engineering, biomedical engineers are in demand.

The Biomedical Engineering Society (BMES) serves as the professional society representing the interests of biomedical engineers. They are a relatively new society having incorporated in 1968. Its purpose is to promote the increase of biomedical engineering knowledge and its utilization

Engineering Science and Engineering Mechanics

Engineering mechanics is a field of study that permeates all major engineering disciplines. Solid mechanics deals with analytical formulation of direct and reaction forces on structures, and the resulting response of the structures to these forces. The discipline of engineering mechanics is based on applied mathematics and large scale computing. Fail-safe design is another dimension in this discipline. Because of the general applicability of this subject to many aspects of engineering design and manufacturing, opportunities are available in a broad spectrum of industries and government laboratories.

The American Society for Mechanical Engineering (ASME International), the professional society representing mechanical engineering, also represents engineering mechanics. The ASME is described in Section 1.4.2.

Agricultural Engineering

Agricultural engineering is in transition with more emphasis currently being placed on biological aspects rather than traditional agriculture topics. As such many programs carry some reference to biology in their titles—for example, **BioResource and Agricultural Engineering**. Agricultural engineers are trained to creatively apply scientific principles in the design and development of new products, systems, and processes for the conversion of raw materials and power sources into food, feed, and fiber. They are also concerned with protecting the environment and worker health and safety. The diversity of knowledge and skills which agricultural engineers possess is valuable in the agricultural and agribusiness industries. The agricultural engineer develops skills in design and problem solving, which are based on fundamental principles in the engineering sciences including mathematics and physics, computer tools, communication, teamwork, instrumentation and biology. Agricultural engineers are distinguished from other engineering disciplines by their commitment to meeting human and animal needs for food, feed, fiber and ensuring a sustainable, safe living and working environment. Biological and economic constraints will continue to make this a challenging career opportunity.

The American Society of Agricultural Engineers (ASAE) is a professional and technical organization dedicated to the advancement of engineering applicable to agricultural, food, and biological systems. Its 9,000 members, representing more than 90 countries, serve in industry, academia, and public service.

Petroleum Engineering

Petroleum engineering is primarily concerned with producing oil, gas, and other natural resources from the earth through the design, drilling, and use of wells and well systems. The petroleum industry develops methods for conveying fluids in to, out of, or through the earth's subsurface for scientific, industrial, and other purposes, remaining mindful of the ecological needs for safety. The curriculum in petroleum engineering provides a proper balance between fundamentals and practice. Graduate engineers are prepared for life-long learning but are capable of being productive contributors immediately. Petroleum engineers are currently in high demand in the industry, and their starting salaries are consistently among the top in the nation. The curriculum includes study of design and analysis of well systems and procedures for drilling and completing wells; characterization and evaluation of subsurface geological formations and their resources; design and analysis of systems for producing, injecting, and handling fluids; application of reservoir engineering principles and practices for optimizing resource development and management; and use of project economics and resource valuation methods for design.

The Society of Petroleum Engineers (SPE), with more than 50,000 professionals from oil and gas-producing regions around the world, represents the technical interests of this discipline. Its mission is to provide the means to collect, disseminate, and exchange technical information concerning the development of oil and gas resources, subsurface fluid flow, and production of other materials through well bores for the public benefit. The Society also provides opportunities through its programs for interested individuals to maintain and upgrade individual technical competence in these areas. The SPE accomplishes this mission through an international schedule of meetings and exhibitions, periodicals, short courses, books, electronic publications and section programs.

Mining Engineering

Mining engineering involves prospecting for mineral deposits; planning, designing, and operating profitable mines; processing and marketing the extracted minerals; insuring safe and healthy working conditions; and protecting and restoring the land during and after a mining project. Mining engineers use technologically advanced equipment, machines, robotics, and computers every day. Those in charge of mine design, plan and test mines using computer simulators before ever breaking ground. Mining engineers in management use mine scheduling software to plan mining activity once operation is underway. Surface mining operations use larger mobile equipment than any other industry in the world. Mining methods and equipment are also applied to the removal of earth and rock outside the mining industry. Each year, an individual requires an equivalent of 40,000 pounds of new minerals, and energy equal to that produced by burning 30,000 pounds of coal. With each different mineral comes a different mine site and a different location, giving the mining engineer a very diverse work environment, and limitless horizons to work anywhere in the world.

The Society for Mining, Metallurgy, and Exploration (SME) is an international society of 16,000 professionals with members in nearly 100 countries. Services available to SME members include: publications, professional registration, peer-review of technical papers, future leader and college accreditation programs, meetings and exhibits, public education, and SME short courses.

Nuclear Engineering

Nuclear engineering is concerned with the science of nuclear processes and their application to the development of various technologies. Nuclear processes are fundamental in the medical diagnosis and treatment fields, and in basic and applied research concerning accelerator, laser and super conducting magnetic systems. Utilization of nuclear fission energy for the production of electricity is the current major commercial application, and radioactive thermal generators power a number of spacecraft. For the longer term, electricity production based on nuclear fusion is expected to become an increasingly important segment of the field. Nuclear engineers are concerned with maintaining expertise in the design and development of advanced fission reactors, performing basic and applied research in the development and ultimate commercialization of fusion energy, developing both institutional and technical options for radioactive waste and nuclear materials management, and in fostering research in nuclear science and applications, with emphasis on bioengineering, detection and instrumentation and environmental science.

The discipline is represented by the American Nuclear Society (ANS). The ANS has a diverse membership composed of approximately 11,000 engineers, scientists, administrators, and educators representing corporations, educational institutions, and government agencies.

Manufacturing Engineering

Manufacturing is a prime generator of wealth and is critical in establishing a sound basis for economic growth. Manufacturing education is important in achieving and maintaining a long-term competitive position for U. S. industry in the current global economy. The program covers concepts such as design for manufacture, flexible manufacturing, new communication and information networks, and the impact of automation on human experience. Achieving and maintaining a long-term competitive economic position requires students to be able to analyze manufacturing systems for overall technical performance, economic performance, and environmental performance. Manufacturing engineers should also be able to quantify trade-offs between these performance characteristics in relationship to the technology deployed in the manufacturing system.

The Society of Manufacturing Engineers (SME) is the world's leading professional society serving the manufacturing industries. Through its publications, expositions, professional development resources and member programs, SME influences more than 500,000 manufacturing executives, managers and engineers. The SME has some 60,000 members in 70 countries and supports a network of hundreds of chapters worldwide

Table 1.5
Web site addresses for the professional societies

Professional Society	Web Site Address
Electrical and Electronics Engineers (IEEE),	http://www.ieee.org/
American Society for Mechanical Engineering (ASME International)	http://www.asme.org/
American Society for Civil Engineering (ASCE)	http://www.asce.org/
IEEE Computer Society	http://www.computer.org/
American Institute of Chemical Engineers (AIChE)	http://www.aiche.org/
Institute of Industrial Engineers (IIE)	http://www.iienet.org/
American Institute of Aeronautics and Astronautics (AIAA)	http://www.aiaa.org/
American Society for Materials (ASM International)	http://www.asm-intl.org/
Biomedical Engineering Society (BMES)	http://mecca.org/BME/BMES/
American Society of Agricultural Engineers (ASAE)	http://asae.org
Society of Petroleum Engineers (SPE),	http://spe.org/
Society for Mining, Metallurgy, and Exploration (SME)	http://www.smenet.org/
American Nuclear Society (ANS).	http://www.ans.org/
Society of Manufacturing Engineers (SME)	http://www.sme.org/

1.5 ENGINEERING FUNCTIONS

Engineers from all disciplines are involved in a wide variety of functions ranging from sales to maintenance. The more common engineering functions will be briefly described in the following subsections.

Design and Development

Companies continuously work on their product lines to maintain or increase their market share. Their ability to release a stream of high-quality, high-performance, reliable products that are competitively priced dictates their success and the profits they generate. **Design and development engineers** work to create new and profitable products. These "new" products are often redesigned models of an older product. Important in these redesigns are improvements in performance and reliability at lower costs. On rare occasions design and development engineers create an entirely new product, which establishes a new market. Cellular phones are a recent example of such a new product.

Design and development engineers interact with marketing personnel to establish the customer's needs. They convert the customer's needs to a product specification and then consider a large number of different design concepts that fulfill this specification. The best design concept is selected from among these options, and then it is developed. During this process, the design and development engineers work with manufacturing engineers, production engineers and test engineers to insure a timely and cost effective approach for producing a safe and reliable product.

Testing

Society today is increasingly litigious. It is a common occurrence for trial lawyers to seek remedies for damages due to what may be perceived as unsafe products. Product liability is a very serious concern. When a company releases a product for the market, it is imperative that it be safe. The only questions are—how safe and under what conditions? **Test engineers**, as the name implies, test products to ascertain if their performance specifications have been met. They conduct tests that enable them to predict the life of the product and the number of failures that will occur with time in service. They provide management with the data needed to determine warranty costs. During the design and development phase, they may test subsystems to provide data helpful in designing the product. Basically test engineers verify analytical methods for predicting performance. When the analytical methods are not sufficient, the test results are essential to prove the adequacy of the product.

Design Analysis

Almost every design and development team includes one or more members with excellent skills in design analysis. **Design analysis engineers** are responsible for modeling where components from the product are converted to simplified models that may be analyzed. The analysis conducted may be in closed form where well-known mathematical formulas are used to produce results that predict performance or insure safety. In many instances, closed form solutions are not available and numerical techniques, such as finite element models, are used to generate the required results. In recent years, a large amount of specialized software has become available that enables the design analysis engineers to solve increasingly complex problems—rapidly and accurately.

The close interaction of the design analysis engineer with the design team during the development cycle is vitally important. If a problem is identified by analysis early in the design cycle, it can be corrected quickly and the costs of the error are minimal. However, if the problem goes undetected until hardware is produced and tested, the costs to correct the error are dramatically higher and the time required correcting the problem may delay introducing the product to market.

Manufacturing

A **manufacturing engineer** serves on the design and development team to provide expertise on tooling and manufacturing methods. As the components of a product are designed, many decisions are made that affect the cost of manufacturing and assembly of the product. It has been established that a major fraction of the total life-cycle cost of a product is committed in the early stages of design [7]. Quality cannot be manufactured or tested into a product; it must be designed into the product. Manufacturing engineers provide the design and development team with the knowledge of manufacturing equipment and processes available within their facilities and available from qualified suppliers. As the design proceeds, they identify manufacturing requirements and begin the design of any specialized tooling required for production. They design the manufacturing cells for producing component parts. They certify vendors who will supply externally purchased components. They design the lines, cells and the tooling used in assembly.

Manufacturing engineers also work with the quality control department to establish procedures to insure the quality of each component manufactured. They select machine tools and instrumentation to insure timely adjustments for manufacturing processes to remain within the control limits established by quality control personnel.

Production

Production engineers also participate on the design and development team. Their role is to ensure the flow of material and components to the manufacturing facilities and to the assembly lines. Often there is a need to purchase components or materials that require long lead times for delivery. In some cases, the lead-time is comparable to the design time. The production engineer must identify these items, anticipate the number required and place the order before the design is complete. After the design is complete, they work with purchasing personnel to insure delivery of components just-in-time to be used in production. They attempt to minimize the work-in-progress inventory. They work with personnel from the sales department to match production to anticipated demand for the product. They work with personnel from shipping to ensure that the product is removed from the production facilities and delivered to customers in a timely manner. Their goal is to meet the demand of the customers with a minimum of work in progress and finished product in inventory.

Maintenance

Most students own an automobile or have one available to them. Have you ever tried to travel from point A to point B and found the car would not start? This problem improves in your understanding of the need for maintenance. Manufacturing facilities often operate on a 24/7 schedule (this schedule implies the equipment operates 24 hours a day seven days a week). **Maintenance engineers** work to keep the equipment in operation. They seek to avoid breakdowns with scheduled maintenance. They plan for periodic lubrication of bearings, replacement of parts subjected to high rates of wear, replacement of belts, and scheduled inspections. They develop a recording system that provides a database for predicting the time between failures and identifies the equipment that will probably fail. They design monitoring instrumentation that enables the facilities to be shut down in a controlled manner prior to failure of any critical parts. They supervise the maintenance crews that perform the repair work.

If you enjoy high stresses associated with the inevitable breakdown, consider this engineering function for a career.

Sales

Sales engineers provide an interface between the customer buying technical products and the company marketing them. They work with the engineers and purchasing agents representing the customer and provide technical information necessary to intelligently purchase a product or a system. Many engineering products and/or systems are complex with the availability of several different options. Sale engineers must know the details of the product and/or the systems, provide an overview of the options, and answer questions from the customer's representatives. People skills are as important as analytical skills in this position because the effectiveness of a sales call is often affected by personalities.

The sales engineers also seek information regarding new products from the customer. In some cases, the product line offered does not satisfy the requirements of the customer. When this occurs, it is important to ascertain the customer's requirements and the price he or she would be willing to pay for the product if it were to become available. This information is passed on the marketing representatives and is used in planning for new product developments.

Management

After demonstrating the ability to perform engineering functions, many relatively young engineers are promoted into **management** positions. If the company you are with is organized along functional lines, the first management position is usually "section head." A section is a small group of engineers (8 to 12) that specializes in a certain area such as motors, controls, transmissions, power, safety, etc. The section manager assigns tasks to the engineers in the group and provides supervision and support to aid them in performing the work. A department manager (2nd level of management) supervises several section managers. The section manager's role is a mix of both technical and administrative skills.

In companies that are organized with product development teams, program managers are responsible for the design and production of a certain product. Often an engineer becomes the program manager. He or she oversees the team, leads team meetings, makes assignments, arbitrates decisions, and is responsible for the development schedule and the budget. The program manager usually reports to a vice president for development. For larger development teams, the program manager's administrative skills are usually more important than his or her technical skills.

Education

Many engineering students continue their studies after completing the B. S. degree and pursue advanced engineering degrees. Those graduating with a M. S. degree may go into industry to serve in assignments demanding a higher level of analytical skills. Others take faculty positions in educational institutions (usually a community college or a four-year college). In these educational institutions, they usually teach three or four undergraduate classes each semester. They also perform many service functions for the college and community.

Some students complete the Ph. D. and take a faculty position with a college of engineering in what are known as research universities. They are responsible for teaching, research and service in these positions. The teaching is limited to one or two courses per semester usually divided between graduate and undergraduate courses. They engage in active research programs and seek funding from external sources (federal and state governments, industry and foundations) to finance their research. They guide the studies of graduate students working in their research topic. Publication in technical journals and writing textbooks is an important part of their scholarly activities. Participation in technical societies, by presenting papers, organizing sessions and serving as officers of the society, are ways for performing community service. Many faculty members are also active consultants to both industry and government.

Consulting

Consulting engineers work on a retainer from government agencies or industry. They are independent businessmen and women who basically offer engineering services at a price. In years past, it was unusual for an engineer to select this career path. Most companies and government agencies hired their own specialists. The downsizing of companies and government has changed this practice. Many companies find that outsourcing engineering work is more cost effective. They call on engineering consultants with a given specialty as required, and pay fixed prices for well-defined engineering tasks. The advances in communications—fax, cell phones, e-mail and the Internet—have enhanced the flow of information essential to conducting a successful consulting engineering business. Being in an office at a company's location eight hours a day five days a week, is not required. A consultant can solve a problem or create an engineering document from a distance while maintaining communication with any client in any country for as long as is necessary to complete an assignment. The availability of high-speed, low-cost computers has enabled the consultant to compete effectively with an in-house analyst if the consultant has access to the same software and the data needed to perform the assignment.

1.6 REWARDS

For many years, engineering graduates have received relatively high starting salaries and most have usually found interesting positions available upon graduation. However, salaries later in ones career depend upon many different factors including:

- Years of experience
- Level of technical and supervisory responsibility
- Type of employer
- Discipline and/or function
- Degree level
- The economy

Salaries also depend on current salary-wage policies, individual mobility, working conditions, and local and regional salary scales. Many significant factors, such as challenge, productivity, creativity, technical and supervisory responsibility, and job commitment are dependent on the individual's initiative. However, other factors affect salaries, which the engineer cannot control. These include company size, company policies, fringe benefits and promotion opportunities. Industry and company growth, labor-management relations, and research and development funding also affect earnings. There is considerable variation, even for engineers with similar position, responsibility and experience.

It is not possible to consider all of the parameters that affect salaries in this discussion. However, the three most important factors affecting average engineering salaries—degree level, years of experience, and management responsibilities are presented in Fig. 1.3 and Fig. 1.4.

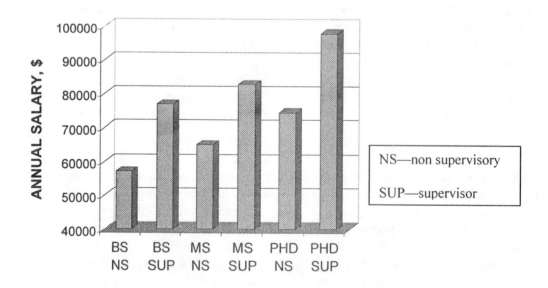

Fig. 1.3 Salaries for all engineers' as a function of degree and management responsibilities. This data is from "Engineers Salaries: Special: Industries Report, Engineering Workforce Commission of the AAES, 1997.

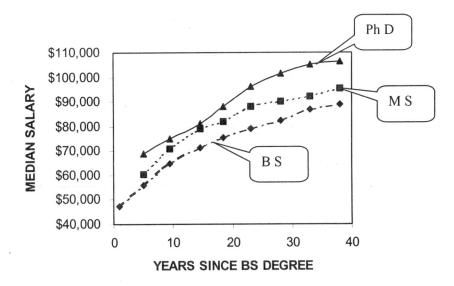

Fig. 1.4 Salaries for all engineers as a function of degree and years of experience.
This data is from Engineering Workforce Commission of the AAES, 2000

The results presented in Fig. 1.3 and Fig. 1.4 are clear. Engineers with advanced degrees command a higher average salary than those with only a B. S. degree. Also, those with Ph. Ds earn more than engineers with a M. S. degree for the same class of work. Finally, for all degree levels, management responsibilities are rewarded with higher salaries. The differences are significant with manager's pay exceeding non-manager's pay by about 35, 28 and 25% for the B. S., M. S. and Ph. D. degrees, respectively.

The role of experience and degree level is illustrated in Fig. 1.5. All engineers gain in salary with years of experience. The gains are more rapid in the early years, with increases of about twice the increase in the cost of living regardless of the degree level. However, after about 20 to 25 years of experience, the rate of increase in annual salary decreases. In the last decade of a typical engineering career, an individual's compensation may not keep up with the increases in the cost of living. These data are averages, and you may be an exception with either higher or lower rewards for your services. But the message of Fig. 1.4 is clear—make your mark early in your career. After you reach the age of 45 to 50, significant salary gains are more difficult for most people to obtain.

The data in Fig. 1.3 and 1.4 are from surveys that were conducted 1997 and 2000, respectively. As such, the salaries offered in 2004 and later may be significantly higher than those reported here[7]. Regardless of the exact numbers, the message is clear. If money is important to you— continue your studies and earn an advanced degree. Also, if you have the talents and the interests, pursue a position in management.

Another factor not considered in a discussion of salaries is the satisfaction derived from completing an engineering assignment. With sufficient experience, many assignments typically given to younger engineers lose their challenge. (Been there, done that.) With an advanced degree, an employer is paying you more and has higher expectations of your capabilities. Accordingly, the assignments tend to be more challenging, more visible and more rewarding. Success in one tough assignment leads to another and continued success leads to promotion and recognition.

---

[7] More current data for salaries in select engineering disciplines are presented in Table 1.3.

1.7 PROFESSIONAL REGISTRATION

Today about 30% of graduate engineers are registered and licensed as "professional engineers." Professional registration is optional for many engineers unlike medicine or law where registration is mandatory before one may practice. When public safety is an issue, registration is required for engineers. Civil engineers often design buildings, highways, bridges and other structures that would endanger the public if they failed in service. Consequently, civil engineers participating in these activities must register and become licensed to practice.

You are encouraged to begin the four-step registration process for the simple reason that you do not know when in the future you will be required to demonstrate your qualifications. Sometimes evidence of your B. S. degree is sufficient for qualification, but in other instances it is not. Also as you approach graduation you are very well prepared to begin the registration process.

State boards of registration control the procedure for registration. There are minor variations from state to state, but the guide shown below is typical of the process required:

- Graduation from a four year engineering program that has been accredited by ABET.
- Achieving a passing grade on the "engineering fundamentals" examination.
- Practicing engineering for a specified number of years to gain engineering experience. The state board of registration specifies the number of years of experience required.
- Achieving a passing grade on the "principles and practice of engineering" examination.

The "engineering fundamentals" examination may be taken before you graduate. In fact, it is recommended that you take it in October of your senior year[8]. The examination questions are designed to test your skills in fundamental subjects. Since engineers of all disciplines take the same examination, it is evident that the coverage must be on subjects common to all disciplines—mathematics, physics, chemistry, and engineering science. The examination is scheduled for eight hours. After passing the examination and graduating, you receive a certificate that designates you as an Engineer-in-Training.

The "principle and practice of engineering" examination is discipline specific. The civil, mechanical, electrical, etc. engineers must pass a discipline-oriented examination. While the topics covered are more focused, the questions probe the depth of your knowledge in the discipline. After passing this examination, you receive a license as a professional engineer and a seal that may be used when certifying your work.

1.8 SUMMARY

Two reasons are cited for classifying engineering as a profession. The first is based on a highly structured curriculum that prepares you for a position as an engineer upon graduation. You are expected to be productive almost immediately after assuming a beginning position with an industrial firm. The second reason is based on the type of work performed after graduation. When practicing in industry or for the government, one is engaged in engineering activities. A professional engineer is almost always salaried and is not compensated directly for overtime. Nor are you docked for time off. Engineers assume responsibility and are committed to the project, development team and the company or agency. They devote the time necessary to complete a project on schedule and on budget.

Statistics are cited for students entering engineering colleges. The students admitted to colleges of engineering are top notch—student leaders and scholars ranking in the top 5 or 10% of their high school graduating class. Average scores for both mathematics and verbal in the SAT

[8] The engineering fundamentals examination is offered twice a year in April and October. It is usually offered at several of the colleges of engineering in each State.

examinations are outstanding. About 21% of those graduating from engineering are women, but the percent of women in a typical class depends strongly on the discipline (see Fig. 1.1).

Pursuing a successful career in engineering depends on your ability to develop a number of different skills and understandings. These include:

- The ability to design.
- The ability to perform well on a development team.
- The ability to communicate in many ways and in different settings.
  - o Writing letters, memos, reports, specifications, etc.
  - o Participating in discussions, and presenting design briefings.
  - o Preparing two and three-dimensional drawings.
  - o Preparing slide presentations.
- Performing design analysis and interpreting the effect of the results on the design.
- Designing and conducting experiments and interpreting the data acquired.
- Understanding professional and ethical responsibilities.
- Understanding the need for life-long learning and actively pursuing this goal.

The most popular of the engineering disciplines are described and a chart of the B. S. degrees awarded by discipline is shown in Fig. 1.2. Starting salaries for beginning engineers by discipline shows some differences with chemical and computer engineers leading civil engineers. The technical emphasis of each of the popular disciplines is described. Professional societies representing each discipline are discussed.

Engineers of all disciplines perform many different functions in their professional activities. The tasks commonly conducted in each of these functions include:

- Design and development
- Testing
- Design analysis
- Manufacturing
- Production
- Maintenance
- Sales
- Management
- Education
- Consulting

Rewards (salaries) for engineers depend on many different factors. However, three factors—degree level, years of experience and management responsibilities are the most important. Data showing salaries as a function of these three factors are given in Fig. 1.4 and 1.5. Advanced degrees and management responsibilities are rewarded with significantly higher salaries. Experience is also important, but the salary gains due to experience are only significant for about the first 20 to 25 years. After that time, salary gains are barely equal to inflation.

Finally, the advantages of professional registration are discussed. A guide to the four-step process involved in becoming a registered professional engineer is provided. All engineering students are encouraged to take the eight-hour "engineering fundamentals" examination late in their junior year or early in their senior year.

REFERENCES

1. Valenti, M. "Teaching Tomorrow's Engineers," Special Report, Mechanical Engineering, Vol. 118, No. 7, July 1996.
2. Zhang, G. Engineering Design and Pro/ENGINEER, 4th Edition, College House Enterprises, Knoxville, TN, 2001.
3. Patterson, J. and P. Kim, The Day America Told the Truth: What People Really Believe about Everything that Really Matters, Prentice Hall, New York, NY, 1991.
4. Sommers, C. H., "Teaching the Virtues," *Public Interest,* No. 111, Spring 1993, pp. 3-13.
5. Grayson, L. P., The Making of an Engineer: An Illustrated History of Engineering Education in the United States and Canada, John Wiley & Sons, New York, NY, 1993.
6. Anon, *ASEE Prism*, Volume 13, No. 2, October 2003, p. 19.
7. Anon, Improving Engineering Design: Design for Competitive Advantage, National Research Council, National Academy Press, Washington, D. C., 1991.
8. Anon, "American Consulting Engineers Council Directory", American Consulting Engineers Council, 10015 15th Street, N. W., Washington, D. C. 20005.
9. Landis, R. B. Studying Engineering: A Road Map to a Rewarding Career, Discovery Press, Burbank, CA, 1995.

EXERCISES

1.1 Write a two-page paper on why engineers should be considered professionals.

1.2 Write a two-page paper comparing the engineering profession to the medical profession.

1.3 Write a two-page paper comparing the engineering profession to the legal profession.

1.4 Write a two-page paper comparing the engineering profession to the accounting profession.

1.5 Write a brief paper giving reasons that you know for why bright well-qualified students for leave engineering after only one semester.

1.6 Why do bright, well-qualified students fail courses in their first year of study?

1.7 Prepare a list of the skills that you currently have and provide a grade (from 1 to 10 with 10 being exceptional) for your achievement level in each skill.

1.8 Add to the list in Exercise 1.7 your goal for an achievement level in each skill at the conclusion of this course.

1.9 Do you believe that you will learn new skills in this course?

1.10 What is the purpose of the ethical codes endorsed by the professional societies?

1.11 Write a two-page paper describing your current ethical standards. Do you believe these standards will change during your tenure at college?

1.12 Are you committed to life-long learning? If so, why? If not, why?

1.13 Have you selected an engineering discipline? If so, state the discipline and give the reasons why you have selected it. If not, state the reasons for your delay?

1.14 If you are still concerned with selecting an engineering discipline, prepare an action plan to gain the information necessary for your decision?

1.15 If you are concerned that your first choice of an engineering discipline was not correct, prepare an action plan to evaluate your choice and to change disciplines if necessary.

1.16 For the engineering discipline of your choice, describe the mission and the activities of the professional society that represents their interests. Visit their web site.

1.17 Most colleges of engineering offer only a limited number of degree programs. What are the degree programs in your college? Do these programs coincide with your career objectives?

1.18   After reviewing the engineering functions, select the two that best suit your interests. Write a paper discussing why these functions interest you.

1.19   Using the data in Fig. 1.3, determine the gain in salary for each degree level that is obtained by serving in a management role.

1.20   Using the data in Fig. 1.3, determine the gain in salary for a non-supervisory position by earning an advanced engineering degree.

1.21   Using the data in Fig. 1.3, determine the gain in salary for a supervisory position by earning an advanced engineering degree.

1.22   Write a paper discussing why engineering salaries begin to plateau after about 20 to 25 years of experience. Include in your paper what you plan to do to avoid this problem.

1.23   Today the retirement age for most engineers is about 65. At what age do you plan to retire? Do you believe the age requirement to receive social security benefits will change prior to your retirement? Why?

1.24   Write a paper arguing for and against professional registration for engineers. Do you plan to take the "engineering fundamentals" examination in a few years?

CHAPTER 2

A STUDENT SURVIVAL GUIDE

2.1 OVERVIEW FOR SURVIVAL

As a new student in the College of Engineering, you are encountering many new challenges and meeting many new friends. You have to arrange for housing, transportation and meals (the basics) in an entirely new setting. You also have to cope with a schedule of classes and find the locations of classrooms, lecture halls and laboratories in a maze of buildings scattered all over campus. Finally, you must learn to study and pass courses in a much more competitive environment than you found in high school.

This chapter will discuss some strategies for coping with the new environment and for procedures that will help ensure your successful completion of your studies for a B. S. degree in engineering. In writing this survival guide, it is recognized that you are bright and have demonstrated your scholastic and leadership abilities—otherwise you would not have been admitted to the College. Success should be insured, but it is a fact that about half of each year's incoming class does not complete the program. Many students leave in the first year because of failing grades or lack of interest in the program. Some procedures to insure your success in completing the program will be discussed in this section. Then several important tools for improving your performance will be described later in this chapter.

Goals and Priorities

Many students fail to establish realistic goals that enable them to control their time and direct their energy in a manner needed to succeed in a competitive environment. Long-term, intermediate-term and short-term goals must be established to focus your activities. Otherwise your efforts are diffused, sufficient progress is not made and one or more goal is not achieved. Goals should pertain to both academic and social activities. The goals should be written in priority order and they should be dated. Your goals should be reviewed and revised periodically with changes made to reflect both progress and realism.

Confidence

Perhaps the most important student characteristic for success is confidence. It is essential that you maintain a positive attitude with a can-do philosophy. If you start to doubt your ability to pass a test, fear will become a significant factor that blocks your ability to think clearly during the examination. It is a well-know fact that fear and/or anger limits a person's ability to think clearly and act promptly. The question is how to gain confidence? The answer is to prepare until you are certain that you understand the course material and become confident in your ability to execute well on competitive examinations under pressure.

Stay Current

The clock is as important in studying engineering as it is in a football or basketball game. The course (game) begins on the first day of class and ends at the last minute allowed for the final examination. Your instructor will try to use all of that time to cover as much material as possible. There is little or no float time in an engineering course. Reading and homework will be assigned for each and every class period. Moreover, the material is cumulative—to understand the second topic it is necessary to master the first topic, etc. If you fall behind in the assignments, it is extremely difficult, if not impossible, for most students to catch up and become current.

Attend Class

As an instructor, I was amazed at the number of students cutting classes. An hour of class time costs the student, or his or her parents, more than a ticket to a rock concert. What student would throw away a ticket to a rock concert? Yet, some students throw away many opportunities to attend class over the semester or quarter. If you cut a class, you do not hear the instructor's interpretation of the topics discussed during that period. You do not participate in class discussion. You do not hear or understand questions asked by your classmates or share in the answers given by the instructor. You make yourself dependent entirely on the textbook or you count on the generosity of your fellow students to share their notes. It is simply common sense to maximize the opportunities for learning, and attending class is a well-established way for you to listen to explanations, ask questions and gain a deeper understanding of the course material.

Managing Time

You are accustomed to schedules because you moved from class to class following a prescribed schedule while attending high school. The schedule in college is similar except that a class period for a specific course is only scheduled for two or three hours per week instead of an hour every day. You have the **illusion** of more free time. A typical load of 16 credit hours for a semester entails only 12 or 13 hours of classroom participation and 6 to 8 hours of laboratory involvement. Unfortunately, this is misleading because the reading, and other assignments require significant amounts of out-of-class time. A rule followed by many instructors is to adjust the material and class assignments so as to require a **time factor of 3 to 4 on the credit hours for the class**. This means that if you are taking a three-credit hour class expect to spend 9 to 12 hours each week to attend class, complete the assignments and study for the examinations. Your 16 credit hour schedule really involves a time commitment of **48** to **64** hours per week. Time management becomes important when considering the actual time needed to succeed in engineering courses.

Developing Good Habits

It is self evident that good study habits are essential to successfully completing an engineering degree program. Yet many successful high school students have never developed good study habits. Competition in high school was not intense and they could meet or exceed their teachers' expectations without much out-of-class work. The academic competition is much more intense in a college of engineering and your instructor's expectations are higher than you can imagine. It is essential that you establish and follow a personalized schedule for study periods for each of your classes. The time allocation must be sufficient for you to complete the class assignments. Your schedule must insure that you remain current in each subject.

In addition to good study habits, you should develop good personal habits. These include eating a well balanced diet with three meals a day. A good workout with several exercises designed to maintain body tone is important to your well-being. Arranging your schedule so that you rest six to eight hours each night and awake feeling ready to face, if not the world, certainly your instructors. Finally, entertain in moderation. If you consume too much alcohol during a party, you will awake the next morning with a hangover. Hangovers not only feel terrible, they destroy your capability to perform academically for a day or two.

Managing People

While you are not yet an engineering manager, you will still interact with a number of people, and will find it necessary to manage your relations with them. A short list of these people includes your instructors, roommate, classmates, teammates, and friends on and off campus. Maintaining relationships requires time, and as you will soon determine, time is a precious commodity. Your instructors should command priority; fortunately they are easy to manage. All you have to do is go to class, and perform well on examinations. Your instructor will respond with good grades. You will share some common interests with some members of each of your classes and they will, with time, become classmates. Small groups of classmates form social entities. You may go to a sporting event together, run or workout as a group or gather together to study. This social bonding is an important part of the educational process. Education should be fun and classmates help make it so.

Teammates differ to some degree from classmates, because a team is a more formal group that has been organized to perform a well-defined task such as the design a prototype of some product. Teammates also bond socially and often remain friends long after the team has been disbanded. Friends on and off campus are important to your social life. Pursuing an engineering degree will take most of your time and energy, but it is essential that you schedule some time for socializing.

Your roommate and dorm neighbors will play a central role in your life on campus. Living in close proximity and sharing space dictates that a set of rules be established regarding community behavior. There are many decisions to be made involving music, visitors, study time, quiet time, lights out, entertaining, etc. You are encouraged to discuss a schedule with your roommate and neighbors that will provide an environment conducive for study. Also important is quiet time to enable you to sleep and rest. Performance and health both depend upon having a good night's rest. Medical experts recommend eight hours sleep to avoid illnesses and to perform at your peak capability.

Seeking Help

Compared to high school, the university is mammoth with tens of thousands of students and countless buildings scattered over a huge area. Orientation held in the summer preceding fall admission is recommended. These orientation sessions will provide some of the answers to your questions. Information will be provided about the engineering programs, the locations of the buildings, dormitory life, availability of food and entertainment, etc. However, after you arrive and begin classes, many unforeseen questions will arise. Where do you seek reliable answers?

If your questions are about dormitory life, find the **dorm manager** and he or she will provide guidance. The dormitory managers are carefully selected; they relate well to new students often anticipating questions. The point is for you to ask questions at your earliest opportunity. The answers will alleviate your concerns.

If your questions are about the engineering program, find the **Office for Freshman Advising**. This office, usually found near the Dean's office, is staffed with professional counselors. They know the details of all the programs in the college, and they understand the needs of the students entering the program. They can provide valuable advice about program requirements that will ease your entry into the program.

If you have questions about a class you are taking, check the instructor's schedule. He or she will have a few hours each week reserved for office hours. You may stop at his or her office during these times and find your instructor. He or she should be willing to take the time to discuss your concerns. Sometimes meeting with the instructor during regularly posted office hours is not possible. In these instances, contact the instructor by phone or e-mail and arrange an appointment. The instructors will help you understand the course material, but you should be prepared to ask clearly phrased questions that define the problems you are encountering. The instructor will be willing to respond to your questions, but he or she will not be willing to tutor you.

2.2 TIME MANAGEMENT

Weekly Schedules

The college schedule is a weekly affair. You have classes Monday, Wednesday and Friday or possibly on Tuesday and Thursday. The process repeats for about 15 weeks before the final examination period. It makes sense to prepare a weekly schedule showing the time and location of your classes similar to the one shown in Fig. 2.1.

WEEKLY SCHEDULE
FALL SEMESTER 2003

NAME		ADDRESS			PHONE NUMBER		
TIME	SUN	MON	TUES	WED	THUR	FRI	SAT
MORNING							
6:00							
7:00							
8:00		ENES 100		ENES 100		ENES 100	
9:00				ENES 100		ENES 100	
10:00		MATH 140		MATH 140		MATH 140	
11:00		CHEM 133		CHEM 133		CHEM 133	
12:00							
AFTERNOON							
1:00		ENGL 101	CHEM 133	ENGL 101		ENGL 101	
2:00			CHEM 133			MATH 140	
3:00			CHEM 133			MATH 140	
4:00							
5:00							
EVENING							
6:00							
7:00							
8:00							
9:00							
10:00							
NIGHT							
11:00-6:00							

Fig. 2.1 Weekly class schedule for first semester engineering students.

The recommended load of 14 credit hours for the first semester schedule is common for all the disciplines in the college of engineering[1]. While the times for classes on the schedules will differ from one student to another, the number of credit hours and the number of contact hours will probably

---

[1] The number of credit hours scheduled in the first semester is lower than the average (15-16) because time is allowed for the first year students to adjust to a new campus and new social conditions.

remain the same. The number of contact hours[2] (19) exceeds the number of credit hours because laboratories and recitation periods are not counted with the same weight as lecture hours.

The next step in time management is to schedule study periods. How much time should you schedule? The general rule most instructors use in designing assignments is that the **average** student spends two hours studying for each hour of lecture and another hour studying for each hour of laboratory or recitation. This rule implies that a student should schedule $3 \times 2 + 3 \times 1 = 9$ hours of out-of-class study time for Chemistry 133. For English 101 an average student would schedule $3 \times 2 = 6$ hours of out-of-class time for study and writing. Adding the study time for all four course required in the first semester gives a total of 27 hours that should be devoted to study for the **average** student. The word **average** is stressed. You may be brilliant and be able to perform well with less than the recommended 27 hours. Or you may be having trouble with one or more courses and find that 27 hours is not sufficient. The weekly schedule shown in Fig. 2.2 provides for 27 hours of study time.

WEEKLY SCHEDULE							
FALL SEMESTER 2003							
NAME		ADDRESS			PHONE NUMBER		
TIME	SUN	MON	TUES	WED	THUR	FRI	SAT
MORNING							
6:00	PREPARE	PREPARE	PREPARE	PREPARE	PREPARE	PREPARE	PREPARE
7:00	BRKFAST	BRKFAST	BRKFAST	BRKFAST	BRKFAST	BRKFAST	BRKFAST
8:00	OPEN	ENES 100	STUDY	ENES 100	STUDY	ENES 100	STUDY
9:00	OPEN	STUDY	STUDY	ENES 100	STUDY	ENES 100	STUDY
10:00	OPEN	MATH 140	STUDY	MATH 140	STUDY	MATH 140	STUDY
11:00	WORSHIP	CHEM 133	STUDY	CHEM 133	STUDY	CHEM 133	STUDY
12:00	LUNCH	LUNCH	LUNCH	LUNCH	LUNCH	LUNCH	LUNCH
AFTERNOON							
1:00	OPEN	ENGL 101	CHEM 133	ENGL 101	OPEN	ENGL 101	SOCIAL
2:00	OPEN	STUDY	CHEM 133	STUDY	OPEN	MATH 140	SOCIAL
3:00	OPEN	STUDY	CHEM 133	STUDY	OPEN	MATH 140	SOCIAL
4:00	OPEN	STUDY	STUDY	STUDY	OPEN	STUDY	SOCIAL
5:00	WORKOUT	WORKOUT	WORKOUT	WORKOUT	WORKOUT	WORKOUT	WORKOUT
EVENING							
6:00	DINNER	DINNER	DINNER	DINNER	DINNER	DINNER	DINNER
7:00	STUDY	OPEN	OPEN	OPEN	OPEN	SOCIAL	SOCIAL
8:00	STUDY	OPEN	OPEN	OPEN	OPEN	SOCIAL	SOCIAL
9:00	STUDY	OPEN	OPEN	OPEN	OPEN	SOCIAL	SOCIAL
10:00	OPEN	OPEN	OPEN	OPEN	OPEN	SOCIAL	SOCIAL
NIGHT							
11:00-6:00	SLEEP	SLEEP	SLEEP	SLEEP	SLEEP	SLEEP	SLEEP

Fig. 2.2 A complete weekly schedule for a first semester student.

Many assumptions were made in developing the schedule presented in Fig. 2.2. The first assumes you are a morning person because all mornings are packed with classes and study except for Sunday. There are two reasons for carrying a heavy schedule in the morning. First, most people think more clearly in the morning after a good night's rest. Second, if you encounter difficulties in completing an assignment during a study period, time is available to recover later in the day.

Every hour in the morning has been filled. If you have a period or two between classes schedule it for study. It is easy to find a quiet desk in the library and complete an assignment in mathematics in an hour. On the other hand, you may decide to work with a small study group in one of the student lounges. Try to avoid spending that hour in idle conversation with a group of friends.

---

[2] Contact hours are the number of hours you attend class, recitation or laboratory periods in a given week.

The weekly schedule shows four of the seven afternoons devoted to classes or study periods. Two afternoons are completely open and Saturday afternoon and evening are scheduled for social events. The open periods (unscheduled time) provide much needed flexibility in the schedule. On some weeks you will find the assignments much more demanding than other weeks. When this situation occurs, you will be able to use the open periods to study. Open time after dinner has been scheduled for four evenings of a typical week. A total of 28 hours of open time is available. There is a significant amount of flexibility to add study periods or even time to work part time if necessary.

All work and no play makes Jack and Jill very dull. Accordingly, schedule some time for a social life, but not too much party time. Friday night and Saturday afternoon and night are reserved for socializing. Some fatherly advice from the author—**do not drink to excess or even think about illegal substances.** There is absolutely nothing to be gained except a headache that impairs your ability to study effectively.

The schedule also reflects the author's concern for your physical well-being. An hour has been set-aside for each of three meals a day, seven days a week. Nutrition is important to your health today and even more important to your health throughout life. Eat regularly, on schedule and avoid junk food. You will feel better and perform better. Sleep is as important as eating to your well being. Seven hours a night is reserved for sleep, but you may need more to stay healthy. You will begin to have frequent colds[3] if the amount of your sleep and rest is not sufficient. Regularity is also important in establishing beneficial sleep patterns. Go to bed at the same time each night and arise at the same time each morning. A sleep period from 11 PM to 6 AM is scheduled.

Attending class and studying is sedentary. You will lose your body tone without interrupting this pattern of class, study and sleep. Try going to the gym and working out at least three times a week. Actually, the schedule provides time for a work out every day from 5 to 6 PM. Working out before dinner will suppress your appetite and help you maintain a healthy weight.

An hour is set aside each morning to prepare you to meet the world. This hour will permit you to shower, shave (or to arrange your hair) and dress. Dress should be casual, but not sloppy. If you have a few minutes after dressing, sit down and prepare a detailed schedule for the day.

You may find that the schedule shown in Fig. 2.2 is not suitable. That's fine. It is not necessary to impose this schedule on you. However, it is imperative that you prepare a weekly schedule that is more suitable, and one that accommodates time for:

- Attending all your classes, laboratories and recitations
- Sufficient study time at suitable times (27 hours is a minimum suggested)
- Adequate uncommitted (open) time to provide needed flexibility in your schedule
- A measure of social time to keep your spirits high
- Enough time for meals, sleep, work outs, and mentally preparing for the day

Daily Schedules

You will prepare a weekly schedule each semester. It is a guide to your daily activity, but it does not contain the detail needed to use your time efficiently. The daily schedule provides the detailed guide for allocating sufficient time to tasks essential to your success in each course. To provide information on preparing a daily schedule, let's consider a typical Monday as defined in the weekly schedule of Fig. 2.2. A new format for the daily schedule is constructed as illustrated in Fig. 2.3.

---

[3] If you are in good health and resting your body sufficiently, you should not experience more than two or three colds a year, and they should be of short duration.

TIME MORNING	MONDAY	AFTERNOON		EVENING	
DAILY SCHEDULE					
MONDAY, OCTOBER 15, 2001					
NAME					
6:00	SHOWER	1:00	ATTEND ENGL 101 CLASS	6:00	DINNER
	DRESS	2:00	STUDY	7:00	OPEN
	PLAN DAY		WORK TWO MATH	8:00	OPEN
7:00	BREAKFAST		PROBLEMS	9:00	OPEN
	ARRANGE STUDY GROUP	3:00	STUDY	10:00	OPEN
8:00	ATTEND ENES 100 CLASS		COMPLETE MATH	NIGHT	
9:00	STUDY		PROBLEMS	11:00-6:00	SLEEP
	COMPLETE ENES 100	4:00	STUDY		
	ASSIGNMENT		READ CHEM		
10:00	ATTEND MATH 140 CLASS		ASSIGNMENT		
11:00	ATTEND CHEM 133 CLASS	5:00	WORKOUT		
12:00	LUNCH		RUN THREE MILES		
			SHOWER AND DRESS		

Fig. 2.3 An example of a daily schedule for a typical Monday.

Let's review this example. Each hour of the day is scheduled and tasks assigned. At 6:00 AM, you rise, shower, etc. and plan the day. Next you walk[4] to the dining hall and enjoy breakfast. At breakfast you arrange for a study group to meet immediately following the ENES 100 class. At 8:00 AM you arrive on time for the ENES 100 class. Following class your study group meets at 9:00 AM in a student lounge and works collaboratively to complete the ENES 100 assignment.

At 10:00 AM, you arrive on time and attend a MATH 140 lecture. Immediately after this lecture you proceed to the Chemistry building to attend the CHEM 133 lecture scheduled for 11:00 AM. After two lectures in a row, lunch is a welcome break.

Following lunch, you cross campus and arrive on time for the ENGL 101 class scheduled for 1:00 PM. After this class, you proceed to the library and find a quiet desk in the stacks. You have planned two hours (from 2:00 to 4:00 PM) to complete the problems assigned in the MATH 140 class. Unless you encounter difficulties, this amount of time should be sufficient. At about 4:00 PM, you are saturated with mathematics and turn your attention to a reading assignment made during the CHEM 133 lecture. Upon completing the reading assignment at about 5:00 PM, you walk over to the gym, dress in running gear and jog for three miles. After showering and dressing, you walk over to the dining hall for dinner.

Your daily schedule indicates that the evening is open. Should you relax or study for a while after dinner? You have efficiently used the four hours of study scheduled during the day to complete assignments in three of your four classes; however, you have not considered the assignment for ENGL 101. To make this decision, examine your class schedule for Tuesday. You have four hours set aside for study in the morning, and a CHEM laboratory in the afternoon. If you are current in all of your subjects, the four-hour study period should enable you to complete the ENGL 101 assignment and to prepare for the CHEM laboratory. You can relax Monday evening or take on a part time job if you are able to consistently complete your assignments during the study periods scheduled during the day.

While the class structure is consistent from week to week, the assignments vary from day to day. Sometimes the assignments can be completed quickly with little effort. Other times term papers are due and the library research and writing requires much more time than the usual assignment. A few times each semester an hourly examination will be scheduled in each course. The time to prepare for the examinations is often significant, particularly for those students who fall behind schedule.

---

[4] Remember to allow for time to walk between buildings in preparing your daily schedule. Total walking time probably approaches an hour each day for students on large campuses with scattered buildings.

Because of the changing demands for time from one course to another, it is vital that several open periods be available in the daily schedule to provide you with an opportunity to put in the extra effort needed for success.

2.3 MONEY MANAGEMENT

Money, or the lack thereof, is often a problem for many students. It is a difficult problem to be discussed in a textbook because of the large variation in the circumstances of the students and their personal spending habits. To simplify the discussion, let's divide a student's money requirements into two categories—mandatory and discretionary. The mandatory category is the funding necessary for tuition, fees, books and room and board. Reference to a university catalog gives an estimate of these mandatory costs as shown in Table 2.1. An example of a typical in-state-student attending a State university is given. The costs increase by about $8,000 to $9,000 for out-of-state students attending a State university. The mandatory costs for attending most privately financed universities often exceed $30,000.

Table 2.1
Estimated mandatory educational costs
Student living on campus

ITEM	AMOUNT
Tuition and Fees (in State)	$5,700
Room	$4,000
Board	$2,800
Books and Supplies	$1,000
Total	$13,500

If you live in the vicinity of the university, you may choose to live at home and commute. If this is the case, you will probably drive, car-pool, or use public transportation to travel from home to the university. In this case, the room and board costs in Table 2.1 will be eliminated (actually absorbed by your parents); however, commuting costs are incurred. Commuting cost vary widely from zero if Mom or Dad drops you off on the way to work to about $3,000 per year[5] if you drive an older used car a long distance to the parking lot. A commuting cost of $3,000 per year is assumed in the estimates for the commuting student in Table 2.2 to arrive at a total annual mandatory cost of $9,700.

Table 2.2
Estimated mandatory educational costs
Student commuting to campus

ITEM	AMOUNT
Tuition and Fees (in State)	$5,200
Books and Supplies	$1,000
Commuting Costs	$3,000
Total	$9,700

The next category is discretionary costs where you have a choice about expenditures. Of course you will need walking around money—the question is how much? If you are on a room and board plan, it probably does not include three meals a day seven days a week. You will find it necessary to purchase a few meals at the local eateries each week. You will also find it necessary to finance your social life. Social costs vary widely depending on your choice of entertainment and

---

[5] This estimate is based on driving 10,000 miles per year at an average cost of $0.30 per mile for gasoline, insurance and maintenance.

tastes. A night at the movies with a soft drink and a slice or two of pizza is a low cost evening. Taking a date to dinner at a good restaurant followed by a rock concert will dent your budget big time. The amount allocated for discretionary costs is a decision you should make with your parents if they are funding a large portion of your college expenses. They will want to help, but their resources are not infinite.

If you are working your way through college, the hours you must work increase with the amount of your discretionary spending. As you commit more hours to the workplace, your study time decreases and you endanger your chances for success. Often working students find it necessary to withdraw from courses during the semester and accept a reduced course load that lengthens the time to complete a B. S. degree from four to five or more years.

Budgeting

Depending on your circumstances attending college is going to cost from about $10,000 to $15,000 per year[6]. For this discussion, assume that you have an agreement with your parents and know the level of support that they can provide and when they will make their contribution. You will find the university is not a benevolent organization. They will demand their money[7] and if it is not forthcoming, you will be declared a persona non grata. Clearly, you have two problems.

1. Raising $10,000 to $15,00 per year.
2. Insuring the cash flow so that the funds will be available at the correct time.

Planning is essential to your money management and developing a budget is the first step in managing your funds. To provide an example of budgeting, let's assume that your parents take care of the cost of tuition and fees. You are to handle the cost of books, commuting expenses, and walking around money. The first step is to prepare an estimated budget of annual expenditures. You list your financial requirements, make many decisions about books and supplies, commuting costs, and estimate out-of-pocket expenditures at $85 per week or $3,000 for the 36 week academic year. Your budget of estimated expenditures is listed in Table 2.3. Clearly, the listing of estimated expenditures indicates a $7,000 problem.

Table 2.3
Budget of estimated expenditures
Student commuting to campus

REQUIREMENTS	AMOUNT
Books and Supplies	$1,000
Commuting Costs	$3,000
Pocket Money	$3,000
Total	$7,000

Let's assume that you worked during the summer and managed to save $3,000 of your after tax earnings. That's great because your shortfall has been reduced to $4,000. How do you raise the $4,000? Your parents are tapped out; they have their hands full with their own financial concerns. You have three options:

[6] Assuming in-state residency and attendance at a typical state-supported university.
[7] Near the beginning of a typical undergraduate catalog, the university describes its charges for tuition and fees. They also clearly indicate that "all charges incurred during a semester are payable immediately. Returning students will not be permitted to complete registration until all financial obligations to the university—including library fines, parking violations, and other penalty fees and service charges—are paid in full."

- A scholarship
- A student loan
- A job

Without a doubt, a scholarship is the best way for helping to resolve your problem. Some scholarships are based on merit (your high school record) and others are based on financial need. Scholarship awards usually do not happen without your actively seeking financial aid. Contact the **Office of Student Financial Aid** at your university and be prepared to fill out a number of forms. There are a large number of scholarships available and you may qualify.

If a scholarship or grant is not possible, you may wish to consider a student loan. The Office of Student Financial Aid administers all types of federal, state and institutional financial aid programs. This office will not magically recognize your needs. Financial aid does not happen automatically; you must make it happen by applying and following up to make certain that all of the forms have been processed.

Your third option is to work and earn money while you pursue your degree. Many students are forced to take a part time job to pay part of their college expenses. If you decide to work, you may be able to arrange a part time position with a local company for say 10 to 15 hours per week. Working much more than this amount probably would seriously cut into your scheduled open and/or social time, and endanger your chances for successful completion of a full load during the semester.

Another option for working to help pay your college expenses is to enroll in the cooperative education program. Students enrolled in this program alternate semesters of full time work with semesters of full time study. The program enables a student to gain professional experience, integrate theory and practice, confirm career choices and help finance their education. Students are eligible to participate in the cooperative program at any time during their studies; however, most employers seek students with sophomore standing or higher.

Let's continue with the budgeting example. Suppose you have been successful in obtaining a $900 scholarship, in arranging a $1,000 student loan, and landing a part time job for 12 hours a week that yields $80 per week ($80/week × 36 weeks = $2880 per academic year) after taxes. Your parents are pleased with the scholarship award and agree to give $900 in cash as a tuition reimbursement. You have arranged for cash flow of $4780 for the academic year. The next step is to budget the flow of cash into and out of your checking account[8]. Both monthly and weekly budgets are recommended to track the flow of your funds. A monthly budget for a hypothetical student commuting from home to the university is shown in Table 2.4.

The monthly budget for February shows expenses exceeding income by $155. Is this cause for concern? If you examine the expenses, you will note that books and supplies total $185. This is much higher than usual because of the beginning of the spring semester. Next month, you do not plan on purchasing any books, and expenditures on supplies will only be budgeted at $15.

The budget provides $50 for miscellaneous because one always encounters unforeseen expenses. However, the miscellaneous items ordinarily should not exceed 10% of the total expenses budgeted for the month.

In an analysis of the budget, you should always try to identify unusual items. On the income side of your budget, you will find a cash gift of $80. The money came from an older brother who responded to your tale of woe about the high price of textbooks. This is an unusual event and should be recognized as such. The gift will not be part of your regular budgeting for next month.

---

[8] It is assumed that you have established a checking account to accommodate cash flow requirements prior to enrolling in college.

Table 2.4
Monthly budget
February 2004

Monthly Income		Monthly Expenses	
Item	**Amount**	**Item**	**Amount**
Balance Checking Acct.	$3,215	Books	$150
Wages	$ 320	Supplies	$ 35
Cash gifts	$ 80	Car Pool Payments	$110
		Meals	$ 90
Total Income	$ 400	Entertainment	$120
		Miscellaneous	$ 50
Income less Expenses	$400 – $555 = ($155)		
		Total Expenses	$555
New Balance	$ 3070		

Considering the unusual items on both the income and expense sides of your budget, you can conclude that the negative cash flow in March will be much smaller, but the cash flow will remain negative. If you seek to balance the monthly budget, it will be necessary to work more hours to increase income or to entertain more economically to decrease expenditures.

Have you ever approached the end of a week (or month) and discovered you were broke. How did you spend that much money? Did the cash evaporate? What happened? To avoid this difficulty, it is suggested that you keep a weekly budget and a daily record. The weekly budget is a short-term version of the monthly budget as shown in Table 2.5. Use a period of one week because many payments are made on a weekly basis.

Table 2.5
Weekly budget
February 15 to 21, 2004

Weekly Income		Weekly Expenses	
Item	**Amount**	**Item**	**Amount**
Balance Checking Acct.	$ 3,120	Supplies	$ 15
Wages	$ 80	Car Pool Payments	$ 25
		Lunch 5@$4 ea	$ 20
Total Income	$ 80	Entertainment	$ 30
		Miscellaneous	$ 10
Income less Expenses	$ 80 – $100 = ($20)	Total Expenses	$100
New Balance	$ 3,100		

The weekly budget shows a small negative cash flow as expected. It appears that the money in your checking account will be sufficient to carry you until the summer when you can take on a full time job and begin to save money again. The question is your ability to survive within the budget, week after week? A daily record of all of your expenditures will help.

To develop a guide for your daily record, let's define discretionary as the sum of the expenditures for lunch, entertainment and miscellaneous which is $60 for the week of February 15 to 21. The expenditures budgeted for the car pool payment and supplies are required and not discretionary. For Friday, let's suppose you record your expenditures as shown in Table 2.6. Your expenditures for the day and a modest evening out were $24.25. This may sound reasonable when compared to the $65 for discretionary spending, but the week has seven days. If you would like more entertainment (and everyone does), it will be necessary to cut back on other expenditures. The daily record gives the data necessary for you to examine what you buy with your money. What could you give up with out great pain and sacrifice? That coffee and pastry for $1.95 tastes good, but it is junk food that does little for you other than adding empty calories. The coke and pizza is in the same category. Take in the movie and avoid the munchies. You will save money, meet your budget and avoid gaining unwanted weight.

Table 2.6
Daily record of expenditures
Friday, February 20, 2004

Item	Amount
Coffee and Danish	$1.95
Lunch	$4.75
Pizza and Coke	$2.80
Share in Fuel Costs	$5.00
Movie with Munchies	$9.75
Total	$24.25

The advantages of the daily record of expenditures are:

1. Provides accurate data on expenditures that reveals your spending patterns.
2. Gives a daily reminder of your budget constraints as you record your purchases.
3. Enables you to compare daily expenditures with your weekly budget.

It is recommend that you maintain the weekly budget and a daily record of expenditures. The budget and record of expenditures will not necessarily relieve your financial difficulties, but they will enable you to deal with them more effectively.

Loans and Credit Cards

Many students are eligible for student loans that are guaranteed by the Federal government. This guarantee is important since it removes the risk of your defaulting from the lending agency. Accordingly, as a student you can arrange a loan with a low interest loan that is repaid after graduation when you should be earning a significant salary. If you must incur debt to finance your college education, a low interest rate, deferred payment, government-sponsored loan is the best alternative.

The worse alternative is to accumulate credit card debt. Credit cards are used by almost everyone today because they offer a convenient way of paying for anything from gasoline to dinner. Carrying around a pocket full of cash is not necessary. You do not need to go the bank to cash a check. Credit cards also offer free **short-term** credit. The **short-term** nature of this credit is emphasized. After receiving the monthly statement that summarizes the purchases charged to the credit card, you have about two weeks to pay the bill. If you do not pay on time, the interest charge is very high—18 to 21% of the outstanding balance is common. Suppose you carry an average credit balance of $1,500 on a credit card for a year at an interest rate of 21%. The cost of this credit card loan for a year is $315. The interest on a student loan for $1,500 at a rate of 6% is only $90.

Use a credit card freely only when you are certain that you have enough money in your checking account to pay off the entire balance, on time, each and every month.

Scholarships

Scholarships were discussed previously in this section. The topic is introduced again because of its importance. There is no scholarship guru who automatically awards scholarships to students on the basis of merit or need. You must actively pursue a scholarship by asking for financial aid. Visit the Office of Student Financial Aid and talk to their staff. There are many scholarships based on merit. If you have an excellent high school record, you may qualify for one of these merit scholarships. If your personal income is limited, you may qualify for an **Institutional Scholarship, a State scholarship or a Federal grant.** The point here is that you must ask for financial aid[9], fill out the forms and submit them on time. Usually these scholarships or grants are for one year, and it is necessary for you to reapply each year.

If you cannot manage to obtain a scholarship or grant, it may be possible to obtain a low interest federally-subsidized loan. While these loans must be repaid, the taxpayer is helping you by deferring interest payments and reducing interest rates significantly. To obtain one of these loans complete a free application for **Federal Student Aid** and submit it to the aid program processor. The deadline for submission is February 1. If you are eligible for financial aid and find it necessary to incur debt to pursue your education, a subsidized loan will significantly reduce the costs of borrowing money needed to support your studies.

2.4 PEER MANAGEMENT

As you move through the program, you will encounter hundreds of fellow students. Some of these students will be no more than faces in a common classroom, while others will become close friends. These close friends may affect your life style and behavior. Peer pressure is a recognized factor in affecting behavior. You will have an established set of behavioral patterns before you arrive on campus. As you develop friendships in the classroom, dormitories, fraternities, gymnasium, etc., your behavioral patterns will change. You will adapt to accommodate with the group. In fact, you will become associated with several groups of friends each having somewhat different behavioral patterns. For example, a group of friends may evolve from a design team in an engineering class. Another group from the dormitory and still another from the pick-up team formed to play basketball at the gym.

These groups of friends make up the social fabric that is so important for you to enjoy the educational experience during your tenure in college. It is healthy to be involved with friends, but there are two dangers that should concern you. First, the time involved in pursuing social activities with your friends should not be excessive. If you find the demands of your social groups cutting into your scheduled study time, you will not be prepared for class. The results are predictable when this happens—your grades will suffer badly and you will miss important opportunities to extend the depth of your understanding.

Second, the behavior of some social groups should cause concern. Does someone in the group break the law? Do a few in the group drink excessively or smoke pot? Do their moral boundaries exceed yours? Does everyone in the group have ethical standards comparable to yours? These are the questions you should address before you become deeply involved with a social group. It is easy to find trouble if you associate with the wrong people.

---

[9] A personal note—out of ignorance, I failed to ask for financial aid until the second semester of my junior year. I worked long hours in a factory to finance my education at a private university. When I did visit the Office for Student Financial Aid and explained my situation, a tuition scholarship for my senior year was made available almost immediately and life became much easier.

2.5 STUDY HABITS

Good study habits are important to successfully completing an engineering degree. A study habit is a routine that you establish to prepare for class. There are several characteristic associated with study habits. First is timing. Many students prefer to study in the morning when they are rested and their minds are fresh. Others prefer to study very late at night when most students have retired, and it is quiet. You should choose the time that is the most effective for you. Establishing several blocks of study time each week is an important scheduling habit to develop.

Organization is another important feature for effective study. You will be carrying loads of four to six courses each semester with different sets of requirements and different types of assignments. Organizing the assignments, assigning priorities and making time estimates and commitments for each course are critical tasks. Time management, as discussed previously, is an important element to organization; however, priorities must be set so that assignments are completed and handed in on time. If you list action items and due dates at the conclusion of each class, it is easy to review your notes at the end of each day and establish priorities for your time.

Where you study is as important as when you study. Single desks in the library stacks are ideal. It is quiet, smoke free and there is little or no chance of interruptions. If you form a small study group, tables in the library will serve your purpose. You may talk quietly and discuss problems and approaches to be used in their solution. You may find your dorm room suitable if your roommate is cooperative. If not—head for the library. Student lounges may be suitable in some instances, but they are often noisy with students talking, eating and moving about.

Developing smart study methods is also important. Most engineering students spend much of their study time working problem assignments. A typical assignment for a class may include three or four problems. The instructor probably intends the students to spend 15 to 20 minutes per problem. Students understanding the material will be able to complete the assignment in an hour or so; however, if you have difficulty with one or more problems you will need to devote additional time to complete the work. How much time should you spend before asking for help? You probably should not devote more than about 30 minutes trying different approaches for the solution to a difficult problem. If you have not solved the problem by then, seek help from your TA, instructor or a study group. Study groups will be discussed later in this chapter.

Prepare neat homework solutions. A paper filled with scrawls and incomplete erasures is a clear indication to the instructor grading your work that you do not understand the material. If you know what you are doing, most solutions can be completed with a dozen lines and a diagram or two. Plan your approach and execute the answer. If you make a mistake, either erase completely or rewrite the solution. Always use a pencil because it is easy to erase after you have made a mistake. Ballpoint pens are not appropriate writing instruments for most students.

2.6 PROBLEM SOLVING METHODS

From your first semester until graduation, you will be required to solve problems that require accurate numerical answers. A large percentage of your grade in science, mathematics, engineering science and engineering courses will be a result of your performance in solving these problems. It is vital that you master a proven technique to solve problems. It is recommended that you follow an eight step plan as indicate below:

1. Write the problem statement as you begin the solution.
2. Carefully read and understand the problem.
3. Identify the unknown quantity that you are being asked to determine.
4. Write the rule or equation you plan on using in your solution.
5. List the information provided and prepare sketches as required.

6. Execute the solution.
7. Add units as required to the final result.
8. Check the solution for numerical errors.

Let's consider an application problem in the first calculus course to illustrate this approach.

Step 1: Write the problem statement.⇒⇒Through how many revolutions does a bicycle wheel with a radius of one-foot turn when the bicycle travels one mile[10].

Step 2: Carefully read the problem and understand the physics involved.

Step 3: It is clear that the unknown quantity is the number of times the wheel turns (without slipping) as the bicycle travels a distance of one mile. Hence, the controlling equation is:

$$S = N (2\pi R). \tag{2.1}$$

Step 4: List the known information pertaining to the problem as:
$S = 1$ mile $= 5280$ ft is the distance traveled.
N is the number of turns made by the wheel, which is the unknown.
$R = 1$ ft. is the radius of the wheel.

Step 5: Execute by solving Eq. (2.1) for N to obtain:

$$N = \frac{S}{2\pi R} = \frac{5280\,\dfrac{\text{ft}}{\text{mile}}}{2\pi\,\dfrac{1\,\text{ft}}{\text{rev}}} = 840.3\,\frac{\text{rev}}{\text{mile}} \tag{a}$$

Step 6: Add the units required to the numerical answer. Checked the units and add rev/mile with the numerical result.

Step 7: Check the numerical result. You should mentally perform an arithmetic estimate of the answer and check if it is reasonable. Also, key the numbers into the calculator a second time to verify the result. Many careless errors are detected with this simple procedure.

With repeated usage, the seven steps will become a ritual that you will follow with little thought. You will also find that checking your results for accuracy and for units is important because errors of both types are common.

2.7 COLLABORATIVE STUDY OPPORTUNITIES

Collaborative study is encouraged in college; however, this statement this does not give you license to blindly copy the work of a fellow student. Collaborative study is defined as two or more people working together sharing ideas and discussing approaches for solving problems. Collaboration enables you to find errors in solutions and then to discuss these errors. All parties involved benefit from this discussion. The individual with the correct solution increases the depth of his or her understanding by explaining the proper approach. The individual with the wrong answer discovers the correct approach from a fellow student. The student works the problem again and has another chance

---

[10] Problem 40, page 57 of <u>Calculus with Analytic Geometry,</u> by Robert Ellis and Denny Gulick, Sanders College Publishing, New York, 1994.

to learn and to succeed. The opportunities for learning are enhanced by students teaching each other in this manner.

Regular meetings of small study groups reinforce study habits. The group bonds with time, and learning becomes a social experience. Respect, confidence and esteem grow as group members become comfortable with one another.

Forming small (two to four members) study groups for each of your classes is suggested. Meet at a scheduled time after all members of the group have had an opportunity to work on an assignment. Compare results, share successes and discuss difficulties. Seek knowledge from one another to determine the correct approach for solving each and every type of problem. If all members of the study group are encountering difficulty, seek help from your teaching assistant (TA) or your instructor.

2.8 WORKING THE SYSTEM

Believe it or not, a system is in place in the College of Engineering that is designed to help you succeed. Well-trained professional counselors housed in the Dean's offices usually provide advising for first year students. They will listen to your problems—academic or personal and will provide valuable advice. However, these counselors are only vaguely aware of your existence. You are one among hundreds of students entering the college each fall, and they may not be aware of your concerns. You must seek them out, arrange an appointment, and frame your questions and appeals for help. They often can find remedies for your academic pain.

Your instructors also want you to succeed in their courses. They have organized the course material so as to maximize your learning opportunities over the duration of the semester. They have scheduled preliminary and final examinations for the course that provide you with several opportunities to demonstrate your level of understanding. They assign, collect and correct many homework assignments that permit you to develop problem-solving skills important in performing well on examinations. But you must engage in the process. To be fully engaged means:

- Prepare in advance for every class
- Attend class
- Arrive on time
- Stay awake and alert
- Ask questions when you become confused
- Complete homework assignments correctly and on time
- Stay current with all assignments

If you participate fully in the educational process and are having difficulties, arrange an appointment to discuss your problems with the instructor. He or she will have office hours posted. Avail yourself of the help offered, but do not go to the instructor's office and say, "I don't understand." Be prepared with questions about specific principles that are causing you difficulty. Show your work and indicate where you went astray. Your instructor will help you with specific questions, but he or she will not become your tutor. Their obligation for your success does not include tutoring you.

2.9 PREPARING FOR EXAMINATIONS

If you attend class, understand the material presented in class, and successfully complete all of the homework assignments, preparing for an examination is not difficult or time consuming. Exam preparation becomes time consuming if you have fallen behind in your daily preparation for the class and find it necessary to cram. There is no need to cram if you **stay current**. Let's assume that you have studied diligently and are reasonably current with the material. What should you do in preparing for the examination?

A confident student with a good set of notes should try to write the examination by anticipating the instructor's questions. Anticipating the type of questions on the examination is not only possible; it is usually easy to do. In his or her presentation to the class, the instructor will emphasize a few topics indicating their importance. The instructor will make statements like:

- "This is really very important."
- "If you remember anything remember this."
- "I really like this derivation."
- "This is the most important principle in the course."

The instructor will usually dwell on the important topics and move quickly through the ones considered of less importance. Review you notes and place a star beside the passages where the instructor signaled either importance or his or her interest. In the typical five or six week period for a preliminary examination you may find about a half a dozen stars. Write a question about the topics you have identified as important and ask one of the members from your study team to solve the problems on your exam. Discuss this process with your study team and arrive at a dozen questions that the team solves. After a few examination experiences, you will gain skill at anticipating the instructor and be able to predict the form, if not the substance, of at least half of the examination questions in advance.

2.10 STRATEGIES FOR TAKING EXAMINATIONS

There are strategies that you may exercise in taking an examination that will enhance your performance. Let's discuss some of them:

Confidence: You must study and learn until you become confident that you understand the material and that you will perform well on the examination. If you believe you will screw up the examination, it could be a self-fulfilling prophecy. Confident does not mean cocky, but rather comfortable with the subject and the instructor.

Calm: To be calm, you are at peace with yourself. You know that you have prepared for the examination. You have demonstrated the ability to perform well on the homework assignments. You are not nervous, and you look forward with anticipation to the opportunity to demonstrate your problem solving skills.

Plan and Select: Read each problem in the entire examination. Be certain that you understand each problem. If not, ask the instructor for clarification. Then arrange the problems in an order—from the easiest to the most difficult. Many instructors will arrange their examinations in this fashion with the easy problem first to give you confidence and the more difficult problems later to give range to the grades. Other instructors will arrange problem chronologically with the first problem covering material studied early in the examination period and the last problem covering much more recent material.

Most hourly examinations are limited to a small number of problems. The instructor has timed the examination and in most cases those with experience and knowledge in the subject can complete it in 15 to 20 minutes. You have 50 minutes, which is usually more than sufficient to complete your solutions, **but not adequate to study during the examination**. Divide one-half of the time for the exam (25 minutes) by the number of problems on the examination. Suppose there are five problems on the examination. They you should plan on spending five minutes to write the solution to each problem, and hold 25 minutes in reserve.

Execute: Begin with the easiest problem because you know the approach for its solution. If your initial assessment of the difficulty of the problem was correct, you successfully complete the solution in a few minutes and can begin the solution of the next easiest problem. You are ahead of schedule and full of confidence. You continue in this manner until, say problem No. 4, when you encounter a snag and become stuck. When you have expended five minutes on problem No. 4, stop your efforts and move to the next problem. You should return to this problem only if time remains at the end of the examination. Continue solving problems and completing solutions until your initial time allotment of 25 minutes has expired. Hopefully, most of your work is complete on all but one or two problems and you have already scored enough points to insure a good grade on the examination.

Next, take 15 of the remaining 25 minutes and return to the problem causing difficulty. Review the problem statement and the solution to the point of the snag. Perhaps you will find an error, a missing formula, or an omission and be able to complete the solution. If not write the instructor a brief note explaining your difficulty. He or she may reward your efforts with generous application of partial credit.

Checks: Use the final ten minutes of the examination for checking your results and insuring that all of the numerical answers are accompanied with appropriate units. There are three different checks that are useful in discovering errors. The first is a simple sanity check. Does the number look reasonable or is it obviously too large or too small? Common sense often is a good guide to finding obvious errors.

The second is a numerical check. Examine the equation and make a mental estimate of the result. With practice you will soon be able to estimate the answer to within ± 20%. Next, run the numbers in the equation through the calculator again. Rerun the numbers until the answer is confirmed. In many instances, a button is pressed inadvertently and errors result. Recalculating eliminates these careless errors.

The third check is for dimensional homogeneity. Sometimes errors occur because you fail to recognize the need for unit conversions or you do not make the conversions correctly. By checking an equation for dimensional homogeneity, it is possible to locate and correct these errors. Equations employed in the solution of engineering problems are dimensionally homogenous—they are independent of the units of measurement. Consider an example from physics where the distance S that a baseball falls after is dropped from a very tall building is to be determined. You seek the distance traveled by the ball after it fallen for 6 seconds. If you assume[11] that the drag due to air resistance is negligible, the equation governing the distance the ball falls is given by:

$$S = gt^2 / 2 \qquad\qquad (2.2)$$

        where g is the gravitational constant
             t is time of the fall

When applying this equation, you must insure that it is dimensional homogenous by using units that are consistent on both sides of the equation. To illustrate the various options you have in dealing with Eq. (2.2), consider the three following cases:

    Case 1: U. S. Customary units using units of feet for length and seconds for time. The gravitational constant in this system of units is 32.17 ft/s^2. Accordingly, you write Eq. (2.2) as:

$$S = gt^2 / 2 = (32.17 \text{ ft/s}^2)(6 \text{ s})^2/2 = (32.17 \text{ ft/s}^2)(36 \text{ s}^2)/2 \qquad (a)$$

---

[11] The drag is not negligible. It is assumed to be negligible to simplify the mathematical solution leading to Eq. (2.2). As your skills develop, drag forces will be considered and you will have the opportunity to solve the more complex mathematical expressions that result when the drag force is introduced.

The time unit "s" cancels out on the right hand side of Eq. (a) and the remaining unit represents a length expressed in feet. Hence:

$$S = 579.1 \text{ ft} \tag{b}$$

Case 2: U. S. Customary units using units of inch for length and seconds for time.

The gravitational constant in this system of units is g = (32.17 ft/s²)(12 in./ft) = 386.0 in./s². Accordingly, you write Eq. (2.2) as:

$$S = gt^2/2 = (386.0 \text{ in./s}^2)(6 \text{ s})^2/2 = (386.0 \text{ in./s}^2)(36 \text{ s}^2)/2 \tag{c}$$

The time unit "s" cancels out on the right hand side of the equation and the remaining unit represents a length expressed in inches. Hence:

$$S = 6948 \text{ in.} \tag{d}$$

Case 3: SI units using units of meter for length and seconds for time.

The gravitational constant in the SI system of units is 9.807 m/s². Accordingly, you write Eq. (2.2) as:

$$S = gt^2/2 = (9.807 \text{ m/s}^2)(6 \text{ s})^2/2 = (9.807 \text{ m/s}^2)(36 \text{ s}^2)/2 \tag{e}$$

The time unit "s" cancels out on the right hand side of Eq. (a) and the remaining unit represents a length expressed in meters. Hence:

$$S = 176.5 \text{ m} \tag{f}$$

A careful check of units on each side of the equation will enable you to locate errors introduced by using inconsistent units in evaluating equations.

2.11 ASSESSMENT FOLLOWING THE EXAMINATION

Errors on an examination can be grouped into three classes.

1. Careless mistakes
2. Execution errors
3. Concept errors

Careless mistakes are omission of units—the correct numbers substituted into the correct equation but the wrong answer for the result, reversing digits in a number, etc. In this case, you clearly understood the material but you lacked focus. It is recommended that you slow down, and increase your concentration on the problem. More time in checking the results also eliminates many of these careless errors.

Execution errors occur when you understand the principles covered on the examination, but lack skill in applying these principles to one or more specific problems. Omissions of a force in a physics problem, or inappropriate compound combinations in an equation representing a chemical reaction are examples of execution error. Execution errors indicate a need to solve more and different homework problems. "Practice makes perfect" is an expression that is as true in the academic world as it is in the sports world. True, you must have talent in both the sports and academic worlds, but you would not have been admitted to the College of Engineering without the necessary academic talents.

Concept errors are much more serious. They occur when you do not employ the correct principle in your attempt to solve a problem. Concept errors indicate that you do not understand the material. These errors signal your instructor that you are in trouble. Hopefully, the instructor will identify this type of error with a comment on the graded examination. If an explicit comment is not made, the instructor may signal his or her concern by giving you zero credit for your futile efforts toward a solution. If you have concept errors it is important that you seek help from the instructor. It will be necessary to devote more time to the course until you have mastered the material.

It is important to carefully review your performance following an examination. The instructor will usually devote class time to show the correct procedure for solving every problem on the examination. You will have the information necessary to study your mistakes. Classify them and adapt your study habits to improve your performance. If you find an execution error, ask questions of your study group or your instructor until you clearly understand the correct solution to the problem.

Unfortunately, many students do not attempt to learn from their errors. It is essential that you understand why you made the error. You lost points on an examination once for this error; it is senseless to lose points again for the same error. Engineering is a discipline that builds from a basic foundation to more advanced subjects. If you have flaws in the foundation, your performance in the more advanced subjects will be impaired. Make it your business to identify your errors on an examination, correct your procedures and **never never never repeat the same error**.

2.12 SUMMARY

An overview for your success as a student in a College of Engineering has been presented. Success as a student of engineering depends on setting goals and priorities, developing confidence in your ability to perform and being dedicated to staying current in each subject. Also important is attending class, managing time, developing good study habits, managing your relationships with many new people and seeking help when the need arises.

Time management is based on schedules with carefully allocated time to all activities including attending class, studying, open or flexible time, social meetings, and maintaining your physical well being. Weekly schedules, so important in college, have been described in considerable detail and a sample schedule is given. A daily schedule is also presented that allocates time for specific tasks for the entire 24-hour day.

Many students find themselves short of funds, and they resort to working to cover the shortfall. Working is great if you can manage the time without endangering your performance in each and every course in your schedule. However, some students cannot manage time well and often drop a course or two before the end of the semester. This practice lengthens their time to complete the B. S. degree[12]. Clearly time management and money management are related. Accordingly, procedures for estimating costs for community and resident students, for both monthly and weekly budgeting and for daily recording of expenditures have been described. The use of credit cards has been discouraged by illustrating the cost of a credit card loan.

A brief discussion has been given on peer management. You will be interacting with several new groups of friends in the next few years. Your relations with some of these groups will be very helpful. Study groups and design teams provide opportunities for collaborative learning. Groups formed to workout in the gym or on the track help maintain your physical well-being. Most social groups help provide you with an enjoyment of the educational experience. Warnings have been given to avoid excessive time committed to social activities and to avoid groups that cause behavioral

[12] The average time to complete a B. S. degree in engineering is about 4.8 years in State supported colleges.

concerns. It is important to compare ethical standards of a group with your own set of standards. Do not participate with any group with which you are not ethically comfortable.

Study habits so essential to your success have been discussed. Suggestions have been made for when and where you study. Advice has also been given on how to study smart and how to prepare papers that are respected when reviewed by your instructor. Study groups have been encouraged because of the many benefits of collaborative study.

An eight-step approach to problem solving was reviewed. While eight steps may seem cumbersome when you are trying to hurry through an examination, experience shows that the process actually saves time by reducing errors. The seven steps include:

1. Write the problem statement as you begin the solution.
2. Carefully read and understand the problem and identify the unknown quantity that you are to determine.
3. Write the rule or equation you plan on using in your solution.
4. List the information provided and prepare sketches as required.
5. Execute the solution.
6. Add units as required to the final result.
7. Check the solution for numerical errors.

An example was provided demonstrating the problem solving approach.

Collaborative study opportunities were described. Study teams or design teams are encouraged for all of your classes. Meet at scheduled times after each team member has had an opportunity to work independently on an assignment, and then compare and discuss approaches and results. Student to student learning is effective. Respect, confidence and esteem grow as the group members bond and become comfortable with one another.

There is a system in place in the College of Engineering to help you succeed in your studies. The counselors, instructors and teaching assistants are ready to help you. However, they usually will not initiate the process. It is your responsibility to seek help. Procedures for helping yourself and for seeking help from others in the College have been described.

Next, techniques for preparing for an examination have been discussed. The most important factor in preparing for an examination is to stay current and avoid the need to cram. Finally, strategies to follow in taking an examination were described. They include:

- Be confident
- Be calm
- Select problems and plan your approach
- Manage your time on each problem
- Execute the solution
- Check the results

Follow this strategy when you take your next examination and determine if it is helpful.

REFERENCES

1. Combs, P. and J. Canfield, <u>Major in Success: Make College Easier, Fire Up Your Dreams, and Get a Very Cool Job</u>, 3rd Ed., Ten Speed Press, Berkeley, CA, 2000.
2. O'Brien, P. S., <u>Making College Count: A Real World Look at How to Succeed In and After College</u>, Graphic Management Corporation, Green Bay, WI, 1996.
3. Hanson, J., <u>The Real Freshman Handbook: An Irreverent and Totally Honest Guide to Living on Campus</u>, Houghton Mifflin, Boston, MA, 1996.
4. Carter, C., <u>Majoring in the Rest of Your Life: Career Secrets for College Students</u>, 3rd Ed., Farrar Straus and Giroux, New York, NY, 1999.
5. Davis, L., <u>Study Strategies Made Easy</u>, Specialty Press, New York, NY, 1996.
6. Dodge, J., <u>The Study Skills Handbook: More Than 75 Strategies for Better Learning</u>, Scholastic Trade, New York, NY, 1995.
7. Luckie, W. R., et al, <u>Study Power Workbook: Exercises in Study Skills to Improve Your Learning and Grades</u>, Brookline Books, Cambridge, MA, 1999.
8. Tufariello, A., <u>Up Your Grades: Proven Strategies for Academic Success</u>, vgm Career Horizons, Lincolnwood, IL, 1996.
9. Hjorth, L., <u>Claiming Your Victories: A Concise Guide to College Success</u>, Houghton Mifflin, Boston, MA, 2000.
10. Jensen, E. and T. Kerr, <u>Student Success Secrets</u>, 4th Ed., Barron's Educational Series, Hanppauge, New York, NY, 1996.

EXERCISES

2.1 Prepare three different lists of goals—short term, intermediate and long term. Define the time for accomplishment of each set of goals and provide your motivation for setting each goal.
2.2 Do you believe that confidence is important to your success in achieving the goals stated in Exercise 2.1? Explain your response to this question.
2.3 Is it possible to raise your confidence level? How?
2.4 Describe your plan for staying current in each and every subject for which you have registered this semester.
2.5 Did you cut class in high school? Do you plan to attend all of your classes this semester? If you cut a class, how do you plan on mastering the material covered in that class?
2.6 How many total clock hours per week hours do you intend to devote to your studies this semester? How many credit hours are you carrying? How many study hours are you planning to spend for each credit hour earned?
2.7 Write a critical assessment of your study habits—good and bad.
2.8 Identify the friend who you find most helpful as you adapt to college life. Identify the friend who causes you the most difficulty in adapting. Write a short paper describing the reasons for your determination. Also indicate your plans for the future of these relationships.
2.9 Locate the Office for Freshman Advising and the offices of each of your instructors. The instructors should have their office hours posted. Did you find these postings and did you make a note of them?
2.10 Prepare a weekly schedule for your activities this semester similar to one shown in Fig. 2.2. Also list time reserved for class, study and social activities. How many open hours were you able schedule?
2.11 Describe the provisions in your schedule for taking care of your physical and mental health.
2.12 Describe the consequences of becoming ill for three weeks in the middle of the semester.

2.13 How many hours do you sleep each night? Do you maintain a regular schedule for retiring and rising?

2.14 Prepare a daily schedule for each day of the week. Identify your open periods and discuss a strategy for using them when an examination is scheduled.

2.15 Discuss your balance between working, studying and playing over the weekend.

2.16 Suppose you are working 16 hours per week, prepare a weekly schedule. Also list what time has been reserved for class, study and social activities. How many open hours were you able schedule?

2.17 What is a strategy for preparing for examinations for students working 16 hours per week?

2.18 How many hours do you believe you could work per week without significantly affecting your performance in class?

2.19 Prepare a listing of your mandatory educational expenses.

2.20 Prepare a listing of your discretionary educational expenses.

2.21 Describe the rationale for each item listed under discretionary educational expenses.

2.22 Prepare a budget for your costs for this academic year. Are you living on campus in a dorm? Are you commuting to school?

2.23 Does your budget of annual expenditures indicate a shortfall or a surplus? If it is a shortfall, how do you plan to raise the funds necessary to balance your budget?

2.24 Prepare a monthly budget showing your income and expenses.

2.25 For the next week, prepare a daily record of expenditures.

2.26 Analyze your expenditures for the past week using the data revealed in your daily records. Is it possible to reduce your spending without severe sacrifices?

2.27 Do you have one or more credit cards? Do you diligently pay-off the outstanding balance each and every month?

2.28 What is the annual interest rate on each of your credit cards?

2.29 Suppose you have an average outstanding balance of $1252 with a credit card that carries an annual interest rate of 21%. Determine the monthly interest charges and the total interest charges accumulated over the year.

2.30 Do you know the procedure to apply for scholarships, grants or financial aid?

2.31 Have you discussed your financial situation with the counselors in the Office for Student Financial Aid?

2.32 Do you have a friend who sometimes behaves in ways that cause you concern? What actions cause you concern? What do you do when you observe this behavior? Do you plan on continuing this relationship?

2.33 How many blocks (two to three hour periods) of study time have you scheduled each week?

2.34 Where do you study? Which location do you find the most conducive? Why?

2.35 Have you established a procedure for controlling the amount of time devoted to a typical assignment? Why not?

2.36 Using the problem solving approach outlined in Section 2.5, solve the following problem: A 36 inch diameter pipeline services two cities 87 miles apart with fuel oil. Determine the number of gallons of fuel oil required to initially fill the pipeline.

2.37 Using the problem solving approach outline in Section 2.5, solve the following problem. A V-notch weir, illustrated in Fig. Ex 2.37, is used to measure the rate of flow of water in an open irrigation channel. The relation used to convert the measured height "h" to flow rate Q is given by:

$$Q = (8/15)(2g)^{1/2}\tan(\theta/2)\, C_D\, h^{5/2}$$

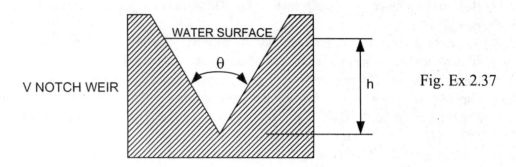

V NOTCH WEIR

Fig. Ex 2.37

Determine the flow rate and express the result in terms of gallons per minute (GPM) if the head h of the flow through the weir is measured at 14 inches and the angle $\theta = 70°$. Note that the drag coefficient for the weir is $C_D = 0.60$, and g is the symbol for the gravitational constant.

2.38 Have you joined a study group (team) for this class? If so, when and where do you study?

2.39 Which course do you consider the most difficult this semester? Have you joined a study group associated with this course?

2.40 Write a short paper describing the benefits of collaborative learning for your study efforts. What are some of the disadvantages?

2.41 What is your role in your favorite study group? What is your role in the study group you least enjoy?

2.42 Write a short paper describing how you "work the system?"

2.43 Have you ever visited your instructor's office during office hours? Is he or she ready to receive you? Did you find the help provided useful? Do you plan on visiting again when you encounter difficulties with the course material?

2.44 Do you stay current in the mathematics course you are taking? If not, why not?

2.45 Which is the most difficult course you are taking this semester? How much time do you allocate for study in this course? Is that sufficient? What other strategies do you employ to improve your performance other than study time?

2.46 What actions do you take in the week before the scheduled date of an examination to prepare?

2.47 What actions do you take in the day before the examination to improve your performance?

2.48 Write a short paper describing your strategies for taking an examination?

2.49 Write a short paper describing your actions after an examination is returned to you with a grade of 70.

2.50 Write a short paper describing your actions after receiving a grade of 78 on an examination.

2.51 Write a short paper describing your actions after receiving a grade of 53 on an examination.

2.52 Write a short paper describing your actions after receiving a grade of 92 on an examination.

2.53 Explain the methods you use to check answers during an examination?

2.54 Do you find yourself rushed when taking an examination? Which subjects cause you the most difficulty in completing the examinations on time? Explain why you are so rushed on these examinations.

2.55 What actions do you plan for managing your time on examinations more effectively?

PART II

PRODUCT AND SYSTEM DEVELOPMENT PROCESSES

CHAPTER 3

DESIGN AND PRODUCT DEVELOPMENT

3.1 THE IMPORTANCE OF PRODUCT DEVELOPMENT

We use hundreds of products every day. Products surround us. This morning I prepared breakfast by toasting bread, frying eggs, and making coffee. How many products did I use in this simple task? The toaster (Philips), the frying pan (T-FAL), the coffee maker (Krups) and the gas cooktop (Kitchen Aid) are all products that are designed, manufactured and sold to customers both in the U. S. and abroad. Some products are relatively simple, such as the frying pan with only a few parts, yet some are much more complex, such as your automobile with several thousand parts. Some products are inexpensive and have a short life prior to disposal. Other products are very expensive and are expected to provide dependable service for many years. Some products require periodic maintenance while others are disposable. Regardless of the expected life or cost all products are expected to perform well and to satisfy the customers' needs.

Corporations worldwide continuously develop their product lines with minor improvements introduced every year or two with more major improvements every four or five years. Product development involves many engineering disciplines and is a major responsibility of engineers entering the work place, as illustrated in Table 3.1. An examination of Table 3.1 shows that designing, manufacturing, and product sales represent more than 80% of the responsibilities of mechanical engineers in their initial position in industry. This distribution of activities is typical of many of the engineering disciplines where engineers are responsible for developing either new products or new manufacturing processes. The viability of many corporations depends on the introduction of a steady stream of successful products to the marketplace. A successful product must satisfy the customer by providing robust and reliable service at a competitive price that completely meets the customer needs.

3.2 DEVELOPMENT TEAMS

Interdisciplinary teams develop products with representation from several disciplines including engineering, marketing, manufacturing, production, purchasing, etc. A typical organization chart for a team developing a relatively simple electro-mechanical component is shown in Fig. 3.1. The team leader coordinates and directs the activities of the designers (mechanical, electronic, industrial, etc.) and is supported by a marketing specialist to ensure that the evolving product meets the needs of the customer. Financial, sales and legal assistance is usually provided on an as-needed basis by corporate staff or outside contractors.

A purchasing agent supports the team leader by establishing close relations with the suppliers who provide materials and component parts used in the final assembly of the product. In the past 10 to 15 years, the relationship between suppliers and end-product producers has changed significantly. Suppliers have become part of the development team; they often provide a significant level of engineering support that is external to the internal (core) development team. In some developments, these external (supplier funded) development teams—when totaled—are larger than the internal development team. For example, the internal development team for the Boeing 777 airliner involved 6,800 employees, and the external teams were estimated to include about 10,000 employees [2].

Table 3.1
Responsibilities of Mechanical Engineers in Their First Position [1]

ASSIGNMENT		PERCENT OF TIME
Design Engineering		40
Product Design	24	
Systems Design	9	
Equipment Design	7	
Plant Engineering/ Operations/ Maintenance		13
Quality Control/ Reliability/ Standards		12
Production Engineering		12
Sales Engineering		5
Management		4
Engineering	3	
Corporate	1	
Computer Applications/ Systems Analysis		4
Basic Research and Development		3
Other Activities		7

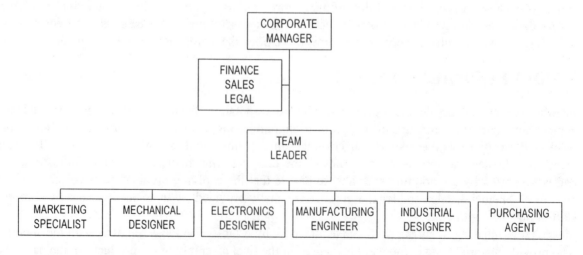

Fig. 3.1 Organization of a product development team for a relatively simple product.

3.3 DEVELOPING WINNING PRODUCTS

There are two important fundamentals involved in developing successful products that win in the competitive marketplace. The first is the quality, performance and price of the product. The second is the time and the cost of the development, and the manufacturing cost of the product in production.

Quality, Performance and Price

Let's discuss the product first. Is it attractive and easy to use? Is it durable and reliable? Is it effective and does it meet the needs of the customer? Is it better than the products now available in the marketplace? If the answer to all of these questions is an unqualified **YES**, then the customer may want to buy the product if its price is right. Next, you need to understand what is implied by product cost and its relation to the price actually paid for the product. Cost and price are distinctly different quantities. Product cost clearly includes the cost of materials, components, manufacturing and assembly. The accountants also include less obvious costs such as the prorated costs of capital equipment (the plant and its machinery), the price of the tooling, the development cost, and even the expense of maintaining the inventory in establishing the total cost of producing the product.

      Price is the amount of money that a customer pays to buy the product. The difference between the **price** and the **cost** to produce a product is **profit**, which is usually expressed on a per unit basis.

$$\text{Profit} = \text{Product Price} - \text{Product Cost} \qquad (3.1)$$

This equation is the most important relation in engineering or in any business. If a corporation cannot make a profit, it soon is forced into bankruptcy, its employees lose their positions, and the stockholders lose their investment. It is this profit that everyone employed by a corporation seeks to maximize while maintaining the strength and vitality of its product lines. The same statement can be made for a business that provides services instead of products. If a business is to make a profit and prosper, the price paid by the customer for a specified service must be more than the cost to provide that service[1].

The Role of Time in the Development Process

Let's now discuss the importance of the development process in producing a line of winning products. Developing a product involves many people with expertise in different disciplines. Also, it takes time, and it costs a lot of money. Let's first consider development time. Development time, as it is used in this context, is time to market, i.e., the time from the product development kickoff to the introduction of the product to the market. This is a very important target for a development team because of the many significant benefits that accrue from being first to market. A corporation realizes many competitive advantages with a rapid development capability. First, the product's market life is extended. For each month cut from the development schedule, a month is added to the life of the product in the marketplace with an additional month of sales revenue and profit. We show the benefits of being first to market on sales revenue in Fig. 3.2. The shaded area between the two curves on the left side of the graph is the enhanced revenue due to the longer sales.

      A second benefit of early product release is increased market share. The first product to market has 100% of market share in the absence of a competing product. For products with periodic development of "new models," it is generally recognized that the earlier a product is introduced to

---

[1] The most recent example of this simple fact is the airline industry with two major airlines currently in bankruptcy proceedings with several other airlines planning for that eventuality. The problem is clearly excessive costs of labor and the effect of work rules, which limit management's flexibility in assigning personnel.

compete with older models—without sacrificing quality and reliability—the better chance it has for acquiring and retaining a large share of the market. The effect of gaining a larger market share on sales revenue is also illustrated in Fig. 3.2. The shaded area between the two curves at the top of the graph shows the enhanced sales revenue due to increased market share.

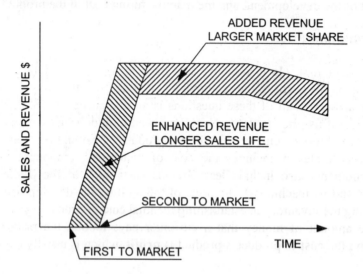

Fig. 3.2 Increased sales revenue due to extended market life and larger market share.

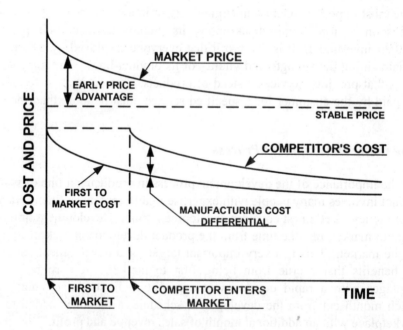

Fig. 3.3 The development team bringing the product to market first enjoys a higher initial price and subsequent cost advantages from production efficiencies.

A third advantage of a short development cycle is higher profit margins. If a new product is introduced prior to availability of competitive products, the corporation is able to command a higher price, which enhances the profit. With time, competitive products will be introduced forcing price reductions. However, in many instances, relatively large profit margins can still be maintained because the company that is first to market has added time to reduce the product's production costs.

The workers learn better methods for producing components and reduce the time needed to assemble the product. The advantage of being first to market, with a product where a manufacturing learning curve exists, is shown graphically in Fig. 3.3. The manufacturing learning curve reflects the reduced cost of manufacturing and assembling a product with time in production. These cost reductions are due to many innovations introduced by the workers after mass production begins. With time and improvements in worker productivity due to experience, it is possible to significantly reduce production costs.

Let's next consider development costs, which represent a very important investment for the companies involved. Development costs include the salaries of the members of the development team, money paid to subcontractors, costs of pre-production tooling, expense of supplies and materials, etc. These development costs can be significant, and most companies must limit the number of developments in which they invest. The size of the investment may be appreciated by noting that the development cost of a new automobile is estimated [2] at about $1 billion, with an additional expenditure of $500 to $700 million for the new tooling required for high-volume production.

We have included this discussion on time and cost of product development to help you begin to appreciate some of the business aspects of developing winning products. Any company involved in the sale of products depends completely on their ability to continuously introduce winning products in the marketplace in a timely manner. To win, the development team must bring a quality product to market that meets the needs of the customer. The development costs must be minimized while maintaining a product development schedule that permits an early (preferably first-to-market) introduction of the product.

3.4 LEARNING ABOUT PRODUCT DEVELOPMENT

The best way to learn about product design is to work on a development team and develop a prototype. For this reason, we have selected the design of a smart deicing system for bridges for your first development effort. The degree of complexity of the system your team will be required to design and build will be defined in detail in Chapter 4. You are to work as a member in a team of four to six students to develop a working model (prototype) of this system. In some instances, companies develop three or four prototypes before finalizing a particular product design. However, time available during this semester will limit each team to a single prototype of the smart system.

We understand that the development of a smart system is a difficult task particularly when the assignment is given so early in your engineering program. However, our experience with more than 5,000 students beginning to study engineering indicates that you will gain immensely from your efforts. We know from previous classes that first year students are very creative, that you are cooperative, work well within a team structure and design and build successful prototypes. Of course, the product selected for your initial development project cannot be overly complex. Since beginning this course 1990 at the University of Maryland, College Park, the students participating in this course have developed playground equipment, windmills for the generation of electricity and pumping water, furniture, human-powered water pumps, wind-powered vehicles, digital and analog bathroom and postal scales, a solar-cooking unit, a solar-desalination facility and a weather station. Extensive student surveys clearly indicate that many hours are devoted to the project. The surveys also show that you have fun working with your hands, learn a lot about engineering and appreciate the experience of working as a member of a development team.

3.5 PROTOTYPE DEVELOPMENT

We often divide the prototype development process into three major phases that include:

1. Designing a model or a prototype.
2. Preparing the assembly kit.
3. Assembling and evaluating the prototype.

We will begin the process by providing the design team with a description of the smart system and its intended application in rather general terms. Also, the team will be provided with design requirements pertaining to the system's specification—the ability of the deicing system to prevent ice from forming in weather conditions where all of the factors are present that will permit ice to form on a bridge. The description will also include the information pertaining to measuring the temperature of the concrete block used to simulate the pavement of the bridge. Information is also provided about the deicing solution and the requirement to disperse this solution to prevent the formation of ice on the surface of bridge. We suggest that you make full use of the library to search for literature and patent files. There is absolutely nothing to be gained by reinventing the wheel (or a temperature sensor in this instance).

After you understand in general terms the design requirements for the smart system, we will introduce the product specification [3] in more detail. The product specification, written relatively early in the development process, is usually prepared in tabular format. It tersely states design requirements and design limits or constraints. Remember engineers always design with constraints and/or limits. The specification also includes all of the performance criteria and provides design targets on weight, size, power, cost, etc.

We will place explicit design constraints that limit your flexibility in designing the model of the deicing system. First, there is a maximum limit of $25.00 per team member for the cost of materials, supplies, and components purchased for the system and the testing of the model. The idea is to keep your out-of-pocket expenses at an affordable level, while giving the team adequate funds to build a model that will demonstrate the design concepts. We will also limit the size of the hardware used to construct the model, so that the laboratory facilities and workshops available in the College can handle the large number of systems that will be designed and constructed this semester.

3.6 DESIGN CONCEPTS

After everyone on your development team understands the design specification and the design limitations, you are ready to generate design concepts. (A much more complete description of the design requirements, constraints, etc. is given in Chapter 4.) Design concepts are simply ideas for performing each function involved in the operation of the deicing system that will help you meet the system specifications. To illustrate design concepts, let's examine the functions (processes) involved in automatically activating a deicing system for a bridge, as depicted in the block diagrams presented in Fig. 3.4.

Design Concept—Design Function Relationship

In an autonomous (smart) deicing system it is necessary to sense the weather conditions that will lead to ice formation on the road surface of a bridge. Typically three weather related conditions exist concurrently when ice forms on the roadway of a bridge.

1. The temperature of the road surface is at or below freezing.
2. The temperature of the air above the road surface is at or below freezing.
3. Water is present on the road surface.

Your design team will be required to develop temperature sensors and a sensor to detect the presence of water. The output from these sensors will be recorded in a controller (RCX) that contains a programmable microprocessor and sufficient memory to store the data necessary to control the deicing system. Your team will write a computer program in ROBOLAB™ or another suitable language to automatically activate the flow of a deicing agent when the weather conditions indicate that icing is imminent.

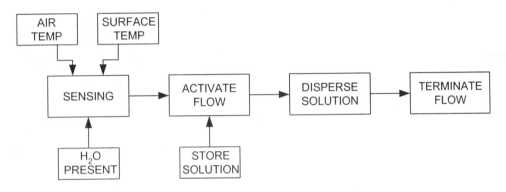

Fig. 3.4 Functions involved in a smart deicing system.

The State Highway Department has specified that a brine solution containing NaCl is to be employed as the deicing solution. Your design team is to develop storage a facility that is sufficiently large to prevent two icing events without the need to replenish the supply of the deicing solution. It will be necessary for your team to develop a mechanism that will initiate the flow of the deicing solution from the storage facility to the road surface on the bridge. This mechanism is to be controlled by the computer program your team has developed and stored in the RCX.

Simply turning on the flow of saline solution will not insure that the bridge's road surface will remain free of ice. It will be necessary for your team to develop a method for dispersing the solution so that a uniform layer of solution covers the road surface. In this case the road surface is simulated with a small concrete block. Finally, when a sufficient quantity of saline solution has been dispersed the controller is to signal the flow control mechanism to terminate flow. This sequence completes a single deicing event. However, the system should be capable of preventing the formation of ice in two independent icing events.

Your team will need to develop many different ideas for each subsystem included in the design of your deicing system. Each concept should be described in considerable detail prior to developing preliminary design proposals for the design concepts. The more completely you develop the design proposals, the better prepared you will be to conduct what is termed a design-trade-off analysis, where you compare several design concepts and select the best one.

Design-trade-off Analysis

To perform a design-trade-off analysis, we consider each design proposal individually and list its strengths and weaknesses. Factors usually considered in this analysis include size, weight, cost, ease of manufacturing, performance, appearance, etc. After the strengths and weaknesses of each design proposal have been listed, you can compare and evaluate each design proposal and select the most suitable for your team's system.

We have briefly discussed design concepts, design proposals and design trade-off analyses for a single function that is involved in the development of an autonomous deicing system. Your team has several functions to consider:

- Sensing the temperature of the surface of the concrete roadway.
- Sensing the temperature of the air above the concrete roadway.
- Sensing the presence of water on the surface of the roadway.
- Writing a program to activate and terminate the flow of the deicing solution.
- Developing a flow control device to initiate and terminate flow.
- Developing a flow dispersal method to provide a uniform layer of saline solution over the simulated surface of the roadway.
- Developing circuits that will permit the RCX to interface with the temperature sensors developed.

The procedure in designing anything is to generate ideas for performing each function, expanding these ideas with more detail until they can be treated as design proposals, and finally to perform a design trade-off analysis where the merits and faults of each proposal are evaluated. When the design proposals are selected for each of the different functions involved in the deicing system, the winning concepts must then be integrated into a seamless system that automatically disperses a sufficient quantity of saline solution over the entire surface of the simulated roadway to prevent the formation of ice.

We often refer to the part of a product that performs some function as a subsystem. The collection of all of the subsystems constitutes a complete system, which when manufactured and assembled, becomes the prototype of the system. The integration of the various subsystems is often difficult, as one subsystem influences the design of the others. The manner in which two subsystems interact is defined as the interface between these two subsystems. In a mechanical application, the power output from an internal combustion engine and the capacity of an automatic transmission is an example of an interface. Frequently, interfaces between the subsystems are troublesome and difficult to manage. It is vitally important that you control the interfaces to permit seamless integration of all of the subsystems without loss of effectiveness and efficiency.

Preparing the Design Documentation

The final step in designing the prototype is the preparation of the design documentation package. The documentation package, which is an extended engineering report, contains:

- Engineering sketches illustrating the design concepts.
- Engineering assembly drawings that describe how the various components fit together to form the complete system.
- A complete parts list.
- Engineering drawings of all the component parts in sufficient detail to permit the component to be manufactured by anyone capable of reading an engineering drawing.
- A written report that describes the design concepts and predicts product performance.

We recognize that many of you may need instruction in graphics, and we have included three chapters in this text to help you prepare three-view and pictorial drawings and to make tables and graphs.

The design documentation package contains a parts list that identifies every unique part employed in the assembly of the system. The quantity required of each component is also included on the line describing the component. For example, if you are going to use one-liter plastic water bottles to store the saline solution, you would:

1. Assign a part number for the bottles to identify that these containers are required to assemble the model of the dispersal system.
2. List the quantity required as "2" in the quantity column. This implies that two containers will be employed in the design—one on each side of the roadway.

3. Describe the bottles in sufficient detail so that another team member can procure exactly the size and type of bottles that are required.

NUMBER	NAME OF ITEM	DESCRIPTION OR DRAWING NO.	QUANTITY	PRICE
PARTS LIST FOR A SOLUTION DISPERSAL SYSTEM				
1	STORAGE TANK	PLASTIC WATER BOTTLE 1 LITER SIZE	2	$ 0.10
2	TUBING	TYGON TUBE 1/4 ID, 3/8 OD, 1/16 IN. WALL	2 m	$ 1.25
3	FACE PLATE	PLASTIC SHEET WITH LABEL FOR SCALE	1	$ 1.15
4	SUPPORTS	1/4 X 1 WOOD STRIPS -------DWG. NO. 100-01	12 m	$ 3.25
5	BASE	1/2 X 4 X 4 PLYWOOD BASE -------DWG. NO. 100-02	1	$ 0.50
6	MOTOR	LEGO GEAR MOTOR B775225-017	2	$ 17.00
7	BEAMS	LEGO BEAMS	4	$ 1.00
8	PINION	LEGO 8 TOOTH SPUR GEAR	2	$ 1.00
9	RACK	LEGO RACK GEAR	2	$ 1.00
10	PLATES	LEGO PLATES	10	$ 1.00
11	BRADS	1/32 SHEET METAL --- DWG. 100-08	1 Box	$ 1.45
12	ADHESIVE	GENERAL PURPOSE STRUCTURAL ADHESIVE	1 Tube	$ 1.25

Fig. 3.5 Example of a parts list for a saline solution dispersal system.

The parts list contains all of the items that are to be purchased for your prototype. It also contains the components that must be manufactured. If the component is to be manufactured, it is identified on the parts list by its name and a drawing number. An example of a parts list is shown in Fig. 3.5.

An engineering report is also included as part of the design package. This report supports the design by describing the key features of your deicing system. Your report should treat each function involved in the system and describe the design concepts that your team considered. A rationale for each design proposal that was adopted based on systematic design trade-off studies, is an essential element. Additionally, the design report contains the analyses used to predict the performance of your prototype. The analysis is conducted before actual testing the model of the system, and both the analysis and the test results are useful in assessing the merits of the product development. More information on preparing of an engineering report is included in a Chapter 14.

Final Design Presentation

The design phase of the product development process is concluded with a final design presentation. This is a formal review of your team's design that is presented to the class (peer review), to the instructor and others in the audience. It is the team's responsibility to describe all of the unique features of the design and to predict the performance of the product. It is the responsibility of the class (peer group) and the instructors to question the feasibility of the design and the accuracy of the predicted performance. If you note a shortcoming or a flaw in another team's design, identify the problem to the team presenting their work. As a peer reviewer of another team's design, be tactful in your critique. Criticism is always difficult to offer to another person. Offer constructive suggestions in good faith and in good taste. If you are on the receiving end of criticism, do not be defensive. The individual offering the critique is not attacking your capabilities; he or she is actually trying to give you a suggestion that might help you achieve a better design.

The purpose of the design review is to locate deficiencies and errors. It is better to correct errors in the paper stage of the process, not later in the hardware phase, when it is much more difficult and costly to fix the problems.

3.7 PROTOTYPE ASSEMBLY AND EVALUATION

Preparing the Assembly Kit

The second major phase of the product development process is to prepare an assembly kit. The assembly kit is a collection of all of the parts, in the appropriate quantities that are required to completely assemble the prototype. The parts list, as described above, is the essential document that directs our procurement and manufacture of the components needed to build the model or the prototype. The parts list identifies everything that we need and indicates the precise quantity of each item required to assemble a single unit.

An efficient method for collecting the required parts is to divide the list into three groups:

- Those parts to be purchased.
- The materials and supplies to be purchased.
- The components to be manufactured.

One or two team members handle purchasing the necessary components, materials and supplies while the other team members work in the student model shop to fabricate the parts according to the detailed drawings prepared in the design phase. It is important to recognize that an engineering drawing defines the part that is being manufactured. Always work from the engineering drawing and not from your memory or understanding of the part geometry.

It is suggested that one team member serve as the "inspector," checking the finished parts against the drawings for the parts that have been manufactured. If the parts have been purchased, the "inspector" should check the purchased items against the parts list to insure that the item meets the description and that the correct quantity has been procured.

Model Assembly

When the assembly kit is complete and checked against the parts list, we can begin the final phase— system evaluation. The first part of the evaluation is to assemble the model of the system. We often call this step the "first article build," because it is the first time that we have attempted to fit together all of the components needed to assemble the prototype. The "first article build" is a success if: all of the parts are available; they all fit together properly; the tolerances on each part and each feature are correct; and the surface finish on all of the parts is acceptable. Prototype design is often judged against the four Fs—form, fit, finish and function. The "first article build" permits us to assess how well the team performed with regard to the first three of the Fs—form, fit and finish.

If the parts do not fit, modifications of one or more parts are required. These modifications require design changes—a dreaded and costly process in the real world. In fact, one of the most important facts used by management to judge the quality of a development team is the number of design changes required both before and after the introduction of a product to the marketplace. Clearly, you want to design each part in a product correctly during your team's first effort and minimize the number of required design changes.

We anticipate that each team will make a few errors in the preparation of the detail design drawings and that changes will be required. The natural tendency is to take the incorrect part to the model shop and to correct (modify) it and move on with the "first article build." This behavior is acceptable only if the team revises the drawing of the incorrect part to reflect the modification made to correct the design deficiency. In the real world, the integrity of the drawing package is much more important than the prototype. The second, and all subsequent assemblies, will be fabricated from the details shown in your engineering drawings and not by examining the prototype. The prototype is often scrapped after it has been built and tested.

System Evaluation

After the assembly of the model of the system is complete, it should be carefully inspected to insure that it is safe. Fortunately, this system is relatively safe since the parts do not undergo significant motion, and the hazards due to extreme pressures and temperatures should not exist. However, be certain that you have eliminated all sharp edges and points on the model, and that you have avoided introducing pinch points. (Pinch points are openings, which close due to motion of one or more parts when a product is operated. Fingers, hair and/or clothing caught in pinch points may represent a hazard).

The final step in this development process is to test your system to determine if it is functional and effective. This is a big day and indeed a big hour. Typically a one-hour time slot has been scheduled for evaluating each team's prototype in the College testing facilities. Your prototype will be judged based on the following criteria:

1. Performance: accuracy over the entire specified range and readability of the display.
2. Capability: ability to meet the specific requirements for the application.
3. Cost: minimize.
4. Design innovation: novelty and creativeness.
5. Quality of parts and assembly: workman (women) ship.
6. Appearance: attractive and consistent with the specified application.

3.8 TEAMWORK[2]

There are three reasons for requiring you to participate on a development team. First, the project is too ambitious for an individual to complete in the time available. You will need the collective efforts of the entire team to develop a working model of the autonomous deicing system during the semester. We plan on pressing the development teams to complete the project on a prescribed schedule. Second, we want you to begin to learn teamwork skills. Experience has shown us that most students entering the engineering program have not developed these skills. From K-12, the educational process has focused on teaching you to work as an individual often in a setting where you competed against other students in your class. We will demand that you begin functioning as a team member where cooperation, following and listening are as important than individual effort. Leadership is important in a team setting, but following is also a critical element for successful team performance. The final reason is that need teamwork experience because you will probably find yourself on a development team early in your career if you take a position in industry. A recent study [4] by the American Society for Mechanical Engineers (ASME), the results of which are shown in Table 2, ranked teamwork as the most important skill to develop in an engineering program. Teamwork was also the first skill, in a list of 20, considered important by managers in industry. We are hopeful that this course will be instrumental in exposing you to team working skills so necessary for a successful career.

As you work within your team in the development of a prototype, you will be introduced to many of the 20 topics listed in Table 3.2. We hope that you will begin to develop an appreciation of the need for these skills and begin to enhance your level of understanding of the design process that requires you to utilize many of them.

---

[2] A chapter on development teams and teaming skills is given in Appendix A.

Table 3.2
Priority Ranking of Skills Considered Important for
New Mechanical Engineers with Bachelor Degrees

1.	Teams and Teamwork	11.	Sketching and Drawing
2.	Communication	12.	Design for Cost
3.	Design for Manufacture	13.	Application of Statistics
4.	CAD Systems	14.	Reliability
5.	Professional Ethics	15.	Geometric Tolerancing
6.	Creative Thinking	16.	Value Engineering
7.	Design for Performance	17.	Design Reviews
8.	Design for Reliability	18.	Manufacturing Processes
9.	Design for Safety	19.	Systems Perspective
10.	Concurrent Engineering	20.	Design for Assembly

3.9 OTHER COURSE OBJECTIVES

While your experience in developing this model of a deicing system is the primary objective of this course, there are several other related objectives. You will quickly recognize a critical need for graphics as you attempt to describe your design concepts to fellow team members. We have included three chapters on graphics to help you learn the basic skills required at this entry level. These chapters include materials on three view and pictorial drawing and graphs and tables.

You will also be required to become familiar with using three additional software application programs: word processing, spreadsheet, and graphics presentation. Our experience indicates that nearly all of you are already proficient with word processing; therefore, we will not cover this topic here. However, many of you are not familiar with how to use a spreadsheet. Accordingly, we have included a chapter describing Microsoft Excel to help you prepare your parts list, perform calculations and prepare graphs. Spreadsheets provide very powerful tools that you will find useful in many different ways for the remainder of your life. We hope that you will take this opportunity to learn how to use this important software tool. We have also included a description of Microsoft PowerPoint to aid you in the preparation of first-rate slides for your design briefings.

As you proceed with the development of the model of your team's system, we frequently will require you to communicate both orally and in writing. Presenting design reviews before the class gives you the opportunity to learn presentation skills such as style, timing and the preparation of visual aids. Preparing the final design report gives you experience in making complete, high-quality engineering drawings and in writing a technical report, which contains text, figures, tables and graphs.

The final objective of the course is design analysis. We understand that your engineering analysis skills have yet to be developed. For this reason, we will present key equations that mathematically model the electrical and mechanical components used in the design of many of the subsystems that you employ in the design of your team's deicing system. We do not expect you to completely understand all of the theoretical aspects of the relatively complicated subsystems involved. Most of you will take several courses later in the curriculum dealing with these subjects in great detail. However, at this stage of your career, we want you to begin to appreciate the relationship between analysis and design. To help you with the analysis, we have included a chapter on electrical circuits, a chapter on sensors and a chapter on programming in ROBOLAB. The coverage is introductory, but it should give you a direction and a start in understanding sensing and programming required to design autonomous systems.

REFERENCES

1. Valenti, M. "Teaching Tomorrow's Engineers," Special Report, Mechanical Engineering, Vol. 118, No. 7, July 1996.
2. Ulrich, K. T., S. D. Eppinger, <u>Product Design and Development</u>, McGraw-Hill, New York, NY, 1995, p. 6.
3. Cross, N., <u>Engineering Design Methods</u>, 2nd Edition, Wiley, New York, NY, 1994, p. 77-81.
4. Anon, "Integrating the Product Realization Process into the Undergraduate Curriculum," ASME Report to the National Science Foundation, 1995.
5. Macaulay, D., <u>The Way Things Work</u>, Houghton Mifflin Co., Boston, 1988.
6. Horowitz, P. and W. Hill, <u>The Art of Electronics</u>, 2nd Edition, Cambridge University Press, New York, 1989.
7. Dally, J. W., Riley, W. F. and K. G. McConnell, <u>Instrumentation for Engineering Measurements</u>, 2nd Edition, Wiley, New York, 1993.
8. Wang, E., <u>Engineering with LEGO Bricks and ROBOLAB</u>, College House Enterprises, 2003.

EXERCISES

3.1 List ten products that you have used today and the companies that manufactured and marketed them.

3.2 Suppose that you do not want to work on product development, manufacturing or sales. What opportunities remain for you in engineering?

3.3 Write an engineering brief describing the characteristics of a winning product.

3.4 Write the most important equation in engineering or business.

3.5 Why is General Motors Corporation reluctant to develop an entirely new model of an automobile?

3.6 What is a design concept? How many concepts will your team generate in developing the model of the deicing system this semester?

3.7 What are the differences between the different temperature sensors available for measuring temperatures near the freezing point of water?

3.8 Why is it important to prepare a parts list during a development program?

3.9 What is an assembly kit and why do we prepare one in the development of the first prototype?

3.10 Why do we invest scarce company funds in developing the model of a complex system?

3.11 What are the safety considerations with which you must be concerned in building and testing your team's system?

3.12 Why do you believe industry representatives ranked teams and teamwork as the most important skill for new engineers?

CHAPTER 4

DESIGNING ANTI-ICING SYSTEMS FOR HIGHWAY BRIDGES

4.1 INTRODUCTION

Preventing snow and ice from forming on a road's surface is more effective than breaking the bond between the ice and the pavement after ice has formed [1]. Anti-icing programs are designed to treat roadways with suitable chemicals prior to a winter storm to prevent icing. Anti-icing is preferred over deicing to maintain road surfaces because delays in implementing deicing strategies are eliminated. Anti-icing improves highway safety and increases driver's mobility during and after snow and/or ice storms. Anti-icing strategies are more economical because these approaches require lighter and fewer applications of chemicals. Thus, savings accrue in cost of chemicals, road crew time and wear on snow removal equipment. Your team's task this semester is to design and develop a road weather information system. This system is critical in predicting the brief window of opportunity for an anti-icing effort prior to the onset of a winter weather event—a snowstorm or a light drizzle turning to sleet.

Two of the important objectives of this textbook are to assist you in learning about the product realization process and to provide an initial and challenging experience in engineering design. This textbook is written to support your efforts as you work within a team structure to develop an engineering system, such as the road weather information system described above. The road weather information system involves a smart anti-icing system that your team will develop for a typical bridge under simulated conditions for freezing rain. In this chapter, we will provide a brief description of the system that is to be designed and the approach that your team will follow in demonstrating that it will be effective when installed in a real application.

For this anti-icing system, your team will be expected to design, build and test a working model of a system that senses the road surface and air temperatures in addition to sensing the presence of water on the bridge's surface. It is anticipated that your team will also develop an algorithm that predicts the onset of the formation of a thin film of ice on the surface of the bridge. Finally, your design team will develop a method for coating the surface of the bridge with a saline solution that will prevent icing from forming on the roadway as the temperature drops to − 10 °C.

This development is intended to be a semester long project[1]. Your instructor will consider your individual performance as well as the team's performance a significant part of your final grade in this course. There are ten deliverables associated with the development of a smart anti-icing system, which include:

---

[1]This course will involve several other requirements and you should expect several assignments drawn from many other chapters included in this textbook in addition to the system design described in this chapter.

1. Prepare a statement of work describing the design specification, the operating principles and the initial design concepts for the smart anti-icing system your team is to develop. (Due the second week of the semester).

2. Present a preliminary design briefing before the class describing your team's initial design concepts, problems encountered, initial plans and the development schedule. We anticipate that you will use a Gantt chart in describing the development schedule. (Due the fourth week of the semester).

3. Present a critical design briefing before the class describing the final design concepts, discussing design trade-offs, predicting performance and updating the development schedule. (Due the sixth week of the semester).

4. Construct models or prototypes of the essential subsystems involved in your design. The purpose of the models or prototypes is to validate a design concept involved in the team's development. Your instructor must approve of this undertaking before the team begins its work. (Due the seventh week of the semester).

5. Submit the first draft of the team's design report. This report includes a description of the sensors and their performance characteristics, a discussion of the operation of the saline solution dispersal system, an analysis of its performance, a discussion of the design trade-offs and a set of high-quality engineering drawings. (Due the tenth week of the semester).

6. Prepare an assembly kit containing all of the parts necessary to build your system. Also write a program in ROBOLAB to initiate and control the anti-icing process. (Due the eleventh week of the semester).

7. Present a model of the team's system for evaluation by instructor. The instructor will also perform a safety inspection and qualify it for testing. (Due the twelfth week of the semester).

8. Conduct test of you team's model and evaluate its performance. Compare its performance with the requirements in the design specification. (Due the next to last week of the semester).

9. Submit the final draft of the team's design report. This report includes a description of the operation of the system, a discussion of problems encountered in the design, a discussion of the design trade-offs, a description of performance during testing, a set of the final engineering drawings and recommendations for design improvements. (Due the final week of the semester).

10. Present a final design briefing before the class describing the final design concepts, discussing design trade-offs, describing test results, explaining failures and successes and making recommendations for design improvements. (This design briefing is part of the final examination for the course).

4.2 THE ANTI-ICING SYSTEM

4.2.1 Temperature Sensors

Your team is to design, build and test two sensors that are to be employed to measure the temperature of the air and the temperature of the concrete surface of a bridge. A description of several temperature-sensing methods is provided in Chapter 7 to assist your team in this development. It is anticipated that the sensor used to measure air temperature will be shielded from the sun so that it provides a reliable measurement of the air temperature. The sensor employed to measure the temperature of the road's surface is to be embedded in a concrete block that is used to simulate the bridge pavement. Surface mounting this sensor is not advisable because of the possibility of damage due to either automotive or pedestrian traffic. Extreme care must be exercised in waterproofing this installation because both temperature sensors will be exposed to the environment for several years.

The temperature of both the air and the pavement will be measured with the RCX controller that is described in Chapter 5. It is anticipated that your team will measure and record the two temperatures at suitable time intervals for a period of one hour during your model test. The data recorded on the RCX is to be downloaded to a PC. A graph of temperature as a function of time for both the atmosphere and the pavement surface is to be prepared using EXCEL and included in your design reports.

4.2.2 Modeling the Bridge Pavement

It is anticipated that your team will model the bridge surface by casting a block of concrete approximately 75 mm deep with lateral dimensions of 300 mm by 400 mm. A small bag of ready mix concrete, available at Home Depot or Loews, should provide a sufficient quantity of material to cast this block. A bag of ready mix contains cement, small gravel and sand. Its contents are mixed with a suitable quantity of water to form thick slurry, which is cast by poring it into an open face mold with appropriate dimensions. One of the temperature sensors should be embedded in this block of concrete to a depth of approximately 20 mm. Then the surface of the concrete should be toweled smooth. After the concrete has solidified and gained sufficient strength, the mould supporting the edges of the block should be removed. This block of concrete with its embedded sensor will serve to simulate the bridge pavement in subsequent experiments that your team will conduct.

4.2.3 Sensing Water on the Bridge Pavement

While pure deionized water is an insulator, the water we drink is a good conductor. Your team can make use of this fact by developing a conductivity sensor to determine whether or not the bridge surface is coated with water. Consider the circuit shown in Fig. 4.1 that consists of two terminals a resistor and a battery for a voltage supply.

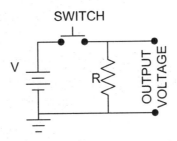

Fig. 4.1 A simple circuit loop contain a switch.

When the switch is open the output voltage is zero; however when the switch closes the output voltage becomes equal to the supply voltage V. If two terminals are placed on either the surface or embedded a small distance below the surface of the concrete block[2], the presence of water will serve as a switch and close the circuit. The use of the RCX in monitoring a continuity (switch) sensor is described later in Chapter 7.

---

[2] Concrete is porous and water will be conducted through its voids some distance below the surface of the block.

4.2.4   The Anti-icing Solution

When the bridge surface is wet and the temperature conditions are below freezing, your team is to coat the concrete surface with a saline solution (brine) that will prevent the formation of ice. It is anticipated that a solution of water mixed with a salt such as NaCl will serve this purpose. The concentration of the salt in the brine is to be determined by the team, but it essential that the salt remain in solution for long periods of time regardless of the temperature. Clearly, the higher the concentration of salt in the solution, the more effective the anti-icing agent will be preventing the formation of ice on the road surface.

4.2.5   Disbursing the Saline Solution

When the sensing system indicates that icing conditions are imminent, your team's system should disperse a thin layer of the anti-icing solution over the surface of the block that simulates the bridge pavement. It is anticipated that the anti-icing solution will cover the entire surface of the block and that the application will provide a layer of brine approximately 3 mm thick. The capability of your system to disperse the solution in a uniform layer over the entire surface of the test block will be a factor in assessing the merits of the design of your team's dispersal system.

    The container for storing the anti-icing system should be sufficiently large to provide a saline solution for a minimum or two anti-icing events. Clearly a valve will be required to open the piping system that carries the saline solution from the storage vessel to the test block. Sufficient pressure will also be required to drive the saline solution through the piping system and to enable dispersal over the surface of the block.

4.3 TESTING THE ANTI-ICING SYSTEM

In addition to simulating the pavement of a bridge, it will be necessary for your team to simulate the atmospheric conditions that lead to icing. Your class instructor will provide a test chamber that will contain a wooden box—150 mm deep with lateral dimensions of 500 mm and 800 mm. This box is sufficiently large to hold your test block with a space of 100 mm around all sides of your test block. Your instructor will also supply a ¼ inch diameter copper tube to serve as a water supply. The copper tube will pass though a box containing dry ice that will cool it to a temperature of + 1 °C. A manual valve operated by your instructor will control the flow rate through this line. A showerhead positioned over your test block will provide a brief period of simulated rainfall that will thoroughly wet the surface of your test block with nearly freezing water.

    Your team will be responsible for simulating the temperature of the air around the test block and the temperature of the concrete block. We anticipate that you will achieve low temperatures (about – 10 °C) by using dry ice. An arrangement suggested for cooling the test block is illustrated in Fig. 4.2. First, pieces of dry ice are placed to form a layer on the bottom of the test frame and then the test block is placed on top of this layer. Next pieces of dry ice are pack around the sides of the test block. The temperature of the test block, which will decrease with time, will be monitored on the display of the RCX. When the temperature drops to – 10 °C or lower, the instructor will at your team's request provide a brief shower of water at a temperature of 1 °C.

    To provide cold air over the test block, it is suggested that dry ice be used to cool a stream of air. An arrangement that your team may consider for cooling the air is presented in Fig. 4.3. A fan with a diameter of about 300 mm is used to force air through a short channel with screens at both ends. Pieces of dry ice fill this channel and serve to cool the air passing through it. The length of the channel, the fan capacity and the amount of dry ice control the temperature of the airflow emerging from this arrangement.

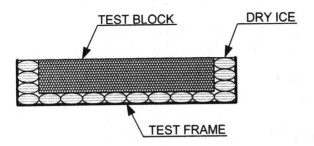

Fig. 4.2 Method for cooling model of highway surface.

Figure 4.3 Arrangement for providing a cold stream of air.

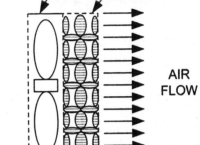

The evaluation of your anti-icing system will be performed in the test chamber with your team members and the instructor present. The test will begin when your team arranges its supply of dry ice in the test frame and the fan channel. Your team will monitor the temperatures and when the temperatures of both the test block and the air above the block are at least − 10 °C you will signal the instructor to provide the simulated rainfall. At this time, the anti-icing system must become completely automatic. All three of your sensors must provide a go signal to activate the anti-icing system. When the go signal is received your team's system must provide a dispersed quantity of saline solution over the test block. When this quantity of saline solution has been dispensed, the flow must be terminated with sufficient solution remaining in storage for another anti-icing event.

We anticipate that you will be able to complete this verification test to determine the effectiveness of the anti-icing system in less than one hour. Your instructor will verify the automatic operation of your system, the temperature when the flow was activated, and the amount of fluid remaining in the storage vessel. He or she will also inspect the test block to determine if any ice formed during the event.

Criteria that will be used to judge the merits of your team's development effort include:

1. Performance
   - Ability of the three sensors to measure temperatures accurately.
   - Effectiveness of the valve to control the flow of the anti-icing solution.
   - The efficiency of the dispersal system.
   - The operation of the program controlling the system actuators.
2. Cost
   - Minimize the cost of materials and components used in the test setup.
   - Minimize overall cost consistent with developing a model of a high performance anti-icing system.
3. Design innovation
   - Simplicity
   - Novel concepts
   - Ease in manufacturing and assembly

4. Quality
   - Reliability of the measurements
   - Perfect programming
   - Effective components
   - Superior workman(women)ship
5. Appearance
   - Consistent with the surroundings
6. Package size
   - See extended description below
7. Safety
   - Avoid all hazards[3]

The criterion—package size—requires additional explanation because it is more involved than the other criteria used in determining the merits of your team's development. The package size is a design constraint. We are limiting the size of each system by requiring that all of its parts fit into a copy paper box that normally contains ten packages (500 sheets each) of 8.5 by 11 in. of paper.

You will be required to assemble your molds and support structures using simple hand tools such as a drill, screwdriver, hammer, pliers, soldering iron, etc. You will be allotted 50 minutes immediately prior to the class period scheduled for the evaluation test to assemble the component parts of your system. You will also be required to disassemble the system in the 15-minute period immediately following the evaluation tests. Unless you receive instructions to the contrary from your instructor, you will have only one opportunity to "officially" test your system. However, the test facility will be available six weeks prior to the scheduled evaluation test, and you are encouraged to check your design concepts well in advance of the critical test. When you have completed the evaluation test, pack the parts from your system in a copy paper box and dispose of it in an environmentally sensitive manner.

While 120 V ac power is available in the test facility, we believe that it represents a safety hazard if used in your system. Accordingly, **the use of the 120 Volt ac utility line power supply is prohibited** except to drive the fan that will be used to supply the cold air surrounding the test block. We recognize that an electrical supply is essential for electronics; however, we anticipate that you will employ a battery to provide the necessary power. **Do not employ more than one nine-volt battery in any sensor or actuator without written permission from your instructor.**

REFERENCES

1. Brink, M. and M. Auen, "Go Light with the Salt, Please: Developing Information Systems for Winter Roadway Safety," TR News, No. 230, p. 3-9, January-February, 2004.

---

[3] A list of hazards is described in detail in Chapter 17 on Safety, Risk and Performance.

By Stanley J. Reeves and James W. Dally

CHAPTER 5

BASIC ELECTRIC CIRCUITS

5.1 INTRODUCTION

This chapter introduces you to the basic quantities, concepts, and physical devices used in electric circuits. Let's begin with definitions of charge, current, and voltage, the basic quantities of electricity and other components—resistance, capacitance, and inductance—that characterize circuit elements. Only the simplest concepts governing circuits will be described in this chapter. The basic passive components—those that do not require an external power source will be described.

5.2 BASIC ELECTRICAL QUANTITIES

Electricity involves charge carriers—either electrons, which are negative charge carriers, or holes which are positive charge carriers. Since we cannot see, smell or touch an electron, electricity has always been somewhat mysterious to most of us. Holes, which are atoms with an electron missing from its outer ring, are even more obscure. Nevertheless, we will try to introduce many of the basic electrical concepts in a simple, straightforward manner and investigate the mystery.

Electricity can be understood most simply by drawing an analogy between charge flowing through a circuit and water flowing through a pipe. The analogy is not perfect, but it provides a great deal of insight into the basic operation of electricity.

5.2.1 Charge

Charge (q) is the fundamental quantity of electricity. It is the substance of which current is made, just as a volume of water is the substance of which water current is made. Charge is measured in units of coulombs (C). One electron has a charge of 1.602×10^{-19} C. We can think of charge as a substance (free electrons) that is contained within a conductor such as a wire. This is analogous to a hose full of water. Charge does not really **do** anything until it moves. Excess charge can create a force, but nothing happens until the charge moves.

5.2.2 Current

Current (I) is charge in motion. The amount of current flow is measured in amperes (A), which are coulombs per second. Just as we speak of the rate at which a volume of water moves through a pipe, we can speak of current as a flow rate. The higher the current, the higher the rate at which charge is moving by a point in a conductor. This concept is expressed in equation format as:

$$I = dq/dt \tag{5.1}$$

5.2.3 Voltage

Voltage is another fundamental electrical quantity. The unit of voltage is volts (V). It is the electrical equivalent of pressure. Just as a pump creates pressure to force water through a pipe, a voltage source creates pressure to push charge through a conductor.

A voltage source is a circuit element that delivers a specified voltage (electrical pressure) regardless of the current flowing. The most basic voltage source is a battery. A flashlight battery pushes charge through a conductor that comprises the light bulb filament, creating friction within the filament, and causing the bulb to glow. A 6-V battery pushes four times as hard as a 1.5-V battery. The symbol for a constant voltage source (called a **direct current**, or dc, source) such as a battery is shown in Fig. 5.1a. A more general voltage source is shown in Fig. 5.1b.

Fig. 5.1 (a) dc voltage source symbol, (b) general voltage source symbol.

5.2.4 Resistance

Resistance measures the difficulty of pushing current from one place to another. The unit of resistance is ohms (Ω). The higher the resistance of a device, the more difficult it is to push current through it. Pumping water through a small diameter pipe requires more pressure to achieve the same flow rate as pumping through a larger diameter pipe. A high-resistance electrical device is analogous to a small diameter pipe, resisting more strongly the flow of charge. Likewise, a low-resistance device is analogous to a large diameter pipe that offers little resistance to flow.

A resistor is a circuit element that has a fixed resistance regardless of the input voltage. A resistor in a circuit loop with a voltage source is shown in Fig. 5.2. The connecting lines between the voltage source and the resistor are wires, which can be thought of as resistors with zero resistance. In our water analogy, wires are extremely large pipes that offer little resistance to water flow.

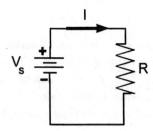

Fig. 5.2 A simple circuit loop containing a voltage source (battery) and a resistor.

The relation between the applied voltage V_s, the resistance R, and the current flow I is given by:

$$V_s = I R \qquad\qquad (5.2a)$$

$$I = V_s / R \qquad\qquad (5.2b)$$

Equations (5.2a) and (5.2b) are both expressions of Ohm's law. When V_s is expressed in volts and I in amperes, R is given in ohms Ω.

The resistance R of a conductor is given by:

$$R = \rho L / A \qquad (5.3)$$

where ρ is the specific resistance of the metal conductor (Ω-cm).
       L is the length of the conductor (cm).
       A is the cross-sectional area of the conductor (cm^2).

We can fabricate resistors by winding very-small-diameter, high-resistance wire on coils or by evaporating thin films of metal on ceramic substrates. The relation in Eq. (5.3) allows us to select the metal and to size the conductor to provide a specified resistance.

A resistor is a dissipative component. When current flows through a resistor, power is lost and heat is generated. The power dissipated by the resistor is given by:

$$p = V_d I \qquad (5.4)$$

where   p is the power expressed in watts (W).
       V_d is the voltage drop across the resistor.

Substituting Eqs. (5.2a) and (5.2b) into Eq. (5.4), we obtain:

$$p = I^2 R = V_d^2 / R \qquad (5.5)$$

We have used subscripts with the voltage in Eqs. (5.2a) and (5.2b) and the subscript d with the voltage in Eq. (5.5). For the simple circuit shown in Fig 5.4, the supply voltage V_s is equal to the voltage drop across the resistor V_d. In this case, the subscripts can be interchanged. Clearly, it is easy for us to determine the power lost if we know the current flowing through a resistor or the voltage drop across the resistor.

5.2.5 Capacitance

Capacitance is a bit more difficult to understand than the previous quantities. However, it also has a water-related analogy. It is used to characterize devices that have a degree of "springiness" – devices in which charge can be pumped in but which tend to push charge back when the external pressure is removed. The ease with which charge can be pushed into the device measures its capacity. The unit of capacitance is farads (F).

A capacitor is a passive electrical component that has a fixed capacitance. The circuit symbol for a capacitor is shown in Fig. 5.3. The capacitor consists of two flat, parallel-plate electrodes separated by a dielectric, which also serves as an electrical insulator. The capacitance C of a flat-plate capacitor is given in units of picofarad and is determined by:

$$C = k \, \varepsilon \, A / h \qquad (5.6)$$

where ε is the dielectric constant.
       A is the area of the plates.
       h is the distance between the two plates as shown in Fig. 5.3.
       k is a proportionality constant; 0.00885 for dimensions in millimeters.

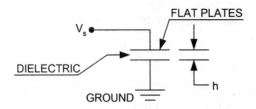

Fig. 5.3 A flat parallel plate capacitor.

When a voltage V is applied across the capacitor, it stores a charge q that is given by:

$$q = C V \qquad (5.7)$$

where C is the capacitance given in farads denoted with the symbol F.
q is the charge on the capacitor given in coulombs.

When the capacitor is charged, it becomes a storage device. The electrical energy stored (w) is determined from:

$$w = C V^2 /2 \qquad (5.8)$$

where w is the stored energy in joules when C is expressed in farads.

When a voltage is first applied across the terminals of a capacitor, current flows and it becomes charged. However, when the capacitor is fully charged and maintained at a constant voltage, no current flows through it. The capacitor conducts current only when the voltage across it changes with respect to time.

The water analogy for a capacitor is a balloon or membrane stretched across the cross-section of a pipe to block the flow of water. As pressure is applied to attempt to pump water through the pipe, the balloon begins to stretch, which allows a small amount of water to be flow. As the balloon stretches, it resists more and more until the backpressure of the balloon resisting the flow equals the forward pressure of the pump. At that point, the water ceases to move.

5.2.6 Inductance

Another passive component is the inductor, whose symbol is shown in Fig. 5.4. We produce an inductor by winding a coil of small-diameter, insulated wire about a core. The inductance L is given by:

$$L = 4\pi N^2 \, A\mu/ \ell \times 10^{-9} \qquad (5.9)$$

where L is the inductance in Henrys
N is the number of turns of wire in the coil.
μ is the permeability of the media within the coil.
A is the cross sectional area of the coil.
ℓ is the length of the coil.

When a step voltage is first applied across an inductor, it acts like an open circuit; it does not allow any current to flow through it. Gradually, current flow through the inductor begins to increase until the inductor offers no resistance at all. In a sense, the inductor has an internal inertia with respect to current.

The inertia must be overcome for current to flow. Once current begins to flow, the inertia of the inductor tends to keep the current flowing even when the external voltage source is removed. Inductance L is a measure of the electrical inertia of the inductor.

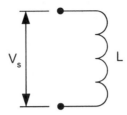

Fig. 5.4 An inductor coil across an alternating voltage source.

A solenoid, which is a coil with an inductance, is commonly used to develop a magnetic field. The solenoid produces a magnetic force on a metal plunger that extends into the core of the coil. A solenoid engages a relay that switches your battery across the starter motor when you start the engine on your auto.

In dc circuits, an inductor coil acts like a small resistor and it has little effect on the current flowing in the circuit. However, when the coil is exposed to a current that fluctuates with respect to time (an alternating current) there is a significant voltage drop across the coil that is given by:

$$V_d = L \ (dI/dt) \hspace{3cm} (5.10)$$

where (dI/dt) is the rate of change of the current with respect to time.

The water analogy for an inductor is a water wheel in a channel through which water flows. At first, the inertia of the stationary wheel prevents the water from moving through the channel and past the wheel. As the water pushes on the wheel, the wheel begins to turn and speed up, allowing more and more water to pass. Eventually, the wheel is turning at the rate the water would flow without the presence of the wheel. If the water pressure on the up stream of the wheel is turned off, the rotational inertia of the wheel keeps it turning. It pulls water from the upstream side and pushes it down stream. With time, the wheel begins to slow since there is no external pressure, until the wheel and the water flow both cease.

5.2.7 Diode

The diode is another important circuit element. While no particular electrical quantity is associated with diodes, they act in some respects like a current switch or a check valve. A diode allows current to flow freely in one direction (ideally, just like a wire) but does not allow current to flow at all (ideally) in the opposite direction. In a sense, a diode is the electrical equivalent of a check valve in a pipe carrying water. The symbol for a diode is shown in Fig. 5.5. The direction of the arrow indicates the direction in which the diode allows current to flow. We will discuss a common use of diodes later in the chapter.

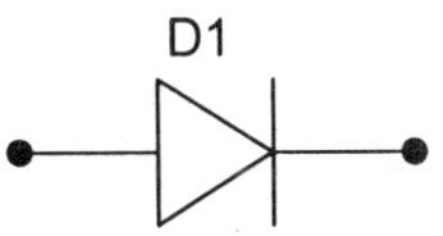

D1

Fig. 5.5 Circuit symbol of a diode

5.3 KIRCHHOFF'S LAWS

5.3.1 Kirchhoff's Voltage Law

We introduced Ohm's law with a simple circuit loop containing a voltage source and a resistor as shown in Fig. 5.2. This is a good beginning; however, what happens when you begin to insert additional components in the circuit? You will need to develop a few more tools useful in analyzing slightly more complex circuits. Let's begin by adding one additional resistor to the simple circuit loop as shown in Fig. 5.6a. Then determine the relation for the current flowing in this slightly more complex circuit.

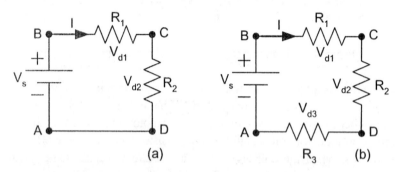

Fig. 5.6 Circuit loops containing two or three resistors.

To approach this problem, we need to use **Kirchhoff's voltage law** that states:

The sum of the voltages changes about a circuit loop is zero.

If we refer to the drawing of the circuit shown in Fig. 5.6a, we can write Kirchhoff's law in an equation format as:

$$+ V_{A-B} - V_{B-C} - V_{C-D} - V_{D-A} = 0 \tag{5.11}$$

In the interpretation of Eq. (5.11), we move in the direction of the current flow (clockwise in this case). Moving from node A to B, we pass through the battery that increases the voltage at B relative to A by an amount V_s. This is the reason for the + sign shown on the term V_{A-B}. Continuing around the circuit loop, we proceed from node B to C and pass through resistor R_1. There is a voltage drop V_{d1} as the current passes through this resistor equal to $I R_1$. The fact that the voltage drops across R_1 is the reason for the minus sign on term V_{B-C}. Moving onto the next leg of the circuit between node C and D, we encounter a second voltage drop $V_{d2} = I R_2$. Because the voltage decreases as we move from node C to D, a minus sign is assigned to the term V_{C-D}. Finally, we consider the leg of the circuit between nodes D - A. We have a perfect conductor between these two points with no loss or gain of voltage. For this reason, $V_{D-A} = 0$. Substituting these four voltage changes into Eq. (5.11) gives:

$$V_s - I R_1 - I R_2 - 0 = 0$$

$$I = V_s / (R_1 + R_2) \tag{5.12}$$

If we apply the same analysis techniques to the circuit with the three resistors shown in Fig. 5.6b, it is easy to add another term to the relation shown above and write:

$$V_s - I R_1 - I R_2 - I R_3 = 0$$

$$I = V_s / (R_1 + R_2 + R_3) \qquad (5.13)$$

Examine Eqs. (5.12) and (5.13), and observe that the resistors in a series arrangement in the circuit add together to give an equivalent resistance R_e given by:

$$R_e = R_1 + R_2 + R_3 = \Sigma R \qquad (5.14)$$

Equation (5.14) leads to a well know resistor rule for resistors connected in a series arrangement. The effective resistance of several resistances in series is the sum of the individual resistances. The equivalent resistance is used to simplify the circuit diagram by replacing the three-resistor circuit with a single equivalent resistor as shown in Fig. 5.7.

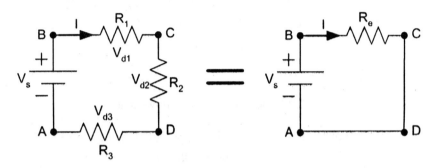

Fig. 5.7 An equivalent circuit with an equivalent resistor replaces a more complex circuit.

5.3.2 Kirchhoff's Current Law

What is the effect if the resistors are not all connected in a series arrangement? What if we have a parallel arrangement of the resistors such as shown in Figs. 5.8a and 5.8b? To analyze this type of a circuit, we need to introduce **Kirchhoff's current law**. This simple law states:

The sum of the currents flowing into a node must be equal to the sum of the currents flowing from that node.

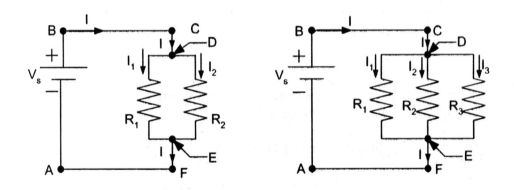

Fig. 5.8 Circuits with parallel arrangements of resistors.

Let's refer to node X, presented in Fig. 5.9, with currents flowing into and out it. Kirchhoff's current law permits us to write:

$$I_1 + I_2 - I_3 - I_4 = 0 \qquad (5.15)$$

To deal with the resistors in parallel, as shown in Fig. 5.8a, we apply Eq. (5.15) to the current flow at node D and write:

$$I - I_1 - I_2 = 0 \qquad \text{(a)}$$

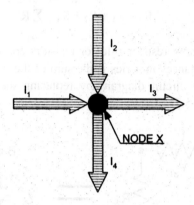

Fig. 5.9 Node X with current flows in and out.

Next, consider Kirchhoff's voltage law Eq. (5.11), and apply it to the circuit in Fig. 5.8. It is clear then that the voltage gain due to the battery is offset by the voltage drop across the two parallel resistors between points D - E. We write this fact in an equation:

$$V_s - V_{D\text{-}E} = 0 \qquad \text{(b)}$$

We know that the current divides at node D with I_1 flowing through R_1 and I_2 through R_2 to produce the voltage drop $V_{D\text{-}E}$. This knowledge permits us to write:

$$V_{D\text{-}E} = I_1 R_1 = I_2 R_2 = V_s \qquad \text{(c)}$$

Equation (c) gives relations for both I_1 and I_2:

$$I_1 = V_s / R_1 \qquad \text{and} \qquad I_2 = V_s / R_2 \qquad \text{(d)}$$

Following the form of Eqs. (d), we write:

$$I = V_s / R_e \qquad \text{(e)}$$

where R_e is the equivalent resistor that replaces the two parallel resistors in Fig. 5.8a to give the same equivalent circuit as shown in Fig. 5.7.

Let's substitute Eqs. (d) and (e) into Eq. (a) to obtain:

$$V_s/R_e = V_s [(1/R_1) + (1/R_2)] \qquad \text{(f)}$$

Eliminating V_s from Eq. (f) gives:

$$1/ R_e = (1/R_1) + (1/R_2) \qquad \text{(5.16a)}$$

or

$$R_e = (R_1 R_2) / (R_1 + R_2) \qquad \text{(5.16b)}$$

If we find three resistors in parallel, we can follow the same procedure in the analysis of the circuit and show that the equivalent resistance for the three parallel resistors is:

$$1/R_e = (1/R_1) + (1/R_2) + (1/R_3) \qquad (5.17)$$

While resistors connected in series were summed to give the equivalent resistance, resistors in parallel follow a different rule. For parallel resistors, the reciprocals are summed and set equal to the reciprocal of the equivalent resistance. This fact implies that the equivalent resistance for a parallel arrangement of resistors is less than that of the smallest individual resistance.

5.3.3 Voltage Divider

A consequence of series resistance is voltage division. Consider the circuit in Fig. 5.10. We want to find the voltage across R_2. From Ohm's law, we have

$$I = V_s /(R_1 + R_2) \qquad (5.18)$$

$$V_{d2} = IR_2 = V_s R_2/(R_1 + R_2) \qquad (5.19)$$

Equation (5.19) shows that the voltage across R_2 is a fraction of V_s. The fraction is $R_2/(R_1 + R_2)$.

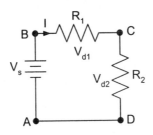

Fig. 5.10 Voltage division circuit.

Another consequence of Kirchhoff's laws is that voltage sources in series add just as resistors do as long as the orientation of the source is taken into account. That is why two 1.5 V flashlight batteries are used in an ordinary flashlight; the voltage of the two batteries adds to provide a total voltage source of 3 V. On the other hand, if one of the batteries is inserted upside down, it will push **against** the voltage of the first battery and give a net zero voltage source. In that case, the flashlight will not work.

At this point, you should recognize that Ohm's law and Kirchhoff's two laws can be employed to analyze circuits. Also, you should be able to take a complex circuit and use equivalent resistances to reduce the circuit to its most elementary form.

5.4 ALTERNATING AND DIRECT CURRENT

5.4.1 Definitions

Most circuit systems are classified as either direct current (dc) or alternating current (ac). A typical dc system is one that is powered by a battery, such as a flashlight or an automotive electrical system. Typical ac systems are those that are powered from a wall outlet in the house, such as a blow dryer or television. Direct current simply means that the current is constant in the circuit; it does not oscillate in amplitude with time. Alternating current oscillates with time usually in the form of a sine wave.

Paradoxically, the terms ac and dc are used to classify voltage sources as well as current, since constant voltage sources ordinary produce constant current and alternating voltage sources ordinary produce alternating current. The rate at which a periodic signal cycles is called the **frequency**, which is measured in cycles per second, or **hertz** (Hz). The **amplitude** is the peak value of the periodic waveform. The **RMS** (root-mean-square) value is the square root of the average squared value. For a sine wave, this is equal to amplitude/$\sqrt{2}$. The 110 volts ac available at a wall outlet is actually 110 volts RMS. The peak value is about 155. In the United States, ac supply is a sine wave that oscillates at 60 Hz. A graph of voltage as a function of time in a household power supply (110 V ac) is shown in Fig. 5.11.

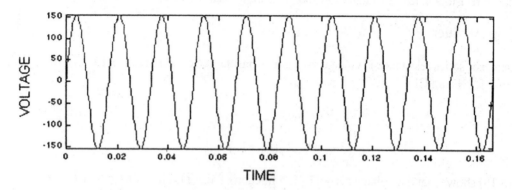

Fig. 5.11 Voltage as a function of time in wall outlet

5.4.2 Converting ac to dc

In many cases, the power is supplied by plugging into an ac outlet, even though the internal circuitry of a product like a PC requires a dc voltage source. In these cases, a circuit must be used to convert ac voltage to dc. Two types of circuits that accomplish this are the **half-wave rectifier** and the **bridge rectifier**. Both rectifiers take advantage of the diode, which allows current to flow one way but not the other.

Half-Wave Rectifier

A half-wave rectifier is shown in Fig. 5.12. It consists simply of a diode in series with an ac power supply and the load (the device to be powered—a resistor in this case). When V1 is positive, the diode acts like a wire and current passes through unimpeded. However, when V1 is negative, no current is allowed to flow backward, so the voltage drop across R1 is zero. The input and output voltages for this circuit are shown in Fig. 5.13.

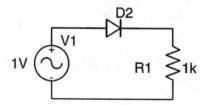

Fig. 5.12 Half-wave rectifier (diode) in series with load R1.

Note that in the graph showing output voltage the negative voltage values are truncated by the action of the diode. The dotted line in each figure shows the average value of the voltage on each graph. The input voltage oscillates equally on either side of zero, so the average value is zero. On the other hand, the output voltage is always nonnegative, so the average output voltage is positive. This output voltage waveform actually has both a dc component and an ac component. For some applications, this output can be used as a dc supply; the ac component (the fluctuations around the average value) can be ignored. In other cases, further filtering is required to smooth out the fluctuations to provide a more constant voltage.

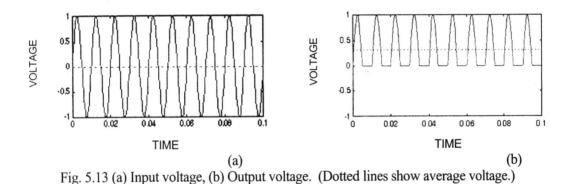

(a)   (b)
Fig. 5.13 (a) Input voltage, (b) Output voltage. (Dotted lines show average voltage.)

Bridge Rectifier

A bridge rectifier, shown in Fig. 5.16, operates on the same principle as the half-wave rectifier. It is a more complex circuit, but the performance is better. The network of four diodes allows the current to flow only in one direction regardless of the sign of the input voltage. Hence, its output voltage waveform is modified as shown in Fig. 5.15.

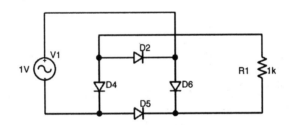

Fig. 5.14 Bridge rectifier in series with a load R1.

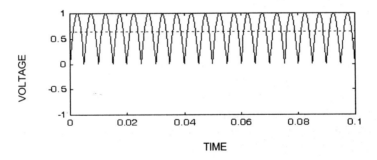

Fig. 5.15 Output voltage for bridge rectifier with the input defined in Fig. 5.13.
(Dotted lines show average voltage.)

It is clear that the bridge rectifier converts the negative-going voltage lobes to positive voltage lobes. Thus, the average value (the dc component) is twice as high as for a half-wave rectifier. This doubling of the average value comes at the expense of using four diodes instead of one.

Capacitor Filter

It is possible to smooth out the bumps in either the half-wave-rectified signal or the bridge-rectified signal by inserting a capacitor in series with the load. A capacitor-filtered half-wave rectifier is shown in Fig. 5.15. The capacitor begins to accumulate charge when the source voltage is positive. When the source is negative, the capacitor begins to discharge and acts like a temporary battery to maintain the voltage across R.

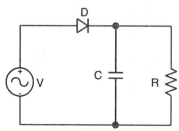

Fig. 5.16 Capacitor-filtered half-wave rectifier.

A graph of voltage as a function of time from a simulated capacitor-filtered half-wave rectifier circuit is shown in Fig. 5.17. The darker trace is the input voltage, and the lighter trace is the output. Notice that the output does not reach the same peak value as the input. This difference is because the capacitor does not charge instantaneously; it is still charging when the peak input begins to drop. However, when the input drops below the output voltage, the capacitor slowly discharges, sending current through the resistor and keeping the output voltage higher than it would be otherwise. Thus, the average value (the dc component) is increased.

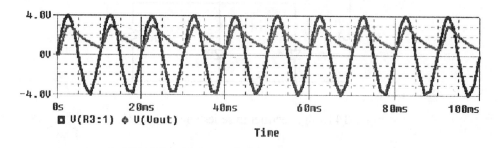

Fig. 5.17 Simulated input (dark) and output (light) voltages of a capacitor-filtered half-wave rectifier.

5.5 SUMMARY

We have described the basic electrical quantities of a typical circuit. In addition, we have defined Ohm's law, resistance, and the power dissipated by current flow through a resistor. The other common passive components—the capacitor, inductor, and diode—have also been introduced.

Circuit analysis is vitally important for all engineers regardless of their discipline. For this reason, we have introduced Kirchhoff's two laws, and given two examples of circuit analysis leading to the equations for the equivalent resistance of circuits with series and parallel-connected resistors.

We have also discussed the differences between ac and dc systems. Circuits that convert from ac to dc with different levels of efficiency, depending on the number and type of components used have been presented.

REFERENCES

1. Irwin, J. D., <u>Basic Engineering Circuit Analysis</u>, 6th ed., John Wiley & Sons, New York, 2000.
2. Irwin, J. D. and Kerns, D. V., Jr., <u>Introduction to Electrical Engineering</u>, Prentice-Hall, 1995.
3. Horowitz, P. and Hill, W., <u>The Art of Electronics</u>, 2nd ed., Cambridge University Press, New York, 1989.
4. Dorf, R. C., <u>Introduction to Electric Circuits</u>, 2nd ed., John Wiley & Sons, New York, 1993.

EXERCISES

6.1   Draw the symbols used on drawings to represent a resistor, a capacitor and an inductor. Also, write the relations for the voltage across these components in terms of the current flow.

6.2   If you design a circuit with a 9 V battery containing an equivalent resistance of 2,000 Ω, find the current drained from the battery. If the battery has a capacity of 2A-h (ampere-hours), determine the life of the battery. Assume that it cannot be recharged. What is the power dissipated by the resistor?

6.3   A size D, 1.5 V, alkaline battery has a capacity of 1200 hours when the battery is discharged through a 100 Ω resistor. Determine the current flow through this resistor and state the capacity in terms of ampere-hours (Ah).

6.4   A 9 V alkaline battery has a capacity of 6 hours when the battery is discharged through a 100 Ω resistor. Determine the current flow through this resistor and state the capacity in terms of ampere-hours (Ah).

6.5   Write Kirchhoff's two laws. Use circuit diagrams to illustrate these two laws.

6.6   Draw a circuit with four series-connected resistors and give the expression for the equivalent resistance. If these resistors have values of 1000, 400, 1500 and 10,000 Ω, find the equivalent resistance. Draw a circuit diagram that contains the equivalent resistor, which is also equivalent to your initial circuit diagram.

6.7   Draw a circuit with four parallel-connected resistors and give the expression for the equivalent resistance. If these resistors have values of 1000, 400, 1500 and 10,000 Ω, find the equivalent resistance. Draw a circuit diagram that contains the equivalent resistor, which is also equivalent to your initial circuit diagram.

6.8   Trace the current path through a bridge rectifier when the source voltage is positive. Trace the current path through a bridge rectifier when the source voltage is negative.

CHAPTER 6

SENSORS

6.1 INTRODUCTION

Transducers are electromechanical devices that convert a mechanical change such as displacement or force into a change in an electrical signal that can be monitored as a voltage after some degree of signal conditioning. A wide variety of transducers are commercially available for use in measuring mechanical quantities such as force, torque, strain, velocity, acceleration, etc. Transducer characteristics include—range, linearity, sensitivity and operating temperatures. Primarily the sensor that is incorporated into a transducer to produce an electrical output determines these characteristics. For example, a set of four strain gages arranged on a tension link provides a transducer that produces a resistance change $\Delta R/R$ in proportion to the load applied along the axis of the link. The strain gages serve as the sensor in this force transducer and play a significant role in establishing the characteristics of the transducer.

Sensors used in transducer design include potentiometers, differential transformers, strain gages, capacitors, piezoelectric and piezoresistive crystals, thermistors, etc. The important features of these different sensors are described in this chapter.

6.2 POTENTIOMETERS

The simplest type of potentiometer, shown schematically in Fig. 6.1, is the slide-wire resistor. This sensor consists of a length L of resistance wire attached across voltage source v_s. The relationship between the output voltage v_o and the position x of a wiper, as it moves along the length of the wire, can be expressed as:

$$v_o = (x/L)v_s \qquad\qquad x = (v_o/v_s)L \qquad\qquad (6.1)$$

Thus, the slide-wire potentiometer can be used to measure a displacement x.

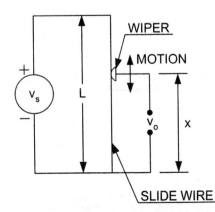

Fig. 6.1 Slide-wire resistance potentiometer.

Resistors fabricated from a short length of straight wire resistors are not feasible for most applications, because its resistance is too low. Very low resistance of a sensor imposes excessive power requirements on the voltage source. To alleviate this difficulty, high-resistance, wire-wound potentiometers are obtained by winding the wire around an insulating core, as shown in Fig. 6.2. The potentiometer illustrated in Fig. 6.2a is used for linear displacement measurements. Cylindrically shaped potentiometers, similar to the one illustrated in Fig. 6.2b, are used for angular displacement measurements. The resistance of a wire-wound potentiometer can range between 10 and 10^6 Ω depending upon the type and diameter of the wire used and the length of the coil.

The resistance of the wire-wound potentiometer increases in a stepwise manner as the wiper moves from one turn to the adjacent turn. This step change in resistance limits the resolution of the potentiometer to L/n, where n is the number of turns in the length L of the coil. Resolutions ranging from 0.05 to 1% are common, with the lower limit obtained by using many turns of very small diameter wire.

The active length L of the coil controls the range of the potentiometer. Linear potentiometers are available in many lengths up to about 1 m. Arranging the coil in the form of a helix can extend the range of the angular-displacement potentiometer. Helical potentiometers are commercially available with as many as 20 turns; therefore, angular displacements as large as 7,200° can be measured quite easily.

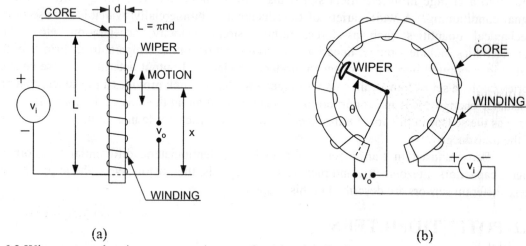

(a)   (b)

Fig. 6.2 Wire-wrapped resistance potentiometer for (a) axial displacements and (b) for angular displacements.

In order to improve resolution, potentiometers have been introduced that utilize thin films of conductive plastic with controlled resistivity instead of wire-wound coils. The film resistance on an insulating substrate exhibits very high resolution together with lower noise and longer life. For example, a resistance of 50 to 100 Ω/mm can be obtained with conductive plastic films that are used for commercially available potentiometers with a resolution of 0.001 mm or less. The frictional force that must be overcome to move the wiper is depended on the construction details of the particular potentiometer; however, 1 to 2 ounces is common. The life expectancy for potentiometers fabricated with conductive plastics is commonly specified at 20×10^6 strokes. An example of a linear motion potentiometer with a single end shaft is presented in Fig. 6.3. Note the three terminals used for electrical connections.

The dynamic response of both the linear and the angular potentiometer is severely limited by the inertia of the shaft and wiper assembly. Because this inertia is large, the potentiometer is used only for static or quasi-static measurements where a high frequency response (bandwidth) is not required.

Fig. 6.3 A linear motion potentiometer.

Potentiometers are used primarily to measure large displacements—10 mm or more for linear motion and 15° or more for angular motion. Potentiometers are relatively inexpensive yet accurate; however, their main advantage is simplicity of operation, because only a voltage source and a digital voltmeter (DVM) to measure voltage are required for a complete instrumentation system. Their primary disadvantages are a limited frequency response that precludes their use for dynamic measurements and relatively high forces required to overcome friction of the wiper.

6.3 DIFFERENTIAL TRANSFORMERS

Differential transformers, based on a variable-inductance principle, are also used to measure displacement. The most popular variable-inductance sensor for linear displacement measurements is the linear variable differential transformer (LVDT). A LVDT, illustrated in Fig 6.4a, consists of three symmetrically spaced coils wound onto an insulated bobbin. A magnetic core, which moves through the bobbin without contact, provides a path for magnetic flux linkage between coils. The position of the magnetic core controls the mutual inductance between the center or primary coil and the two outer or secondary coils.

When ac excitation is applied to the primary coil, voltages are induced in the two secondary coils. The secondary coils are wired in a series-opposing circuit, as shown in Fig. 6.4b. When the core is centered between the two secondary coils, the voltages induced in the secondary coils are equal but out of phase by 180°. Because the coils are in a series-opposing circuit, the voltages v_1 and v_2 in the two coils cancel and the output voltage is zero. When the core is moved from the center position, an imbalance in mutual inductance between the primary and secondary coils occurs and an output voltage, $v_o = v_2 - v_1$, develops. The output voltage is a linear function of core position, as shown in Fig. 6.5, providing the motion of the core is within the operating range of the LVDT. The direction of motion can be determined from the phase of the output voltage relative to the input voltage.

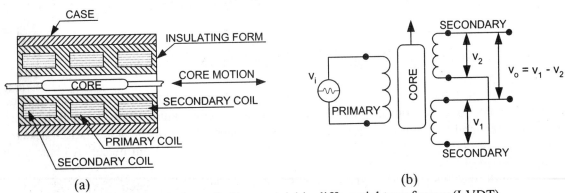

(a)   (b)

Fig. 6.4 (a) Sectional view of a linear variable differential transformer (LVDT).
(b) Schematic circuit diagram for the LVDT coils.

The frequency of the voltage applied to the primary winding can range from 50 to 25,000 Hz. If the LVDT is used to measure transient or periodic displacements, the carrier frequency should be 10 times greater than the highest frequency component in the dynamic signal. The highest sensitivities are attained with excitation frequencies between 1 and 5 kHz. The input voltage can range from 3 to 15 V rms. and the power required is usually less than 1 W. Sensitivities of different LVDTs vary from 0.02 to 0.2 V/mm of displacement per volt of excitation applied to the primary coil. At rated excitation voltages, sensitivities vary from about 30 to 350 mV per V_i − mm of displacement. The higher sensitivities are associated with short-stroke LVDTs, with an operating range of ± 0.13 mm; the lower sensitivities are for long-stroke LVDTs, with an operating range of ± 254 mm.

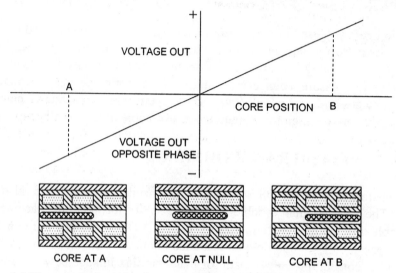

Fig. 6.5 Phase-referenced voltage as a function of LVDT core position.

Because the LVDT, a passive sensor, requires ac excitation at a frequency different from common ac supplies, signal-conditioning circuits are needed for its operation. A typical signal conditioner, shown in Fig. 6.6, provides a power supply, a frequency generator to drive the LVDT and a demodulator to convert the ac output signal from the LVDT to an analog dc output voltage. Finally, a dc amplifier is incorporated in the signal conditioner to increase the magnitude of the output voltage.

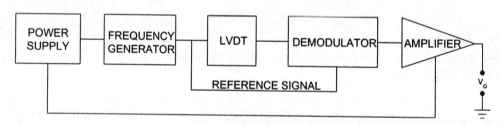

Fig. 6.6 Block diagram of the signal conditioning circuit for a LVDT.

Many different types of LVDTs are commercially available today. They are designed for normal applications, for rugged environments, for high temperature (220 °C) and high-pressure measurements and to monitor motion sensitive mechanisms. Most of the LVDTs are packaged in a stainless steel tube that encases the coils and shields the electronics from noise. A photograph of a general purpose LVDT is presented in Fig. 6.7.

With the development of microelectronics, circuits have been developed that permit miniaturization of the signal conditioner shown in Fig. 6.6. These miniaturized circuits are packaged within the case of an LVDT to produce a small self-contained sensor known as a direct current differential transformer (DCDT). A DCDT operates from a battery or a regulated power supply and provides an amplified output signal that can be monitored on either a digital voltmeter (DVM) or an oscilloscope. The output impedance of a DCDT is relatively low (about 100 Ω).

Fig. 6.7 General purpose LVDTs.

Both the LVDT and the DCDT are used to measure linear displacement. An analogous device, known as a rotary variable differential transformer (RVDT), has been developed to measure angular displacements. As shown in the schematic diagram in Fig. 6.8, the RVDT consists of a primary coil and two secondary coils wound symmetrically on a large-diameter insulated bobbin. A rotor, fabricated from a magnetic material, is mounted on a shaft that extends through the bobbin and serves as the core. As the shaft rotates and turns the core, the mutual inductance between the primary and secondary windings varies and produces an output voltage versus rotation response that resembles a modified sinusoid.

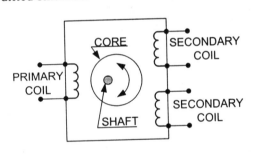

Fig. 6.8 Schematic diagram of a rotary variable differential transformer (RVDT).

Although the RVDT is capable of a complete rotation (360°), the range of linear operation is limited to ± 30° although a maximum range of ± 60 is possible with compromised linearity. The linearity of a typical RVDT, with a range of ± 30°, is about 0.5% of the range. Reducing the operating range improves the linearity, and an RVDT operating within a range of ± 5° exhibits a linearity of about 0.1% of this range.

The LVDT, DCDT, and RVDT have many advantages as sensors for measuring displacement. There is no contact between the core and the coils; therefore, friction and hysteresis are eliminated. Because the output is continuously variable with input, resolution is determined by the characteristics of the voltage recorder. Non-contact also ensures that life will be very long with no significant deterioration of performance over this period. The small core mass and freedom from friction give the sensor a limited capability for dynamic measurements. Finally, the sensors are not damaged by over travel; therefore, they can be employed as feedback transducers in servo-controlled systems where over travel may occur due to accidental deviations beyond the control band. Typical performance characteristics for DCDTs, and are listed in Table 6.1.

In more recent years, the DCDT has been equipped with an analog to digital converter (A/D) to provide a digitized output[1]. Their input voltage may be varied from 8.5 to 30 V, and their output signal has a resolution of at least 15 bits or 1 part in 32,768.

---

[1] For example, the Schaevitz Sensor model HC485 displacement sensor has a two-wire addressable RS-485 output.

Table 6.1
Performance characteristics of DCDTs
Courtesy of Schaevitz Sensors

DCDT Model No.	Linear Range (mm)	Scale Factor V/mm	Response − 3 Db (Hz)
050	± 1.25	8.00	500
125	± 3.00	3.20	500
250	± 6.00	1.60	500
500	± 12.5	0.80	200
1000	± 25.0	0.40	200
2000	± 50.0	0.20	200
3000	± 75.0	0.13	200
5000	± 125	0.08	200
10000	± 250	0.04	200

6.4 RESISTANCE STRAIN GAGES

Electrical resistance strain gages are thin metal-foil grids (see Fig. 6.9) that can be adhesively bonded to the surface of a component or structure. When the component or structure is loaded, strains develop and are transmitted to the foil grid. The resistance of the foil grid changes in proportion to the load-induced strain. The strain sensitivity of metals, first observed in copper and iron by Lord Kelvin in 1856, is explained by the following simple analysis.

The resistance R of a uniform metallic conductor can be expressed as:

$$R = \rho L/A \qquad (6.2)$$

where ρ is the specific resistance of the metal (Ω - cm).
   L is the length of the conductor.
   A is the cross-sectional area of the conductor.
Differentiating Eq. (6.2) and dividing by the resistance R gives:

$$dR/R = d\rho/\rho + dL/L - dA/A \qquad (a)$$

The term dA represents the change in cross-sectional area of the conductor resulting from the applied load. For the case of a uniaxial tensile stress state, recall that

$$\varepsilon_a = dL/L \qquad \text{and} \qquad \varepsilon_t = -\nu\varepsilon_a = -\nu dL/L \qquad (b)$$

where ε_a is the axial strain in the conductor.
   ε_t is the transverse strain in the conductor.
   ν is Poisson's ratio of the metal used for the conductor.

If the diameter of the conductor is d_o before application of the axial strain, the diameter of the conductor d_f after it is strained is given by:

$$d_f = d_0\left(1 - \nu\frac{dL}{L}\right) \qquad (c)$$

From Eq. (c) the change in area for small strains can be approximated by:

$$\frac{dA}{A} = -2v\frac{dL}{L} + v^2\left(\frac{dL}{L}\right)^2 \approx -2v\frac{dL}{L} \qquad (d)$$

Substituting Eq. (d) into Eq. (a) and simplifying yields:

$$\frac{dR}{R} = \frac{d\rho}{\rho} + \frac{dL}{L}(1+2v) \qquad (6.3)$$

that can be written as:

$$S_A = \frac{dR/R}{\varepsilon_a} = \frac{d\rho/\rho}{\varepsilon_a} + (1+2v) \qquad (6.4)$$

where S_A is defined as the sensitivity of the metal or alloy used for the conductor.

Fig. 6.9 An electrical resistance strain gage.

It is evident from Eq. (6.4) that the strain sensitivity of a metal or alloy is due to the changes in dimensions of the conductor, as expressed by the term $(1 + 2v)$ and the change in specific resistance as represented by the term $(d\rho/\rho)/\varepsilon$. Experimental studies show that the sensitivity S_A ranges between 2 and 4 for most metallic alloys used in strain-gage fabrication. Because the quantity $(1 + 2v)$ is approximately 1.6 for most of these materials, the contribution due to the change in specific resistance with strain varies from 0.4 to 2.4. This increase in specific resistance is due to variations in the number of free electrons and their changing mobility with applied strain.

A list of alloys commonly employed in commercial strain gages together with their sensitivities is presented in Table 6.2. The most commonly used strain gages are fabricated from the copper-nickel alloy known as Advance or Constantan. The response curve for this alloy ($\Delta R/R$ as a function of strain) is shown in Fig 6.10. This alloy is widely used because its response is linear over a wide range of strain (beyond 8%), it has a high specific resistance and it has excellent electrical stability with changes in temperature.

Table 6.2
Strain sensitivity S_A for common strain-gage alloys

Material	Composition (%)	S_A
Advance or Constantan	45 Ni, 55 Cu	2.1
Nichrome V	80 Ni, 20 Cu	2.1
Isoelastic	36 Ni, 8 Cr, 0.5 Mo, 56.5 Fe	3.6
Karma	74 Ni, 20 Cr, 3 Al, 3 Fe	2.0
Armour D	70 Fe, 20 Cr, 10 Al	2.0
Platinum-Tungsten	92 Pt, 8 W	4.0

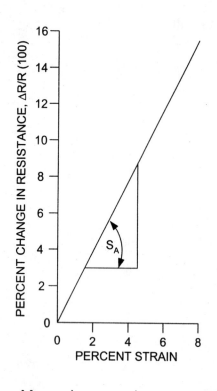

Fig. 6.10 Change in resistance $\Delta R/R$ as a function of strain for Advance alloy.

Most resistance strain gages are of the metal-foil type, where a photoetching process is used to form the grid configuration. Because the process is very versatile, a wide variety of gage sizes and grid shapes can be produced. The shortest gage available is 0.20 mm; the longest is 102 mm. Standard gage resistances are 120 and 350 Ω; however, special-purpose gages with resistances of 500, 1000 and 5000 Ω are also available.

The etched metal-film grids are very fragile and easy to distort, wrinkle or tear. For this reason, the metal grid is bonded to a thin plastic film that serves as a backing or carrier before photoetching. The carrier film, shown in Fig. 6.9, also provides electrical insulation between the gage and the component after the gage is bonded.

A strain gage exhibits a resistance change $\Delta R/R$ that is related to the strain ε in the direction of the grid by the expression:

$$\Delta R/R = S_g \, \varepsilon \qquad\qquad (6.5)$$

where S_g is the gage factor or calibration constant for the gage. The gage factor S_g is always less than the sensitivity of the metallic alloy S_A because the grid configuration of the gage with the transverse conductors is less responsive to axial strain than a straight uniform conductor.

The output $\Delta R/R$ of a strain gage is usually converted to a voltage signal with a Wheatstone bridge, as illustrated in Fig. 6.11. If a single gage is used in one arm of the Wheatstone bridge and equal but fixed resistors are used in the other three arms, the output voltage is given by:

$$v_0 = \frac{v_s}{4} \frac{\Delta R_g}{R_g} \qquad\qquad (6.6)$$

Substituting Eq. (6.5) into Eq. (6.6) gives:

$$v_0 = \frac{v_s}{4} S_g \varepsilon \qquad\qquad (6.7)$$

The input voltage is controlled by the gage size (the power it can dissipate) and the initial resistance of the gage. As a result, the sensitivity, $S = v_o / \varepsilon = v_s S_g / 4$, usually ranges from 1 to 10 $\mu V/(\mu \varepsilon)$.

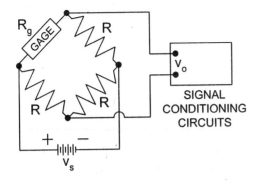

Fig. 6.11 Wheatstone bridge circuit used to convert resistance change $\Delta R/R$ of a strain gage to an output voltage v_o.

6.5 CAPACITANCE SENSORS

The capacitance sensor, illustrated in Fig. 6.12a, consists of a target plate and a second plate called the sensor head. These two plates are separated by an air gap of thickness h and form the two terminals of a capacitor that exhibits a capacitance C given by:

$$C = \frac{kKA}{h} \tag{6.8}$$

where C is the capacitance in picofarad (pF).
A is the area of the sensor head ($\pi D/4$).
K is the relative dielectric constant for the medium in the gap (K = 1 for air).
k = 0.225 is a proportionality constant for dimensions specified in in.
k = 0.00885 for dimensions given in mm.

If the separation between the head and the target is changed by an amount Δh, then the capacitance C becomes:

$$C + \Delta C = \frac{kKA}{h + \Delta h} \tag{a}$$

Equation (a) can be written as:

$$\frac{\Delta C}{C} = \frac{\Delta h / h}{1 + (\Delta h / h)} \tag{6.9}$$

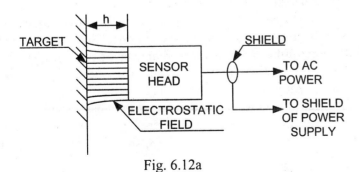

Fig. 6.12a

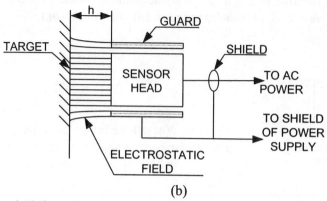

(b)

Fig. 6.12 Capacitor sensors: (a) without a guard ring, where edge effects in the electrostatic field affect the range of linearity, and (b) with a guard ring to extend the range of linearity.

This result indicates that ($\Delta C/C$) is non-linear because of the presence of Δ/h in the denominator of Eq. (6.9). To avoid the difficulty of employing a capacitance sensor with a non-linear output, the change in the impedance due to the capacitor is measured. Note that the impedance Z_C of a capacitor is given by:

$$Z_C = -j/(\omega C) \tag{b}$$

With a capacitance change ΔC, it can be shown that:

$$Z_c + \Delta Z_c = -\frac{j}{\omega}\left[\frac{1}{C + \Delta C}\right] \tag{c}$$

Substituting Eq. (b) into Eq. (c) and solving for $\Delta Z/Z$ gives:

$$\frac{\Delta Z_c}{Z_c} = -\frac{\Delta C/C}{1 + \Delta C/C} \tag{6.10}$$

Finally, substituting Eq. (6.9) into Eq. (6.10) yields:

$$\frac{\Delta Z_c}{Z_c} = -\frac{\Delta h}{h} \tag{6.11}$$

From Eq. (6.11) it is clear that the capacitive impedance Z_C is linear in Δh and that methods of measuring ΔZ_C will permit extremely simple plates (the target as ground and the sensor head as the positive terminal) to act as a sensor to measure the displacement Δh. Cylindrical sensor heads are linear and Eq. (6.11) is valid provided $0 < h < D/4$, where D is the diameter of the sensor head. Fringing in the electric field produces non-linearity if (h + Δh) exceeds D/4. The linear range can be extended to h ≈ D/2 if a guard ring surrounds the sensor as shown in Fig. 6.12b. The guard ring essentially moves the distorted edges of the electric field to the outer edge of the guard, significantly improving the uniformity of the electric field over the sensor area and extending its linear range. An example of a button type head for a capacitance gage, presented in Fig 6.13, show the guard ring that surround the sensing head.

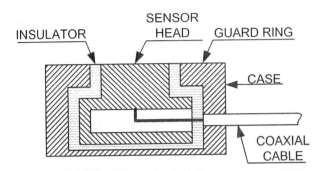

Fig. 6.13 A button capacitor gage showing the guard ring and other construction details..

The sensitivity of the capacitance probe is given by Eqs. (b), (6.8), and (6.11) as:

$$S = \frac{\Delta Z_C}{\Delta h} = \left| \frac{Z_C}{h} \right| = \left| \frac{1}{\omega\, Ch} \right| = \left| \frac{1}{\omega KkA} \right| \qquad (6.12)$$

The sensitivity can be improved by reducing the area of the probe; however, as noted previously, the range of the probe is limited by linearity to about D/2. Clearly there is a range-sensitivity trade-off. Of particular importance is the circular frequency ω in Eq. (6.12). Low frequency improves sensitivity but limits frequency response of the instrument, which represents another trade-off. It is also important to note that the frequency of the ac power supply must remain constant to maintain a stable calibration constant.

The capacitance sensor has several advantages. It is non-contacting and can be used with any target material provided the material exhibits a resistivity less than 100 Ω-cm (any metal). The sensor is extremely rugged and can be subjected to high shock loads (5000 g's) and intense vibratory environments. Their use as a sensor at high temperature is particularly impressive. They can be constructed to withstand temperatures up to 2000 °F and they exhibit a constant sensitivity S over an extremely wide range of temperature (74 to 1600 °F). Examination of the relation for the sensitivity S in Eq. (6.12) shows that the dielectric constant K is the only parameter that can change with temperature. Because K is constant for air over a wide range of temperature, the capacitance sensor has excellent temperature stability.

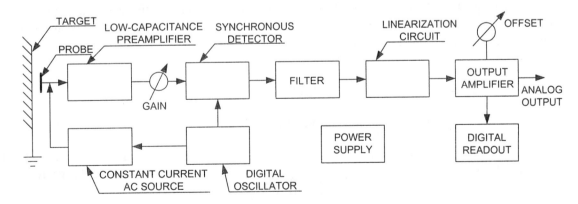

Fig. 6.14 Schematic diagram of an instrument system employed with a capacitance sensing system.

The change in the capacitive impedance Z_C is usually measured with the circuit illustrated schematically in Fig. 6.14. The probe, its shield, and guard ring are powered with a constant current ac supply. A digital oscillator is used to drive the ac supply and to maintain a constant frequency at 16.6 kHz. This oscillator also provides the reference frequency for the synchronous detector. The voltage drop across the probe is detected with a low capacitance preamplifier. The signal from the preamplifier is then amplified again with a fixed-gain instrument amplifier. The high voltage ac signal from the instrument amplifier is rectified and given a sign in a synchronous detector. The rectified signal is filtered to eliminate high frequency ripple and give a dc output voltage related to Δh. A linearizing circuit is used to extend the range of the sensor by accommodating for the influence of the fringes in the electrostatic field. Finally, the signal is passed through a second instrument amplifier where the gain and zero offset can be varied to adjust the sensitivity and the zero reading on a DVM display. When the gain and zero offset are properly adjusted, the DVM reads Δh directly to a scale selected by the operator.

6.6 EDDY CURRENT SENSORS

An eddy current sensor measures distance between the sensor head and an electrically conducting surface, as illustrated in Fig. 6.15. Sensor operation is based on eddy currents that are induced at the conducting surface as magnetic flux lines from the sensor intersect with the surface of the conducting material. The magnetic flux lines are generated by the active coil in the sensor, which is driven at a very high frequency (1 MHz). The magnitude of the eddy current produced at the surface of the conducting material is a function of the distance between the active coil and the surface. The eddy currents increase as the distance decreases.

Changes in the eddy currents are sensed with an impedance (inductance) bridge. The two coils in the sensor are used for two arms of the bridge. The other two arms in the bridge are housed in an associated electronic package illustrated in Fig. 6.16. The first coil in the sensor (active coil), which changes inductance with target movement, is wired into the active arm of the bridge. The second coil is wired into an opposing arm of the same bridge, where it serves as a compensating coil to balance and cancel the effects of temperature change. The output from the impedance bridge is demodulated and becomes the analog signal, which is linearly proportional to distance between the sensor and the target. This signal is then amplified prior to recording on some analog recorder or conversion to a digital signal.

The sensitivity of the sensor is dependent upon the target material with higher sensitivity associated with higher conductivity materials. For aluminum targets, the sensitivity is typically 100 mV/mil (4 V/mm). Thus, it is apparent that eddy current sensors are high-output devices if the specimen material is non-magnetic; however, the sensitivity decreases significantly if the specimen material is magnetic. While the sensitivity is dependent on the resistivity (conductivity) and the magnetism of the target material, this fact does not affect the accuracy of the sensor because it is easily calibrated for a specific target material.

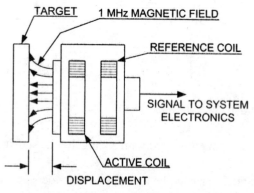

Fig. 6.15 Schematic diagram for an eddy current sensor.

For nonconducting, poorly conducting or magnetic materials, it is possible to bond a thin film of aluminum foil to the surface of the target at the location of the sensor to improve its sensitivity.

Because the penetration of the eddy currents into the material is minimal, the thickness of the foil can be as small as 0.2 mm (ordinary kitchen type aluminum foil).

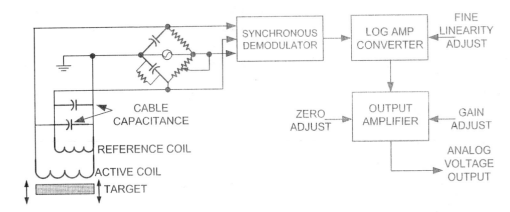

Fig. 6.16 Block diagram of the signal condition circuits used in converting impedance changes due to target displacement to an analog output voltage.

The effect of temperature on the output of the eddy current sensor is small. The sensing head with dual coils is temperature compensated; however, a small error can be produced by temperature changes in the target material because its resistivity is a function of temperature. For instance, if the temperature of an aluminum target is increased by 500 °F, its resistivity increases from 3 to 6 $\mu\Omega$-cm. and the bridge output is reduced by about 2% for this change in resistivity. For aluminum, the temperature sensitivity of the eddy current sensor is 0.004% °F.

The range of the sensor is controlled by the coil diameter with larger sensors exhibiting the larger ranges. The range to diameter ratio is usually about 0.26. Linearity is typically better than ± 0.5% and resolution is better than 0.05% of full scale. The frequency response is typically 20 kHz, although small-diameter coils can be used to increase this response to 50 kHz.

The fact that eddy current sensors do not require contact for measuring displacement is quite important. As a result of this feature, they are often used in transducer systems for automatic control of dimensions in fabrication processes. They are also applied extensively to determine thickness of organic coatings that are nonconducting.

6.7 PIEZOELECTRIC SENSORS

A piezoelectric material, as its name implies, produces an electric charge when it is subjected to a force or pressure. Piezoelectric materials, such as single-crystal quartz or polycrystalline barium titanate, contain molecules with asymmetrical charge distributions. When pressure is applied, the crystal deforms and there is a relative displacement of the positive and negative charges within the crystal. This displacement of internal charges produces external charges of opposite sign on the external surfaces of the crystal. If these surfaces are coated with metallic electrodes, as illustrated in Fig. 6.17, the charge q that develops can be determined from the output voltage vo because:

$$q = v_o\, C \qquad\qquad (6.13)$$

where C is the capacitance of the piezoelectric crystal.

The surface charge q is related to the applied pressure p by:

$$q = S_q\, A_p \qquad\qquad (6.14)$$

where S_q is the charge sensitivity of the piezoelectric crystal.
A is the area of the electrode.

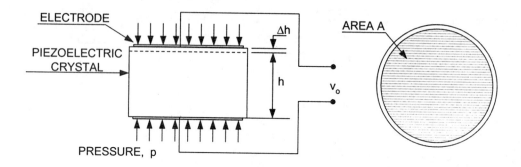

Fig. 6.17 Piezoelectric crystal deforming under the action of an applied pressure.

The charge sensitivity S_q is a function of the orientation of the sensor (usually a cylinder) relative to the axes of the piezoelectric crystal. Typical values of S_q for common piezoelectric materials are given in Table 6.3.

The output voltage v_o developed by the piezoelectric sensor is obtained by substituting Eqs. (6.8) and (6.14) into Eq. (6.13). Thus:

$$v_0 = \left(\frac{S_q}{kK}\right) h \, p \qquad \text{(a)}$$

The voltage sensitivity S_v of the sensor can be expressed as:

$$S_v = \left(\frac{S_q}{kK}\right) \qquad (6.15)$$

The output voltage v_o of the sensor is then:

$$v_0 = S_v h \, p \qquad (6.16)$$

Again, voltage sensitivity S_v of the sensor is a function of the orientation of the axis of the cylinder relative to the crystallographic axes. Typical values of S_v are also presented in Table 6.3.

Table 6.3
Typical charge S_q and voltage S_v sensitivities of piezoelectric materials

Material	Orientation	S_q (pC/N)	S_v (V-m/N)
Quartz SiO$_2$ Single Crystal	X-cut, length longitudinal	2.2	0.055
	X-cut, thickness longitudinal	− 2.0	− 0.05
	Y-cut, thickness shear	4.4	0.11
Barium Titanate BaTiO$_3$ Ceramic Poled	Parallel to polarization	130	0.011
	Perpendicular to polarization	− 56	− 0.004

Most piezoelectric transducers are fabricated from single-crystal quartz because it is the most stable of the piezoelectric materials and is nearly loss free both mechanically and electrically. Its properties are: modulus of elasticity 86 GPa, resistivity 10^{14} Ω-cm, and dielectric constant 40.6 pF/m. It exhibits excellent high-temperature properties and can be operated up to 550 °C. The charge sensitivity of quartz is very low when compared to barium titanate; however, with high-gain charge amplifiers available for processing the output signal, the lower sensitivity is not a serious disadvantage.

Barium titanate is a polycrystalline material that can be polarized by applying a high voltage to the electrodes while the material is at a temperature above the Curie point (125 °C). The electric field aligns the ferroelectric domains in the barium titanate and it becomes piezoelectric. If the polarization voltage is maintained while the material is cooled well below the Curie point, the piezoelectric characteristics become permanent and relatively stable after a short aging period.

The mechanical stability of barium titanate is excellent; it exhibits high mechanical strength and has a high modulus of elasticity (120 GPa). It is more economical than quartz and can be fabricated in a wide variety of sizes and shapes. While its application in transducers is second to quartz, it is frequently used in ultrasonic applications as a driver. In this application, a voltage is applied to the electrodes and the barium titanate deforms and delivers energy to a work piece or test specimen.

Most sensors exhibit relatively low output impedance (in the range of 100 to 1000 Ω). However, when piezoelectric crystals are used as the sensing elements in transducers, the output impedance is usually extremely high. The output impedance of a small cylinder of quartz depends the geometry of the crystal and on the frequency ω associated with the applied pressure. Because the sensor acts like a capacitor, the output impedance is given by:

$$Z_C = \frac{1}{j\omega C} = -\frac{j}{\omega C} \qquad (a)$$

Clearly, the impedance ranges from infinity for static applications to about 10 kΩ for very high-frequency applications (100 kHz). With this high output impedance, care must be exercised in monitoring the output voltage; otherwise, very serious errors can occur.

A circuit diagram of a typical system used to measure a voltage produced by a piezoelectric sensor is shown in Fig. 6.18. The piezoelectric sensor acts as a charge generator. In addition to the charge generator, the sensor is represented by parallel components including a capacitor C_p (about 10 pF) and a leakage resistor R_p (about 10^{14} Ω). The capacitance of the lead wires C_L must also be considered, because even relatively short lead wires have a capacitance larger than the sensor. The amplifier is either a cathode follower or a charge amplifier with sufficient input impedance to isolate the piezoelectric sensor in spite of its very large output impedance. If a pressure is applied to the sensor and maintained for a long period of time, the charge q developed by the piezoelectric material decays with time because a small current flows through both R_p and the amplifier resistance R_A. The time available for readout of the signal depends upon the effective time constant τ_e of the circuit that is given by:

$$\tau_e = R_e C_e = \frac{R_p R_A}{R_p + R_A}(C_p + C_L + C_A) \qquad (6.17)$$

where R_e is the equivalent resistance of the circuit.
C_e is the equivalent capacitance of the circuit.

Time constants ranging from 1000 to 100,000 s can be achieved with quartz sensors and commercially available charge amplifiers. These time constants are sufficient to permit measurement of quantities that vary slowly with time or measurement of static quantities for short periods of time.

Problems associated with measuring the output voltage diminish as the frequency of the mechanical excitation increases and the output impedance Z_C of the sensor decreases.

The inherent dynamic response of the piezoelectric sensor is very high, because the resonant frequency of the small cylindrical piezoelectric element is so large. The resonant frequency of the transducer depends upon its mechanical design as well as the mass and stiffness of the sensor. However, the most significant advantage of the piezoelectric sensor is its very high-frequency response.

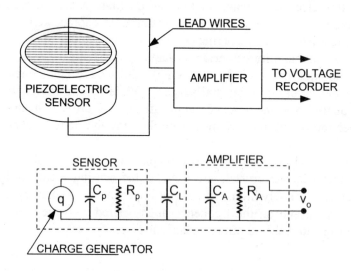

Fig. 6.18 Schematic diagram of a measuring system with a piezoelectric sensor.

6.8 PIEZORESISTIVE SENSORS

Piezoresistive sensors, as the name implies, are fabricated from materials that exhibit a change in resistance when subjected to a pressure. The development of piezoresistive materials was an outgrowth of research on semiconductors by the Bell Telephone Laboratories in the early 1950s that eventually led to the transistor. Piezoresistive sensors are fabricated from semiconductive materials usually silicon containing boron as the trace impurity for the P-type material and arsenic as the trace impurity for the N-type material. The resistivity ρ of a semiconducting material can be expressed as :

$$\rho = 1/(eN\mu) \tag{6.18}$$

where e is the electron charge, which depends on the type of impurity.

N is the number of charge carriers, which depends on the concentration of the impurity.

μ is the mobility of the charge carriers, which depends upon strain and its direction relative to the crystal axes.

Equation (6.18) shows that the resistivity of the semiconductor can be adjusted to any specified value by controlling the concentration of the trace impurity. The impurity concentrations commonly employed range from 10^{16} to 10^{20} atoms/cm^3, which permits a wide variation in the initial resistivity. For example, the resistivity for P-type silicon with a concentration of 10^{20} atoms/cm^3 is 50,000 $\mu\Omega$-cm, which is about 30,000 times higher than the resistivity of copper. This very high resistivity facilitates the design of miniaturized sensors. Another advantage of using single-crystal semiconducting silicon as a sensor is its high strength and stiffness (modulus). The elastic modulus of silicon is 190 GPA, which is nearly equal to that of steel (207 GPa). The yield strength of silicon—7.0 GPa—is nearly five times as large as 4340 steel with a yield strength of about 1.5 GPa. In addition, of single crystal silicon exhibits relatively high fatigue strength because dislocations produced during cyclic loading move to the surface and dislocation pinning, which produces flaws, does not occur.

Considerable progress has been made in applying fabrication techniques used in the microelectronics industry to the development of micro-miniature sensors. These solid-state sensors incorporate silicon as the mechanical element and piezoresistive sensors. In a piezoresistive pressure transducer, a diaphragm is etched from silicon to form the mechanical element of the transducer. Resistances that form the arms of a Wheatstone bridge are diffused or ion implanted directly into the silicon diaphragm to provide the sensing element.

Extremely small sensors may be produced using micromachining techniques that involve photolithography methods. An accurate pattern of the mechanical element is transferred to a photoresist coating on a silicon wafer. Anisotropic etchants, which etch at different rates along different crystal axes, are used to produce three-dimensional shapes in exposed silicon areas.

6.9 PHOTOELECTRIC SENSORS

In many applications, where contact cannot be made with the object being examined, a photoelectric sensor can often be employed to monitor changes in the intensity of light that sense the quantity being measured. When light impinges on a photoelectric sensor, it either creates or modulates an electrical signal. Most photoelectric devices employ semiconductor materials and operate by either generating a current or by changing resistivity. These devices are photodetectors that respond quickly to changes in light intensity.

6.9.1 Photoconducting Sensors

Photoconductive cells, illustrated in Fig. 6.19, are fabricated from semiconductor materials, such as cadmium sulfide (CdS) or cadmium selenide (CdSe), which exhibit a strong photoconductive response. When a photon with sufficient energy strikes a molecule of, say CdS, an electron is driven from the valence band to the conduction band and a hole or vacancy remains in the valence band. This hole and the electron both serve as charge carriers, and with continuous exposure to light, the concentration of charge carriers increases and the resistivity decreases. A circuit used to detect the resistance change ΔR of the photoconductor is also shown in Fig. 6.19. The resistance of a typical photoconducting detector changes over about 3 orders of magnitude as the incident radiation varies from dark to very bright.

When a photoconductor is placed in a dark environment, its resistance is high and only a small dark current flows. If the sensor is exposed to light, the resistance decreases significantly (the ratio of maximum to minimum resistance for R_d in Fig. 6.19 ranges from 100 to 10,000 in common commercial sensors); therefore, the output current can be quite large. The sensitivity depends on cell area, type of cell (CdS or CdSe), and the power limit for the cell. If the maximum voltage v_s is used to supply the cell at its maximum power limit, then sensitivities of up to 0.2 mA/lx result for CdS type cells. A typical cadmium sulfide photoconductor exhibits a maximum dark resistance of about 1 to 2 $M\Omega$. When subjected to a light intensity of about 2 foot-candles the resistance decreases to about 1 to 5 $k\Omega$. The change of about three orders of magnitude is large enabling many applications with very simple control circuits.

Photoconductors respond to radiation ranging from long thermal radiation through the infrared, visible, and ultraviolet regions of the electromagnetic spectrum. The sensitivity S, changes significantly with wavelength and drops sharply at both short and long wavelengths; consequently, photoconductive cells exhibit the same disadvantage as most other photodetectors because its calibration constant depends on the wave length of the impinging light.

The photocurrent requires some time to develop after the excitation is applied and some time to decay after the excitation is removed. The rise and fall times for commercially available photoconductors is usually about a second. Because of these delays, the CdS and CdSe photoconductors are not suitable for dynamic measurements. Instead, their simplicity, high sensitivity and low cost lend them to applications involving counting and switching based on a slowly varying light intensity.

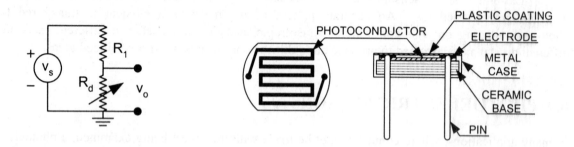

Fig. 6.19 When a photoconductor is used as R_d, the output voltage will vary with the light intensity.

6.9.2 Photodiode Sensors

Photodiodes are semiconductor devices that are responsive to high-energy particles and photons. Photodiodes operate by absorption of photons or charged particles and generate a current that flows in an external circuit, which is proportional to the incident power. Photodiodes can be used to detect the presence or absence of minute quantities of light and can be calibrated for extremely accurate measurements from intensities below 1 pW/cm^2 to intensities above 100 mW/cm^2. Silicon is the most common semiconductor material used in fabricating planar diffused photodiodes. These photodiodes are employed in such diverse applications as spectroscopy, photography, analytical instrumentation, optical position sensors, beam alignment, surface characterization, laser range finders, optical communications and medical imaging instruments.

Planar diffused silicon photodiodes are P-N junction diodes. A P-N junction can be formed by diffusing either a P-type impurity (anode), such as boron, into a N-type bulk silicon wafer, or a N-type impurity, such as phosphorous, into a P-type bulk silicon wafer. The diffused area defines the photodiode active area. To form an ohmic contact it is necessary to diffuse another impurity into the backside of the wafer. The impurity is an N-type for P-type active area and P-type for an N-type active area. Then contact pads are deposited on the front active area on defined areas, and on the backside, completely covering the device. The active area is then covered with an anti-reflection coating to reduce the reflection of the light for a specific predefined wavelength. The non-active area on the top is covered with a thick layer of silicon oxide. A schematic illustration of a planar diffused photodiode fabricated from N-type silicon is presented in Fig. 6.20.

By controlling the thickness of bulk substrate, the speed and responsivity of the photodiode can be controlled. Note that the photodiodes, when biased, are operated in the reverse bias mode, i.e. a negative voltage applied to anode and positive volt-age to cathode.

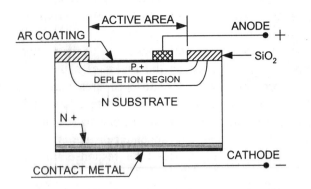

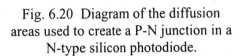

Fig. 6.20 Diagram of the diffusion areas used to create a P-N junction in a N-type silicon photodiode.

Electrical Characteristics of Photodiodes

A silicon photodiode can be represented by a current source in parallel with an ideal diode as shown in Fig. 6.21. The current source represents the current generated by the incident light, and the diode represents the P-N junction. In addition, a junction capacitance (C_J) and a shunt resistance (R_{SH}) are in parallel with the other components. A resistance (R_s) is connected in series with all components in this model.

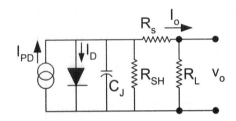

Fig. 6.21 Representative circuit for a photodiode.

Physically the shunt resistance is the slope of the current-voltage curve of the photodiode at the origin, i.e. $v_0 = 0$. Although an ideal photodiode should have an infinite shunt resistance, actual values range from 10 MΩ to several 1000 MΩ. Experimentally the shunt resistance is determined by applying $\pm$ 10 mV across the diode, measuring the resulting current and calculating the resistance from Ohm's law. Shunt resistance is used to determine the noise current in the photodiode with no bias (photovoltaic mode). For superior photodiode performance, a very high shunt resistance should be specified.

The series resistance of a photodiode arises from the resistance of the contacts and the resistance of the undepleted silicon shown in Fig. 6.20. The series resistance is given by:

$$R_s = \frac{(h_s + h_d)\rho}{A} + R_C \qquad (6.19)$$

where h_s is the thickness of the substrate and h_d is the thickness of the depleted region
A is the diffused area of the junction
ρ is the resistivity of the substrate
R_C is the contact resistance.

The series resistance is used to determine the linearity of the photodiode in photovoltaic mode[2]. Although an ideal photodiode should have zero series resistance, typical values ranging from 10 Ω to as much as 1000 Ω are measured.

The junction capacitance C_J is due to the geometry of the planar structure of the photodiode. The boundaries of the depletion region, shown in Fig. 6.20, act as the plates of a parallel plate capacitor. The junction capacitance is directly proportional to the diffused area and inversely proportional to the thickness of the depletion region. Higher resistivity substrates have lower junction capacitance because of the larger voltage drop across the substrate. Furthermore, the capacitance is dependent on the magnitude of the reverse bias voltage. The junction capacitance is very important in dynamic measurements of light intensity because it markedly affects the response time of the photodiode.

The rise or fall time of a photodiode is defined as the time for the signal to rise or fall from 10% to 90% or from 90% to 10% of the final value, respectively. This parameter can be expressed in term of the frequency response f_r (the frequency at which the photodiode output decreases by 3dB). The rise time may be approximated by:

$$t = 0.35/f_r \qquad (6.20)$$

There are three factors defining the response time of a photodiode:

1. t_{drift}—the charge collection time of the carriers in the depleted region of the photodiode.
2. $t_{diffuse}$—the charge collection time of the carriers in the undepleted region of the photodiode.
3. t_{RC} —the RC time constant of the diode and the associated circuit.

The factor t_{RC} is given by:

$$t_{RC} = 2.2\ RC \qquad (6.21)$$

where $R = R_s + R_L$ and $C = C_J + C_s$. Note that C_s is the stray capacitance due to lead wires etc.

Because the junction capacitance C_J is dependent on the diffused area of the photodiode and the applied reverse bias, faster rise times are obtained with smaller diffused area photodiodes, and larger applied reverse biases. In addition, using short leads and careful arrangement of the electronic components involved in the circuit containing the photodiode can minimize stray capacitance.

The total rise time is determined by:

$$t_r = \sqrt{t_{drift}^2 + t_{diffuse}^2 + t_{RC}^2} \qquad (6.22)$$

Generally, in photovoltaic mode of operation without a bias voltage, rise time is dominated by the diffusion time for diffused areas less than 5 mm^2 and by RC time constant for larger diffused areas. When operated in photoconductive mode with a reverse bias voltage, if the photodiode is fully depleted, the dominant factor is the drift time. In non-fully depleted photodiodes, however, all three factors contribute to the response time.

Optical Characteristics of Photodiodes

The responsivity R of a silicon photodiode is a measure of the sensitivity to light, and it is defined as the ratio of the photocurrent I_{PD} to the incident light power P at a given wavelength:

$$R = I_{PD}/P \qquad (6.23)$$

---

[2] Photodiodes act as photovoltaic deices when no bias voltage is applied.

The responsivity is a measure of the effectiveness of the conversion of the light power into electrical current. It varies with the wavelength of the incident light, as indicated in Fig. 6.22, the reverse bias voltage and the temperature. Responsivity increases with reverse bias voltage because it improves the charge collection efficiency in the photodiode.

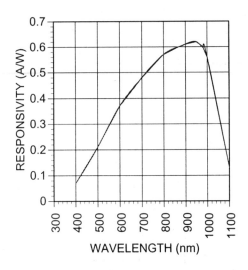

Fig. 6.22 Spectral responsivity of a typical planar diffused photodiode.

The linearity of a photodiode is usually represented by the output current versus light intensity relationship illustrated in Fig. 6.23. The lower limit is determined by the noise level and is defined as the point where the noise signal $v_n = v_o$. The linear range of the silicon photodiode shown in Fig. 6.23 extends from the this level to an upper limit that depends on the load resistance in the external circuit. Typically, this linear range covers 8 to 10 orders of magnitude.

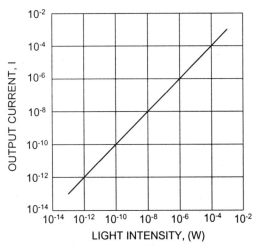

Fig. 6.23 Output current as a function of light intensity for a silicon photodiode.

The choice of using a semiconductor diode as a photovoltaic detector or as a photoconductive detector depends primarily on the frequency response required in the measurement. For light intensity fluctuations at the lower frequencies (less than 100 kHz), a photovoltaic circuit exhibits a lower noise voltage than a photoconduction circuit with reverse bias voltage. For higher frequencies required for measuring high-speed light pulses or high frequency modulation of a continuous light beam, the reverse bias serves to accelerate the electron/hole transition times, which improves the frequency response. Photoconductive diodes operate over a frequency range from DC to 100 MHz and are capable of measuring the intensity of light pulses with rise times in the 3 to 12 ns range.

Semiconductor photodiodes are small, rugged, and inexpensive and because of these advantages they have replaced the vacuum tube detectors in most applications. The photodiodes may be used in either mode with operational amplifiers to give a responsivity that is exceptionally high.

Circuits showing photodiodes operating in the photoconduction mode and in the photovoltaic mode are presented in Fig. 6.24.

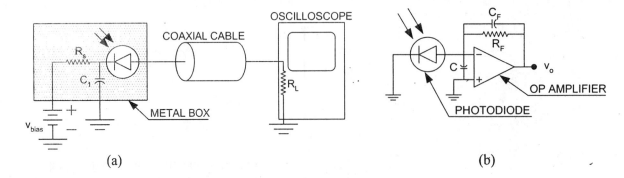

Fig. 6.24 Circuits used with photodiodes. (a) High intensity light at high frequency.
(b) Low intensity light with low frequency.

6.10 RESISTANCE TEMPERATURE DETECTORS

The change in resistance of metals with temperature provides the basis for a family of temperature sensors known as resistance temperature detectors (RTDs). The sensor is simply a conductor fabricated either as a wire-wound coil encased in a glass or ceramic tube or as a grid formed from a thin film adhered to a ceramic substrate. The change in resistance of the conductor with temperature is given by the expression:

$$\Delta R/R_o = \gamma_1 (T - T_o) + \gamma_2 (T - T_o)^2 + \ldots\ldots + \gamma_n (T - T_o)^n \qquad (6.24)$$

where T_o is a reference temperature.

R_o is a reference resistance at temperature T_o.

$\gamma_1, \gamma_2, \ldots\ldots\gamma_n$ are temperature coefficients of resistance.

Resistance temperature detectors are often used in ovens and furnaces where inexpensive but accurate and stable temperature measurements and controls are required.

Platinum is widely used for sensor fabrication—it is the most stable of all the metals, is the least sensitive to contamination and is capable of operating over a very wide range of temperatures (4 °K to 1064 °C). Because platinum provides an extremely reproducible output, it has been selected for use in establishing the International Temperature Scale (1990) over the range from 13.8033 °K to 961.78 °C.

The performance of platinum RTDs depends strongly on the design and construction of the package containing the sensitive wire coil. The most precise sensors are fabricated with a minimum amount of support; hence, they are fragile and often fail if subjected to rough handling, shock or vibration. Most sensors used in transducers for industrial applications have platinum coils supported on ceramic or glass tubes. These fully supported sensors are quite rugged and will withstand shock levels up to 100 g's. Unfortunately, the range and accuracy of such sensors are limited to some degree by the influence of the constraining package. An example of a RTD fabricated with a thin film of platinum deposited on a ceramic substrate then encapsulated in glass is presented in Fig. 6.25.

The sensitivity S of a platinum RTD is relatively high (S = 0.390 Ω/°C at 0° C); however, the sensitivity varies with temperature because of the nonlinear response of the sensor to temperature, as indicated by Eq. (6.28). The sensitivity S decreases to 0.378, 0.367, 0.355, 0.344, and 0.332 Ω/°C at temperatures of 100, 200, 300, 400 and 500 °C, respectively.

The dynamic response of an RTD depends almost entirely on construction details. For large coils mounted on heavy ceramic cores and sheathed in stainless steel tubes, the response time may be several seconds or more. For film or foil elements mounted on thin polyimide substrates, the response time can be less than 0.1 s.

Fig. 6.25 Thin film platinum resistance temperature detectors (Pt-RTD) fabricated with a thin film platinum deposited on a ceramic substrate then encapsulated in glass.
Courtesy of U. S. Sensor Corp.

6.11 THERMISTORS

A second type of temperature sensor based on resistance change of the sensing element with temperature is known as a thermistor. A thermistor differs from a resistance temperature detector (RTD) because the sensing element is fabricated from a semiconducting material instead of a metal. The semiconducting materials, which include oxides of copper, cobalt, manganese, nickel and titanium, exhibit very large changes in resistance with temperature. As a result, thermistors can be fabricated in the form of extremely small beads as shown in Fig. 6.26.

Resistance change with temperature for a thermistor can be expressed by an equation of the form:

$$\log_e \rho = A_0 + \frac{A_1}{\theta} + \frac{A_2}{\theta^2} + \cdots + \frac{A_n}{\theta^n} \qquad (6.25)$$

where ρ is the specific resistance of the material.
A_1, A_2, ... A_n are material constants.
θ is the absolute temperature.

The temperature-resistance relationship as expressed by Eq. (6.25) is usually approximated by retaining only the first two terms. The simplified equation is then expressed as:

$$\log_e \rho = A_0 + \frac{\beta}{\theta} \qquad (6.26)$$

Using Eq. (6.26) is convenient and acceptable when the temperature range is small and the higher-order terms in Eq. (6.25) are negligible.

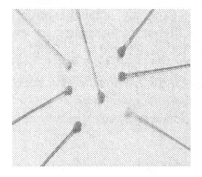

Fig. 6.26 Miniature bead-type thermistors.
Courtesy of U. S. Sensors Corp.

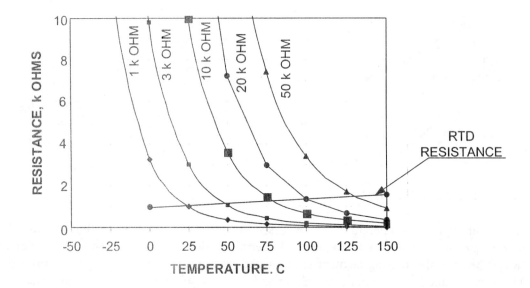

Fig. 6.27 Resistance temperature characteristics of thermistors and a RTD.

Thermistors have many advantages over other temperature sensors and are widely used in industry. They can be small (0.005-in. diameter) and, consequently, permit point sensing and rapid response to temperature change. Their high resistance minimizes lead-wire problems and their output is more than 10 times that of a resistance temperature detector (RTD) as shown in Fig. 6.27. Finally, thermistors are very rugged, which permits use in industrial environments where shock and vibration occur. The disadvantages of thermistors include nonlinear output with temperature, as indicated by Eqs. (6.25) and (6.26) and a limited range. Significant advances have been made in thermistor fabrication methods and it is possible to obtain stable, reproducible, interchangeable thermistors that are accurate to 0.5% over a specified temperature range.

6.12 THERMOCOUPLES

A thermocouple is a temperature sensor that consists of two dissimilar materials joined together to form both thermal and electrical contact. A potential develops at the interface of the two dissimilar materials as the temperature changes. This thermoelectric phenomenon is known as the **Seebeck effect**. Thermoelectric sensitivities (μV/°C) for a number of different materials in combination with platinum are listed in Table 6.4. Data from Table 6.4 can be used to determine the sensitivity of any thermocouple junction by noting, for example, that:

$$S_{Chromel/Alumel} = S_{Chromel/Platinum} - S_{Alumel/Platinum} = +26.8 - (-13.6) = 39.4 \ \mu V/°C$$

Table 6.4
Thermoelectric sensitivities for different materials in contact with platinum

Material	Sensitivity ($\mu V/°C$)	Material	Sensitivity ($\mu V/°C$)
Constantan	− 35	Copper	+ 6.5
Nickel	− 15	Gold	+ 6.5
Alumel	− 13.6	Tungsten	+ 7.5
Carbon	+ 3	Iron	+ 18.5
Aluminum	+ 3.5	Chromel	+ 26.8
Silver	+ 6.5	Silicon	+ 440

Common thermocouple material combinations include iron/constantan, chromel/alumel, chromel/constantan, copper/constantan, and platinum/platinum-rhodium.

The output voltage from a thermocouple junction is measured by connecting two identical thermocouples into a circuit as shown in Fig. 6.28. The output voltage v_o from this circuit is related to the temperature at each junction by an expression of the form:

$$v_o = S_{A/B} (T_1 - T_2) \tag{6.27}$$

where $S_{A/B}$ is the sensitivity of material combination A and B.
   T_1 is the temperature at junction J_1.
   T_2 is the temperature at junction J_2.

In practice, junction J_2 is a reference junction that is maintained at a carefully controlled reference temperature T_2. Junction J_1 is placed in contact with the body at the point where a temperature is to be measured. When a meter is inserted into the thermocouple circuit, junctions J_3 and J_4 are created at the contact between the wires (material B) and the terminals of the meter. If these terminals are at the same temperature ($T_3 = T_4$), the added junctions (J_3 and J_4) do not affect the output voltage v_o.

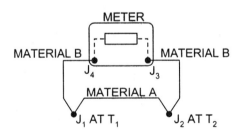

Fig. 6.28 A typical thermocouple circuit.

Designed as a sensor, the thermocouple can be made quite small (0.0005-in. diameter wire is available) to reduce its mass and size. These miniaturized thermocouples have a rapid response time (milliseconds) and provide essentially point measurements of temperature. A thermocouple can cover a wide range of temperatures; however, the output is nonlinear, and linearizing circuits are required to relate the output voltage v_o to the temperature T. In addition to nonlinear output, thermocouple sensors suffer the disadvantage of very low signal output and the need to carefully control the reference temperature at junction J_2.

6.13 INTERFACING SENSORS WITH THE RCX

Ports 1, 2 and 3 on the RCX are used to interface with both active and passive sensors. Recall that passive sensors do not require power from the RCX to function. Moreover, the output from a passive sensor is a voltage drop across the sensor terminals due to a resistance or a voltage source placed across the terminals of the RCX. For example, the voltage source can be from an externally powered Wheatstone bridge, a solar cell or a thermocouple. The resistive loads may be from potentiometers, thermistors or fixed value resistors.

Active sensors require power, supplied by the RCX, to function. The LEGO light sensor is an example of an active sensor because it requires power drawn from the RCX to operate the light emitting diode (LED) that illuminates the scene. The light reflected from the scene passes through a lens and impinges on a photodiode producing an output voltage that is monitored by the RCX and converted to a reading of light intensity. The supply voltage to the sensor and the output voltage from the sensor are transmitted from the same port over the same wire. The RCX is able to use the same wire for two different functions by multiplexing. First, the RCX connects the Port to a voltage source (about 8 V) for 3 ms providing power to the sensor. Next, it monitors the output voltage from the sensor for 1 ms. By rapidly switching back and forth between supplying voltage and monitoring output voltage, the RCX performs these two function with a single port and a single pair of lead wires.

This discussion will be limited to passive sensors because dealing with the complexities of multiplexing is beyond the scope of this textbook. It is possible to use LEGO sensors that are active because they have been specifically designed for the RCX and incorporate circuits compatible with multiplexing.

6.13.1 Measuring Voltage with the RCX

The RCX can handle input voltages of 5 V across the terminals of its input Ports. The voltage measuring circuit in the RCX is shown in Fig. 6.29. The shaded box represents the RCX with its digital voltmeter (DVM). A 10 kΩ resistor in series with a 5 V voltage source, internal to the RCX, are across the terminals of the DVM. External to the RCX is an unknown voltage v_u in series with an external 10 kΩ resistor.

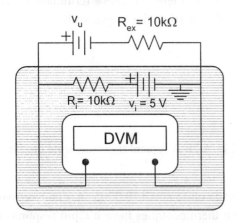

Fig. 6.29 Internal and external wiring required to measure voltage.

Let's perform a circuit analysis to determine the output voltage v_0 that is measured by the DVM. First, we will redraw the circuit shown in Fig. 6.29 to better identify the circuit loop that will be considered in the analysis. The new circuit diagram presented in Fig. 6.30, shows loop 1 that includes the two voltage sources and the two resistors. We also show the current flow counterclockwise around this loop. Note the polarity of the two voltage sources, which implies that v_u

> 0. Recall, Kirchhoff's voltage law, which states that the sum of the voltages around a loop must equal zero. Beginning at point A and applying this law permits us to write:

$$\Sigma V = + v_i - IR_i - v_u - IR_{ex} = 0 \qquad \text{(a)}$$

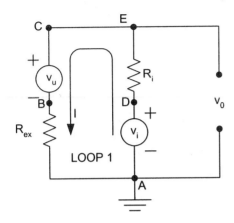

Fig. 6.30 Circuit diagram for the RCX showing the internal voltage source and resistor together with the external voltage source and resistor.

Solve Eq. (a) for the current flow I to obtain:

$$I = \frac{v_i - v_{ex}}{R_i + R_{ex}} \qquad \text{(b)}$$

The voltage v_0 across the DMV is the voltage between points A and E, which is given by:

$$v_0 = v_i - IR_i \qquad \text{(c)}$$

Substituting Eq. (b) into Eq. (c) and simplifying yields:

$$v_u = \frac{v_i R_i + v_{ex} R_{ex}}{R_i + R_{ex}} \qquad \text{(d)}$$

If $R_i = R_{ex} = R$, then Eq. (d) reduces to:

$$v_0 = \tfrac{1}{2}\,(v_i + v_{\underline{u}}) \qquad \text{(6.28)}$$

Set v = 5 V and Eq. (6.28) reduces to:

$$v_u = \tfrac{1}{2}\,(5 + v_u) \qquad \text{(6.29)}$$

There is always a danger of damaging the RCX if an accidental over voltage (more than 5 V) is applied to the terminals of one of its ports. To prevent damage a pair of diodes is placed across the external circuit arm as illustrated in Fig. 6.31. The 1N751 is a Zener diode that begins to conduct at 5.1 V and prevents the voltage on the internal circuitry of the RCX from exceeding its break down voltage. Similarly, a second Zener diode is reversed to prevent the negative voltage applied to the terminals of the RCX from exceeding − 5.1 V.

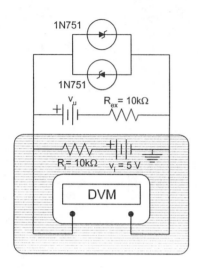

Fig. 6.31 Voltage limiting Zener diodes protect the RCX from large positive and negative voltages.

RCX Characteristics

The DVM on the RCX does not look like a typical digital voltmeter, because it is incorporated in the display. The RCX contains a 10-bit analog to digital converter (ADC) that converts the voltage to a digital code that varies from 0 to 1023. The number displayed, to the left of the little man, may be in digital code (some number between 0 and 1023), binary (0 or 1), or proportional (0 to 100), depending upon the programming of the sensor ports.

The DVM can be used to monitor the three input (sensor) ports or the three output (actuator) ports. Selection of the port to be monitored is made with the View button. As the button is depressed a caret ($\wedge$) appears below the input port being monitored. The display then provides a reading that is related to the voltage applied to this port. The relation between the voltage and the reading will depend on the sensor selection in programming this port. We will discuss programming the ports for specific sensors later in Section 16.3.

The input and output ports on the RCX provide the input and output terminals where connections of sensors and actuators are made. On a casual inspection these ports look like standard 2 × 2 LEGO plates. However, on closer inspection metallic rings encircle about 90° around each of the four studs. Connections are made to the RCX by pressure contact through these rings. The four studs are not electrically independent. As shown in Fig. 6.32, the two left hand studs are connected together on the underside of the port and the two right hand studs are also connected together. The left hand studs are the positive terminals and the right hand studs negative.

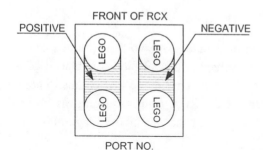

Fig. 6.32 Input port configuration showing electrical connections.

6.13.2 Designing and Building Sensors

There are four different sensors commercially available from LEGO—touch, light, temperature and angle of rotation. These are excellent sensors designed specifically to interface with the RCX. The touch and temperature sensors are passive and the light and rotation sensors are active. However, it is possible to design and build your own sensors using components that are available from local electronic supply stores. In fact with Internet orders, it is possible to obtain almost any electronic device or component from regional supply houses in a few days.

Touch Sensor

Let's begin with the touch sensor because it is the simplest of all the sensors. It consists of a simple momentary switch, a LEGO support block to facilitate mounting in a LEGO design environment, and a connector that is compatible with the RCX. A schematic diagram of a momentary switch[3] that can be used for the touch sensor is shown in Fig. 6.33. . As shown in Fig. 6.33, it is in the normally open mode because a spring keeps it open until the button is pressed bringing the switch bar down on the contacts closing it. When the button is released, the switch returns to the open position.

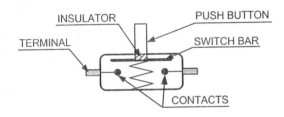

Fig. 6.33 A normally open momentary switch used in the design of a touch sensor.

If the touch sensor is to be employed with the RCX, the electrical connects must be compatible with the ports on the RCX. The easiest way of connecting the switch to the RCX is to take a LEGO lead wire with port compatible connectors on both ends and cut it into two pieces. Take one of the two pieces and strip the insulation for about 6 to 8 mm from the ends of the two wires. Then use a soldering iron to solder U shaped connectors to the two LEGO lead wires. Next prepare the ends of a piece of # 24 dual-lead, stranded-wire about 4 to 6 inches long. Solder U terminals to the leads on one end of the wire and then solder the leads on the other end to the to the two terminals on the momentary switch. Connect the four U terminals to the barrier strip[4] as shown in Fig. 6.34. The other end of the lead wire has a LEGO connector plate that is compatible with the LEGO terminals on the RCX.

At this stage of the development of the touch sensor, check if the RCX responds when the push button on the switch is depressed. Set the RCX to Program 1 (but do not run the program[5]), connect the sensor to Port 1 and depress the View Button until the ∧ mark is pointing to Port 1. The display should indicate 0. Press the button closing the switch and note that the display reads 1. This reading indicates that the switch is functioning properly and will act as a touch sensor. For this simple

---

[3] Momentary switches are available from a large number of companies through electronic suppliers. For this example, we selected a Panasonic detector switch part number ESE-11HS1 that was available from Digi-Key for $0.78. This switch, which is rated at 5 VDC and 10 mA, is actuated with a force of 35 g applied to its push button.

[4] The barrier blocks provide a convenient means for connecting sensors to the RCX. Only one dual lead wire with a LEGO connector is needed to connect a number of different sensors to the RCX when the barrier block is used as an intermediate connection device.

[5] Program 1 is locked into the RCX and was loaded with the firmware. If you run the program before taking the readings, the display will indicate the raw value that varies for 0 to 1023.

sensor no calibration is required—it is either open with a zero reading on the RCX or closed with a reading of 1.

If the touch sensor is to be used in a LEGO design environment, it should be mounted on a LEGO plate or brick. The selection of either a plate or a brick will depend on the size of the switch and the method you choose to attach the switch to then LEGO component. We show an example of a mounting arrangement with a very small switch attached to a 1 × 2 LEGO brick and a 2 × 8 plate in Fig. 6.34. Before the switch was attached to the brick, its two studs were cut off to form a smooth top surface. The switch was mounted to the brick with a fastener and quick drying cement (auto body repair compound).

Fig. 6.34 Left: Mounting the touch sensor on a 1 × 2 LEGO brick.
Right: Test assembly with the touch sensor RCX and barrier block.

Angle Sensor

The angle sensor is also easy to design and build because it consists of a rotary potentiometer (see Fig. 6.2b), a LEGO support block to facilitate mounting in a LEGO design environment, and a connector that is compatible with the RCX. Use the remaining end of the LEGO lead wire that was cut to provide a connector for the touch sensor to solve the connector problem. The most significant design problem is selecting a potentiometer. For this example, we have selected single turn potentiometer that is encased in a very small rectangular case[6]. This single turn potentiometer is designed with an effective electrical angle of 270° ± 5°. Thus, it will serve as a sensor capable of measuring angles that vary from about 0 to 270°. The total resistance of the potentiometer is specified as 10 kΩ ± 10%. The potentiometer used in this experiment was checked and its resistance was measured as 10.8 kΩ. The element material over which the wiper moves is a conductive cermet.

This potentiometer has a threaded mounting shank that is ¼ in. in diameter with ¼ - 32 threads. It is fastened to a 2 × 2 LEGO angle as shown in Fig. 6.35. It was necessary to enlarge the hole in the LEGO angle to accommodate the threaded shank. We enlarged the hole by using several different drills with diameters of 13/64, 7/32, 15/64 and ¼ inch. The plastic used in fabricating the LEGO pieces is so soft that the drills can be turned into the preexisting hole by hand. We also trim off the four studs by holding the angle in a vise and cutting off the studs with a box cutter. This operation

[6] We have selected a 1 W - 10 KΩ ± 10% Bourns Series 50 precision potentiometer with a linear taper. The Digi-Key catalog number for this component is 51CAD-E24-A15-ND and its cost was $7.39.

produced a flat surface so that the nut could be tightened to firmly fix the potentiometer to the LEGO angle. The potentiometer's shaft, which turns the wiper, is 1/8 in. (3.18 mm) in diameter. We have adapted a 40 tooth LEGO gear to the shaft so that it can be used in calibrating the angle sensor. The LEGO angle, with the potentiometer, was mounted with a number of LEGO components to form a calibration assembly as shown in Fig. 6.36. Note that the teeth on the spur gear are close to the edge of the 2 × 4 LEGO plate.

Fig. 6.35 The potentiometer mounted on a LEGO 2 × 2 angle with a 24 tooth spur gear attached to its shaft.

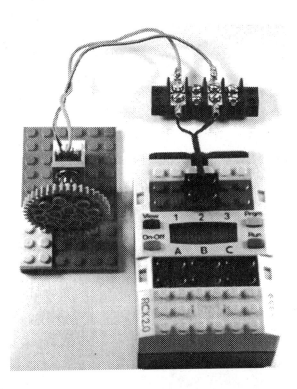

Fig. 6.36 The assembly used to calibrate the angle sensor.

To wire the potentiometer it is necessary to understand that the center terminal (lead) connects with the wiper. The outer terminals (leads) connect to the ends of the resistor element. If you measure the resistance across the outer terminals with an ohmmeter the reading should be 10 K ± 10% regardless of the angular position of the shaft. Next, connect the ohmmeter across the terminal on the left and the center terminal. Rotate the potentiometer in the clockwise direction as far as possible, and observe that the resistance decreases as you rotate the shaft. The resistance across these two terminals approaches zero when the potentiometer wiper is in the full clockwise position.

Now that we understand the position of the wiper relative to the left hand terminal, let's solder lead wires to the left and center terminals. We have connected the lead wires from the potentiometer to a barrier block as shown in Fig. 6.36. The lead wire with the LEGO connector on one end is also attached to the barrier block. This wiring arrangement permits us to use the wire with the LEGO connector for a number of different sensors in these initial experiments.

Select Program #1 on the RCX, click on the View button until the indicator in pointing to Port 2, and then connect the angle sensor to this Port. On the RCX and observe the reading on its display as you rotate the shaft. This reading varies from 1 to 532 as we rotate the shaft from the fully clockwise position to the fully counterclockwise position. Clearly, this reading is not directly related to the angle indicated by the wiper position. The reading is what is known as a raw value. The raw value varies from 0 for an input voltage of 0 V across the DVM to 1023 for a voltage of + 5V. The fact that we are recording raw values ranging from 1 to 532 implies that the voltage across the 10 kΩ potentiometer is varying from about 0 to about 2.5 V.

We are now ready to calibrate the sensor and to prepare a table showing the angle of rotation versus RCX readings. We will use the 40-tooth LEGO gear to indicate the angle of rotation by manually turning the gear one tooth per increment of rotation. We begin with the potentiometer near the full clockwise position with a tooth in alignment with the edge of the 2 × 8 plate and make the initial reading. The we advance the gear by one tooth, which is equivalent to an advance of $360°/40 = 9°$, and repeat the reading. This process is continued by rotating in increments in the counterclockwise direction until we reach the end of the travel of the potentiometer. Then we reversed the process and made the sequence of readings again while returning to the initial zero position. The raw value readings taken from the RCX display during the calibration process are presented in the spreadsheet shown in Fig. 6.36. The readings were taken as we indexed the potentiometer's shaft 9° increments from 0 to 270°. A graph showing the RCX reading as a function of angle is also presented in Fig. 6.37.

Number	Reading 1	Reading 2	Position	Average
1	23	26	0	24.5
2	60	61	9	60.5
3	91	91	18	91.0
4	121	122	27	121.5
5	141	143	36	142.0
6	172	172	45	172.0
7	196	199	54	197.5
8	221	221	63	221.0
9	242	243	72	242.5
10	262	264	81	263.0
11	281	281	90	281.0
12	298	299	99	298.5
13	315	315	108	315.0
14	329	330	117	329.5
15	346	346	126	346.0
16	362	363	135	362.5
17	377	378	144	377.5
18	392	392	153	392.0
19	406	406	162	406.0
20	419	419	171	419.0
21	434	433	180	433.5
22	444	445	189	444.5
23	456	456	198	456.0
24	468	467	207	467.5
25	479	480	216	479.5
26	489	490	225	489.5
27	500	501	234	500.5
28	509	509	243	509.0
29	517	518	252	517.5
30	527	527	261	527.0
31	532	532	270	532.0

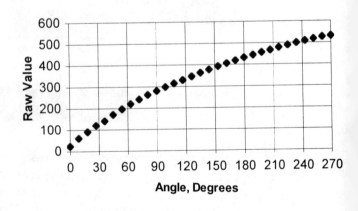

Fig. 6.37 Calibration curve for the 10 kΩ potentiometer used as an angle sensor.

An examination of the calibration curve in Fig. 6.37 indicates that the output from the angle sensor is not linear. Let's explore the reason for the non-linearity by performing a circuit analysis to determine the output voltage v_0 measured by the DVM. First, we prepare the circuit diagram, shown in Fig. 6.39, which includes the internal voltage source with $v_i = 5.00$ V, the internal resistor $R_i = 10$ MΩ and the external resistor R_s. Note the external resistor is a variable resistor that represents the angle sensor. We also show the current flow counterclockwise around this loop. Beginning at point A and applying Kirchhoff's voltage law permits us to write:

$$\Sigma V = + v_i - IR_i - IR_s = 0 \qquad \text{(a)}$$

Solve Eq. (a) for the current flow I to obtain:

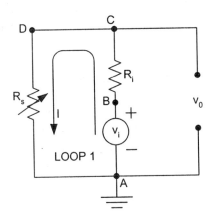

Fig. 6.38 Circuit diagram representing the angle sensor and the internal components in the RCX.

$$I = \frac{v_i}{R_i + R_s} \qquad \text{(b)}$$

The voltage v_0 across the DMV is the voltage between points A and C, which is given by:

$$v_0 = v_i - IR_i \qquad \text{(c)}$$

Substituting Eq. (b) into Eq. (c) and simplifying yields:

$$v_0 = v_i - \frac{v_i R_i}{R_i + R_s} = v_i \left[1 - \frac{1}{1 + R_s/R_i} \right] = v_i \left[\frac{R_s/R_i}{1 + R_s/R_i} \right] \qquad (6.30)$$

Let's consider the maximum and minimum values for the sensor resistance R_s. If $R_s = 0$, its minimum value, Eq. (6.30) gives $v_0 = 0$. Recall the raw value displayed on the RCX was 1, which is nearly[7] 0. If $R_s = 10.8$ MΩ and $R_i = 10.0$ MΩ, then $R_s/R_i = 10.8/10.0 = 1.08$ and Eq. (6.30) gives $v_0 = 2.596$ V. The equivalent raw value is $(2.596/5.00)(1023) = 531$. It is encouraging to note that we recorded 532 for the raw value.

The reason for the non-linearity is the form of Eq. (6.30). The ratio R_s/R_i in the denominator causes the non-linearity. Non-linearity of sensors is not a significant problem today, because microprocessors incorporated with the instrumentation for the sensor can be programmed to accommodate non-linear functions.

_____

[7] The value of R_s is not quite zero, because resistance is introduced in the circuit due to lead wires, and contact resistance at the potentiometer's wiper and the Lego terminals.

Temperature Sensor

A temperature sensor it also easy to fabricate because, like the angle sensor, it is a variable resistor. We have two choices for the sensor—a resistive temperature detector (RTD), described in Section 6.10, or a thermistor, described in Section 6.11. We have selected the thermistor because it is less expensive and more readily available at electronic supply stores. There are several types of thermistors available—some are shaped like disks with pigtail leads and others are small cylinders with axial leads. Some have positive thermal coefficients (PTC) of resistance and others have negative thermal coefficients (NTC). We selected a small cylindrical thermistor, with a diameter of 1.91 mm and a negative thermal coefficient (NTC)[8]. The active element of the thermistor was encased in glass. In each series of thermistors, there are several resistances available as indicated in Fig. 6.27. The thermistor selected[9] for the temperature sensor described here exhibited a resistance of 5 kΩ.

The active element of the thermistor is extremely small and encased in glass; thus it is electrically and thermally insulated. The thermal insulation is not an issue, in this sensor development, because we are measuring temperatures that change relatively slowly with time. The electrical insulation is a desirable characteristic because the sensor must be waterproof. Unfortunately, the leads from the thermistor are not insulated nor is the solder joint made to connect lead wires to the thermistor leads.

We waterproofed the thermistor and its leads using shrink tubing. Shrink tube slightly larger than the wire or wires to be insulated are slipped over the assembly. Hot air from a heat gun is directed at the tubing. As the tubing is heated it shrinks and tightly encases the wire or wires. We used two pieces of shrink tubing in waterproofing the temperature sensor. The first piece was applied over one of the leads from the thermistor. Its purpose was to insulate one thermistor lead from the other when the leads were folded to run parallel to one another. The second, larger diameter, shrink tube encased the entire assembly.

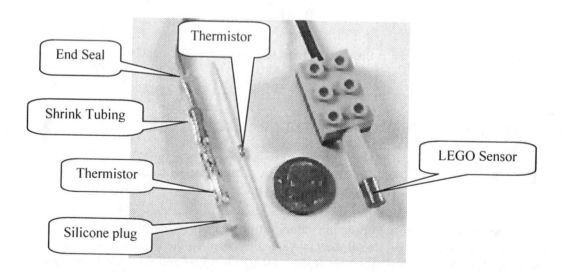

Fig. 6.39 The packaged thermistor, the prepackaged thermistor and the LEGO temperature sensor.

---

[8] The thermistor was a GE76 manufactured by Keystone Thermometrics and distributed by Digi-Key as part number KC008G-ND for $2.66 each.

[9] We purchase four thermistors with nominal resistances at 25 °C of 1k, 2k, 3k and 5kΩ. We finally selected the 5kΩ thermistor because it had resistance temperature characteristics that more closely matched the 10kΩ thermistor use in the LEGO temperature sensor.

The shrink tubing prevents water or other foreign material from entering the temperature sensor from the radial direction; however, one end of the tubing is completely open and a gap between the lead wires and the tubing exists at the other end. To completely seal the tubing, we used a silicone adhesive commonly employed to seal seams on counter tops and shower stalls. A small quantity of silicone adhesive was inserted into the end of the shrink tubing and in the gap around the lead wires. A photograph of this temperature sensor, the prepackaged thermistor and the LEGO temperature sensor is presented in Fig. 6.39.

We are now ready to calibrate the sensor and prepare a table showing the reading on the RCX as a function of temperature. There are two approaches to follow in the calibration process; both use a water bath that is heated or cooled to provide a temperature source that varies from about freezing to 100 °F. In the first approach, our temperature sensor is treated as if it were a LEGO sensor. We record a pseudo reading on the RCX and compare it with the temperature indicated by a LEGO sensor. We recognize that the two readings will be different because the thermistors are not the same. However, we can use the calibration data to correct the output from our 5 kΩ thermistor.

To program the RCX for this calibration approach, wrote a simple program in ROBOLAB, Investigator Level 2 and downloaded it into Program slot 5. In this program, we selected the LEGO temperature sensor for both Ports 1 and 3. After downloading this program, we ran the RCX to prepare Ports 1 and 3 to receive signals from temperature sensors and to display results in °F.

We prepared a water bath by placing a ceramic cup on a temperature controlled hot plate. Initially we filled the cup with ice cubes and water before immersing the two sensors in the bath. Our temperature sensor was connected to Port 1 and the LEGO sensor was connected to Port 3.

After waiting about 20 minutes for the water bath to stabilize and the temperature sensors to come to equilibrium with their surroundings, we began to record temperatures. Temperatures were recorded periodically as the ice cubes melted. When the water approached room temperature, the hot plate was turned on and adjusted to a temperature of about 125 °F. Additional readings were taken as the water temperature slowly increased to 90 °F. At this point, the reading from our temperature sensor remained fixed at 158 °F while the temperature of the water bath continued to increase. At this time, the calibration run was terminated[10]. The data from this calibration experiment was entered into a spreadsheet and the graph shown in Fig. 6.40 was generated.

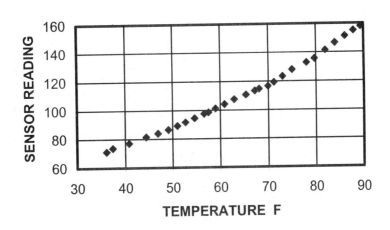

Fig. 6.40 Sensor reading as a function of temperature.

---

[10] The LEGO temperature sensor is rated at temperatures from −20 to +50 °C or −4 to + 122 °F. At Port 1, the RCX was indicating a temperature well outside the range of the LEGO sensor and the program in the firmware that converts the resistance from the LEGO sensor to temperature cut off at an apparent temperature of 158 °F.

Inspection of the results shows a clear trend line that establishes the relationship between the results from the two sensors. The trend line is non linear as expected because both sensors are non linear. Moreover, our temperature sensor is not well matched to the 10 kΩ LEGO temperature sensor. Some scatter in the data is evident. These deviations are largely due to thermal gradients that occurred in the simple temperature bath used in this calibration process. In a more rigorous calibration, the container for the water bath would be thermally insulated with a cover and crushed ice would be used instead of ice cubes.

The second approach to calibration recognizes that the thermistor is a variable resistor similar to the potentiometer. We rotate the wiper of the potentiometer to change its resistance; for the thermistor, we change its temperature. Because the thermistor acts like a variable resistor, we can calibrate it using raw values to represent its output on the RCX. To prepare the RCX, we write a program in ROBOLAB, Investigator, Level 2. In this program, we select the generic sensor for Port 1 and the LEGO temperature sensor for Port 3. After downloading the program and running the RCX to configure the Ports 1 and 3, we connect our temperature sensor to Port 1 and again connect the LEGO sensor to Port 3.

After waiting for the water bath to stabilize and the temperature sensors to come to equilibrium with their surroundings, we began to record the raw values from Port 1 and the temperatures from Port 3. Temperatures were recorded periodically as the ice cubes melted, and when the water approached room temperature, the hot plate was turned on. The water temperature was increased to 110 °F before the calibration run was terminated. The data from this calibration experiment was entered into a spreadsheet and the graph shown in Fig. 6.41 was generated.

Inspection of the results shows a clear trend line that establishes the relationship between the two sensors. The trend line is non linear as expected because thermistors are non linear as indicated in Fig. 6.27. As was the case with the first calibration, scatter in the data occurred because of thermal gradients in the simple temperature bath used in this calibration process. The scatter was more evident at the low temperatures where the presence of ice cubes near the sensors produce errors.

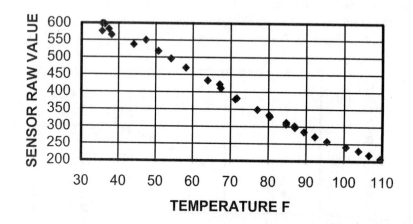

Fig. 6.41 Raw values from our temperature sensor as a function of temperature.

It is possible to determine the resistance of the thermistor at any temperature from the raw value displayed at Port 1. Recall that the raw value RV is given by:

$$RV = \frac{v_0}{v_i}[1023] \qquad (6.31)$$

Substitute Eq. (6.30) into Eq. (6.31) and eliminate v_0 to obtain:

$$RV = \frac{R_s/R_i}{1 + R_s/R_i}[1023]$$ (a)

Let R_i 10×10^3 Ω and $v_i = 5.00$ V and simplify Eq. (a) to obtain the resistance R_s of the thermistor as:

$$R_s = 10\left[\frac{RV/1023}{1 - RV/1023}\right]$$ where R_s is given in kΩ (6.32)

Substituting the raw value of RV = 349 at a temperature of 75 °F into Eq. (6.32) gives the thermistor's resistance as $R_s = 5.18$ kΩ, which is within a few percent of its nominal value of 5 kΩ.

It is evident that an accurate temperature sensor with waterproofing can be constructed using small inexpensive thermistors, shrink tubing and a small quantity of silicone adhesive. The RCX can be used to read the output either in terms of degrees or in raw values depending on the programming of its input ports.

Light Sensor

A relatively complex light sensor is available from LEGO, which interfaces with the RCX to measure light intensity on a scale of 0 to 100. This sensor incorporates a light emitting diode (LED) to illuminate the scene, a photodiode to measure the reflected light and circuitry to power both of these components. This is an active sensor that requires power from the RCX to operate.

In many applications a simple light sensor fabricated from a Cadmium Sulfide (CdS) photoconductor is sufficient. The characteristics of CdS photoconductors were discussed previously in Section 6.9.1. They are variable resistors with the resistance changing from very high values when the sensor is dark to nearly zero resistance when the sensor is exposed to light and the CdS becomes a good conductor. We have discussed the design of angle of rotation and temperature sensors using components (potentiometer and thermistor) that were variable resistors. The development of a CdS light sensor is similar; however, the mounting of the photoconductor is different.

Only two parts are required to build this sensor—a CdS photoconductor (photocell) and an electric plate. We purchased a package of five photocells from Radio Shack and a set of three electric plates from Pitsco[11]. The electric plates, shown in Fig. 6.42, are 2 × 8 LEGO plates with electrical contacts on each of the 16 studs. On the underside, a conducting strip is embedded on each side of the plate. The size of the electric plate poses a problem, because the input ports for the RCX accept a 2 × 2 plate. It is necessary to cut a 2 × 2 plate from one end of the 2 × 8 plate.

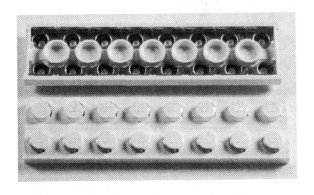

Fig. 6.42 Top and bottom surfaces of the 2 × 8 electric plates used for mounting the CdS light sensor.

A standard 2 × 2 LEGO plate was fitted over one end of the electric plate and used as a guided in the cutting operation. The electric plate was cut using a hacksaw fitted with a fine blade (32 teeth/inch). A very light pressure was exerted on the saw as both the plastic and the copper conductors are soft. After cutting, rough edges of the 2 × 2 electric plate were smoothed with a fine tooth file. The leads of the photoconductor were trimmed and then soldered to the conductors on each side of the electric plate as shown in Fig. 6.43. Care must be exercised in the soldering operation. Excess solder on the conductors on the underside of the 2 × 2 electric plate will prevent it from fitting over the studs on the input ports of the RCX. Excess heat from the soldering iron will melt the thermoplastic from which the electric plate is fabricated. Be quick and limit the amount of solder used in making the connections to the photoconductor's leads.

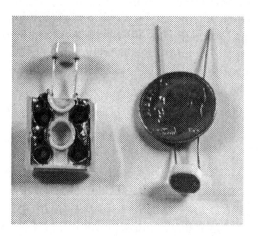

Fig. 6.43 Sensor soldered to electric plate and a sensor anchored with a dime.

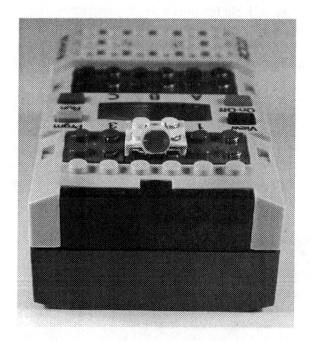

Fig. 6.44 Light sensor mounted on Port 2 of the RCX.

---

[11] The five CdS photocells (part number 276-1657) were purchased from a Radio Shack store for $2.69. The three 9 V electric plates (part number 779886) were purchased from the Pitsco LEGO Educational Division for $6.50.

Let's check to determine some of the characteristics of our light sensor after it is connected to the RCX as illustrated Fig. 6.44. Select Program 1, which is locked into the firmware of the RCX, connect the light sensor to Port 1, Run the program, and press the view button until the reading from Port 1 is displayed. The raw value is shown for the light sensor. In normal room light, the sensor gives a raw value that ranges from about 200 to 400 depending on its orientation. When exposed directly to a bright light source, the raw value decreases to about 15. If we cover the sensor with a black piece of paper, the raw value increases to about 900.

Usually a low cost photocell is employed as a sensor to turn lights on or off and to detect significant changes in light intensity. To check this application, select Program 2 on the RCX, connect the sensor to Port 1, run the program and view the output from the light sensor. In normal room light, the reading is 1. Shield the sensor from the light and the reading goes to 0. The sensor is acting like a switch, turning on when exposed to light and turning off when shielded from light.

It is possible to use the CdS sensor to measure light intensity; however, the sensor is affected by light coming from many different directions. To improve its performance as a sensor to measure light intensity, the sensing element should be placed in a tube to shield it from stray light. Also a lens should be fitted to the front of the tube to focus the incoming light on the sensing element.

6.14 SUMMARY

Several basic sensors have been described in this chapter. These sensors are sometimes used directly to measure an unknown quantity (such as the use of an electrical strain sensor to measure strain); however, in many other instances the sensors are one component of a more involved transducer such as a piezoelectric crystal in an accelerometer or a number of strain gages in a force transducer. Important characteristics of each sensor, which must be considered in the selection process, include:

1. **Size** - with smaller being better because of enhanced dynamic response and minimum interference with the process or event.
2. **Range**—with extended range being preferred so as to increase the latitude of operation.
3. **Sensitivity**—with the advantage to higher output signals that require less amplification.
4. **Accuracy**—with the advantage to devices exhibiting errors of 1% or less after considering zero shift, linearity and hysteresis.
5. **Frequency response**—with preference for wide-bandwidth sensors that permit application in both static and dynamic loading situations.
6. **Stability**—with very low drift in output over extended periods of time and with very small changes signal output with variations in temperature and humidity is considered essential.
7. **Temperature limits**—with the ability to operate from cryogenic to elevated temperatures considered important.
8. **Economy**—with reasonable costs preferred.
9. **Ease of application**—with reliability and simplicity always considered a significant factor.

The use of the RCX in measuring voltage was described. Kirchhoff's voltage law was employed to derive the relation for the voltage recorded by the DVM in the RCX and the voltage drop across a variable resistor. The methods employed to design and build a touch, angle of rotation, temperature and light sensors were described in considerable detail. Examples were described and calibration methods for several sensors were illustrated.

REFERENCES

1. Brindley, K.: Sensors and Transducers, Heinemann, London, 1988.
2. Foster R. L. and S. P. Wnuk Jr.: "High Temperature Capacitive Displacement Sensing", Instrument Society of America, Paper No. 85-0123, 1985, pp. 245-252.
3. Geyling, F. T., and J. J. Forst: "Semiconductor Strain Transducers," Bell Syst. Tech. J., vol. 39, 1960.
4. Herceg, E. E.: Handbook of Measurement and Control, Schaevitz Engineering, Pennsauken, NJ, 1976.
5. Kinzie, P. A.: Thermocouple Temperature Measurements, Wiley, New York, 1973.
6. Macklen, E.: Thermistors, Electrochemical Publications, Ayr, Scotland, 1979.
7. Mason, W. P., and R. N. Thurston: "Piezoresistive Materials in Measuring Displacement, Force, and Torque," J. Acoust. Soc. Am., vol. 29, no. 10, 1957, pp. 1096-1101.
8. Neubert, H. K. P.: Instrument Transducers: An Introduction to Their Performance and Design, 2nd ed., Clarendon Press, Oxford, 1976.
9. Sachse, H.: Semiconducting Temperature Sensors and Their Applications, Wiley, New York, 1976.
10. Sanchez, J. C., and W. V. Wright: Recent Developments in Flexible Silicon Strain Gages, pp. 307-346, in M. Dean and R. D. Douglas (eds.), Semiconductor and Conventional Strain Gages, Academic Press, New York, 1962.
11. Schaevitz, H,: The Linear Variable Differential Transformer, Proceedings of the Society for Experimental Stress Analysis, vol. IV, no. 2, 1947, pp. 79-88.
12. Sedra, A. S. and K. C. Smith: Microelectronic Circuits, 3rd ed., Holt, Rinehart, and Winston, New York, 1991.
13. Smith, C. S.: "Piezoresistive Effect in Germanium and Silicon," Phys. Rev., Vol. 94, 1954, pp. 42-49.
14. Yang, E. S.: Microelectronic Devices, McGraw Hill, New York, 1988.
15. Wang, E., Engineering with LEGO Bricks and ROBOLAB, College House Enterprises, Knoxville, 2003.
16. Baum, D, M. Gasperi, R. Hempel, and L. Villa, Extreme Mindstorms: An Advanced Guide to LEGO Mindstorms, Apress, Berkeley, 2000.
17. Erwin, B., Creative Projects with LEGO Mindstorms, Addison-Wesley, New York, 2001.

EXERCISES

6.1 Describe the differences between a sensor and a transducer. Give an example of a transducer incorporating a displacement sensor. Give another example of a sensor that is also a transducer.

6.2 A slide-wire potentiometer having a length of 100 mm is fabricated by winding wire with a diameter of 0.10 mm around a cylindrical insulating core. Determine the resolution limit of this potentiometer.

6.3 If the potentiometer of Exercise 6.2 has a resistance of 2000 Ω and can dissipate 2 W of power, determine the voltage required to maximize the sensitivity. What voltage change corresponds to the resolution limit?

6.4 A 20-turn potentiometer with a calibrated dial (100 divisions/turn) is used as a balance resistor in a Wheatstone bridge. If the potentiometer has a resistance of 20 kΩ and a resolution of 0.05%, what is the minimum incremental change in resistance ΔR that can be read from its calibrated dial?

6.5 Why are potentiometers limited to static or quasi-static applications?

6.6 List several advantages of the conductive-film type of potentiometer.

6.7 A new elevator must be tested to determine its performance characteristics. Design a displacement transducer that utilizes a 10-turn potentiometer to monitor the position of the elevator over its 100-m range of travel.

6.8 Compare the potentiometer and LVDT as displacement sensors with regard to the following characteristics: range, accuracy, resolution, frequency response, reliability, complexity and cost.

6.9 Prepare a block diagram representing the electronic components in a LVDT. Describe the function of each component.

6.10 Prepare a sketch of the output signal as a function of time for an LVDT with its core located in a fixed off-center position if:

      (a) The demodulator is functioning.
      (b) The demodulator is removed from the circuit.

6.11 Prepare a sketch of the output signal as a function of time for an LVDT with its core moving at constant velocity from one end of the LVDT through the center to the other end if:

      (a) The demodulator is functioning.
      (b) The demodulator is removed from the circuit.

6.12 Describe the basic differences between an LVDT and a DCDT.

6.13 Design a 50-mm strain extensometer to be used for a simple tension test of mild steel. If the strain extensometer is to be used only in the elastic region and to detect the onset of yielding, specify the maximum range. What is the advantage of limiting the range?

6.14 Compare the cylindrical potentiometer (helipotentiometer) and the RVDT as sensors for measuring angular displacement.

6.15 What two factors are responsible for the resistance change $\Delta R/R$ in an electrical resistance strain gage? Which is the most important for gages fabricated from constantan?

6.16 What function does the thin plastic film serve for an electrical resistance strain gage?

6.17 A strain gage with an initial resistance R_g and a gage factor S_g is subjected to a strain ε. Determine ΔR and $\Delta R/R$ for the conditions listed below:

	R_o (Ω)	S_g	ε $(\mu m/m)$
(a)	120	2.02	1600
(b)	350	3.47	650
(c)	350	2.07	650
(d)	1000	2.06	200

6.18 For the conditions described in Exercise 6.17, determine the output voltage v_o for an initially balanced bridge if the input voltage v_s is:

(a)	2 V	(c)	7 V
(b)	4 V	(d)	10 V

6.19 From Exercise 6.18 we note that increasing v_i increases v_o. What would happen if we increase v_s to 50 V to improve the output v_o.

6.20 For the gages specified in Exercise 6.17, monitored in a single-arm Wheatstone bridge with $v_s = 5$ V, determine the strain ε if the output voltage v_o is:

(a)	1.5 mV	(c)	4.8 mV
(b)	3.3 mV	(d)	6.7 mV

6.21 A short-range displacement transducer utilizes a cantilever beam as the mechanical element and a strain gage as the sensor. Derive an expression for the displacement δ of the end of the beam in terms of the output voltage v_0.

6.22 Prepare a graph of the sensitivity S of a capacitance sensor as a function of frequency ω. Assume the dielectric in the gap is air and consider probe diameters of 1, 2, 5 and 10 mm.

6.23 Write an engineering brief describing the advantages and disadvantages of capacitance sensors.

6.24 Describe the operating principles of the instrument, shown in Fig. 6.16, which is used to monitor the output of a capacitance sensor.

6.25 Can eddy current sensors be employed with the following target material?
(a) Magnetic materials
(b) Polymers
(c) Non-magnetic metallic foils
Indicate procedures that permit usage of the sensor in these three cases.

6.26 Determine the charge q developed when a piezoelectric crystal having $A = 15$ mm^2 and $h = 8$ mm is subjected to a pressure $p = 2$ MPa if the crystal is:
(a) X-cut, length-longitudinal quartz
(b) Parallel to polarization barium titanate

6.27 Determine the output voltages for the piezoelectric crystals described in Exercise 6.26.

6.28 Compare the use of quartz and barium titanate as materials for:
(a) Piezoelectric sensors
(b) Ultrasonic signal sources

6.29 Compare the characteristics of a piezoresistive sensor with those of a piezoelectric sensor.

6.30 What advantages does the piezoresistive sensor have over the common (metal) electrical resistance strain gage? What are some of the disadvantages?

6.31 Design a circuit to turn on outside lights at your home as it begins to get dark. Use a photoconduction cell in the circuit and provide for an adjustment to control the intensity level for switch activation.

6.32 Sketch the circuit for a photovoltaic cell and write a paragraph explaining its operation. Write another paragraph stating the advantages and disadvantages of this light sensor.

6.33 Sketch the circuit for a photodiode used in the photoconduction mode and write a paragraph explaining its operation. Write another paragraph stating the advantages and disadvantages of this light sensor.

6.34 Prepare a graph showing the sensitivity S of a platinum RTD with a 100 Ω resistance as a function of temperature T. For the circuits shown Figs. E6.34 (a), (b), and (c), plot a graph of v_0 versus T.

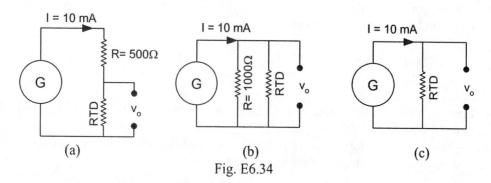

Fig. E6.34

6.35 Write an engineering brief describing the effects of the three circuits in Exercise 6.34 on the output voltage v_0 as a function of temperature T.

6.36   Compare resistance temperature detectors (RTD) and thermistors as temperature sensors. Give an example that is best suited for each sensor.

6.37   Determine the sensitivity of the following thermocouples.

           (a) chromel-alumel

           (b) copper-constantan

           (c) iron-constantan

           (d) iron-nickel

           (e) gold-silver

6.38   Write your own summary of the important topics in this chapter.

CHAPTER 7

PROGRAMMING IN ROBOLAB™

7.1 INTRODUCTION

The RCX incorporates a microprocessor capable of storing programs that provide instructions for controlling the actuators used in your team's deicing system. In this application, the requirements to control the actuator were simple—turn on the power to the actuator (a motor or solenoid activated valve) and maintain this power until the valve was open and flow of the deicing solution initiated. After a sufficient quantity of the saline solution was dispensed the process is reversed, the valve is closed and the flow of brine is terminated with sufficient solution in storage to handle another icing event.

We will be working with two types of control—sequential and decision-based. If sequential control is employed, the operator programs the controller to perform a sequence of tasks with the timing and power level associated with each task decided in advance. Programming for sequential control is relatively easy as it is simply a line of symbols without forks, jumps or loops. We start the program, sequentially perform a specified number of tasks and then stop.

Decision based control requires sensors in the system that to provide feedback signals which give information essential in controlling the system. When the system is in operation, the sensors provide continuous signals that are monitored by the RCX. When the level of a feedback signal deviates sufficiently from a specified level, the power to one or more actuators is adjusted to maintain the sensor signal within a control band centered about the control level. For decision-based control to be effective, the microprocessor must be provided with instructions that establish the upper and lower levels of the control band and that energize actuators, which adjust the parameters affecting the system so as to maintain control. Programming for decision-based control is more challenging as the program entails a more structured program involving forks, jumps and loops and sometimes loops within loops.

7.2 FLOWCHARTS

A flowchart is a programming tool that is often used to graphically illustrate each step in a program. Flowcharts are usually prepared prior to writing a structured program or even assembling a program with a graphical interface. Before writing a program, it is essential that the strategy for executing the steps that control the mechanisms involved in the deicing system be clearly established. The primary advantage of using a flowchart to plan the strategy for a program is that it provides a graphical representation of the program making the program logic easier to visualize.

The American Standards Institute (ANSI) has developed several flowchart symbols that are presented in Fig. 7.1

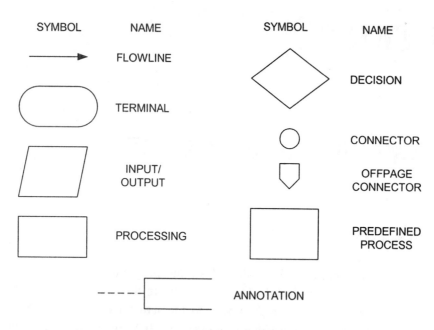

Fig. 7.1 Standard symbols used in preparing flowcharts.

The name and use of each ANSI symbol is listed in Table 7.1.

Table 7.1
Name and use of the flowchart symbols.

Symbol Name	Use
Flowline	For connecting symbols and showing the flow of the logic.
Terminal	For representing the beginning (start) and the end (stop) of the program.
Input/Output	For input and output operations such as reading sensor signals or displaying data.
Processing	For arithmetic and data manipulations operations.
Decision	For any comparison operation. The decision symbol has one input and two outputs.
Connector	For connecting two or more flowlines.
Off page Connector	For indicating that the flowchart continues on the next page.
Predefined Process	For representing a group of statements that perform a specific task or a subroutine.
Annotation	For providing additional information pertaining to one of the flowchart symbols.

Let's consider a few examples to demonstrate the use of flowcharts in programming.

EXAMPLE 7.1

When dinning in a nice restaurant, it is customary to tip the person serving your food by about 15%. Rather than counting pennies in determining amount of the final payment, it is common practice to add the tip to the initial bill and round the sum upwards to the nearest dollar. Let's prepare a flowchart for a program that will determine the amount of the final payment.

Solution: We begin with the terminal symbol with a command to start the program. Next a flowline connects the terminal symbol to an input/output symbol with a command to read the amount of the bill into the program. Flowlines connect to three processing symbols where sequential commands are given for the program to calculate the tip, add the amount of the tip to the initial bill and to round this amount upwards to the nearest dollar. A flowline connects the last processing symbol to an input/output symbol where the program is directed to display the amount of the payment. Finally, a terminal symbol is placed at the end of the top down listing of sequential steps to stop the program.

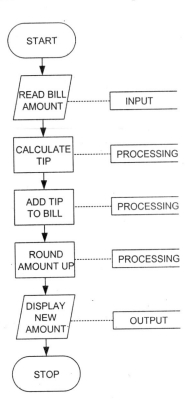

Fig. 7.2 Flowchart to determine the payment that includes an appropriate tip for an attentive waiter or waitress.

The flowchart in Fig. 7.2 is representative of a sequential program in which we move from one step to the next without skipping any blocks (or lines of code). Decisions were not a factor; consequently, it was not necessary to introduce one or more branch points (forks) in the structure of the program. However, in many applications decisions are essential, and a decision-structured program with one or more forks is mandatory. The flowchart for a decision structure is illustrated in Fig. 7.3.

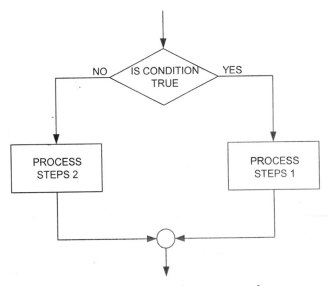

Fig. 7.3 Flowchart for a decision-structured program.

In a decision-structured program, we pose a question or a condition that is either true or false (yes or no). This gives rise to writing lines of code that state **IF** the specified condition is true **THEN** proceed to Process Steps 1 **ELSE** proceed to Process Steps 2.

EXAMPLE 7.2

Downtown San Jose, Costa Rico is laid out on a Cartesian grid. The Avenidas run east and west and the Calles run north and south. Avenida Central and Calle Central divide the city into quadrants. To the north of Avenida Central the avenidas have odd numbers and to the south they have even numbers. Similarly, Calles to the east of Calle Central have odd numbers and to the west they have even numbers. Prepare a flowchart for a program to determine if an avenida is on the north or south side of the city.

Solution: Again we begin the program with a terminal symbol with the command to start. A flowline connects this symbol to an input/output symbol with a command to read the number of the avenida as shown in Fig. 7.4. A flowline connects to a decision symbol, with a command to test if the number is even. **If** the number is even, **then** the flowline connects to an input/output symbol that displays the result "south". **Else** (if the number is odd, the decision condition is not satisfied), the flow line connects to the input/output symbol that displays "north". The flowlines from both input/output symbols go to a connector symbol. The flow line from the connector symbol goes to the terminal symbol with a command to stop the program.

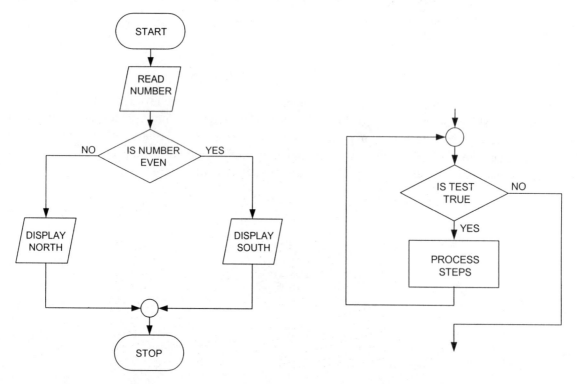

Fig. 7.4 Flowchart for the San Jose avenidas locations.   Fig. 7.5 Flowchart for a loop structure.

The third programming structure is the loop, which executes a set of instructions one or more times. The loop incorporates a test (decision) to control the number of times the instructions in the loop are repeated. When the test is not satisfied the program exits the loop. The flowchart symbols for a loop structure are illustrated in Fig. 7.5.

EXAMPLE 7.3

Suppose a class with 36 students takes an examination that is graded on a scale of 0 to 100. Prepare a flowchart for a program to determine the class average and to display this result.

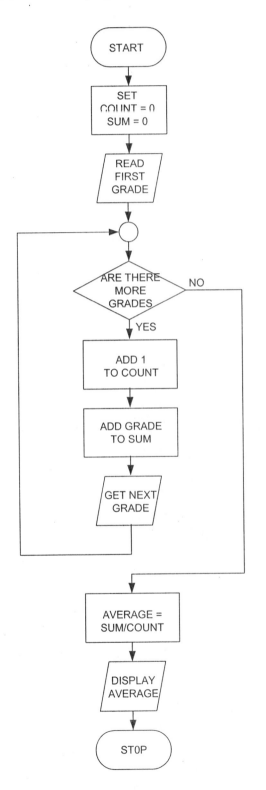

Solution: Again we draw a terminal symbol with the start command to begin the program. Next, a process symbol is employed with instructions to set two variables defined as the count and sum both equal to zero as indicated in Fig. 7.6. Next in line is an input/output symbol with a command to read the first grade in a list of 36 grades. A connector symbol is inserted to indicate a junction point in the program. The flowline then proceeds to a decision symbol with logic to ascertain if there are more grades to be considered. If the answer is yes the flowline connects to two sequential process symbols where the program adds one to the count and then adds the grade to the sum. An input/output symbol with the command to read the next grade follows. At this point, the program loops back to the connector symbol and the process is repeated until each grade is read into the program and processed. When all the grades have been accommodated, the test of the specified condition on the count in the decision symbol is negative. Then the flowline exits the loop and connects with another processing symbol. The average is then determined by taking the sum and dividing by the count. An input/output symbol is used with a command to display the result. Finally, the flowline connects with a terminal symbol to stop the program

Fig. 7.6 Flowchart representing a decision and loop structure.

7.3 PROGRAMMING IN ROBOLAB™

ROBOLAB™ is an icon oriented programming language that is used to provide a list of instructions for the RCX microprocessor. It is based on LabVIEW™—a popular programming tool used by engineers for measurement and control. LabVIEW was developed by National Instruments in Austin, TX and is based on a language known as G. As a programming language, G is similar to Basic, Fortran or the various versions of C. The major difference is that LabVIEW or G is based on graphics (icons) instead of many lines of written text. Hence, a LabVIEW program is comprised of a logical sequence of icons and is essentially independent of text except for comment statements made to explain various aspects of the program to a reviewer.

Chris Rogers and several others at Tufts University developed ROBOLAB to permit people not yet trained in writing structured software to program the RCX. With ROBOLAB, it is possible to create complex programs to control robots to perform many challenging tasks without spending a semester learning a structured programming language with several hundred commands.

The Opening Screen

This section describes the main screens in ROBOLAB, Version 2.5 that is available from Pitsco LEGO Educational Division, P. O. Box 1707, Pittsburg, KS. The opening screen for ROBOLAB, shown in Fig. 7.7, identifies the three professional groups that cooperated in the development of the icon based programming language.

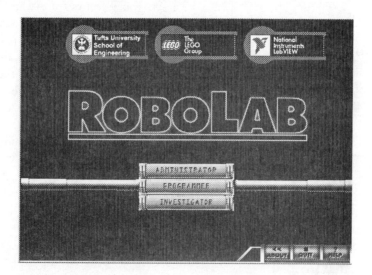

Fig. 7.7 The opening screen for ROBOLAB.

In the center of the screen, observe three bars (buttons) labeled **Administrator**, **Programmer** and **Investigator**. The **Administrator** bar opens a program that enables you to adjust the settings on the RCX, define locations for your files, tests the communication from the IR tower to the RCX and hide (lock) the **Administrator** button. The **Programmer** bar activates the Main Menu screen that in turn enables you to select either the **Pilot** or **Inventor** levels of programming. The **Investigator** bar activates the Menu screen for the programming necessary to record measurements taken from LEGO sensors that include touch, light level, temperature and angle of rotation sensors.

The purpose of the three buttons in the lower right hand side of the screen is obvious. Click on them to make certain you understand their functions.

The Administrator Screen

The opening screen that appears when the **Administrator** bar is activated is presented in Fig. 7.8. The opening screen has three tabs positioned along its lower edge that switch the display as shown below.

1. Administrator screen is used to establish a method for communicating between your computer and the RCX using the IR tower.
2. ROBOLAB Settings screen is used to arrange file locations and to save or delete files.
3. RCX Settings screen is employed to monitor the battery level, firmware version, change power settings, unlock programming slots and adjust time before the power is shut-off.

The screen displayed when the Administrator tab is activated is presented in Fig. 7.8. Five icons (buttons) are displayed. The **Select Com Port** button enables you to select the communication port where the IR tower is connected to your computer. The **Download Firmware** button is to facilitate the downloading of firmware from your computer to the RCX. The downloading is accomplished with IR signals that are transmitted from the IR tower to the RCX. The firmware remains stored in memory in the RCX until the batteries are removed. If you are too slow when changing batteries, the RCX will prompt you to restore the firmware by repeating the downloading process.

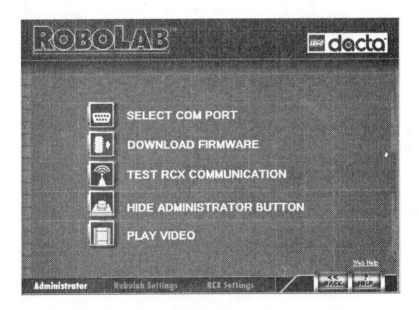

Fig. 7.8 The Administrator Screen

After the firmware has been installed successfully, it is possible to check if the IR tower and the RCX are communicating by clicking the **Test RCX Communications** button. If the test is successful, the RCX emits a confirmation signal. The **Hide Administrator** button locks the settings in place and the Administrator button on the opening screen is disabled. **To unlock the Administrator, press the F5 function key.** A tutorial video may be accessed from the Administrator screen by clicking the **Play Video** button; however, it is necessary to insert the ROBOLAB compact disk in the CD drive of your computer to activate the video. Directions for playing the video are provided when the CD is activated.

ROBOLAB Settings

Clicking on the **Robolab Settings** tab at the bottom of the **Administrator** screen activates the display presented in Fig. 7.9. The path to the hard drive (usually C) for the programs stored in ROBOLAB is shown as well as the path for the user programs that are stored in the **Program Vault**.

Clicking on the icon with the + sign located on the left hand side below the Theme widow creates a new folder that you can use to store any program that you write. The icon with the garbage can is used to delete programs selected from the theme window. Note that programs that are included with the ROBOLAB software are locked and cannot be deleted.

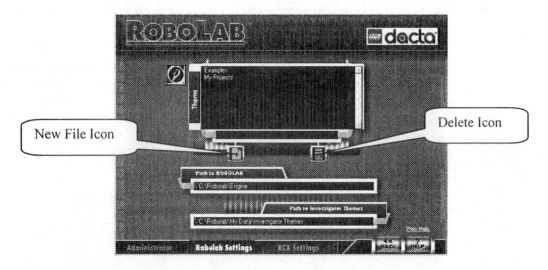

New File Icon   Delete Icon

Fig. 7.9 Display for ROBOLAB Settings.

RCX Settings

Clicking on the **RCX Settings** tab at the bottom of the **Administrator** screen activates the screen presented in Fig. 7.10. To activate this screen, the IR tower must be connected to your computer and the RCX must be turned on and in a position to communicate with the IR tower. The settings stored in the RCX are interrogated and displayed on this screen.

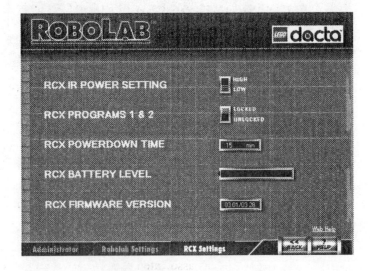

Fig. 7.10 Display for RCX Settings.

The **IR Power Settings** adjusts the power from the RCX to the communication tower. Low power is normally used to conserve battery power; however, if the distance between the RCX and the tower is greater than about 300 mm it may be necessary to use the higher power level. Normally programs in slots #1 and #2 are locked to retain the programs preloaded by the LEGO programmers. If you want to download a program, we recommend that you use one of the higher numbered slots (#3, #4 and #5 are available).

The **Power Downtime** is the time the RCX can be idle prior to automatically tuning itself off. The range of the downtime can be adjusted from 0 to 255 minutes. We suggest a setting of 5 to 10 minutes. Setting the downtime to zero overrides the shutoff command and the RCX will remain on until the batteries are exhausted. The **Battery Level** indicator is self-evident as is the identification of the **Firmware Version** loaded into storage on the RCX.

7.4 PROGRAMMING WITH PILOT IN ROBOLAB™

Programming in ROBOLAB can be accomplished within three different options—**Pilot, Inventor** and **Investigator**. Pilot is the easiest to learn and to use, although it is limited to a sequence of tasks. It is not possible to incorporate forks, jumps or loops in your program when working with the **Pilot** option in ROBOLAB.

Pilot employs a simple interface that involves clicking on an icon and selecting one or more options from a graphical dialog box. The choices you have for tasks (processes) are limited and depend on the level of the **Pilot** that you select. Pilot Level #1 is the easiest and Pilot Level #4 provides more flexibility. Both Pilot Levels #3 and #4 have preloaded themes with sample programs. We will demonstrate level #4 that has all of the functions available in the Pilot mode.

Let's begin by clicking on the **Programmer** button on the opening screen (See Fig. 7.7). The display that appears, shown in Fig. 7.11, enables you to select from either the **Pilot** or **Inventor** modes of programming. We have selected Pilot 4, the most advanced level because it is probably consistent with your capabilities.

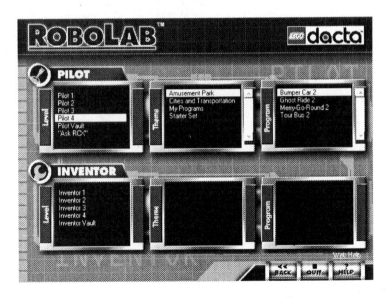

Fig. 7.11 Display for both Pilot and Inventor programming modes.

To open a sample program that already exists in ROBOLAB, click on Pilot 4 in the **Level** window, click on **Amusement Park** in the **Theme** window and finally double click on **Bumper Car 2** in the program window to open the first of six screens associated with the program controlling **Bumper Car 2** as shown in Fig. 7.12.

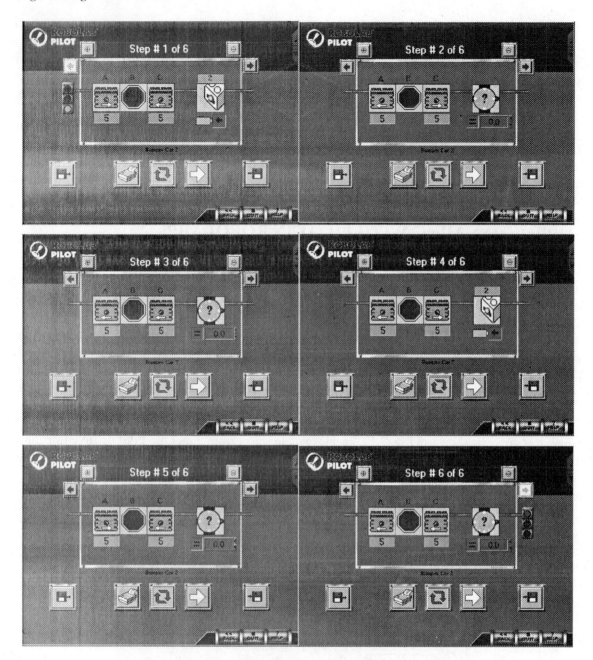

Fig. 7.12 The six-step program in Pilot 4/Amusement Park that controls Bumper Car 2.

Let's examine each step of the program shown in Fig. 7.12.

Step 1: The program begins with the **Traffic Light** icon on the left side of the programming window in Step 1. A pink line connects to the next program element that shows three icons—two **Motor** icons and a **Stop** icon. These output icons control the power connections A, B and C on the RCX. Accordingly, the RCX turns on the motor connected to terminal A in the forward direction at power level 5 (full power), stops (prevents) voltage from being applied at terminal B, and turns on the motor connected to terminal C in the reverse direction at full power. The two motors will run until the next program element in the sequence is activated. In this case, the second element is a **Touch Sensor** icon connected to port 2 on the RCX. As indicated by the arrow in the box below this icon, the program will advance to Step 2 when the button on the touch sensor is depressed.

Step 2: The first element in this screen controls the direction of the motors and the power levels. The direction of the motor connected to terminal A is changed (reversed) and full power is maintained. Power to terminal B is still blocked (disconnected). The direction of the motor connected to terminal C is also changed (reversed) and full power is maintained. The pink line then leads to an icon that resembles a clock face with a question mark. This is the **Clock** icon that indicates the time interval that the program will wait before advancing to the next step. The time shown in the box below the **Clock** icon indicates zero seconds; however, the time interval can be adjusted by clicking on the scroll indicators adjacent to this box.

Step 3: This programming window shows that the icons in Step 3 are identical with those in Step 2 except for the change in direction of rotation of the motor connected to terminal C.

Step 4 This programming window shows that the icons in Step 4 are identical with those in Step 1 except for the change in direction of rotation of both of the motors.

Step 5 This programming window shows that the icons in Step 5 are identical with those in Step 2 except for the change in direction of rotation of both of the motors.

Step 6 This programming window shows that the icons in Step 6 are identical with those in Step 3 except for the pink line that extends beyond the programming window and connects to the **Traffic Light** icon. This icon displays a red light that instructs the program to end.

Pilot Programming Icons.

When examining the program for the **Bumper Car 2**, we identified five of the seven icons accessible in the Pilot mode of ROBOLAB. The seven icons include:

1. **Traffic Light**
2. **Motor**
3. **Lamp On**
4. **Stop**
5. **Touch Sensor**
6. **Light Sensor**
7. **Clock**

The **Traffic Light** icon always appears outside the programming window and is automatically placed to Start and End the program. The **Motor**, **Lamp On** and **Stop** icons control the output ports on the RCX. The icons related to the touch and light sensors and the clock provide different methods for controlling the **Wait For** or **Wait Until** commands.

Output Control

Let's consider the icons controlling the output from ports A, B and C of the RCX. There are three different output icons—**Motor**, **Lamp On** and **Stop**. When you click on a **Motor** icon associated with one of the output ports a pop-up window appears that provides you with a choice of four options for that port as indicated in Fig. 7.13a. These options include: motor reverse, motor forward, lamp on and stop. The **Stop** icon placed at a specified port turns off the power (voltage) supplied to that port regardless of the device, which is connected to that terminal.

The **Motor** icon with the arrow pointed to the right will rotate the motor's shaft to drive a vehicle forward if the orientation of the wiring is correct. Unfortunately, there are 16 different ways to attach the lead wire from the RCX to the motor and errors are common. We suggest that you perform Laboratory Exercise 1 to determine lead wire orientations that produce forward and reverse directions of rotation for motors mounted on either the left or the right side of a wheeled vehicle.

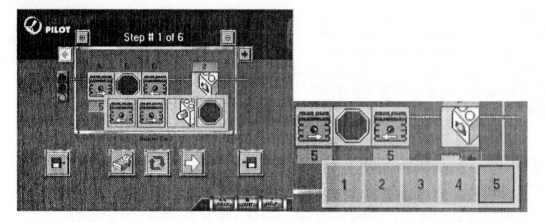

Fig. 7.13a Selections available for the output ports.   Fig. 7.13b Power level selections.

     The number in the box below the **Motor** or the **Lamp On** icons refers to the power level applied to the device. We can change the power level[1] by clicking on this number. A pop up window appears giving you five choices for the power level as shown in Fig. 7.13b. Click on a number from 1 to 5 and the power level changes accordingly—1 corresponds to the lowest level and 5 the highest.

The Wait For Command

In ROBOLAB, the motors run or the lamp shines until some command changes their settings. One of these is the **Wait For** command. The **Clock** icon provides the **Wait For** command and the number in the box below the clock face indicates the wait time interval in seconds. A similar command is the **Wait Until** command, which instructs the program to wait until a specified sensor is activated. The three icons associated with the wait commands, illustrated in Fig. 7.14a, are the **Touch** and **Light** sensors and the **Clock**. You may select among these three options simply by clicking on your choice. If you select a sensor, the number above the icon identifies the input port for connecting the sensor. If you select the clock, the wait time is specified in the box below the icon. You may specify an exact time in seconds with the scroll tabs or with your keyboard. You may also select a random time generator by clicking on the die depicted in Fig. 7.14b.

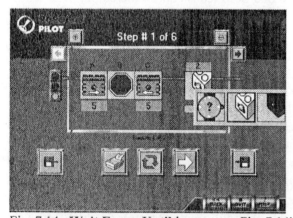 

Fig. 7.14a **Wait For** or **Until** icons.   Fig. 7.14b The die selection provides a random time interval.

---

[1] The power level applied to the LEGO motor or any actuator connected to an output port is controlled by pulse width modulation. The RCX turns the power on and off very rapidly to create a series of voltage pulses that energize the coils in the motor. The width of the pulse is adjusted to increase or decrease the power level. For the lowest power level, the power is on for 1 count and off for 7 counts. For power level 4, the power is on for 7 counts and off for 1 count. At the highest power level, the power is on all of the time.

The sensors are attached to one of the three input ports. The port number is specified in the program in the box above the icon as shown in center illustration in Fig. 7.15. If the number in the box is not correct, click on it and select the correct value from numbers 1, 2 or 3 on the pop up window. If a touch sensor is associated with the **Wait Until** command, it is necessary to specify if the program waits until the button on the sensor is depressed or until it is released before proceeding to the next program element. The arrow direction relative to the button shown on the left in Fig. 7.15 indicates the command will be executed when the button is depressed. Of course, simply clicking and selecting the opposite direction will change it.

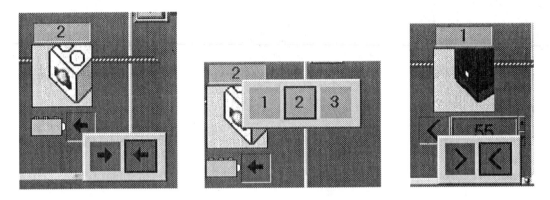

Fig. 7.15 Illustrations depicting methods for changing sensor settings for program control.

The light sensor measures the intensity of light entering its lens with a range from 1 to 100. In the illustration shown on the right side in Fig. 7.15, the light intensity is set at 55 and the < sign indicates the measured light intensity must be less than 55 before the **Wait Until** command has been satisfied and the program can advance to the next element. Of course you can adjust the intensity level or change the comparison criterion from less than < to greater than >.

Defining Pilot Features

At this stage of our discussion, we have focused on the programming window in Pilot Level 4. Let's now explore some of the other features offered by the Pilot Level 4 interface. There are seven buttons that facilitate using this approach to programming, which are identified in Fig. 7.16.

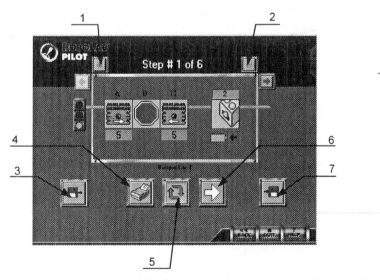

Fig. 7.16 Additional icon buttons to facilitate programming in the Pilot mode.

Let's identify each of these seven buttons by their number in Fig. 7.16 and describe their functions.

1. Clicking on the button with the + sign introduces a new step in the program.
2. Click the button with the − sign deletes the step showing in the programming window.
3. Click on the diskette icon to load a program from the Pilot Level 4 Program folder.
4. Click on the print icon to print the screen associated with a single step in the program.
5. The circling red arrow icon is a toggle that switches from the program executing once and stopping or executing continuously until the program is terminated with the stop button on the RCX.
6. Clicking on the white arrow icon downloads the program from your computer to the RCX if the IR tower is in place and in line-of-sight to the RCX.
7. Clicking on the second diskette icon enables you to save the file to the My Programs folder within the Theme section of Pilot Level 4. However, you must name the program.

Programming in Pilot 4

The previous discussions in this section have prepared you to program at the most advanced level in the Pilot mode of ROBOLAB. Let's consider two examples to demonstrate programming to control two different vehicles so that they perform several predetermined tasks. The first example pertains to a vehicle without sensors. The internal clock in the RCX is used to control the time of the maneuvers required for the vehicle to complete its assigned mission. The second example pertains to the same vehicle except that it is equipped with two touch sensors and the light sensor.

EXAMPLE 7.4

Let's write a program to control a vehicle with a four-wheel drive as shown in Fig. 7.17. Note that the vehicle is not equipped with either the touch or light sensors. The program will control of the vehicle as it executes the following maneuvers:

*   Travel in a straight line from the starting point and stop at a line 5 ft from this point.
*   Turn left 90 degrees and travel 2 ft.
*   Turn right 90 degrees and travel 1 ft.
*   Turn 180 degrees and travel 4 ft and stop for 5 seconds.
*   Return to the starting point.

After you have completed writing the program, download it to the RCX and verify that the wheeled vehicle successfully executes the assigned maneuvers.

Fig. 7.17 A robot with four wheel drive.

Solution: We assume here that you have access to a vehicle similar to the one shown in Fig. 7.17 so that it can be adapted to either of these two examples. In Example 7.4, we must program without feedback from sensors; hence, our strategy for controlling the vehicle must be based on the time to execute the required maneuvers. Before beginning to program let's make a diagram of the route our vehicle will travel in executing the assigned maneuvers as shown in Fig. 7.18. Then we prepare a list of commands presented in Table 7.2 that will be required for the vehicle to traverse this track. In preparing the list of commands, we did not number the start and stop commands because they are not included in the programming window in Pilot Level 4. Also, we conducted simple experiments to determine that the speed of the vehicle was 150 mm/s when operating at a power level of 5. In these experiments, we observed that our vehicle did not travel in a straight line, but drifted to the right. Switching the wheels around did not alleviate the problem. Performance was improved by reducing the power level from 5 to 3 on motor A. We also conducted experiments to determine that the time to execute a 90° turn was 0.9 s. Scaling the turn time linearly with the angle turned gave about 0.5 s for 45° and 1.8 s for 180°. The results you determine in your measurements with your vehicle may be different because of the many factors that affect the speed and the time to turn through a specified angle. You may find in your experiments that the values used for speed and turning times may need corrections, but changes are easy to make in programming with Pilot Level 4.

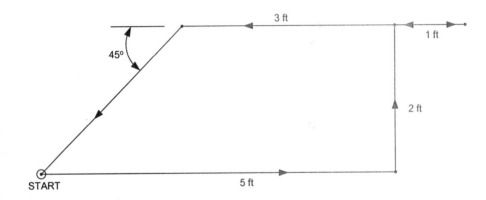

Fig. 7.18 Track layout for Example 4.

Table 7.2
List of commands for controlling the vehicle.

Step No.	Command	Motor Settings	Clock Setting
	Start	Traffic Light (Green)	
1	Straight	A($\Rightarrow$,3); B(Stop); C($\Rightarrow$,5)	10 s
2	Turn left 90°	A($\Leftarrow$,3); B(Stop); C($\Rightarrow$,5)	0.9 s
3	Straight	A($\Rightarrow$,3); B(Stop); C($\Rightarrow$,5)	4 s
4	Turn right 90°	A($\Rightarrow$,3); B(Stop); C($\Leftarrow$,5)	0.8 s
5	Straight	A($\Rightarrow$,3); B(Stop); C($\Rightarrow$,5)	2 s
6	Turn 180°	A($\Leftarrow$,3); B(Stop); C($\Rightarrow$,5)	1.8 s
7	Straight	A($\Rightarrow$,3); B(Stop); C($\Rightarrow$,5)	8 s
8	Stop	A(Stop); B(Stop); C(Stop)	5 s
9	Turn left 45°	A($\Leftarrow$,3); B(Stop); C($\Rightarrow$,5)	0.5 s
10	Straight	A($\Rightarrow$,3); B(Stop); C($\Rightarrow$,5)	7 s
	Stop	Traffic Light (Red)	

In preparing Table 7.2, we identified the port A and C for the motor connections and then used parentheses to contain the arguments that describe the motor settings. For instance, A($\Rightarrow$,3) indicates that we will program the motor wired to port A to turn in the forward direction and to operate at a power level of 3. C($\Leftarrow$,5) indicates that the motor wired to port C is operating in the reverse direction at full power. Our logic is simple for the turns. Because we have not designed the vehicle with steering, it is necessary to reverse one wheel while driving the other wheel forward to execute a turn. The distance traveled and the turning angles are both established by the time settings provided.

Let's use the results shown in Table 7.2 to prepare our 10-step program in Pilot Level 4. We will begin by clicking on Starter Set from the Theme window and then double clicking on Car Step 2. This four-step program will not control our vehicle to perform the required maneuvers; however, it does provide a starting point. We will modify each of the existing steps and add six more steps to produce the program that follows the listing of commands presented in Table 2. This program is shown in Fig. 7.19.

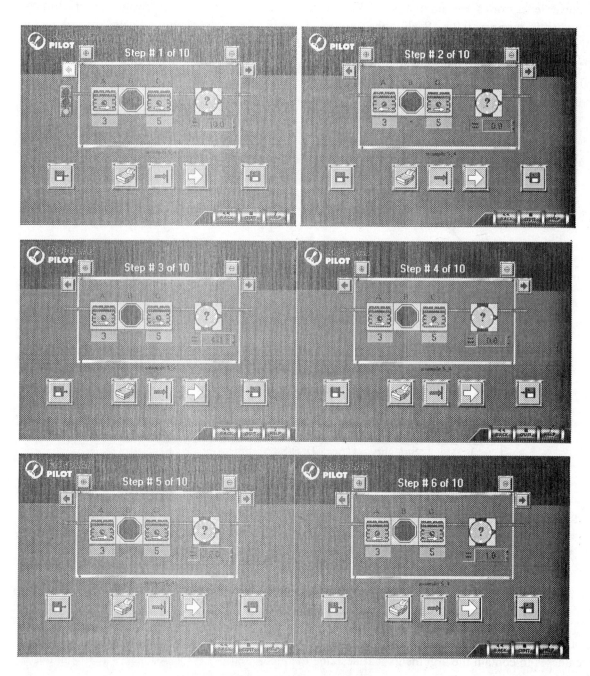

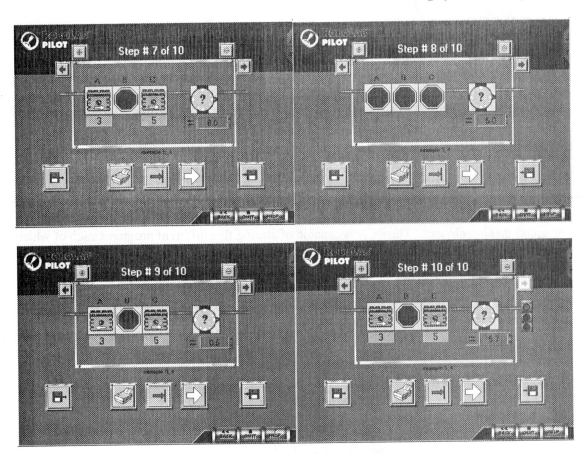

Fig 7.19 Program for maneuvering the vehicle about the track defined in Fig. 7.18.

We save the program to our folder (or disk) under the name Example 7.4. Next, plug the IR tower into your computer, turn on the RCX, select program slot #3 and line up the RCX to receive the IR signal. When you click on the download button (the one with the large white arrow), a task indicator shown in Fig. 7.20a will show the progress in transferring the program from your computer to the RCX. When the transfer is complete another view of the display window of the RCX, presented in Fig. 7.20b, is shown on the monitor indicating the program slot where the program now resides. The RCX also emits a chirp to indicate successful completion of the downloading process.

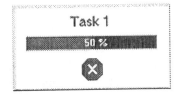

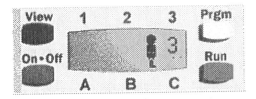

Fig. 7.20a Task bar showing downloading status.   Fig. 7.20b RCX display showing storage slot.

We suggest in Exercise 7.6 that you duplicate this example with your vehicle operating in your environment. You should not expect the same results because your vehicle design may differ from the author's, the friction of the surface upon which the vehicle operates may be different and your motors may perform differently. You will find that accurate control by time alone is difficult to achieve because of the many variables that affect the performance of our vehicles.

7.5 LEGO RCX™

The RCX (Robotic Command Explorer), shown in Fig. 7.21, is a controller that incorporates a Hitachi H8 microcontroller with 32K of external random access memory (RAM). This microcontroller is used to control motors, sensors and an infrared (IR) serial communications port. The RCX also holds 16K of read only memory (ROM), which contains the driver that becomes operational when the RCX is first started. Downloading an additional 16K of firmware from the Robotics Invention System 2.0 to the RCX enhances this driver. The driver accepts and executes commands that are transmitted from your PC to the RCX through an infrared (IR) communications port.

The RCX also contains a 10-bit analog to digital converter (A/D) and can acquire and store 2000 measurements at frequencies up to 200 Hz. The RCX has three input ports that can receive analog signals from sensors. It also has three output ports to provide power (voltage pulses) to motors, lights or sound making devices. The RCX is capable of storing five different programs each of which can execute up to ten simultaneous tasks. The programs are downloaded to the RCX by an infrared (IR) signal generated by an IR tower connected to a computer's serial port. The RCX can also send signals with its infrared transmitter. Four control buttons are available to turn the RCX on and off, select a program, view the operation and run the selected program. A liquid crystal display provides information to the operator about the status of the RCX. The unit is powered with six AA batteries.

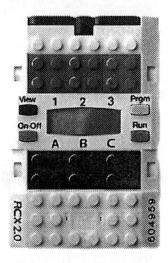

Fig 7.21 The RCX acts to control the actions of autonomous robots and vehicles.

Input Ports

There are six ports on the RCX—the three input ports located above the display and numbered 1, 2, and 3 receive the sensor signals. They each appear to be 2 × 2 plates but all of the studs have electrodes attached to their perimeter. Each input port can accommodate the four standard LEGO sensors—touch, light, rotation and temperature as well as the custom designed sensors described previously in Chapter 6. The sensors are wired to the input ports using the standard LEGO connecting leads.

Sensors are classified as either active or passive. The passive sensors do not require power from the RCX to operate. The LEGO touch sensor (a simple momentary switch) and the LEGO temperature sensor (a variable resistor) are examples of passive sensors. On the other hand active sensors require power from the RCX to function. The LEGO light and the LEGO rotation sensors are examples of active sensors. The RCX is initially programmed so that input ports 1 and 3 are passive and input port 2 is active. However, it is possible to change the programming and configure any port to be either active or passive.

In programming the RCX, there are two different settings that must be configured—**sensor type** and **sensor mode**. The **sensor type** identifies whether an active or passive sensor will be connected to the port. The **sensor mode** pertains to the readout of the sensor output. There are eight different sensor modes, which include:

1. The **raw mode** gives the actual reading of the sensor-induced voltage across the DVM in the RCX unit after it has been converted to a digital number from 0 to 1023.
2. When operating in the **Boolean mode**, the Boolean value 0 or 1 is displayed. For a touch sensor (switch), the value is 0 for an open switch and one for a closed switch.
3. In the **edge count mode**, the RCX displays the number of times the Boolean value changes either from 0 to 1 or from 1 to 0.
4. The **pulse count mode** is similar to the edge count mode. The RCX displays the number of times the Boolean value changes from 1 to 0.
5. When operating in the **percentage mode**, the raw value is converted to a number from 0 to 100 with 100 corresponding to a raw value of 1023.
6. The **rotation mode** is used only with the LEGO rotation sensor. The output from this sensor is decoded using an algorithm specifically written for this sensor. The RCX displays a number from 0 to 16 with each increment corresponding to a rotation of 22.5°.
7. The **Celsius mode** is used with the LEGO temperature sensor. In this case the output from the sensor has been programmed to readout on the RCX display in degrees Celsius.
8. The Fahrenheit **mode** is used with the LEGO temperature sensor. In this case the output from the sensor has been programmed to readout on the RCX display in degrees Fahrenheit.

Output Ports

The three output ports located below the display and labeled as A, B, and C, as illustrated in Fig. 7.22, supply power to the motors or other actuators. Each port may be operated in one of three modes—**on**, **off** and **floating**. When a port is programmed in the **on** mode, a motor connected to that port would receive power. When a port is programmed in the **off** mode, an actuator attached to that port would be braked to stop abruptly. When a port is programmed in the **floating** mode, a motor connected to that port would not receive power, but it would be permitted to freewheel.

Care must be exercised regarding the orientation of the lead wire connection because of the polarity of the power supply. Because the motor is operated on direct current (dc), the polarity affects the direction of rotation. However, polarity is not a serious issue when dealing with LEGO components. If the motor is not turning in the correct direction, disconnect a lead wire and rotate its terminal plate by 90° and replace it. This action reverses the polarity and reverses the direction of rotation of the motor.

The Controller Buttons

Near the center of the RCX, you will notice a liquid crystal display (LCD) and four buttons as shown in Fig. 7.23a. The red button is an on/off toggle switch. When the RCX is turned on numbers and a figure appear in the display window and the other three buttons become functional. The gray button enables you to select the program that controls the actuators and sensors used in your deicing system. As you press the program button, the number of the program that is activated cycles from 1, 2, 3, 4 to 5 and then back to 1, etc. The program selected is indicated in the display window. The random memory in the RCX is capable of storing five different programs. Some of these programs written by LEGO personnel are stored permanently in the RCX and others you will write and download to the RCX using program slot #4.

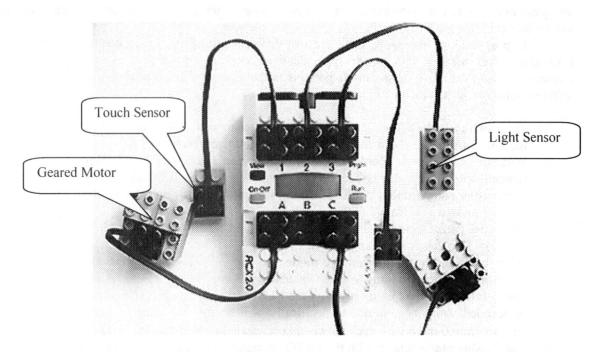

Fig. 7.22   Lead wire connections of the sensors and geared LEGO motors to the RCX.

When pressed, the green or run button starts the program. When the program is running, the little figure in the display window (the run indicator) moves as if it the little man was running. Depressing the run button a second time stops the program.

The black view button permits you to monitor the motors and the sensor outputs if the RCX is on. When the black button is depressed a pointer ∧ appears at the edge of the display pointing to one of the three sensor outputs—1, 2 or 3. The symbol moves (advances) each time the black button is pressed. When the pointer ∨ appears, it will point downward to one of the three output ports A, B or C. The status of either the input or output is monitored in the window to the left of the little man. At positions 1 or 3 the display will indicate 0 if the limit switch is open and 1 if it is closed. At position 2 the number indicates the intensity of the light measured by the light sensor. When the motors are running the number 8 is displayed when the symbol is at either position A or B. If the motor is not running the number is 0. The direction of the motor is also indicated by symbols ▷ and ◁ that appear beside the pointer. The symbol ▷ indicates the motor is running in the forward direction and the symbol ◁ shows the motor is running in reverse as shown in Fig. 7.23b.

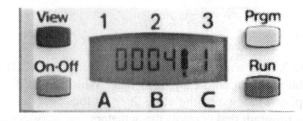

Fig. 7.23a Control buttons and display window on the RCX.

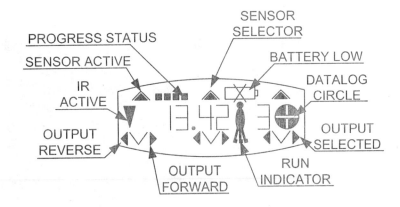

Fig. 7.23b Information displayed on the LCD window on the RCX.

The RCX display shows the output of its internal clock in hour.minutes since it was last turned on. Turning the RCX on and off resets its internal clock. There is also an icon that indicates the status of the internal data buffer that is used when the RCX functions as a data acquisition device to collect and store data. The RCX is also equipped with an internal speaker for playing sounds and music.

7.6. PROGRAMMING IN THE INVESTIGATOR MODE

Introduction

Investigator follows **Pilot** and **Inventor** as the third mode of programming within ROBOLAB. It is used to enable the RCX to acquire, store and transmit data taken from sensors to your PC. Let's explore the screens encountered when opening ROBOLAB in the Investigator mode. The first screen shown in Fig. 7.7 provides three buttons, which enable you to open the Pilot, Inventor or Investigator modes. Select the lower button and open Investigator and the screen presented in Fig. 7.24 appears.

Fig. 7.24 The opening screen in the Investigator mode.

Let's click on the new project button and note the two windows that appear as indicated in Fig. 7.25.

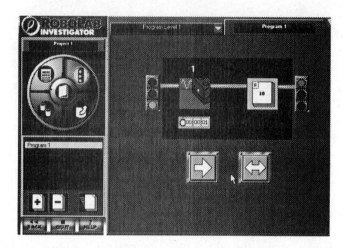

Fig. 7.25 The **Navigator Window** (left) and **Programming Window** (right) in Investigator.

For the moment, let's focus on the **Navigator Window** that controls the programming. Investigator is more complex than either Pilot or Inventor because Investigator divides the programming task into five different areas each of which is accessed by buttons in the **Navigator Window**. These buttons as well as others are defined in Fig. 7.26. Because of the need to switch from one window to another while programming, comparing data, analyzing results or documenting a project, the Navigator Window remains open as you switch from window to another.

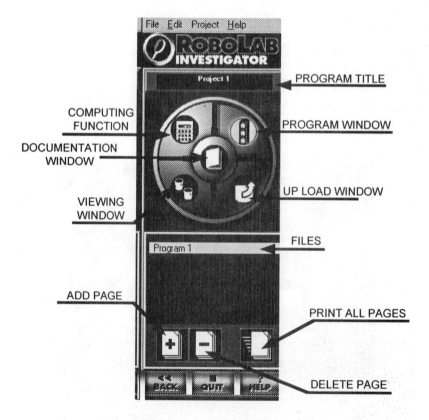

Fig. 7.26 The Navigator Window in the Investigator mode.

Each of the five windows that you will use as you program the RCX to acquire, store and manipulate data are defined in the paragraphs below:

The **Program Window** provides a **programming area** in which you insert icons to acquire the data for a predefined period. You are basically providing instructions to the RCX to acquire measurements using its data acquisition system. You may chose from five different levels of complexity for programming. This window also contains a large white arrow for downloading the program from a PC to the RCX.

The **Upload Window** enables you to **transfer data** stored on a page (file) in the RCX to a PC. Remember that each **Data Set** is stored on an individual page and is transferred one page at a time from the RCX to a PC.

The **Compute Window** permits you to perform **mathematical operations** on your data after it has been transferred to a PC. The operations can be simple such as addition, subtraction, multiplication and division, or more involved operations such as differentiation or integration. Many mathematical and graphical operations are available depending on the level of computing selected. Five levels are available.

The **View Window**, as the name implies permits you to **view** your data before or after manipulation. However, there are four operations possible when working in the View Window. In addition to simply viewing the data, it is possible to **compare** one set of readings with another. When you select the **measure** option, statistical operations become available to you as well as the ability to determine slopes and the area under curves. Finally, it is possible to print out the graphical data from this window.

The Documentation Window enables you to describe your project and the individual pages of data that you have recorded. The description can be embellished with imported JPEG photographs, or if you have access to a UBS digital camera, photographs can be taken and displayed in this window.

The Programming Window

Click on the signal-light-ion on the Navigator wheel and the screen associated with Program Level 1 appears as shown in Fig. 7.27. In this illustration, we have identified the program title, which can be edited to provide a description of your project. Also shown is the program level. Five different programming levels are embedded in the ROBOLAB code. We will begin at Level three in subsequent discussion. The programming window shown in the center of Fig. 7.27 contains the icons selected to form your program. These icons will be described in the next subsection. Finally, the download button used to transfer the program from your PC to the RCX by IR transmission is identified.

Programming in Level 3

Programming in Level 3 is at an intermediate level. It affords the ability to record data from two different sensors at different sampling rates. You can also power devices such as motors or lights that are attached to the output Ports of the RCX. Programming in Level 3 is like programming in the Pilot Mode because wiring (connecting icons together) is not necessary. In Level 3, the icons are arranged in a pre-wired, linear sequence. Loops, forks and jumps in the program are not viable options in Level 3.

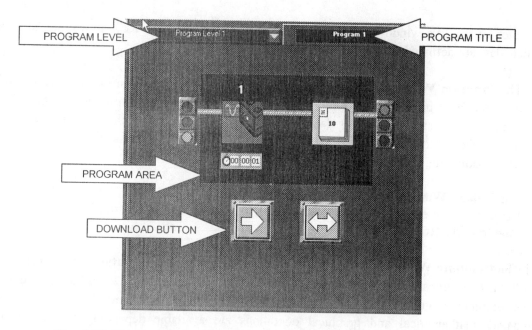

Fig. 7.27 The Programming Window associated with Level 1 programming.

Let's open the Programming Window associated with Programming Level 3 by clicking on the ∇ symbol to the right side of the Program Level display to obtain the screen shown in Fig. 7.28.

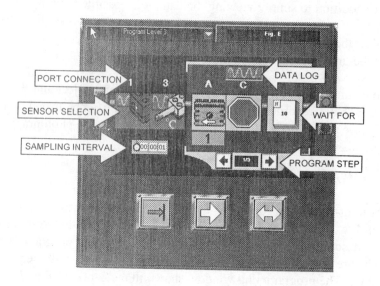

Fig. 7.28 The Programming Window associated with Level 3 Programming.

Specifying the Sensors

First consider sensor selection. It is clear from Fig. 7.28 that the sensors are connected to input Ports 1 and 3. The choice of sensors differs slightly as indicated by the icons shown on the pull down menus that are presented in Fig. 7.29. The sensors in the top two rows are identical for both Ports 1 and 3. The icons identify (in order) the following sensors: touch or contact, light intensity, temperature °C, temperature °F, angle and rotation. The bottom row for Port 1 identifies a generic sensor, a clock and a digital (USB) camera. The two choices in the bottom row for Port 3 include a generic sensor and an icon that indicates no sensor is to be attached to Port 3.

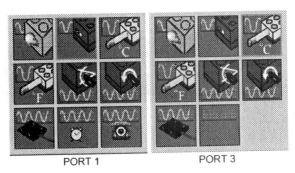

PORT 1   PORT 3

Fig. 7.29 Admissible sensor selections at Ports 1 and 3.

Click on the sensor icon for port 3 and from the pop-up menu select the generic icon. A button labeled "GENERIC" appears under the icon. Click on this label and the pull down menu shown in Fig. 7.30 appears. This listing enables you to adapt port 3 to acquire data from a wide variety of sensors. We have selected the voltage option from this listing and the Generic icon is labeled as such in Fig. 7.31. With this setting, the output voltage from the sensor will be measured and stored in the memory (RAM) of the RCX.

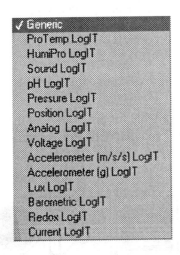

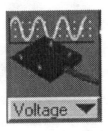

Fig. 7.30 Pull down menu listing types of sensors that can be accommodated by the RCX.

Fig. 7.31 Specifying the signal input for Port 3 as voltage.

Specifying the Data Logging Interval

The sampling rate is specified with the clock icon located below the sensor icons (See Fig. 7.28). Click on this icon a row of five icons that provide a wide choice of data logging rates appears as shown in Fig. 7.32. The first three of these icons (left to right) permit you to sample the output from the sensor every second, minute or hour. The fourth icon enables you to specify the time between readings in seconds. The highest sampling rate that the RCX can handle is 20 acquisitions per second or a sampling interval of 0.05 seconds. The last icon in this row represents the touch sensor. When this selection is made, the output from the sensor will be recorded each time the contact on the touch sensor is depressed. Note that the touch sensor is connected to Port 2 when operating in this mode for data acquisition.

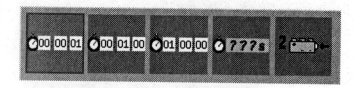

Fig. 7.32 Icons employed to specify data logging rate.

Programming the Output Ports

Level 3 programming in ROBOLAB requires a three step program to control the output Ports A and C and to specify a **Wait For** command. The output Ports and the **Wait For** command are programmed in the area shown in Fig. 7.33. Click on the icon under either the designations for Port A or C and a pop up menu showing motors operating in the forward and reverse directions, a light bulb and a stop sign.

The number below the motor icon shown in Fig. 7.33 adjusts the power level. Click on this number and a pop up menu appears as shown in Fig. 7.34. You have a choice of five power levels—1 to 5. The voltage on the motors is always constant. The power level is adjusted by using pulse width modulation where the voltage to the motors is turned on and off at a high frequency. At a power setting of 5, the voltage is applied continuously. At a power setting of 1 the voltage is applied for only 1/5 th of a cycle.

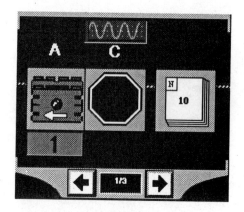

Fig. 7.33 Icons available for programming the output Ports A and C and the **Wait For** command.

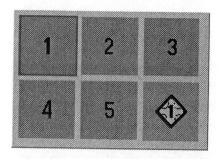

Fig. 7.34 Icons used to control the power level applied to the motors.

In the lower right hand corner of Fig. 7.34, there is an icon that identifies Port 1. When selecting this icon, you are commanding the motor to operate at a power level controlled by the output of the sensor connected to Port 1.

Icon for adjusting power level to
correspond to sensor output.

The light bulb, shown in Fig. 7.33 can be switched on and off if it is connected to either Port A or C. On the other hand if the stop sign is selected, the output voltage will be turned off.

Programming the Wait For Command

The third and last icon in Fig. 7.33 is the **Wait For** command. Click on this icon and a pop up menu appears with 16 different icons as illustrated in Fig. 7.35. These icons permit you to **Wait For** the following events:

- The number of measurements recorded—10, 100, 500 or a specified number.
- The number of seconds between readings—6, 8, 10 or a specified number of seconds.
- Until touch sensor is activated—contact released or depressed.
- Until the light sensor detects a higher or lower light intensity.
- Until the temperature sensor measures a temperature greater or less than a specified value.
- Until the rotation sensors measure an angular displacement greater or less than a specified value.

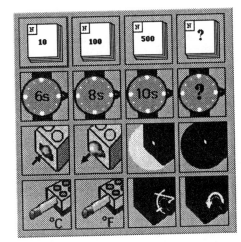

Fig. 7.35 Icons used to specify the
Wait For command.

Data Logging

In some situations logging data may not be required. In these situations it is possible to avoid recording the sensor output by clicking on the icon located above Port C in Fig. 7.33. Clicking on this icon, shown in Fig. 7.36, provides you with two options. The display with a sine wave indicates that you will record data during this step in the program. The display without the sine wave is selected if data logging is not required.

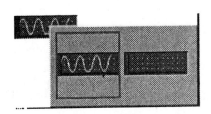

Fig. 7.36 Data logging icons for turning the
logging process on or off.

Three Step Programming

The Level 3 programming mode in Investigator requires three programming steps—whether you need them or not. The program step is identified below the program icons for the output Ports as indicated in Fig. 7.37. The 1/3 shown in this illustration indicates that we are programming the first step in a sequence of three. We advance from one step to another by clicking the right arrow and decrease the program step by clicking the left arrow. You can effectively eliminate a step by programming the output Ports and the time delay as indicated in Fig. 7.37.

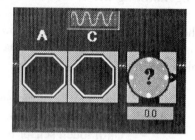

Fig. 7.37 Left—Program step identification and control buttons.

Right—Programming to effectively eliminate a step.

Downloading and Running the Program

Buttons that control downloading and running the program located below the Programming Window are shown in Fig. 7.38. The white single ended Run arrow is to download the program from a PC to the RCX using the IR transmission link. To run the program in the RCX, it is necessary to depress its green Run button. The double-ended white button also downloads your program in the direct mode. When operating in the direct mode, the program is transferred from the PC to the RCX and it immediately begins to run. It is not necessary to depress the green run button.

Fig. 7.38 Control buttons for downloading and running your program.

When you click on the red arrow button, a pop up menu, shown on the right side of Fig. 7.38, provides you with two possibilities for running your program. The red arrow button on the left side is for running the program once and stopping. The double circular arrows on the right side are for running the program continuously.

Handling Several Programs

In some applications, it may be necessary to write several programs in Investigator. It is possible to create a new program by clicking on the page icon with the + sign that is located near the bottom of the Navigator Window. When you create a new page for a program it is numbered and displayed in the file listing area as indicated in Fig. 7.39. More descriptive titles can be added by typing new program names in the title block of the Programming Window. It is possible to move from one program to another simply by clicking on the program title in the file listing area shown in Fig. 7.39. Similarly, you can delete a page (or program) by marking it and clicking on the page icon with the − sign.

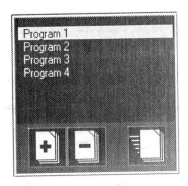

Fig. 7.39 A listing of program files on the area
below the Navigator wheel.

Uploading Data

Suppose that you have programmed your RCX to acquire data from a temperature sensor immersed in a pan filled with water as it is heated on a stovetop. If you measure the temperature every 10 seconds for 10 minutes, you should have 60 data points stored in RAM in your RCX. Let's transfer this data set to a PC. Click on the Upload icon on the Navigator wheel and the window shown in Fig. 7.40 appears. Six features associated with this window are identified in Fig. 7.40 including: (1) The title bar to identify the data set, (2) Upload tools, (3) the Upload button, (4) graphing tools, (5) the graph area and (6) the page (data set) listing area.

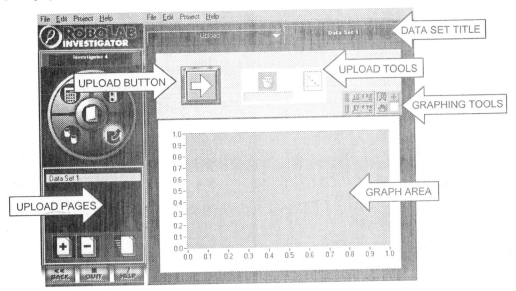

Fig. 7.40 The Upload Window.

Click on the Upload button (white arrow) and the data set is uploaded from the RCX to a PC via the IR tower. During the uploading process, you observe the progress of the data transmission with the uploading bar shown in Fig. 7.41. When the data transfer is complete you are notified with a confirmation that is also presented in Fig. 7.41.

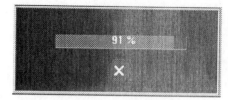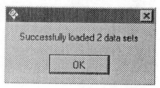

Fig. 7.41 Uploading progress meter and confirmation notice.

To generate the illustrations in Fig. 7.41, we uploaded two sets of data—one from a light sensor and another from a temperature sensor. In the uploading process, ROBOLAB stored the data in two different files and labeled them as Data Set 1 (light intensity) and Data Set 1a (temperature). Data Set 1 was identified with the red bucket and Data Set 1a was placed in a blue bucket as shown in Fig. 7.42.

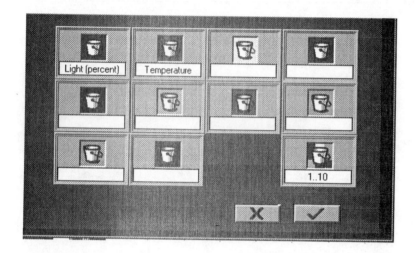

Fig. 7.42 Bucket identifiers for different data sets.

By clicking on the bucket icon in the Upload Tools area (See Fig. 7.40), a pop up screen with all of the buckets appears as indicated in Fig. 7.42. You can edit the title by marking it and typing a new title.

If you plan to collect several sets of data, it is necessary to add a page (open a new file for each uploading event). If you are uploading data from two different sensors, ROBOLAB will automatically open a second page for the second set of data. After the uploading process is complete a bucket color is selected for each data set by clicking on a bucket icon. The bucket color is selected from the ten colors presented in Fig. 7.43.

Fig. 7.43 Bucket colors correspond to the color of the curves when the data is presented in graphs.

Graphing the Data

A graph of the data set (e.g. temperature versus time) is automatically displayed in the graph area of the Upload Window. In the default mode, ROBOLAB automatically plots the points and draws a curve through them. However, several options are available for graphing the data set. Click on the graph icon in the Upload tools area and a pop up menu with these options appears as illustrated in Fig.

7.44. The first three options are self evident from the icons. The fourth option produces a bar chart instead of a curve and the final option is for a tabular display of the data that is also presented in Fig. 7.44.

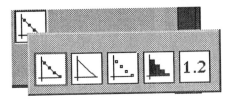

	Time [sec]	Temperat
0	0.0	27.9
1	1.5	27.9
2	3.0	27.9
3	4.5	27.9
4	6.0	27.8
5	7.5	27.8
6	9.0	27.8
7	10.5	27.8
8	12.0	27.7
9	13.5	27.7
10	15.0	27.7
11	16.5	27.7
12	18.0	27.6

Fig. 7.44 Icons for selecting different methods for representing data sets.

Formatting the Graph

The graph is formatted using the graphing tools located on the right side of the Upload Window (See Fig. 7.40). These tools, presented in Fig. 7.45, are described below:

Fig. 7.45 Palette of tools used in formatting the graphs representing the data set.

 The auto-scaling icons for the X and Y-axes together with the locks for fixing the scale.

 Format axis by adjusting scale precision, mapping mode, grid color, etc.

 Zoom control enables you to adjust the size of the graph and to crop (trim) the graph.

 Pan control permits you to move the graph in any direction.

 Cursor (+) icon and full screen icon.

An example of the full screen display of the temperature data together with two cursors is shown in Fig 7.46. The cursor readout displays are located below the full screen graph.

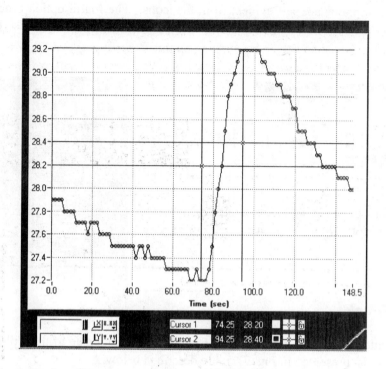

Fig. 7.46 Full screen display of the graph with two cursors for determining intermediate temperatures.

The View Window

When operating in the **View Window** it is possible to: (1) View the graph representing each Data Set. (2) Compare one Data Set against another. (3) Measure the geometric and statistical characteristics of a Data Set. (4) Print out the results. Of these four functions, compare and measure are the most useful. Viewing and printing may be performed in the Upload Window. The View Window is shown in Fig. 7.47.

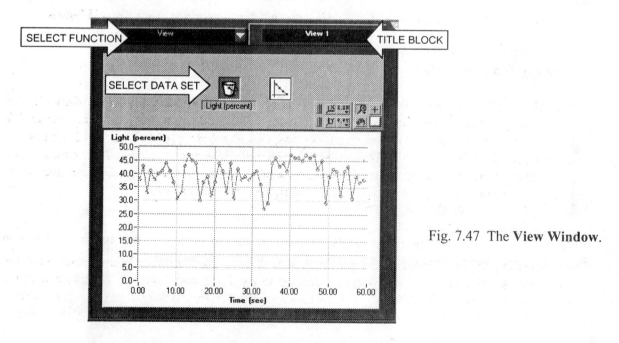

Fig. 7.47 The **View Window**.

Compare Data Sets

The **View Window** is used in the compare mode when the results of two experiments are shown on the same graph. A comparison is made by clicking on the ∇ symbol on the View bar. A menu drops down from which you select the function to be used. We selected a comparison of light intensity versus temperature and obtained the results presented in Fig. 7.48.

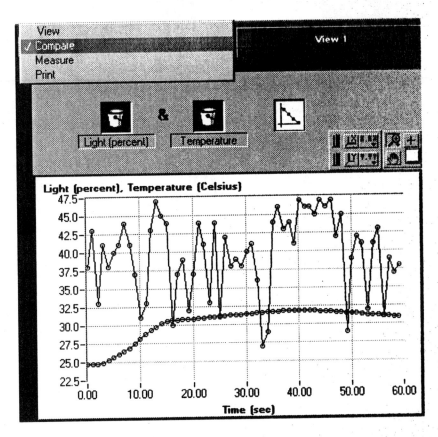

Fig. 7.48 A comparison of one data set against another is obtained in the **View Window**.

Measuring Data Characteristics

When selecting the measure function from the menu on the **View Window** the screen shown in Fig. 7.49 appears. Clicking on the icon with the oscillating curve produces an array of icons that enable you to select from six different measurement options, which include: (1) Minima, (2) Maximum, (3) Mean, (4) Standard Deviation, (5) Slopes and (6) Area under the curve. In Fig. 7.49, which shows random changes in light intensity, we have represented the mean (average) of the light intensities. Note that the mean value is printed in the box to the right of the measure icon.

The Computing Window

The **Computing Widow** is used to manipulate or process data. There are five different tools available for this purpose; however, this description is limited to tools 1, 2 and 3 that are consistent with the Level 3 of Investigator in ROBOLAB. Tools 4 and 5 are more complex, and like the Inventor mode in ROBOLAB, employing these tools requires the user to select functional icons from a palette and to wire them together to form a program. Tools 1, 2 and 3 are pre-wired in a simple series sequence with a limited number of pre-selected operations (functions).

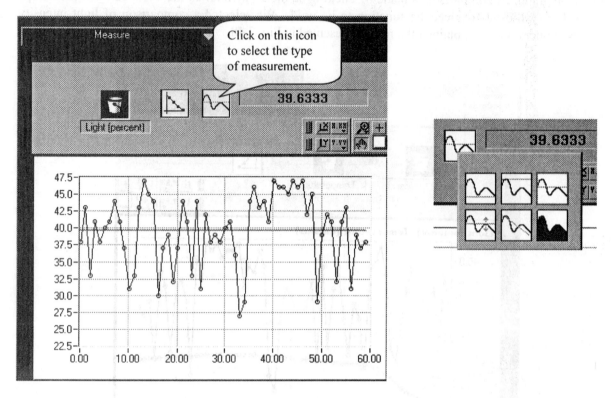

Fig. 7.49. Working in the **Measure** function in the **View Window**.

Computing with Tool 1

Let's begin with Tool 1 by clicking on the ∇ symbol on the tool bar. The screen shown in Fig. 7.50 appears. There are six sequential icons associated with Tool 1. The first icon is used to select the Data Set that you want to manipulate. Click on the second or the fourth icon (from the left side) and the pop up menu, shown in Fig. 7.51, appears that provides a choice of nine mathematical operations. These include adding, subtracting, multiplying and dividing as well as determining the sine, cosine and tangent using each data point as an argument of these trigonometric functions. Also, you can determine the natural log of each data point, and raise the exponential e by the value of each data point.

Click on the third or fifth icon from the left and the pop up menu that is also shown in Fig 7.51 appears. This menu provides icons for 10 buckets or bins and one floating-point number N. The buckets or bins are file locations that may contain data or be open ready to receive a new set of measurements or to receive a file of data that has been generated by using one or more mathematical operations. The last icon in this row (a bucket) represents the file location in which the manipulated data set resides. After the mathematical operation has been performed, the transformed data is displayed in the graphical representation.

New pages can be added so that several different computations can be performed on one or more data sets. In each case the transformed data is displayed in a graphical format.

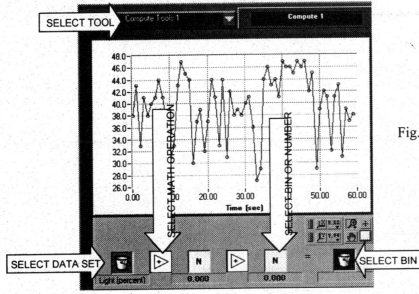

Fig. 7.50 Mathematical operations associated with Tool 1.

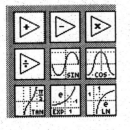 

Fig. 7.51 Mathematical operations to the left, and buckets or bins and a floating point number to the right.

Computing with Tool 2

Let's load Tool 2 by clicking on the ∇ symbol on the tool bar and select the second of the five tools. The screen shown in Fig. 7.52 appears. The icons for manipulating the data are shown at the bottom of this screen. The two bucket (bin) icons provide us with the opportunity to load one data set in the Y-axis and another in the X-axis providing the opportunity to plot one data set against another. Click on the ∇ symbol to the right of the bucket associated with the X-axis and the drop down menu appears that provides a choice of three quantities—X, Y and N. These three quantities are measured when data is acquired by the RCX. X represents the time in the event when the data point was recorded, Y is the quantity measured in appropriate units or percentages and N is the number of the data point. We have shown Y versus N in Fig. 7.52. Note that 60 data points were recorded (0 to 59).

Computing with Tool 3

The window associated with Tool 3 is presented in Fig. 7.53. This tool enables the user to perform several different mathematical operations on one set of data per page. The data set to be operated on is selected by chosing the bucket (bin) containing the results of an experiment. Next two math icons are available that permit sequential mathematical operations on this data set. The icons, shown in Fig. 7.53, enable the following operations:

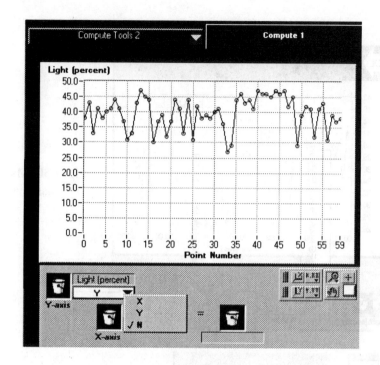

Fig. 7.52 The window associated with Tool 2.

First row:

- No change (multiplying the value of each data point by 1.000).
- Minima (determine the data point with the minimium value).
- Maxima (determine the data point with the maximum value).
- Mean (determine the average value of all the data points).

Second row:

- Standard deviation (determine the statistical quantity known as the standard deviation).
- Slopes (determine the average slope of the curve generated by the data set).
- Areas (determines the area under the curve).
- Differentiate (differentiates the data set at each point).

Third row:

- Integrate (integrates the data set point by point).
- Average lines (provides an average curve for two or more curves).
- Fit lines (fits a line through the data points and provides the equation for this line).

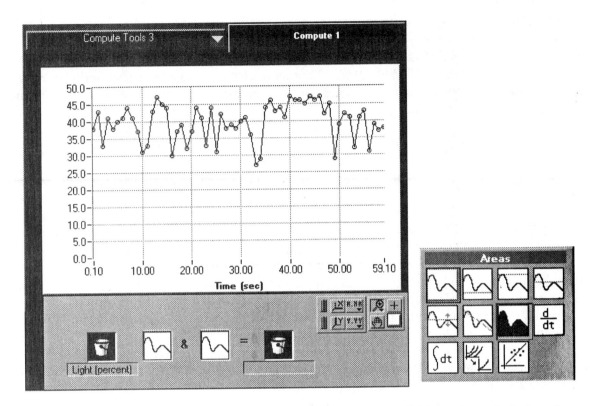

Fig 7.53 The window for Tool 3 and the icons describing the available mathematical operations.

The Journal or Documentation Window

The Journal Window, presented in Fig. 7.54, is employed to prepare a report for your project. The report usually includes a statement of the problem or alternatively the objective of the experiment that you are conducting. Next, you describe the expected outcome of the experiment, and then discuss the sensors used to collect the data and the sampling rate. Digital photographs of the experimental apparatus may be inserted into one of the journal pages. Results drawn from one or more data sets are described and conclusions drawn. Several pages may be used in documenting the project.

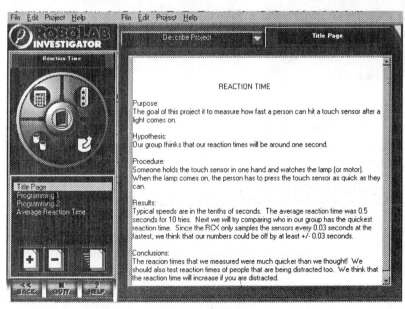

Fig. 7.54 The Journal Window that is used to describe and document the results of your experiments.

EXAMPLE 7.5

To illustrate the use of the RCX to collect and analyze data, let's consider a simple experiment. Heat a cup of water to a temperature of about 70 °C. Then measure the temperature decay of the water as a function of time. Prepare a graph of the temperature relationship with time. Analyze the data and describe the mathematical function that best describes the temperature decay with respect to time.

Solution: Step 1 Programming

We load ROBOLAB and select the Investigator mode Level 3. On the opening Investigator screen, we select new project and name it cooling water. Next we click on the Program icon and select icons for the first step in the program as shown in Fig 7.55. Note in Step 1 of the three-step program we have programmed Port 1 to record data from a LEGO temperature sensor in °C. Port 3 is without a sensor and data will not be recorded using this port.

The output Ports A and C are programmed with stop sign icons, which indicates that power will not be supplied to these Ports. The sign wave in the small display above port C indicates that we will be recording data during step 1 of the program. The selection of the N = 100 icon implies that we will record 100 data points during step 1 of the program. We have defined the sampling interval at 30 seconds that gives a data acquisition rate of 2 measurements per minute. Hence the first step of the program will require 100 × 30 = 3000 seconds = 50 minutes to run. We have programmed the second and third steps of the program to do nothing. The icon arrangement to effectively eliminate these two steps is also shown in Fig 7.55. Finally we place the RCX near the IR tower and click the download button to transfer the program from our PC to the RCX.

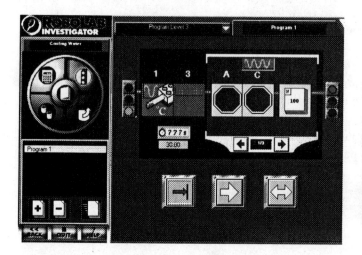
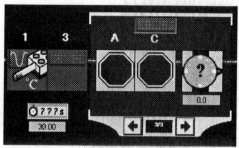

Fig. 7.55 Programming to record 100 temperature measurements at 2 samples per second over a 3,000 second period.

Step 2 Conducting the Experiment

We begin the experiment by connecting a temperature sensor to Port 1 of the RCX. Then we heat a mug of water in a microwave oven adjusting the time and power level to bring the water to about 70 °C. Next we insert the temperature sensor into the hot water and press the run button on the RCX. We note the small figure on the LCD display of the RCX is running thus indicating that data is being recorded. We check the start time of the program and return 50 minutes later. The small figure is stationary indicating that the data collection process is complete.

Step 3 Uploading the Data

We click on the Upload icon to gain access to the Upload Window. Then we click on the white arrow to upload the temperature-time data set. The window that appears is illustrated in Fig. 7.56.

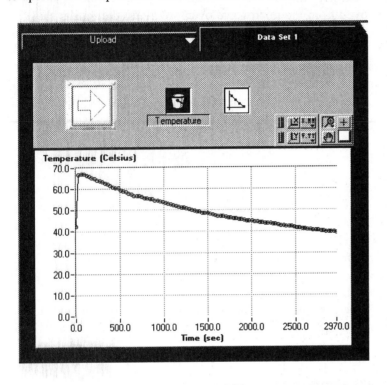

Fig. 7.56 The initial format of the graph of temperature Y(T) versus time X(t).

The graph formatting tools in the Upload Window have been used to improve the visual display of the data as shown in Fig 7.57. We used the automatic scaling features for both the temperature and time axes. Then we selected the number of decimal points used in the captions for T and t on the Y(T) and X(t) axes.

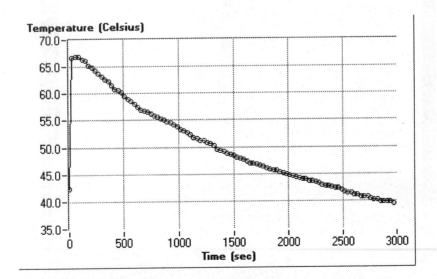

Fig. 7.57 Formatted graph to improve the data presentation.

Step 4 Understanding the Data

Inspection of the graph in Fig. 7.57 shows that the temperature appears to be increasing for about the first minute or so of the experiment. Clearly, the water temperature was decreasing during this interval because the mug had been removed from the microwave oven. Why do the results show an increase in temperature? The reason is due to the mass of the temperature sensor. The sensor was at room temperature (about 24 °C) when it was inserted into the hot water, and the RCX was activated within a few seconds of inserting the sensor. We recorded the actual temperature of the sensor (not the water) as it achieved thermal equilibrium with the surrounding water. After achieving thermal equilibrium, the sensor followed the decreasing water temperature with a negligible temperature differential.

Step 5 Trimming the Data

To correct for the initial lag time of the temperature sensor, we open the Compute Window and change the title from Compute 1 to "Trim and Analyze". We plan to cut out the data recorded during the first minute or so of the experiment because it is in error. To eliminate this erroneous data, we select the trim tool that enables us to cut off data points during either the beginning or end of the experiments or both ends. After clicking on this icon, we move the cursor to the graph area. A new cursor that looks like a miniature magnifying glass appears. Using this cursor, we mark the area of the graph that we wish to retain. The new graph produced in this manner is shown in Fig. 7.58 together with the trim tool. We have eliminated the first 97 seconds of data. The graph does not begin at time t = 0; however, the curve does provide a clear representation of the temperature decay of the hot water due to heat losses.

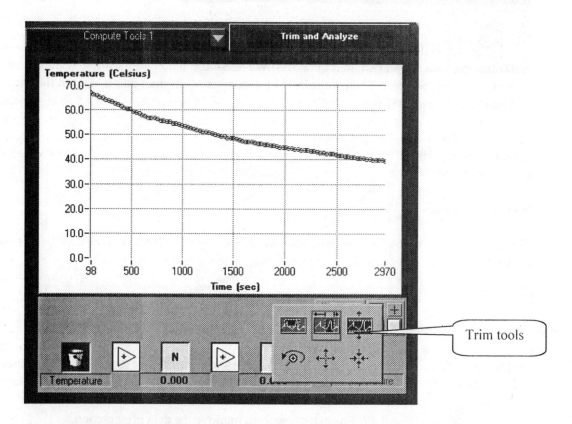

Fig. 7.58 Graph of temperature data after trimming off the initial data points.

Step 6 Analyze the Data

Tool 1 has been selected from the Compute Window because we think the temperature decay behaves like an exponential function. To test this hypothesis, we will use one of the exponential functions that are available with Compute Tools 1. We select the temperature data (the red bucket) and then select the natural log function from the mathematical functions available with Tool 1. For the second mathematical function, we select addition and then add the number 0 to basically eliminate the second mathematical operation. The result, stored in a blue bucket, is displayed in Fig. 7.59.

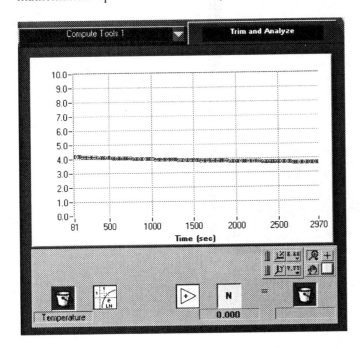

Fig. 7.59 Results obtained by taking the natural log of the data set for temperature.

Examination of Fig. 7.59 shows that when the $\log_e$ (T) is plotted with respect to time t the result is nearly a straight line. We determine the equation of a best fit of a line through these points by using Tool 3. We title a new page "Curve Fit" and select the blue bucket to load the data representing $\log_e$ (T) into the program. We then select the line fit icon as shown in Fig. 7.60. The results are stored in a green bucket and displayed in equation format below the graph in Fig. 7.60.

Step 7 Interpreting the Analysis

The data set in the blue (first bucket) is the $\log_e$ (T) where T is the temperature. The best-fit equation for the line generated when $\log_e$ (T) is plotted with respect to time t is given by:

$$Y = 4.157474 - 0.000172X$$

$$\log_e (T) = 4.157474 - 0.000172t \tag{a}$$

In this case Y is Log_e (T) and X is the time t. Does this make sense? Recall that the temperature of a warm body that is cooling by convection over some time t is given by an exponential equation of the form:

$$T = Ae^{-Rt} \tag{b}$$

where A is a constant and R is the thermal resistance of the heat dissipation system.

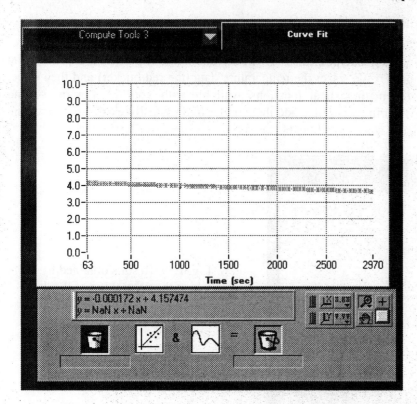

Fig. 7.60 Best fit of a line through the blue data set.

Taking the natural log of both sides of this exponential equation gives:

$$\log_e (T) = \log_e (A) + \log_e (e^{-Rt}) \qquad\qquad (c)$$

or
$$\log_e (T) = \log_e (A) - Rt \qquad\qquad (d)$$

Comparing Eq. (d) and (a) shows that

$$\log_e (A) = 4.157474 \quad\text{and}\quad R = 0.000172 \qquad\qquad (e)$$

Clearly, we have verified our hypothesis and have determined the constants A and R that control the heat dissipation from the cup of warm water used in our experiment.

EXAMPLE 7.6

Determine the angular velocity in revolutions per second (RPS) of a typical LEGO 9-volt motor with gear reduction when operating at power levels of 1 and 4.

Solution:

Step 1. This example requires a strategy for the measurement of angular velocity. In the previous example, it was a simple matter to insert a temperature sensor in the hot water and record the signal output as a function of time over a suitable interval. In this example, we will need a more involved approach. After some thought, we decided to use a light sensor with an encoding disk to measure angular velocity. We also decided to reduce the speed of the encoding disk relative to the motor speed by using a gear reducer.

Step 2. We prepared an encoder disk by using a CAD program. The circular encoder disk is designed with triangular shaped regions. The triangular regions are colored white or black in an alternating sequence as shown in Fig.7.61. In this instance, the disk was divided into eight regions—four white and four black. We will learn later that this dividing the disk into eight regions was a serious error.

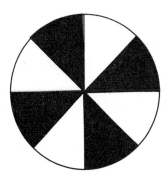

Fig. 7.61 Alternating black and white regions are placed on an encoding disk.

After printing the encoding disk, we cut it from the sheet of paper and glued it to a large (40 tooth) LEGO gear. Using this approach, we were able to develop an encoding disk that could be mounted easily onto a LEGO shaft.

Step 3. We prepared a motor and gear drive assembly that was designed to reduce the speed of the shaft driving the encoder. this arrangement permitted us to mount a LEGO light sensor in close proximity to the encoding disk. The motor and gear drive is shown in Fig. 7.62. Examination of this illustration indicates that the output shaft from the motor is fitted with an 8-tooth pinion. It is mated with a 24-tooth gear that is fixed to one end of the encoder shaft. A gearbox supports the encoder shaft. The 40-tooth gear with the encoding disk is fixed to the other end of the encoder shaft. A LEGO light sensor is mounted in close proximity (0.6 LEGO units) to the encoding disk. As the disk rotates the output signal from the light sensor will oscillate between a low when the black region sweeps by the photodiode to high when the white region passes the photodiode. The output signal should produce a waveform that resembles a series of step functions.

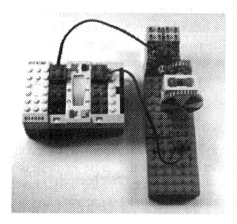

Fig. 7.62 The experimental arrangement for measuring the angular velocity of a LEGO motor.

Step 4. Let's perform a preliminary analysis to determine the number of light pulses that will be generated per revolution of the LEGO motor. We have two factors that affect the output from the light sensor. First—the effect of the gear reducer on speed of the encoder shaft. Second—the number of light pulses generated per revolution of the encoder shaft. The angular velocity of the encoder shaft is given by:

$$RPS_e = (N_p/N_G)\ RPS_m = (8/24)\ RPS_m = (1/3)\ RPS_m \qquad\qquad (a)$$

where RPS_e is the revolutions per second of the encoder shaft
RPS_m is the revolutions per second of the motor.
N_p and N_G are the number of teeth on the pinion and gear, respectively.

Because of the design of the encoder disk with its four black and four white regions, the number of light pulses (N_{LP}) generated by the number of revolutions (N_e) of the encoder shaft during some run time (t) is given by:

$$N_{LP} = 4\ N_e \qquad\qquad (b)$$

Suppose we operate the motor for t = 12 seconds and record data from the light sensor during this run time. Then it is clear from Eq. (a) that:

$$N_e = (1/3)\ (RPS_m)t = (1/3)(RPS_m)(12) = 4(RPS_m) \qquad\qquad (c)$$

Combining Eqs (b) and (c) yields:

$$RPS_m = (1/16)\ N_{PL} \qquad\qquad (d)$$

Using Eq. (d) requires that we fix the run time at t = 12 seconds. If you prefer to run the motor for more or less time, it will be necessary to modify Eqs. (c) and (d).

Step 5. We are now ready to use ROBOLAB to program the RCX to conduct the experiment. Load ROBOLAB and select the Investigator mode Level 3. Click on new project and title the project "Angular Velocity of Motor". Next select the Programming icon and open the Program Window. In step 1 of the program, we have assigned the light sensor to Port 1. The assignment for Port 3, shown in Fig. 7.63, makes it inactive during the experiment. We have specified the sampling interval at 0.1 seconds. This specification implies that we will record 120 data points in the 12-second sampling interval. For the output ports, the LEGO motor is connected to Port A, and the Stop Sign icon at Port C indicates it is not powered. The power level of the motor at Port A is set at level 1. Selecting the Wait For icon and setting its time as 12.0 seconds specify the duration of Step 1 of the program. The program as described above is shown in Fig. 7.63. Programming, as described previously in Fig. 7.37, effectively eliminates step 2 and 3 of the program. Finally, the program is downloaded from our PC to the RCX.

Step 6. Connect the light sensor to Port 1 and the motor to Port A on the RCX. We are not concerned with polarity because the encoder functions the same regardless of the direction of the motor's rotation. Then press the green run button and observe the operation of the motor and gear drive for the run time of 12 seconds. Also note the changing light intensity that is shown on the LCD of the RCX.

Step 7. After conducting the experiments, we click on the Upload icon and upload the data. We changed the title of this page from Data Set 1 to Power Level 1 to reflect the experimental conditions under which the data was obtained. Also, we have decided to show the graph with the data points. The result for this first run of the motor is shown in Fig. 7.64.

Inspection of the results, presented in Fig. 7.64, indicates that we have encountered a problem. The data is showing a beat phenomenon. However, the light intensity monitored by the light sensor should vary like a step function oscillating between maximum and minimum values as the image on the encoder disk goes from white to black. The fact that we observe nine regions where the light intensity did not achieve either the maximum or minimum expected values indicates that we have an aliasing problem in sampling the data from the encoder. Aliasing occurs when the sampling frequency f_s is less than twice the frequency of the waveform we are attempting to record. When the sampling frequency is too low, the waveform we are recording takes on a false identity. It is clear that we have recorded false data and must modify the experimental arrangement and our sampling frequency.

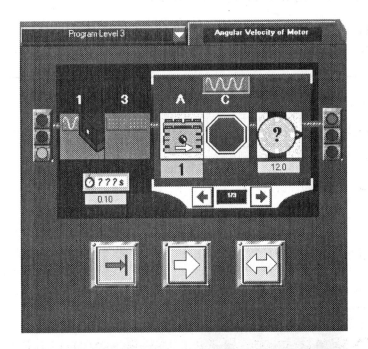

Fig. 7.63 Programming to acquire light sensor data for 12 seconds.

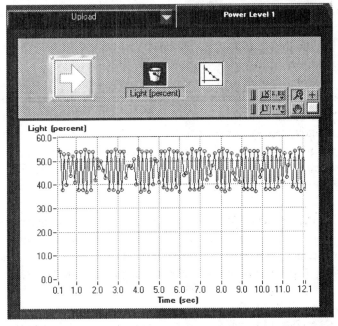

Fig. 7.64 Uploaded data for the motor that represents a false waveform.

Step 7. It is clear that we must increase our sampling frequency and decrease the frequency of the signal monitored by the light sensor to eliminate the aliasing problem evident in Fig. 7.64. To increase the sampling frequency, we reprogram by decreasing the sampling interval to 0.05 seconds, which is the minimum sampling interval for the RCX. We also redesign the encoder and change the pattern on its disk to that shown in Fig. 7.65.

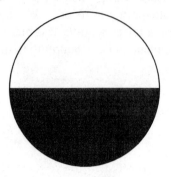

Fig. 7.65 The new encoder pattern will generate one oscillation of the light intensity signal for each rotation of the encoder shaft.

Step 9. After modifying the program and downloading it, we changed the encoder disk and conducted the experiment a second time with a power level of 1. We clicked on the Upload icon and when the Upload Window appeared we added a new page and labeled it "Power Level 1 with New Encoder Disk." We then down loaded the data that is shown in Fig. 7.66.

A comparison of the graphs presented in Fig. 7.64 and Fig. 7.66 shows a marked difference in the results. The beat phenomenon recorded in our first experiment has been eliminated. In Fig 7.66, the light intensity varies between a maximum of 55% and a minimum of 35%. Several data points are evident at both the minimum and maximum values. It is evident from the characteristics of the waveform that we have resolved the aliasing problem and have accurately recorded the output signal produced by the encoder-light sensor device.

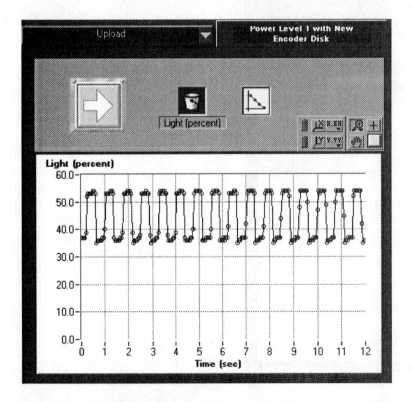

Fig. 7.66 Light intensity with time from the modified experimental arrangement. Power Level 1.

Step 10. The encoder design has been modified; hence it will be necessary to modify the equations developed to convert oscillation count to the angular velocity of the shaft. Repeating the analysis we note that the gear drive has not changed so that

$$RPS_e = (N_p/N_G) \, RPS_m = (8/24) \, RPS_m = (1/3) \, RPS_m \qquad (e)$$

where RPS_e is the revolutions per second of the encoder shaft
RPS_m is the revolutions per second of the motor.
N_p and N_G are the number of teeth on the pinion and gear, respectively.

We changed the design of the encoder disk to one with a single black and a single white region. Accordingly, the number of light pulses (N_{LP}) generated by the number of revolutions (N_e) of the encoder shaft during some run time (t) is given by:

$$N_{LP} = N_e \qquad (f)$$

Suppose we operate the motor for t = 12 seconds and record data from the light sensor during this run time. Then it is clear from Eq. (e) that:

$$N_e = (1/3) \, (RPS_m)t = (1/3)(RPS_m)(12) = 4(RPS_m) \qquad (g)$$

Combining Eqs (f) and (g) yields:

$$RPS_m = N_{PL}/4 \qquad (h)$$

Equation (h) is valid only for a run time of 12 seconds.

   A count of the oscillations in the light intensity waveform for a power level of 1 gives N = 16 for the 12 second run time. Hence, from Eq. (h), it is clear that RPS_m = 16/4 = 4 and that the revolutions per minute $RPM_m = 60 \times 4 = 240$.

Step 11. We modify the program by increasing the power level to 4, down load the program and conduct the experiment again. We click on the Upload icon. When the Upload Window appears, we add a new page. After labeling this page "Power Level 4 with New Encoder", we upload the data. The results are presented in Fig. 7.67. Inspection of the waveform shows a number of data points at each maximum and minimum indicating that the data represents the waveform generated by the signal from the light sensor.
   Counting the number of oscillations gives $N_{PL} = 20$. Hence, from Eq. (h), it is clear that RPS_m = 20/4 = 5 and that $RPM_m = 60 \times 5 = 300$.

   If you were to conduct a similar experiment to measure the speed of your LEGO motors, it is recommended that you document your procedures using the Documentation or Journal Window in ROBOLAB. We have assigned an exercise for you to characterize a LEGO motor and to document the procedure.

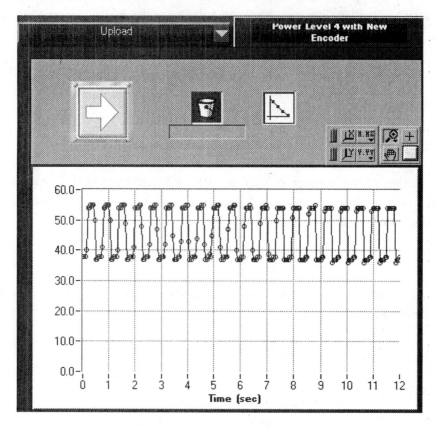

Fig. 7.67 Light intensity with time from the modified experimental arrangement—Power Level 4.

7.7 SUMMARY

The RCX incorporates a microprocessor that can be programmed to control the operation of your deicing system. The degree of control possible is dependent upon the number and type of sensors that you connect to the RCX. Programming consists of downloading a series of commands in a compatible language that direct the microprocessor to execute a number of specific actions. This chapter deals in detail with programming in a graphical language known as ROBOLAB. However, before going into the details of ROBOLAB, the use of flowcharts was described. Flowchart symbols were defined and flowcharts for three different programming examples were presented.

The software used with the RCX for controlling our sensors and actuators is an icon based language known as ROBOLAB. Tufts University, National Instruments and the LEGO Group developed it. It is based on the popular program LABVIEW™—software used by many engineers for laboratory measurements and process control. ROBOLAB is partitioned into three main divisions— **Administrator**, **Programmer** and **Investigator**. The **Administrator** division enables you to download firmware, establish file locations, specify a communication port, etc. The **Programmer** division allows you to program the RCX in either the **Pilot** or **Inventor** modes. Programming in **Pilot** is the easiest to learn and use; however, the capabilities are limited to sequential tasks. Programming in **Inventor** is more challenging, but this approach enables you to structure more complex programs with forks, jumps, loops, events, tasks, etc.

A description of the many of programming icons in **Pilot** has been provided. The **Context Help** window provides excellent information pertaining to the functional icons and the technique for wiring them to modifiers and other types of icons. Two examples were presented to show techniques for programming a wheeled vehicle in the **Pilot** mode. Time, touch and light intensity were used in these examples to control this vehicle.

Significant emphasis on programming involved the **Investigator** mode because of its inherent capability for data logging essential to this semester's project. We described programming at Level 3 in **Investigator**. This level involves relatively simple programming with linear sequential structures. However, it is sufficiently complete to enable you to record the data essential for the successful completion of your project.

Two examples were included in the treatment of **Investigator**. The first example involved measuring temperature during a cooling process. By using the mathematical tools in Investigator, we were able to demonstrate that cooling is an exponential process. The second example concerned the measurement of the angular velocity of a 9-volt, geared LEGO motor. We demonstrated the problem of aliasing when the sampling rate is not sufficient to capture the waveform. The aliasing problem was resolved and the correct waveform was recorded at two different motor speeds.

REFERENCES

1. Wang, E. L., <u>Engineering with LEGO Bricks and ROBOLAB</u>, College House Enterprises, Knoxville, TN, 2002.
2. Baum, D., <u>Definitive Guide to LEGO Mindstorms</u>, Apress, San Francisco, CA, 2000.
3. Ferrari, M., Ferrari, G. and R. Hempel, <u>Building Robots with LEGO Mindstorms</u>, Syngress Publishing, Rockland, MA, 2002.
4. Erwin, B., <u>Creative Projects with LEGO Mindstorms</u>, Addison-Wesley, Boston, 2001.
5. Dally, J. W., Riley, W. F. and K. G. McConnell, <u>Instrumentation for Engineering Measurements</u>, John Wiley & Sons, New York, NY, 1993.
6. Baum, D., Gasperi, M., Hemple, R. and L. Villa, <u>Extreme Mindstorms: An Advanced Guide to LEGO® Mindstorms</u>, Apress, San Francisco, CA, 2000.

LABORATORY EXERCISES

7.1 Wire one motor to port A of the RCX and the other to port C. Select the program in slot #2 of the RCX that will turn both motors on in the forward direction. Prepare a table showing the 16 possible orientations of the lead wires (4^2) leading from ports A and C to the two motors. Run the motors with each possible combination and record the direction of rotation of the motors.

7.2 Write a program to measure the voltage on Ports A and C with all five power level settings. Download the program to the RCX and using a voltmeter measure the voltage at each Port for each power level. Did you note a significant difference in voltage from one power level to another? If not, how is the power level adjusted?

7.3 With your instructor's assistance, measure the voltage output from Port A as a function of time for each of the five power levels. A storage oscilloscope should be used in making these measurements to display the time voltage trace associated with the pulse modulation. Measure the width of the pulses and establish the duty cycle for each power level.

7.4 Write a program that will enable you to determine the forward speed of a wheeled vehicle as a function of the power level applied to the geared motors. Download the program to the RCX and then conduct a test to measure the velocity of your robot at each of the power settings. Prepare a graph of speed versus power level. Is this a linear function? Record the results in your laboratory journal.

7.5 Write a program that will enable turning a wheeled vehicle as quickly as possible. Download the program to the RCX and then conduct a test to measure the time required for your robot to rotate 360° in either direction when the robot is operating at full power. Repeat the exercise to determine the time required to turn through 45°, 90°, 180° and 270° angles.

7.6 Using the meter in the LCD display of the RCX and a LEGO light sensor measure the light intensity reflecting from a white surface. Also measure the light intensity reflecting from black tape. Is this difference between the two measurements significant? Why?

7.7 Design a temperature sensor using a resistive temperature device (RTD). Procure a RTD and construct the sensor.

7.8 Design a temperature sensor using a thermistor. Procure a thermistor and construct the sensor.

7.9 Design and construct a circuit for interfacing the sensor from Laboratory Exercise 7.7 or 7.8 with the RCX.

7.10 Calibrate the temperature sensor and the RCX instrumentation system developed in Laboratory Exercise 7.7 or 7.8.

7.11 Procure a momentary push button switch and adapt it to serve as a touch sensor.

7.12 Design and construct an instrument for measuring wind direction. The signal output from this instrument should be compatible with the RCX.

7.13 Design and construct an instrument for measuring wind speed. The signal output from this instrument should be compatible with the RCX.

EXERCISES

7.1 State, county and city governments collect taxes from travelers visiting their area when they rent hotel rooms. Prepare a flowchart for a program to compute a hotel bill that accommodates the taxes collected by these three different governments.

7.2 Prepare a flowchart for a program to determine if a calle is east or west of Calle Central in San Jose, Costa Rico.

7.3 Suppose a rich uncle deposits $10,000.00 in a Roth Account (tax free) for you to use after its value reaches $1,000,000. The account earns 6% interest compounded annually. Prepare a flowchart for a program that will determine when you will become a millionaire and be able to withdraw funds from this account.

7.4 Write a program for measuring the temperature of a cup of water as it is heated over a low powered heat source. Down load the program, conduct the experiment and upload the data. complete document the experiment and its results.

7.5 Devise an experiment to measure wet bulb temperature of the classroom using a temperature sensor, fan and the RCX. Record your measurements of temperature with respect to time over a two-minute period. Prepare a graph of the temperature time relationship.

7.6 Describe the experiment and the results achieved in Exercise 7.5 using the Journal Window in ROBOLAB.

7.7 Design another experiment to measure the speed of the fan used in Exercise 7.5. We suggest employing a light sensor together with the RCX. Based on the signals from this sensor, derive an equation to convert the pulses recorded to the speed of the fan.

7.8 Conduct the experiment described in Exercise 7.7, and measure the angular velocity of the fan in terms of RPM. Did you check to make sure that the data represents the true waveform and is free of aliasing effects?

7.9 Prepare a Journal window that documents the experiment and its results that were presented in Example 7.5.

7.10 Prepare a Journal window that documents the experiment and its results that were presented in Example 7.6.

7.11 Determine the angular velocity in revolutions per second (RPS) of a typical LEGO 9-volt motor with gear reduction when operating at power levels of 2 and 5. Prepare a Journal window that documents this experiment and its results.

PART III

ENGINEERING GRAPHICS

CHAPTER 8

TABLES AND GRAPHS

8.1 INTRODUCTION

On many occasions you will collect numerical data that must be presented at a design briefing or in an engineering report. At other times, you may have extensive results from a mathematical relationship that must be expressed in a meaningful format. You have two choices in presenting large quantities of numerical data—show the numbers in tabular form or on a graph. Both methods of presentation, the table or the graph, have advantages and disadvantages. The method that you select will depend on your audience, the message and the purpose of the presentation.

In this chapter, the technique for arranging data in a table using Microsoft's Word as the word processing program will be described. In Chapter 11, data entry into a spreadsheet, which is also a convenient method for preparing tables, will be discussed. Five common methods for representing data in the form of graphs will be described including:

1. Pie charts
2. Bar charts
3. X - Y graphs.
4. Semi-log graphs
5. Log - log graphs

Each type of graph is used for a different purpose and selection of the correct type is essential in communicating effectively. For instance, the pie chart is used to show distributions—the person receiving the most and the least. The bar chart is used to compare one set of data with another. X – Y graphs, the most frequently used type of graphic in engineering, shows the variation in Y with X. When Y varies by very large amounts with small changes in X, semi-log graphs are usually employed. The semi-log scales enable you to cover a very large range of Y values. When both X and Y vary over very large ranges, the results are displayed graphically by using a log-log representation.

8.2 TABLES

In Chapter 3 the responsibilities of Mechanical Engineers in their first position were discussed and the statistical data were presented in a table. Let's reproduce this data in a tabular format to show you the ease with which a table may be prepared in Word. First prepare the table heading by selecting the center alignment, identifying the table number and typing the title. You may choose a bold font for the table heading to attract the reader's attention.

Table 8.1
Responsibilities of Mechanical Engineers in Their First Position

---

After typing the heading, you are now ready to prepare the body of the table. With the mouse, point to the menu row and click on "Table" and a drop-down menu appears. You have a choice of whether or not to show grid lines in the table. Click on the "Gridlines" icon and use grid lines when constructing the table. The gridlines aid you in entering the data and they are also helpful to the reader. If you prefer, they can be removed after the data is entered. Next, click on the "Insert" then "Table" shown on the drop down menu. A dialog box that appears on the screen, prompts you to enter information defining the table. It is helpful if you understand the arrangement of the table (how many columns and rows). To represent the data for the responsibilities of mechanical engineers, it is necessary to establish three columns and 15 rows in the arrangement of the table. The initial arrangement is not critical, because it is easy to add or delete either columns or rows later. Finally, select auto-fit in the dialog box permitting Word to control the initial width of the columns. Click the OK button, and an outline of the table appears on the monitor. If the grid lines are light gray when the table first appears on the monitor, they will not print as they are intended only to guide your typing. The result of your first trial after having typed the data into the correct cells should appear as shown below.

ASSIGNMENT	% TIME	% TIME
Design Engineering		40
Product Design	24	
Systems Design	9	
Equipment Design	7	
Plant Engineering/ Operations/ Maintenance		13
Quality Control/ Reliability/ Standards		12
Production Engineering		12
Sales Engineering		5
Management		4
Engineering	3	
Corporate	1	
Computer Applications/ Systems		4
Basic Research and Development		3
Other Activities		7

The result of your first attempt is fair, but it could be improved. Your auto-format selection divided the width of the page into thirds for the column width. This partitioning is not the best choice considering the wasted space associated with the numerical entries and the double rows needed to accommodate two of the assignments. The first column, titled **ASSIGNMENT**, needs more width and the 2nd and 3rd columns are too wide. Modify the table by changing the column widths. You may change the column width by clicking on Table in the menu row and selecting "AutoFit" from the pull down menu. From the adjacent pull down menu that appears, select "To Contents". The appearance of the table has improved and it requires less space on the page. Unfortunately, it is not centered. To center it, click Table from the menu row, click on "Table Properties", and select the "Table" tab from the dialog box. From the Table dialog box, under the choice for alignment, select "Center" and click the OK button. The improved presentation is presented in Table 8.1. As a result of your modifications, the table looks very professional.

Table 8.1
Responsibilities of Mechanical Engineers in Their First Position

ASSIGNMENT	% TIME	% TIME
Design Engineering		40
Product Design	24	
Systems Design	9	
Equipment Design	7	
Plant Engineering/ Operations/ Maintenance		13
Quality Control/ Reliability/ Standards		12
Production Engineering		12
Sales Engineering		5
Management		4
Engineering	3	
Corporate	1	
Computer Applications/ Systems Analysis		4
Basic Research and Development		3
Other Activities		7

Tables are most effective in presenting precise numerical data. When using a table a reader is not required to crudely estimate a number from a curve on an X-Y graph. If you have a valid reason to show your results with six significant figures, it is very easy to do so. A government agency that probably produces the most widely read tables in the United States is the Internal Revenue Service. They prepare tax tables each year that precisely show a person's tax obligations. Another example will be presented later that shows the advantage of conveying numerical data with many significant figures in a tabular format. The application of spreadsheets to produce both graphs and tables will be presented later in Chapter 11.

8.3 GRAPHS

Graphs are used to visualize the data in both reports and presentations. Graphs are not intended to give precise results because tables are employed for that purpose. Graphs show an audience a trend, a comparison, or a distribution—quickly and effectively. There is no need for the audience to study the data to develop an understanding. A graph shows the bottom line and eliminates the time and effort required for the reader to analyze the data and draw logical conclusions.

There are many different ways of presenting numerical data in a graphical format. The distribution of goodies is usually shown with pie charts, because everyone quickly relates to receiving a good size share of the pie. Comparisons among several quantities are made with bar charts with the height or length of each bar indicative of the magnitude of the quantity being compared. Linear graphs, where a dependent parameter Y is plotted as some function of an independent parameter X, are frequently employed in engineering presentations. The data can be generated with mathematical equations when the function Y(X) is known. However, if the mathematical relation is not known, experiments are conducted and measurements of Y are made as X is systematically varied over the range of interest. In both cases, the X-Y graphs represent the numerical data and show a trend—increasing, neutral or flat, decreasing, oscillating, etc. In some instances, the variations in X, Y or both of these quantities is extremely large; therefore, it is not possible to clearly show the trends over the entire range of either X or Y on a graph with linear scales. In these cases, a non-linear format, namely semi-log or log-log is used to present large variations in the data. Let's consider five types of graphs and learn manual techniques for preparing them. Later in Chapter 11, a spreadsheet program, used to prepare these same graphs, will be demonstrated.

Pie Charts

A pie chart is used to show how some quantity is distributed. In Table 8.1, the various assignments for mechanical engineers starting their careers were presented. Let's represent this same data in a pie chart as shown in Fig. 8.1. You begin by drawing a circle to represent the pie, and then decide what fraction of the pie corresponds to each assignment. This determination is easy if you recall that the whole pie contains an included angle of 360°. The slice of pie representing design engineering is a fraction of the whole pie, i. e. $(360°)(0.40) = 144°$. The size of the pie slices in terms of degrees for all of the engineering assignments are listed below:

1. Design engineering $(360°)(0.400) = 144°$.
2. Plant engineering, operations and maintenance $(360°)(0.130) = 47°$.
3. Quality Control, reliability and standards $(360°)(0.121) = 43°$.
4. Production engineering $(360°)(0.121) = 43°$.
5. Sales engineering $(360°)(0.050) = 18°$.
6. Management $(360°)(0.040) = 14°$.
7. Computer applications and systems analysis $(360°)(0.040) = 14°$.
8. Basic research and development $(360°)(0.030) = 11°$.
9. Other activities $(360°)(0.070) = 25°$.

Construct the pie chart by using a protractor to layout these angles on the circle. The pie slices are then labeled as shown in Fig. 8.1. When examining the pie chart, you can visualize the importance of design, because the design slice is huge. You can also see the importance of producing products since these activities—plant operations, quality control, and production—combine to produce still another huge slice. The remainder of the pie, which is less that a quarter, is divided between, sales, management, computers and systems, research and development, and other activities. Of course, these activities are important, but they represent opportunities for a much smaller fraction of the engineers beginning their careers. Compare the effectiveness of the data as presented in Table 8.1 and the pie chart of Fig. 8.1. The pie chart, with a visual representation, quickly leads the reader to the conclusion that design and production are where most opportunities exist for entry-level positions in mechanical engineering.

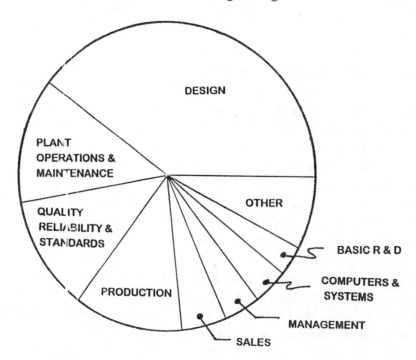

Fig. 8.1 A pie chart showing initial assignments in industry for beginning engineers.

Bar Charts

Bar charts can also be used to show distribution, but they are better suited for illustrating comparisons. As an example of a bar chart, consider the results of a study by the Council of Chief State School Officers on the percentage of high school students graduating with three or more years of secondary mathematics and science in 1982 and 1994. A bar chart showing the comparison is illustrated in Fig. 8.2. The visual effect enhances the comparison. At a glance, the reader can see that many more students were taking mathematics and science in 1994 than in 1982. The second glance indicates that only about a third of the students were in math and science in 1982 while more than half of them were taking math and science in 1994. Finally, it is evident that math was slightly more popular than science in both 1982 and 1994. Three glances and the reader understands the data, because you have presented it in a visual format that effectively carries the message.

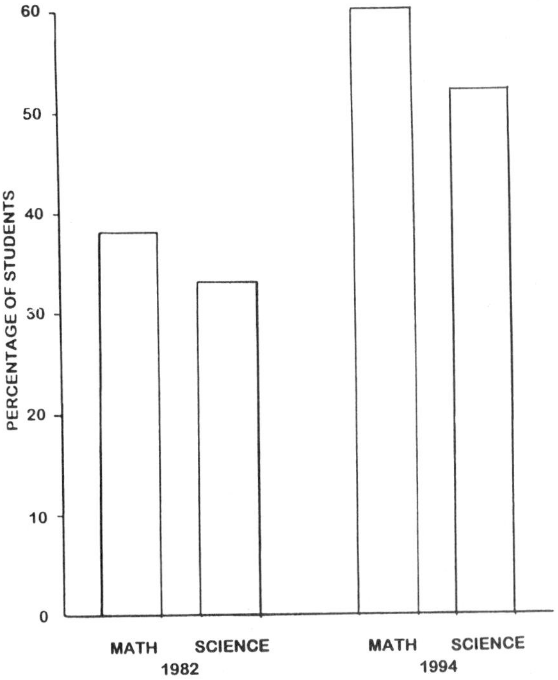

Fig. 8.2 A comparison of the percentage of high school students taking mathematics and science in 1982 and 1994.

The preparation of the bar graph is easy. In constructing Fig. 8.2, you should use an engineering paper with a barely visible grid that facilitated establishing a suitable scale along the ordinate. Only a straight edge and a pencil is necessary in drawing the bars; the grid lines visible on the engineering paper facilitate placing the scale each bar. The time required to prepare a bar chart of this type is less than 10 minutes.

Linear X-Y Graphs

X-Y graphs are probably the most frequently employed graph used in engineering because they illustrate trends. The curves are produced by plotting Y, the dependent variable, along the ordinate as a function of X, the independent variable, which is displayed along the abscissa. Connecting the points with a curve or line segments indicates the trend in Y with changes in X. Graphing X-Y can be accomplished using linear scales for both X-Y, a linear scale for X and a log scale for Y, or log scales for both X and Y.

The numerical results necessary for the following examples of graphing Y as a function of X will be generated with a very important equation that relates to money, and how you may choose to accumulate a fortune or spend it. Begin by assuming that you are going to accumulate some money. Sounds good! Suppose your rich uncle purchased a mutual fund for you at your birth, and plans to give you the proceeds of the fund on your 21st birthday. How much will you have accumulated when the big day arrives? The following relation gives the sum S accumulated:

$$S = P(1 + i)^n \qquad (8.1)$$

where S is the sum accumulated.
P is the amount of your uncle's initial investment.
i is the interest rate for the compounding period.
n is the number of compounding periods over which the interest accumulates.

Suppose your rich uncle invested P = $5000 in a mutual fund that guaranteed an annual interest rate I = 10%, and the interest is compounded semi-annually[1]. It is also assumed that uncle avoided the tax collectors[2] (Federal, State and Local). Let's calculate exactly how much money you can expect to collect on your 21st birthday. You will compute the amount S from Eq. (8.1) substituting uncle's initial investment as P = $5000. The number of compounding periods that occur from your birth until you are 21 years of age is 42, because the fund compounds the interest earned twice each year. The interest rate I = 10% on an annual basis, but the interest rate i is only 5% per the semi-annual compounding period. Substitute these numbers into Eq. (8.1) to obtain:

$$S = \$5000(1 + 0.05)^{42} = \$5000(7.761587) = \$38,807.94$$

This is wonderful. You can buy a very nice car on the good uncle, or defer this option and continue to compound interest accumulating an even larger sum. Another sum accumulated for a total of 80 periods was evaluated using a spreadsheet program (EXCEL) to determine the results from Eq. (8.1). The sums accumulated during the 40-year period are shown in Table 8.2.

In examining Table 8.2, you find that the uncle's principal of $5,000 has grown over the years. The table gives accurate values at the end of each compounding period. For example, at the end of the 20th period, your tenth birthday, the principal in the mutual fund was $13,266.49. Note that the table conveys the exact numerical value with as many significant figures as required. If you maintain the fund until you are 40 years old, accumulating interest for 80 periods, you would have a total of $247,807.21. Compounding interest has produced a significant growth from the $5,000 seed-money planted by your friendly uncle.

Let's represent the data shown in Table 8.2, as an X-Y graph. Take the sum S as the dependent variable represented along the ordinate (the Y axis). Your age is the independent parameter displayed along the abscissa (the X axis). Draw the X and Y-axes on a sheet of engineering paper (which has a faint grid) and apply a scale to each axis, as shown in Fig. 8.3. Scaling is very important because it determines the size of your graph. If the scale is too small, the graph looks like a postage stamp and you have wasted an opportunity to show the graph to full advantage. However, if the scale is too large, you cannot plot all of the data on the paper. To scale the graph correctly, examine the range of the data to be covered for both X and Y. In this case, the range for X varies from 0 to 21 and Y varies from $5,000 to $38,807.94. Scales of 1 inch equal to 5 years along the X-axis, and 1 inch equal to $5000 along the Y-axis were selected. The range displayed for X was from 0 to 25, and for Y from 0 to $40,000. You add numbers to each axis, beginning with zero at the origin and increment in steps of 5 along both axes. Captions are placed adjacent to both axes to remind the reader of the definitions of the variables X and Y. This scaling results in the graph shown in Fig. 8.3; it is approximately 5 by 8 inches in size, which fits nicely on a single page with sufficient space for the figure caption and margins.

---

[1] Interest rates vary significantly from one year to the next. At the time of this revision, the 10% annual interest rate is well above market rates even for long-term investments.

[2] Avoidance of federal taxes is possible, if the funds are invested in a Roth account.

Table 8.2
Accumulated Sum from an Investment of $5000.00
Interest Rate of 10% Compounded Semi-annually.
No Taxes Paid on Interest Earned

Periods	Multiplier	Sum	Periods	Multiplier	Sum
0	1.0000	$ 5,000.00	40	7.0400	$ 35,199.94
1	1.0500	$ 5,250.00	41	7.3920	$ 36,959.94
2	1.1025	$ 5,512.50	42	7.7616	$ 38,807.94
3	1.1576	$ 5,788.13	43	8.1497	$ 40,748.33
4	1.2155	$ 6,077.53	44	8.5572	$ 42,785.75
5	1.2763	$ 6,381.41	45	8.9850	$ 44,925.04
6	1.3401	$ 6,700.48	46	9.4343	$ 47,171.29
7	1.4071	$ 7,035.50	47	9.9060	$ 49,529.86
8	1.4775	$ 7,387.28	48	10.4013	$ 52,006.35
9	1.5513	$ 7,756.64	49	10.9213	$ 54,606.67
10	1.6289	$ 8,144.47	50	11.4674	$ 57,337.00
11	1.7103	$ 8,551.70	51	12.0408	$ 60,203.85
12	1.7959	$ 8,979.28	52	12.6428	$ 63,214.04
13	1.8856	$ 9,428.25	53	13.2749	$ 66,374.74
14	1.9799	$ 9,899.66	54	13.9387	$ 69,693.48
15	2.0789	$ 10,394.64	55	14.6356	$ 73,178.15
16	2.1829	$ 10,914.37	56	15.3674	$ 76,837.06
17	2.2920	$ 11,460.09	57	16.1358	$ 80,678.92
18	2.4066	$ 12,033.10	58	16.9426	$ 84,712.86
19	2.5270	$ 12,634.75	59	17.7897	$ 88,948.50
20	2.6533	$ 13,266.49	60	18.6792	$ 93,395.93
21	2.7860	$ 13,929.81	61	19.6131	$ 98,065.73
22	2.9253	$ 14,626.30	62	20.5938	$ 102,969.01
23	3.0715	$ 15,357.62	63	21.6235	$ 108,117.46
24	3.2251	$ 16,125.50	64	22.7047	$ 113,523.34
25	3.3864	$ 16,931.77	65	23.8399	$ 119,199.50
26	3.5557	$ 17,778.36	66	25.0319	$ 125,159.48
27	3.7335	$ 18,667.28	67	26.2835	$ 131,417.45
28	3.9201	$ 19,600.65	68	27.5977	$ 137,988.32
29	4.1161	$ 20,580.68	69	28.9775	$ 144,887.74
30	4.3219	$ 21,609.71	70	30.4264	$ 152,132.13
31	4.5380	$ 22,690.20	71	31.9477	$ 159,738.73
32	4.7649	$ 23,824.71	72	33.5451	$ 167,725.67
33	5.0032	$ 25,015.94	73	35.2224	$ 176,111.95
34	5.2533	$ 26,266.74	74	36.9835	$ 184,917.55
35	5.5160	$ 27,580.08	75	38.8327	$ 194,163.43
36	5.7918	$ 28,959.08	76	40.7743	$ 203,871.60
37	6.0814	$ 30,407.03	77	42.8130	$ 214,065.18
38	6.3855	$ 31,927.39	78	44.9537	$ 224,768.44
39	6.7048	$ 33,523.76	79	47.2014	$ 236,006.86
40	7.0400	$ 35,199.94	80	49.5614	$ 247,807.21

The data—X and Y coordinates shown in Table 8.2—locate the points that are plotted on the graph. A drop compass was used to draw the small points, but the points can be placed by free hand if you do not have this instrument. Only seven points were plotted to establish this curve. The number of points plotted depends on the smoothness of the function being graphed. In this case, Y varies monotonically with respect to X. It is possible to draw the curve accurately with only six to eight points because the function is relatively smooth and without peaks or valleys.

Inspecting Fig. 8.3 clearly shows the trend of the accumulation sum S with respect to time (age). The sum increases continuously with time and the rate of the increase (the slope of the curve) is also increasing. The visual effect of the X-Y graph is dramatic. At a glance, it is apparent that you are becoming rich. The only question is how rich? By examining the ordinate and abscissa, you can estimate the sum accumulated for a given age, but the estimate may be in considerable error. The faint grid lines used in plotting the data do not show in the copy presented in Fig. 8.3. Accuracy in reading the graph without these grid lines is lost.

If you search for a technique to prepare an X-Y graph that is effective in showing trends, while retaining some accuracy in reading the scales, a good quality graph paper with a fine pitch grid should be employed. An example of this graph, plotted to the same scale on a graph paper with 20 grid lines to an inch, is shown in Fig. 8.4. With these fine pitch grid lines, you can read the scale to:

$$\pm (1/20)(5) = \pm 0.4 \text{ years of age}$$
$$\pm (1/20)(\$5,000) = \pm \$400 \text{ for the sum S.}$$

The resolution of the scale, possible with good quality graph paper, is a significant improvement over the graph without visible gridlines. However, the best accuracy in reporting numerical data is obtained with a tabular format.

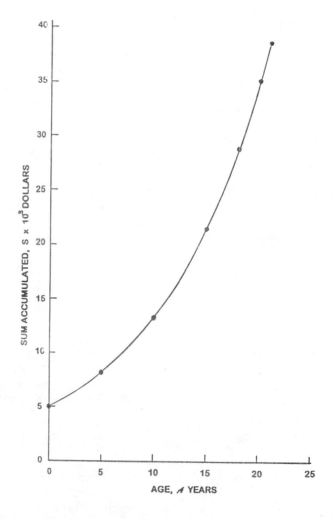

Fig. 8.3 A X-Y graph showing the sum accumulated with time from an initial investment of $5,000.

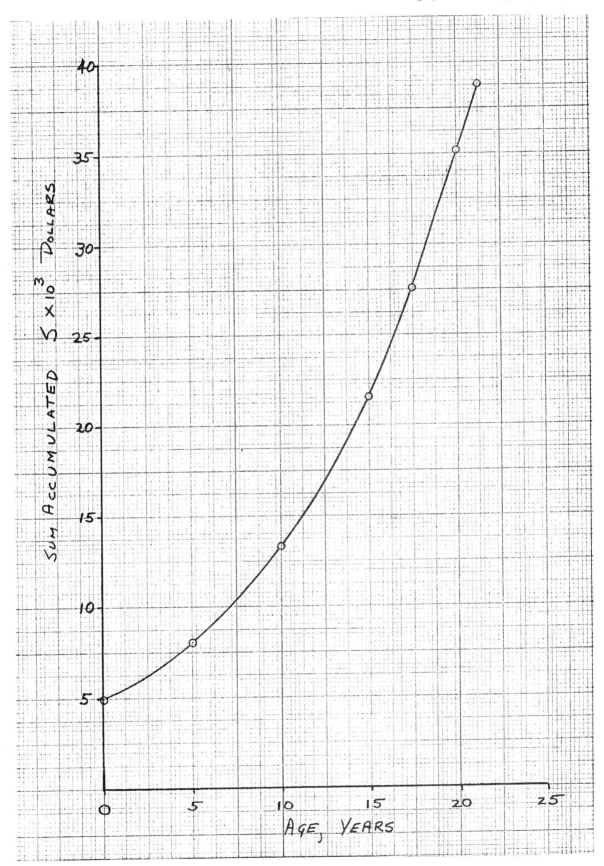

Fig. 8.4 An X-Y graph with fine pitch gridlines used for showing data trends and numerical values.

X-Y Graphs With Semi-log Scales

The results presented in Table 8.2 indicate that you can accumulate some very serious money from a relatively small initial investment if it is invested for a sufficiently long time. If you attempt to plot the larger numbers associated with the sum S in Table 8.2 on the linear graph of Fig. 8.4, a number of data points would go off-scale. You could re-scale, but would lose resolution along the ordinate where the range in S is very large. A better approach is to use semi-log paper in preparing the graph. Semi-log paper, shown in Fig. 8.5, has a linear scale (in this example along the X axis) and a log scale (along the Y axis). Looking closer at the Y-axis indicates that the graph paper is scaled with three log cycles over the 10 inches of the paper that are ruled. Paper with three log cycles was selected because three cycles gives the range needed to cover the large values for the accumulation presented in Table 8.2. Semi-log graph paper is commercially available at engineering supply stores with different numbers of cycles along the log axis.

Let's scale both axes on the semi-log graph paper. The X-axis is scaled with one inch equal to 10 years covering the range from 0 to 50 years. The choice of Y-axis scale is dictated by the selection of the paper. (A three-cycle paper was selected to cover a range of three decades). The first decade is in thousands (10^3 to 10^4), the second in tens of thousands (10^4 to 10^5), and the third in hundreds of thousands (10^5 to 10^6). Zero does not exist on a log scale because the $\log_{10}(0)$ is not defined. In the graph in Fig. 8.5, the origin was placed at Y = $2000. Space for a caption and margin is available at the bottom of the semi-log paper. It is evident that the maximum value of Y, which can be shown on three-cycle paper, is $1,000,000 or 10^6.

The numbers on the Y scale are placed at locations of 1 and 5 as designated on the log scale of the preprinted graph paper. This number is multiplied by 10^3, 10^4, or 10^5 depending on the cycle where it is located. Next, place points on the graph at the coordinate locations defined in Table 8.2. In this example, it is possible to draw a straight line through the data points. The reason for the linearity of the relation for the sum S on a semi-log graph paper is evident, if you take the log of both sides of Eq. (8.1) to obtain:

$$\log_{10}(S) = \log_{10}(5000)(1 + 0.05)^n$$

$$\log_{10}(S) = \log_{10}(5000) + n\,[\log_{10}(1 + 0.05)] \qquad (8.2)$$

Examination of Eq. (8.2) shows that $\log_{10}(S)$ is linear in the number of periods of accumulation (n). The intercept of the straight line with the Y-axis (when n = 0) is $5000. The slope of the straight line is the coefficient of n, namely $\log_{10}(1 + 0.05)$.

The use of semi-log paper has two advantages. First, it permits you to cover a very large range in the quantity represented along the Y (log) axis. A three-cycle semi-log paper may cover a range from 1,000 to 1,000,000. If the need existed, you could cover even a larger range by selecting a semi-log paper with more log cycles. The second advantage of semi-log representation is that it converts certain non-linear mathematical functions, such as Eq. (8.1), into linear relations that are much easier to interpret and extrapolate.

X-Y Graphs With Log-log Scales

When you encounter power functions of the form:

$$Y = AX^k \qquad (8.3)$$

it is advantageous to represent both X and Y on log scales in preparing an X-Y graph. Take the log of both sides of this power function, to obtain:

$$\log_{10} Y = \log_{10} A + \log_{10} X^k = \log_{10} A + k \log_{10} X \qquad (8.4)$$

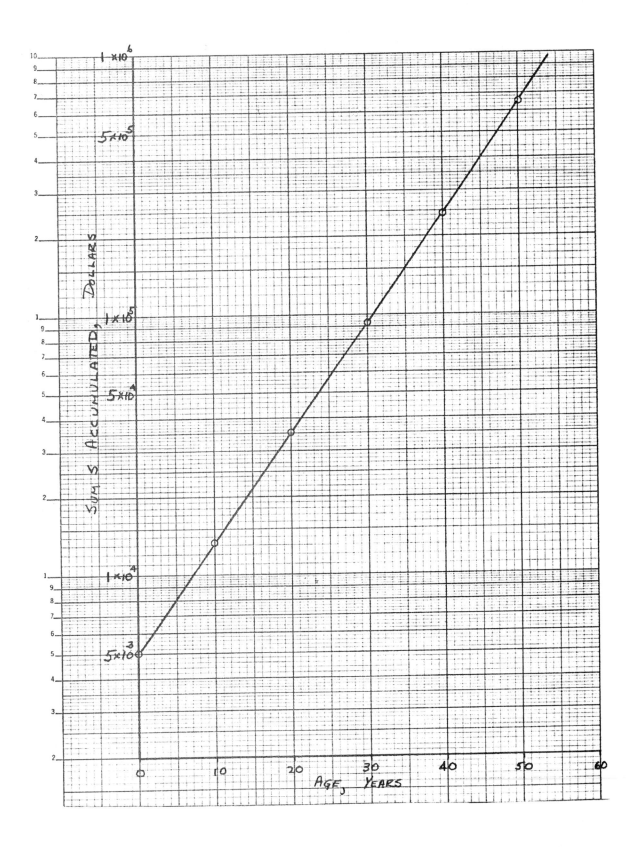

Fig. 8.5 A semi-log representation of the sum S accumulated with time.

For the logarithmic form of the function $Y = AX^k$, the use of log-log graph paper is ideal. Representing both X and Y on a log scale gives you the opportunity to visualize the constants A and k in the power function. To illustrate this fact, consider as an example the function $Y = 2X^{1.4}$ which is described graphically in Fig. 8.6. In this graph, X varies from 0.1 to 100 and Y from 1 to 1000. The scales for $\log_{10} Y$ and $\log_{10} X$ are also shown in Fig. 8.6 to emphasize the difference between X and $\log_{10} X$ and Y and $\log_{10} Y$.

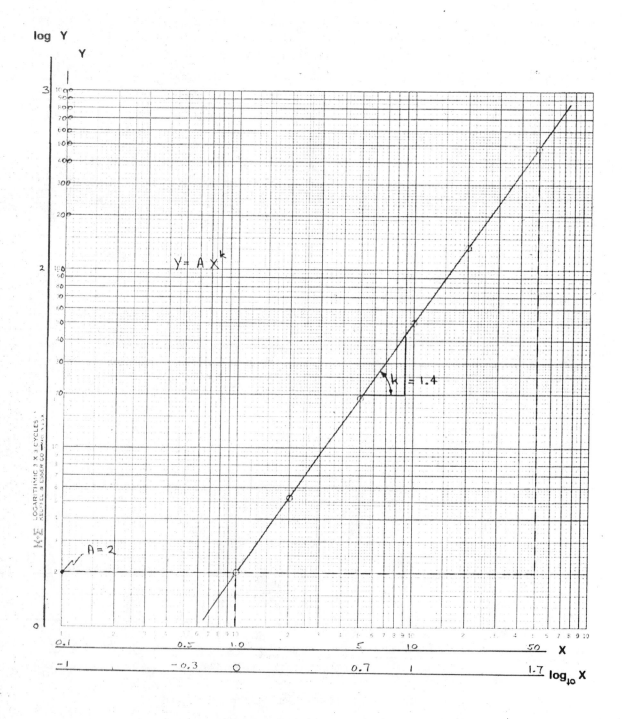

Fig. 8.6 A log-log graph of a power function $Y = AX^k$.

There are three advantages for using log relations and log-log scales on graphs:

1. The capability of linearizing non-linear power functions that often arise in engineering.
2. The ability to plot wide ranging numerical data.
3. Power functions graphed on log-log paper produce a straight line that is easy to interpret and extrapolate.

If the data comes from experimental studies, it is often necessary to establish the constants A and k in a power relation like that shown in Eq. (8.3). The graphical procedure for determining these constants is given in Fig. 8.6. The constant A is determined by setting X = 1 and reading the corresponding Y value as indicated with the dashed lines in Fig. 8.6. The constant k is established by determining the slope of the line. In finding this slope, it is necessary to work with the numbers taken directly from the log scales. Accordingly, the slope k is given by:

$$k = (\log_{10} Y_2 - \log_{10} Y_1) / (\log_{10} X_2 - \log_{10} X_1)$$

$$k = (2.68 - 0.30)/(1.7 - 0) = 1.40. \tag{8.5}$$

The log or log-log graphs are very important because many processes, widely used in engineering, exhibit non-linear behavior, which may be modeled with either exponential or power functions. The behavior of these processes is usually best represented by graphs with either one or both scales expressed in terms of logarithms.

Special Graphs

There are many special graphs that have been developed to help the reader quickly absorb and understand numerical data. This brief chapter on Tables and Graphs does not provide a complete description of the many special and clever graphs that have been devised. However, let's consider one of these—the geographical graph, which is based on a map of the world, the U. S. or some state. Superimposed on the map are contour lines or colored areas dividing the entire geographical region into sub-regions. Either the contour lines or sub-regions are labeled to show the geographical distribution of some quantity across the mapped region. An example of a geographical graph, presented in Fig. 8.7, shows the weather prediction by the National Weather Service for the U. S. for April 9, 2004.

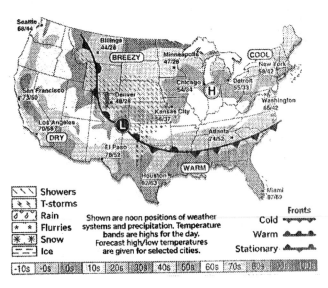

Fig. 8.7 A map showing a weather prediction by the National Weather Service for April 9, 2004.

8.4 SUMMARY

Tables represent an excellent method for reporting data. They have the advantage in presenting precise numerical data. If you need six-figure accuracy, it is easy to obtain in a tabular representation of the data. The procedure for producing very presentable tables using Word has been described.

Many different types of graphs are used to aid in visualizing data. Graphs are much less precise than tables, but they more effectively show trends, comparisons, and distributions. The pie chart, which as the name implies, is a circular area that is divided into pie like segments was introduced. The pie chart is used to indicate distribution of some quantity. Techniques for determining the size of the slices of pie were described.

The bar chart, used to compare the magnitudes of two or more quantities, was introduced. The bar chart was demonstrated and the methods used in its construction were discussed.

Trends are best illustrated using line graphs including the linear X-Y graphs, X-Y graphs with semi-log scales, and X-Y graphs with log-log scales. Linear X-Y graphs are used to show the trend of one variable with respect to another when the range of both variables is not too great. X-Y graphs are presented on semi-log graph paper when one variable has a very large range while the other variable has a more limited range. Log-log graph paper is employed to represent data fields where both variables change over very large ranges. Examples have been provided for these three different types of graphs and techniques used in their construction have been discussed.

Finally, special graphs that make use of maps to convey the concept of geography in the presentation of data were introduced. Weather maps are usually printed in daily newspapers. The Knoxville New Sentinel prepared the map presented in Fig. 8.7; it demonstrate the use of color (shading) to divide America into different temperature zones.

EXERCISES

8.1 Write a memo for a group of young engineers starting their careers with your company that explains the policy of the engineering department relative to reporting data with tables or with graphs.

8.2 Prepare a table using a word processing program that shows the SAT scores for entering freshman at your College of Engineering for the time period from 1982 to 2002. Report both the math and verbal scores for women and men. Also tabulate the number of women and men in each class. The Office of the Dean of Engineering should be able to provide you with the data that is needed.

8.3 Use the data from Exercise 8.2 to construct a bar chart comparing the scores of women and men for the years 1990 and 2000.

8.4 Use the data from Exercise 8.2 to construct three pie charts showing the percentage of men and women enrolled as freshmen in the College of Engineering in 1982, 1992 and 2002.

8.5 Prepare a table similar to the one presented in Table 8.2 showing the sum S with time for the following investment parameters. Use P = $10,000, I = 5%, with interest compounded quarterly. Determine the sum after each compounding period for a total of 40 years.

8.6 Prepare an X-Y graph showing the sum with respect to time using the data generated in Exercise 8.5.

8.7 Prepare a semi-log graph showing the sum determined in Exercise 8.5 on the log scale and the time on the linear scale.

8.8 Evaluate the power function $Y = 2.2\ X^{1.4}$ and plot the result on a log-log graph. Let X vary from 1 to 1000. Try to find log-log paper with a suitable number of cycles for both axes. Confirm that A = 1.8 and k = 1.4 from your graph.

8.9 Evaluate the power function $Y = 5.4 \, X^{2.35}$ and plot the result on a log-log graph. Let X vary from 1 to 100. Try to find log-log paper with a suitable number of cycles for both axes. Confirm that $A = 5.4$ and $k = 2.35$ from your graph.

8.10 You plan on depositing $300 quarterly in a Roth savings account to generate your income after retirement. If you invest for 40 years into this account and it compounds earnings quarterly at an annual rate of 5.75 %, determine the sum you will have accumulated when you retire.

8.11 Shop around and find a better interest rate than the one described in Exercise 8.10. If the rate you negotiate is 6.1% annually with earning compounded semi-annually, determine when you will be able to retire if you seek to have an accumulation of $250,000. Recall that you are to deposit $300 quarterly in a Roth savings account.

8.12 Using the Internet print a copy of today's weather map for the U. S.

CHAPTER 9

THREE-VIEW DRAWINGS

9.1 INTRODUCTION

Engineering graphics is a broad term used to describe a visual means of communication. Most people normally think of communication in terms of writing and speaking because they are more commonly employed in the normal course of life. However, when trying to communicate design ideas, writing and speaking are not always the best way to express complex concepts. A more visual way to communicate is needed. It is more effective to present design concepts and details by means of drawings, sketches, pictures, and many different types of graphs. Visuals aids, such as drawings, convey ideas quickly and with remarkable precision. Also, it is easier to transmit and to receive design information if it is conveyed in the form of drawings, sketches and graphs.

The objective of this part of the book is to introduce you to visual methods of communication. You should understand the advantages of presenting information in drawings, sketches and graphs because they are very common techniques used in communicating engineering concepts. In this part of the textbook, methods for preparing orthographic projections, three-dimensional drawings, sketches and several different types of graphs are presented.

There are two general approaches used in preparing drawings and graphs. The first is a manual approach where the visuals are drawn by hand using a few simple drawing instruments. The second approach makes use of a computer and suitable software programs, which greatly facilitate the preparation of the visual representations. The manual methods for preparing drawings will be described in this chapter. In a different textbook, a computer-aided design (CAD) software program used in preparing engineering drawings is introduced. Microsoft EXCEL, a spreadsheet program used in computation and in preparing several different types of different graphs, is described in Chapter 11. Finally, a graphics presentation program, Microsoft PowerPoint employed to prepare slides and overhead transparencies for oral presentations, is discussed in Chapter 12.

9.2 VIEW DRAWINGS

Let's begin by considering the block-like object shown in Fig. 9.1. You could describe this object as a rectangular block with a slot cut from its top. However, this written description is vague because you have not conveyed the relative proportions of the block, the precise location of the slot, or the size of its features. In preparing an engineering drawing, it is essential to communicate proportion, exact location, and size of every feature of the object, which you are describing. The objective of the drawing is to convey sufficient information for the object to be produced by anyone capable of reading a drawing. You may use pictorial drawings and/or multi-view drawings to quickly and accurately convey this information to the reader. The drawing shown in Fig. 9.1 is a pictorial that is used to improve visualization of features specified in multi-view drawings.

The pictorial (isometric) drawing in Fig. 9.1 is shown with three arrows, which represent the directions of observation of the object showing the front, top and side views. When considered individually, each view is a two-dimensional rendering of the three-dimensional object. Of course, two-dimensional drawings are incomplete, because they show only the information observed in one of many possible views. Nevertheless, the two-dimensional views are important because it is possible to show objects to a true scale on a sheet of ordinary paper or a computer monitor. The problem of incomplete information inherent in two-dimensional drawings is avoided by presenting a sufficient number of views to completely locate and size each feature of the object.

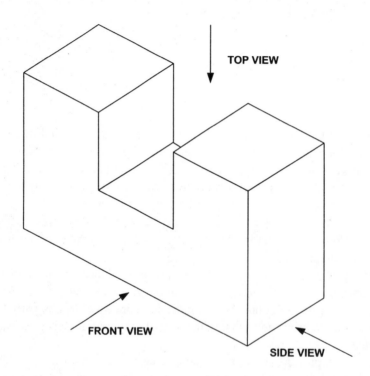

Fig. 9.1 Pictorial drawing showing a rectangular block with a slot. The three viewing directions define the front, top, and side views.

A three-view drawing of the rectangular block presented in Fig. 9.1 is illustrated in Fig. 9.2. The single pictorial drawing of the block is now represented with three different two-dimensional drawings representing the front, top, and side views. These three drawings completely define the proportions of the rectangular block, its size, and the location and size of the slot.

The arrangement of the views is important. The front view is placed in the lower left corner of the paper, the top view is directly above it, and the right side view is to the right of the front view. This arrangement is important because it permits one to prepare the three-view drawing using orthographic projection. With orthographic projection, dimensions are projected from one view to another. For example, you measure width and height of the block, draw the front view, and project construction lines upward to the top view. The width of the object is identical in both the top and front views. You also project construction lines to the right from the front view to define feature heights in the side view. The front and side views both show the height dimensions of the block. You may also project from the top view to the side view by drawing a 45°-construction line from the right hand corner of the front view as shown in Fig. 9.3. Then you project construction lines to the right from the top view until they intersect the 45°-construction line. Next, turn the lines downward to show the depth of the block on the side view. The construction lines used to prepare Fig. 9.2 are illustrated in Fig. 9.3. Orthographic projection is very important because it saves you time in preparing your drawings. It reduces the number of time-consuming measurements that must be made in preparing the different views necessary to completely describe any given part. The construction lines also help you in forming visual images of the locations of each feature on the adjacent views.

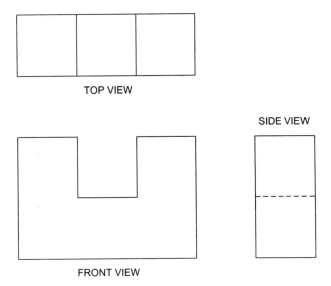

Fig. 9.2 A three-view drawing of the object shows the front, top, and side views.

One more point to consider before leaving Fig. 9.2. Did you notice the dashed line on the side view? Why were all of the lines solid except that one? A convention is followed for the use of line styles and weights in preparing engineering drawing. Solid lines represent edges of features that are observed in a particular view. As you observe the top of the block shown in Fig. 9.1, you can see six lines—the four sides of the rectangle and the two edges of the slot. Because all of these edges are visible, you use solid lines to represent all six edges in the top view. However, when the object is viewed from the side only the four edges outlining the rectangle are visible. When viewing the block from the right side, the slot cannot be observed. If the pictorial or the top and front views are examined, it is evident that a slot exists. To show the slot on the side view, a dashed line representing a hidden line is employed. Drawings differ from photographs because hidden features can be shown on the appropriate view with dashed lines.

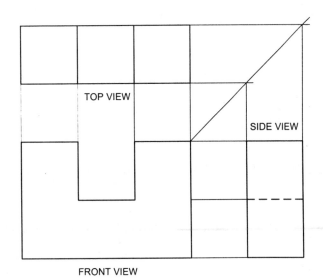

Fig. 9.3 Construction lines used to convey dimensions from one view to another.

9.3 LINE STYLES

Different line styles and different line weights (thickness and darkness of the lines) are used in preparing engineering drawings as illustrated in Fig. 9.4. The weight and the thickness of the lines are controlled by the selection of the pencil lead and the sharpness of the point. The degree of hardness is governed by variable amounts of clay that is mixed with graphite to form pencil lead. Hard leads have a small amount of clay while soft leads have higher clay content. The scale from hard and light to soft and dark is given below:

Hard and Light ⇐ 6H–5H–4H–3H–2H–H–F–HB–B–2B–3B–4B–5B–6B ⇒ Soft and Dark

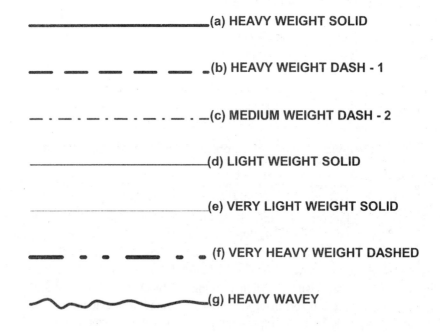

Fig. 9.4 Different line styles used in view drawing.

    a.   Heavyweight, solid-lines representing edges
    b.   Heavyweight, dashed-lines representing hidden lines.
    c.   Medium-weight, long-dash then short-dash lines for centerlines.
    d.   Lightweight solid-lines for dimensioning.
    e.   Very thin, light-lines for construction.
    f.   Very heavyweight, dashed-lines for section cuts.
    g.   Heavy wavy-lines to indicate a break in the view.

Pencils with 4H to 6H leads and a very sharp point are used to draw construction lines. These lines are so light and thin that they will not show when the drawing is reproduced. Light-weight dimension lines are drawn so they will remain visible when the drawing is copied. F lead with a sharp point and a very slightly rounded tip is used to draw dimension lines with the appropriate width and darkness. Medium-weight centerlines are drawn with a HB or B lead with a rounded point. Heavy-weight lines, both solid and dashed, are drawn with B or 2B lead with a rounded point.

    The very soft leads—3B to 6B are usually employed by artists to prepare sketchings, but not by engineers. Engineers handle a completed drawing much more than an artist. In handling, the soft lead smears degrading the appearance of the drawing and detracting from the quality of the copies usually made from the originals.

It is recognized that many of you will be short on pencils, but perhaps team members can share resources. Do not use ballpoint pens for drawing. There are two problems with ball-point pens. First, you have no control over the width of the line because the line width is determined by the diameter of the ball in the pen. Second, you may make errors in preparing the drawing. The lines drawn with a ballpoint pen cannot be erased; your drawings are often sloppy and difficult to read. Work in pencil and use a clean, white, vinyl-eraser to eliminate all traces of your errors. Strive hard to produce an accurate, clean and error-free drawing.

While on the topic of pencils and erasers, below is a list some other simple tools that you should consider acquiring.

1. Transparent ruler with scales in both inches and millimeters.
2. Two triangles—30/60 and 45/45 degrees.
3. Protractor for measuring angles.
4. Compass for drawing arcs and circles.
5. Templates for drawing circles, ellipses, etc.

This is a short list. A complete set of drawing instruments or a drafting board with a T-square have not been included. The purpose of the course is not to teach drafting, but to introduce you to the essentials of engineering graphics. A complete set of drafting tools is not necessary to learn the early lessons in graphics. You are encouraged to invest the small sum needed to acquire the items listed above. You will find them useful in many other courses during your program in engineering.

Finally, a suggestion is made for the paper to use in preparing your drawings. National Brand engineering paper 8-1/2 by 11 in. in size is recommended. It is light green in color to relieve the strain on your eyes. It also has a grid (five squares to the inch) on the reverse side of each sheet. The grid lines show through to the front side of the sheet providing guidelines helpful in preparing many different drawings, sketches and charts. The paper erases well, produces good copies and is pre-punched for a three-ring binder. Equipped with the proper pencils, eraser, drafting tools and paper, you are ready to draw several different features normally encountered in preparing multi-view drawings of engineering components.

9.4 REPRESENTING FEATURES

The techniques for drawing various features in three-view drawings will be demonstrated in this section. The examples include a block with a slot and a step, a block with a tapered slot, and a block with a step and a hole.

Block with Slot and Step

The rectangular block in Fig. 9.1 has one feature—a slot. Many different geometric features are required in depicting parts that incorporate holes, curved boundaries, steps, and tapers. A pictorial of an object with both a slot and a step, presented in Fig. 9.5, serves as the next example. Let's prepare a three-view drawing of this object using the methods of orthographic projection. Examine the pictorial starting with the front view.

As you examine the front view in Fig. 9.5, visualize the outline of the block. Even with the step and the slot, the outline of the block (the outside edges) is a rectangle. Draw the rectangle, representing the outline of the front view, in the lower left hand corner of your quadrille paper. Do not worry about exact dimensions of the rectangle at this stage; however, try to maintain the proportions shown in Fig. 9.5. When drawing the rectangle in the front view, extend the vertical lines upward into the region of the top view and the horizontal lines to the right into the region of the side view. These are construction lines; you should keep them thin and light. Now examine the step in the upper right of the front view of the block. Note the intersection of the planes defining the step with the front plane of the block. These intersections produce two new edges that are visible in the front view. Draw the location of these edges with the horizontal and vertical lines shown in the upper right hand corner of the rectangle (see Fig. 9.6). Extend the vertical line used to locate the step in the front view upward and the horizontal line to the right. Finally, locate the slot on the left side of the pictorial drawing in Fig. 9.5. Again note the planes defining the slot and observe that they intersect the front plane to form two vertical edges. Draw two vertical lines in the correct location to represent these edges. Project these lines upward into the region of the top view with construction lines. You have now completed the front view and are ready to draw the top view.

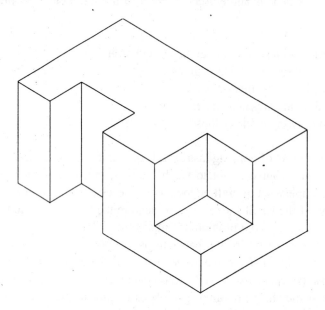

Fig. 9.5 A pictorial drawing of a block with a slot and a step.

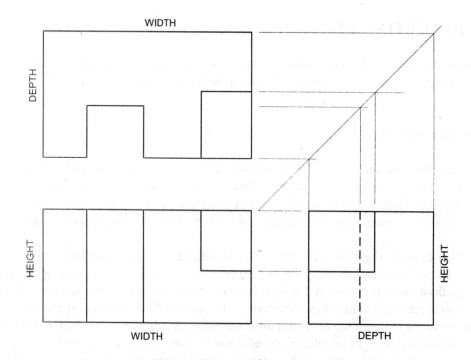

Fig. 9.6 A three-view drawing of the block with a slot and a step.

Look downward at the top of the pictorial drawing. Again, the outline observed is rectangular in shape; however, the rectangle is not perfect since the slot interrupts it. Looking downward, you will observe that the edge defining the outline of the rectangle across the width of the slot is missing. It is possible to observe the bottom plane at the location of the slot. Draw the outline of the rectangle in the location of the top view of Fig. 9.6. Use the construction lines that exist at the location of the top view to establish the width dimensions. An open section in the rectangle is evident at the location of the slot. Next, show the depth of the slot in the top view by drawing the three lines that represent the edges formed by the three vertical planes of the slot with the top plane of the block.

Next, examine the step. The step has two vertical planes that intersect the top plane to form the defining edges. Draw two lines representing those edges in the drawing to complete the top view. If you made full use of the construction lines projected from the front view into the region of the top view, it was easy to draw the top view. It was only necessary to locate the position of three horizontal lines that defined the depth of the slot, step, and rectangular block. These depth dimensions are carried to the right with construction lines.

The final view is of the right side. It is also possible to draw a left side view, but the accepted convention is to draw the right view. After having completed the front and top views, with the orthographic projection of construction lines, the side view is easy to prepare. Draw a construction line at 45° so that it intersects the horizontal construction lines from the top view. Then draw a second set of construction lines downward from these intersection points into the region of the side view. The side view now contains seven construction lines. The three horizontal lines projected from the front view give the height of the step and the rectangular block. The four vertical lines projected from the top view give the depth of the slot, step, and block. The location and dimensions of all of the lines in the side view have been established from your orthographic projections. No measurements are necessary. To complete the side view, darken select portions of the construction lines that represent edges in the side view. Again, examine the side of the pictorial drawing in Fig. 9.5 and observe that the rectangular outline of the block is completed. Also, note the edges produced by the intersections of the planes forming the step with the plane of the right side of the block. Draw these two edges in the upper left corner of the side view (see Fig. 9.6). From a visual perspective, your drawing of the right side is complete; however, a multi-view drawing often carries more information than you can directly observe. You know from the front and top views that a slot exists. Your projection lines from the top view to the side view show the depth of the slot. Even though you cannot "see" the slot in the side view, its presence and its location is represented with a dashed (hidden) line. The three-view drawing of a rectangular block containing a slot and a step is complete as illustrated in Fig. 9.6.

Block with Tapered Slot

Let's consider drawing still another feature—a block incorporating a tapered slot as illustrated pictorially in Fig. 9.7. Begin a three-view drawing with the front view in the lower left corner of your drawing paper as shown in Fig. 9.8. Examine the front view in Fig. 9.7 and observe the edges produced by all the planes intersecting the front plane. Drawing these edges gives four vertical lines and three horizontal lines. The tapered surface intersects the front plane giving an edge shown as a horizontal line in the front view. The top surface is represented by still another edge visible in the front view. All of these lines are projected into the remaining two views with construction lines.

Examine the top view and observe that the rectangular outline of the block remains intact. Observe three edges inside the rectangular outline. The edge due to the intersection of the taper plane with the top surface produces a horizontal line, and the two vertical planes, defining the width of the slot, intersect the top surface to produce two vertical lines in the top view. Project the horizontal lines in the top view to the right so that they intersect the 45° construction line shown in Fig. 9.8. Then project these three lines downward from the intersection points on the 45° line into the side view.

The six construction lines projected into the side view provide the dimensions needed to complete this view. Clearly, it is evident from Fig. 9.7 that the rectangle outlining the side view is intact. In fact when you view the right side, only a simple rectangle formed by the four edges is visible. The information regarding the tapered surface, evident in the pictorial drawing and the front view, is not visible in the side view. Information, showing the location of the tapered surface on the side view, is added to the right-side view by using the dashed (hidden) lines.

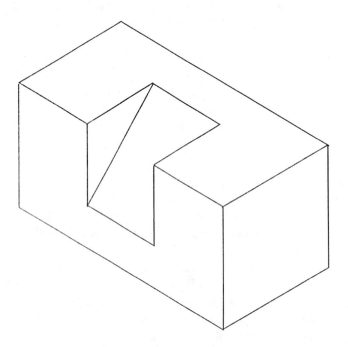

Fig. 9.7 A pictorial drawing of a block with a tapered slot.

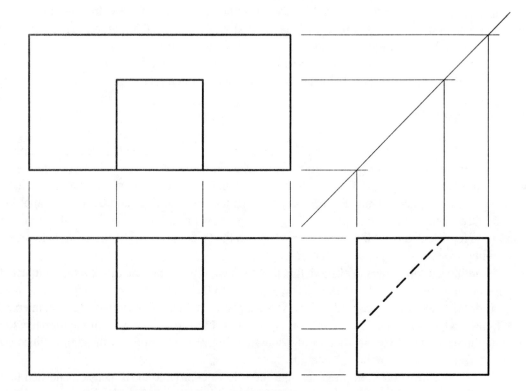

Fig. 9.8 A three-view drawing of the block with a tapered slot.

Block with Step and Hole

A pictorial drawing of a block with a step and a hole is illustrated in Fig. 9.9, and a three-view drawing of this object is presented in Fig. 9.10. The purpose in showing the fourth example of a three-view drawing is to illustrate the method for representing holes and curved boundaries in engineering drawings. Assume in this discussion that you understand the techniques used to draw the outlines of the three-views. The only question remaining concerns drawing the curved boundary that defines the step and the hole that is drilled through it.

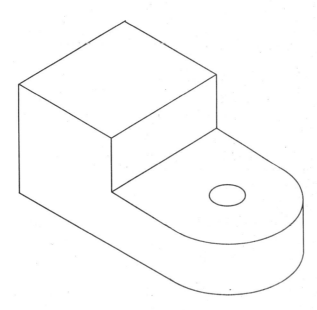

Fig. 9.9 A pictorial drawing of a block with a step and a hole.

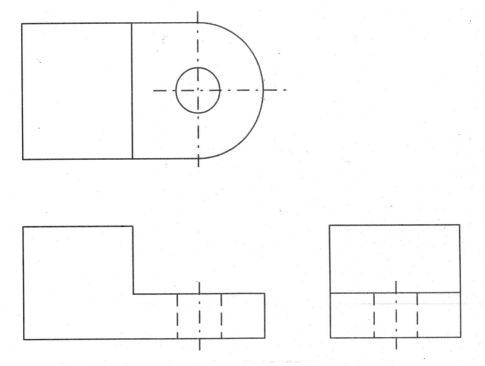

Fig. 9.10 A three-view drawing of the block with a step and a hole.

The circular boundary outlining the step is evident only in the top view (see Fig. 9.10). The front and side views give no evidence of the presence of a curved boundary because the curved surface does not produce intersections (edges) on the front or side planes. The curved surface in the top view is produced by using a compass with its point located in the center of the hole. The radius is set on the compass so that the circular arc is tangent to the horizontal lines forming the outline of the block in the top view. It is advisable to draw the arc prior to drawing the horizontal lines that are tangent to the ends of the arc.

A compass is also employed to draw the hole in the top view. Note the two orthogonal lines defining the center of the hole. These are centerlines that locate the position of the hole relative to other features on the block. The presence of the hole is also depicted in the front and side views even though it is not visible in either of these views. Use dashed (hidden) lines to locate the edges of the holes. The centerline of the hole in the front and side views is also shown. Note the long-dash, short-dash line used to denote the centerlines in all three-views. The centerlines are very important when you dimension the drawings, because holes are located relative to their centerlines and not their edges.

9.5 DIMENSIONING

In the previous section, three-view drawings were prepared maintaining proportionality, but without concern for the dimensions. This approach was a simplification taken to clarify the discussion of three-view drawings. In actual practice, the dimensions are critical because they control the size of the component and the location of all of its features. Let's examine dimensions from two different points of view. First, as the individual preparing the drawing, you must know (or decide) upon the dimensions required to completely define the component. If you are the designer of a new component, it is necessary to start with a blank sheet of paper and assign the dimensions that you believe will optimize the design of the component. If you are redesigning an existing component, it is often possible to start with the original drawing of that component and modify existing dimensions to refine the design. The person preparing the drawing must know all of the dimensions. It is your responsibility to include on your drawing all dimensions necessary for some stranger, perhaps living in another country, to fabricate the component.

Next consider the second point of view—as the person using the drawing. This person may manufacture the part, may assemble the product that incorporates the part or may repair the product. Engineering drawings serve many purposes and are used by many people not involved in the design. The three-view drawing defines this component without ambiguity. The dimensions give its size precisely. The component as defined by the drawing and its dimension can be made by anyone in the world. The component must be interchangeable with another one manufactured previously by a different plant in another country.

Let's begin to learn about dimensioning by adding dimensions to the drawing of the block with a slot shown in Fig. 9.2. This drawing has been copied and dimensions added as indicated in Fig. 9.11. Examine the front view of this drawing note the width of the block has been defined as 3. This dimension is inserted in a break in the dimension line. No units are given except for a notation in the drawing block that all dimensions are in inches, mm, ft, etc. Arrowheads that point to the two extension lines terminate the dimension line. The arrowhead is long (about 1/8 in.) and thin. The extension lines are separated from the view drawings by a small gap (about 1/16 in.).

In the example presented in Fig. 9.11, the block was dimensioned using inches as the unit of measure. The fact that the width was defined as 3 indicates to the person manufacturing the block that the width may be 3 ± 1/64 in. The tolerance on the width is implied or is given in the drawing block. If the designer finds it necessary to specify a block with a higher dimensional accuracy, the dimension would be specified using decimals. For example, if you provided the dimension as 3.00, the person manufacturing the part would understand that the width of the block would have to be between 3.00 ± 0.01 in. If you require still more accuracy, specify the dimension as 3.000; the block must be produced with a width between 3.000 ± 0.005 in. Remember—there is a great difference between specifying the dimension as 3, 3.00 and 3.000 in the accuracy of the block and tooling required to manufacture it. There is also a significant difference in the cost. As you decrease (tighten) the tolerances, the cost of manufacturing a component increases markedly.

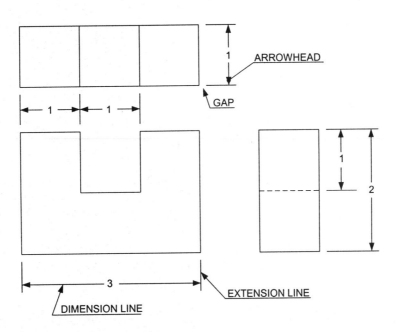

Fig. 9.11 Dimensioning example showing the dimension line, extension line, arrowhead and gap.

Let's return to Fig. 9.11 and examine the dimensioning of the top and side views. The width on the front view has already been shown, but the dimensions for the depth and height of the block and the location and size of the slot must be provided. Note, the top view is used to give dimensions for the width and depth, and the side view is used to give dimensions for the height and depth. In the top view of Fig. 9.11, the depth of the rectangular block has been specified as 1. The dimensions locating the slot and defining its width on this view have also been given. The dimensions of the height of the block and the height of the slot are provided on the side view.

The dimensioning of the block and the slot has been completed. Every dimension necessary to manufacture the component has been specified in the drawing. Also the drawing has not been over-dimensioned. (The same dimension has not been specified more than one time). The choice of where you place the dimensions is somewhat arbitrary. You may dimension the width on either the front or top views, the depth on the top or side views and the height on the front or side views. Two choices for the location of each dimension are possible. The selection between the two views in defining a particular dimension is usually made to reduce drawing clutter and maintain clarity.

Dimensioning Holes and Cylinders

To demonstrate techniques for dimensioning either holes or cylindrical boundaries, let's add dimensions to the drawing shown in Fig. 9.10. Again start with the front view and locate the centerline of the hole as 3.20 from the left edge of the block. A hole is always located by dimensioning between suitable edges to its centerline. Use centerlines for dimensioning because the drill employed to produce the hole is inserted into the component at the location of its centerline.

In this case the dimension for the width is not shown in the front view because it is preferable to specify the radius of the curved (cylindrical) boundary in the top view. Examine the top view and note the radius of the curved boundary is specified as 1.20 R. The R is added to the dimension to indicate that it is given in terms of the radius and not the diameter. In specifying the radius R, you have in effect defined the width of the block as the sum of 3.20 + 1.20 = 4.40. If you had specified the total width of the block in the front view, the drawing would have been over-dimensioned.

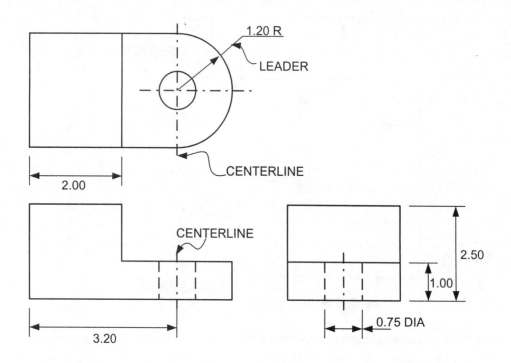

Fig. 9.12 Dimensioning example showing centerlines, leader, and radius and diameter indicators.

The location of the step from the left edge of the block has been specified as 2.00 in the top view. The height dimensions of the features are given in the side view. The total height of the block is given as 2.50 and the height of the step as 1.00. Both heights are measured from the bottom edge of the block. The diameter of the hole is dimensioned on the side view as 0.75 DIA. Holes are dimensioned in terms of their diameter because the drills used in drilling the holes are specified by diameter and not radius.

Are the dimensions as shown in Fig. 9.12 complete? You might question if the depth of the block has been described because it has not been shown in the side view. However, the depth was specified with the radius of the curved boundary in the top view. The depth is equal to two times the radius R or 2.40.

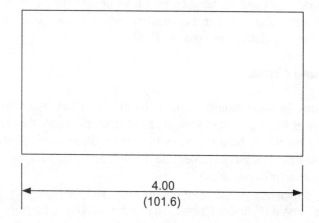

Fig. 9.13 Illustration of a convention used for dual dimensioning.

Again, units for the dimensions have not been shown directly. The units are defined in the drawing block along with other information pertaining to manufacturing the component. The use of two standards for units, the U. S. Customary and the SI systems may cause some confusion in preparing and reading drawings. Some drawings are prepared using U. S. Customary units of in. or ft, and others are prepared using the SI system with units given in mm, cm or m. Sometimes, when the drawing is to be used by many people from different countries, both systems are employed in the dimensioning. An example of dual dimensioning is given in Fig. 9.13. In this case, a break the dimension line is not made. The U. S. Customary unit is placed above the line and the SI unit is placed below the line and enclosed with parentheses.

9.6 DRAWING BLOCKS

The drawing block serves a very important function on an engineering drawing. In the previous section, reference was made to the fact that the units for the dimensions are specified in the drawing block. The unit of measurement used in preparing the drawing is only one of the many facts presented in the drawing block. As illustrated in Fig. 9.14, the drawing block is located in the lower right hand corner adjacent to the border of the drawing. A typical drawing block conveys important information pertaining to all drawings produced by a certain company and to the individual drawing. Information commonly shown in the drawing block includes:

1. The name of the company issuing the drawing.
2. The name of the part that the drawing defines.
3. The scale used in preparing the drawing.
4. Tolerances to be employed in manufacturing the part.
5. Date of the completion or the release of the drawing.
6. Material to be used in manufacturing the part.
7. Heat treatment of the part after manufacturing.
8. Units of measurement to be used in manufacturing the part.
9. Initials of the individual preparing the drawing.
10. Initials of the individual checking the drawing.
11. A unique drawing number to identify the drawing.

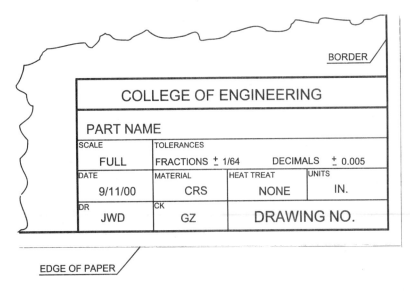

Fig. 9.14 An example of the information presented in a drawing block.

Let's examine each of these items. The name of the company is self-evident; it is illustrated in Fig. 9.14 as the College of Engineering. In designing a part, identify it with some name such as bracket, support, shaft, etc. and include this name on your drawing. The part name should be brief (no more than two or three words).

Next you will indicate the scale used in preparing the drawing. The example shown in Fig. 9.14 indicates that full scale was used in drawing the three-views of the part. The scale that you select depends on the size of the part and the size of the paper that is used for the drawing. In a typical college class, you are usually constrained to standard size copy paper (8-1/2 by 11 inches) because of size limitations imposed by low-cost laser printers. Suppose the part that you are designing is 12 inches wide, 6 inches high and 1 inch deep. Will the three-view drawing of this part fit on this standard size paper? The answer to this question is a clear NO. You need to scale down the part dimensions on the drawing so that it will fit on a single sheet of paper. In this case, you would probably use ¼ scale. With ¼ scale, the front view of the part would be 3 inches wide by 1-½ inches high. You would dimension the drawing using the actual size of the part (i.e. you would show the width as 12 inches although it is only 3 inches on the scaled drawing). You indicate to those individuals reading the drawing that the size of each view has been reduced by a scaling factor of ¼ with a notation in the drawing block.

Tolerances that are to be employed in manufacturing the part are specified in the drawing block. Many companies have standardized their tolerances and these limits are preprinted in drawing blocks that are incorporated on the company's drawing paper. The date of completion of the drawing is also shown. In some instances, the drawing release date is used instead of the completion date. The release date identifies when the design engineers turn over the ownership of the design of a product to the operations (production) function, which is responsible for producing the product.

The material from which the part is to be fabricated is often identified in the drawing block. In the example shown, the abbreviation CRS indicates that cold rolled steel will be used in manufacturing the component. Sometimes the heat treatment of the component, if required, is identified in the drawing block. Heat treatment is a multi-step process employed to enhance the strength and/or hardness of a component fabricated from certain metals. Because heat treatment is usually performed after the part is machined; it is standard practice to indicate whether or not heat treatment is required. This information assists those in charge of operations in controlling the flow of parts in the production process.

The units used in the drawing are given in the drawing block. You may use U. S. Customary units, SI units, or dual units. You cannot use mixed units. (Note the difference between mixed and dual units). U. S. Customary units are in terms of inch (in.) or foot (ft). SI units are in terms of millimeter (mm) or meter (m).

Those responsible for the drawing are identified. The individual who has prepared the drawing signs the block with his or her initials. The drawing is checked for completeness and accuracy, and the individual checking the details also initials the drawing block. Finally, the drawing is given a number. In a large company, which may release thousands of drawings each year, the engineering records department controls the drawing numbers. This department maintains a numbering system that insures that each part or component has a unique drawing number. The engineering records department also organizes the numbering system to group together all of the drawings needed to build a specific product.

9.7 ADDITIONAL VIEWS

Some of the parts used in products are complex and additional views may be needed to ensure that the drawings are properly interpreted. You are not restricted to using only three-views in preparing detail drawings. For very simple parts, you can adequately describe the object with one or two views. There is no need to waste time drawing additional views. For complex parts that are more difficult to visualize, you can provide additional views to clarify the drawing. These extra views may show either external surfaces or internal sections. If it helps to visualize the object, you may add back, bottom and left side views to the more commonly employed front, top and right side views. To draw these additional external views, simply rearrange the layout of the views on the drawing as shown in Fig. 9.15. This drawing, which shows five views of a multiply tapered block, illustrates the proper arrangement to show additional external views. It also maintains the advantages of orthographic projection. Clearly, the extra views help in visualizing the complex shape of this block.

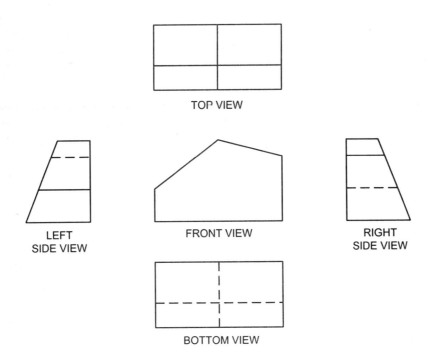

Fig. 9.15 A five-view drawing of a complex block.

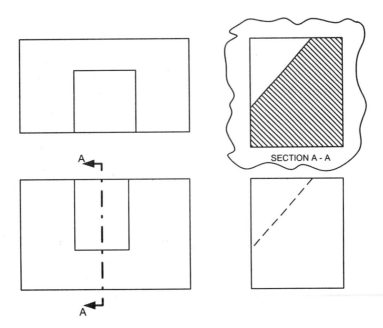

Fig. 9.16 A section cut and the corresponding section view.

An addition internal (sectional) view is often more useful in clarifying a drawing than an additional external view. The concept of a sectional view requires you to mentally slice the object into two pieces with a cutting plane. You then open the part and view the internal section revealed by the cut. An example of the sectioning technique is illustrated in Fig. 9.16 where adding a sectional view has modified Fig. 9.8. On the front view drawing in Fig. 9.16, a very heavy dashed line is shown to indicate the location of the cut. The cut line is turned through 90° at both ends and arrowheads are drawn to indicate the direction of the view. The arrowheads are each labeled with the letter A. In this case, you are viewing the section cut from the right side. Accordingly, you should expect the section view to closely resemble the right side view. You then draw the section that is revealed by the cut.

The section view may be placed at any convenient location on the drawing. In Fig. 9.16, it has been placed it in the open space in the upper right hand corner of the drawing. The section view is usually encircled with a wavy line to distinguish it from the usual three-views. The section view is identified with the section cut by using the letters A-A in a notation. In some drawings, you might make several cuts each revealing a different section with each sectional view identified by A–A, B–B, C–C, etc. In this view, the section revealed by the cut is shown with cross-hatching. The remaining part of the view (outside the cut section) is drawn without cross-hatching. Prepare a section view of the component shown in Fig. 9.12. Make the section cut along the centerline through the hole in the top view, and then examine the section from the right side.

9.8 SUMMARY

Engineering graphics have been introduced by describing techniques for preparing multi-view drawings. Multi-view (usually three-view) drawings are used to communicate the relative proportions of a component and the exact size of all of its features. Drawing techniques are covered in detail with a few simple examples. It is important that you invest the time necessary to learn these techniques because acquiring skills to draw and read three-view drawings is an essential means of communicating in engineering.

Dimensioning is an important task in preparing an error free drawing. All of the dimensions must be included so that the part can be manufactured from the drawing. Often the part is manufactured in another city, or state or even another country. The person manufacturing the part does not know you and cannot ask questions to clarify the ambiguities in the drawing. The drawing and all of its dimensions must completely define the component. An important implication of the dimensioning is the tolerances required in fabricating the part. As you write the numbers for the dimensions, you implicitly assign the tolerances. There is a significant difference between 3, and 3.000 in the precision required in manufacturing and in the cost of the component.

Drawing blocks were described to illustrate the type of information that they normally convey. Remember drawing blocks are not unique and you will observe differences in the information presented by different companies. The concept of scale was also introduced in this section. The scale used in preparing a drawing is your decision. Select the scale factor so that the three-views fit the paper without crowding. Sometimes you will scale down the view drawings so they will fit on the sheet. Other times you will scale up the views of a very small part so that very small features may be clearly represented.

The coverage of three-view drawings presented in this chapter is very brief. If you need additional information, more complete textbooks are listed in the references.

REFERENCES

1. Earle, J. H., <u>Engineering Design Graphics</u>, 4th edition, Addison Wesley, Reading, MA 1983.
2. Bertoline, G. R., and E. N. Wiebe, <u>Fundamentals of Graphics Communication</u>, 3rd Ed., McGraw Hill New York, NY, 2002
3. Lockhart, S. D. and C. M. Johnson, <u>Engineering Design Communication</u>, Prentice Hall, New York, NY, 1999.
4. Jensen, C. H. and J. D. Helsel, <u>Engineering Drawing and Design</u>, Glencoe McGraw Hill, New York, NY, 1997.
5. Sorby, S. A., Manner, K. J., Baartmans, B. J. and S. S. Sorby, <u>3-D Visualization for Engineering Graphics,</u> Prentice Hall, New York, NY, 1998
6. Giesecke, F. E. et al, <u>Engineering Graphics</u>, 7th Ed., Prentice Hall, New York, NY, 2001.
7. Krulikowski, A. Geometric Dimensioning and Tolerancing, Delmar Publishers, Delmar Publishers, Clifton Park, NY, 1998.

EXERCISES

9.1 Prepare a three-view drawing similar to the one shown in Fig. 9.2 except change the width of the notch from 1 inch to 1–3/8 inch.

9.2 Prepare a three-view drawing similar to the one shown in Fig. 9.2 except change the width of the notch from 1 inch to 1–5/8 inch.

9.3 Prepare a three-view drawing similar to the one shown in Fig. 9.2 except change the depth of the object from 1 inch to 1 ½ inch.

9.4 Take a piece of clay or foam plastic and use a razor knife to manufacture a block with the shape shown in Fig. 9.6.

9.5 Prepare a three-view drawing of a block with a taper notch like that shown in Fig. 9.8. Select the dimensions yourself, but be consistent from one view to another with these dimensions.

9.6 Prepare a three-view drawing similar to that shown in Fig. 9.10 except increase the diameter of the hole to 1 ½ inch.

9.7 Prepare a three-view drawing similar to that shown in Fig. 9.10 except decrease the diameter of the hole to ¾ inch.

9.8 Prepare a three-view drawing similar to that shown in Fig. 9.10 except increase the depth of the object from 2.4 to 2.7 inch.

9.9 Dimension the drawing that you prepared for Exercise 9.1.

9.10 Dimension the drawing that you prepared for Exercise 9.2.

9.11 Dimension the drawing that you prepared for Exercise 9.3.

9.12 Dimension the drawing that you prepared for Exercise 9.5.

9.13 Dimension the drawing that you prepared for Exercise 9.6.

9.14 Dimension the drawing that you prepared for Exercise 9.7.

9.15 Dimension the drawing that you prepared for Exercise 9.8.

9.16 Design a drawing block that your team will employ to identify their drawings.

9.17 Prepare a five view drawing of the object shown in Fig. 9.5.

9.18 Prepare a five view drawing of the object shown in Fig. 9.7.

9.19 Prepare a five view drawing of the object shown in Fig. 9.9.

9.20 Prepare a drawing with a section view for the block defined in Fig. 9.1.

9.21 Prepare a drawing with a section view for the block defined in Fig. 9.5.

CHAPTER 10

PICTORIAL DRAWINGS

10.1 INTRODUCTION

Pictorial drawings are three-dimensional illustrations of a machine component, structure or an object. For a person trying to visualize an object with some complexity, the pictorial drawing is the most effective means to convey its size and shape. Some people have difficulty placing the three standard (front, top and side) views together to "see" the object. Pictorials drawings assemble the three-views on a single sheet providing a three-dimensional rendering facilitating visualization. Because pictorials are so easy to visualize, they are often used to illustrate catalogs, maintenance manuals and assembly instructions.

Three different types of pictorials are in common usage:

- Isometric
- Oblique
- Perspective

A simple cube with three different types of pictorials is illustrated in Fig. 10.1. The isometric pictorial is drawn with its three axes spaced 120° apart. The term isometric means "equal measurement" indicating that the three sides are all scaled by the same factor relative to their true length. Parallel lines defining the edges of the cube are also parallel on the isometric drawing. Drawing paper with isometric axes is available in well-stocked office and/or drafting supply stores. You are encouraged to use this special paper because it greatly facilitates the preparation of an isometric pictorial.

Oblique pictorials are drawn with the front view in the x–y plane. The oblique lines, which represent the z-axis (depth), are projected at some angle—often 45°; however, the angle used for the oblique lines may vary from 0 to 90°. Parallel lines defining two or more edges on the cube are also parallel on the oblique drawing. If the true length of the lines is employed in scaling all three sides, it is known as a cavalier oblique pictorial. The cavalier oblique style is frequently used, but the resulting pictorial is distorted. The distortion is due to the depth dimension, which appears to one's eye to be too long.

The perspective is a pictorial drawing that represents what you appear to "see". Artists usually draw or paint using the perspective style. Engineers sometimes represent their designs in this style. Unfortunately, perspective is the most difficult of the three types of pictorials to master. In perspective drawing, you do not have a well-defined coordinate system. Parallel lines tend to converge to a vanishing point as they recede from the observer. Scales used on the different axes are different in order to foreshorten the lines located some distance from the picture plane. The use of converging lines instead of parallel lines and the foreshortening of select dimensions give the drawing perspective. The prospective drawing is similar to a photograph of an object.

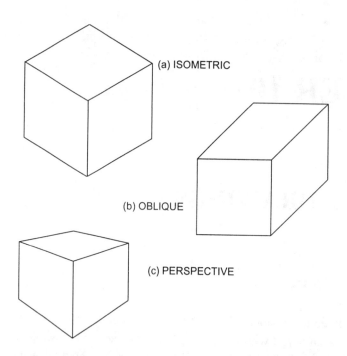

(a) ISOMETRIC

(b) OBLIQUE

(c) PERSPECTIVE

Fig. 10.1 A cube represented with isometric, cavalier oblique and perspective pictorials.

10.2 ISOMETRIC DRAWINGS

In this discussion of the three forms of pictorial drawings, let's consider the following three characteristics of three-dimensional, isometric drawings:

1. The axes used to frame a drawing.
2. The direction of viewing a three-dimensional object.
3. The dimensions used to draw width, height and depth.

Let's begin with the isometric axes shown in Fig. 10.2. The isometric pictorial employs axes, which make 120° angles with each other. The axes divide the paper into the three zones utilized to present three-views. If the axes form the letter Y with the vertical line oriented downward from the two branches, you are looking downward toward the object. From this perspective, you visualize the top view in the region between the branches of the Y as shown in Fig. 10.2. The front view is displayed in the region to the right of the vertical axis, and the left-side view is located in the region to the left of this axis.

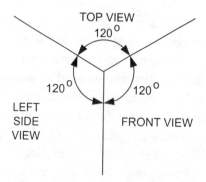

Fig. 10.2 Isometric axes at 120° angles give three regions for the front, top and left-side views.

The axes and the direction of viewing the object have been defined for isometric coordinates in Fig. 10.2. Next consider the dimensions used on an isometric drawing for the width, height and depth. The word isometric means equal measurement; hence, it is clear that the same scale is used along all three axes. To illustrate the techniques used to prepare isometric drawings, consider a simple rectangular block with width W, height H, and depth D. An isometric pictorial of this rectangular block is presented in Fig. 10.3. To draw an isometric pictorial of a rectangular block, follow the procedure outlined below:

1. Use a 30°–60°–90° triangle to draw the isometric axes identified with the numbers 1, 2, and 3 as shown in Fig. 10.3.
2. Measure the length of H from the origin down the vertical axis to establish point A.
3. From point A, draw two additional lines (numbers 4 and 5) parallel to lines number 1 and 2.
4. Along line 5, measure the width W locating point B. Similarly measure the depth D along line 4 to locate point C.
5. From points B and C, draw the vertical lines 6 and 7 that intersect lines 1 and 2, and locate points E and F.
6. From point F, draw line 8 parallel to line 2, and from point E draw line 9 parallel to line 1.
7. Lines 8 and 9 intersect at point G to complete the isometric drawing.

The isometric pictorial that is illustrated in Fig. 10.3 shows the left-side view, the top view and the front view because the block is being viewed from above with the vertical edge bisecting the left side and the front side views. The origin of the isometric coordinates is positioned at the upper left hand corner of the rectangular block.

The procedure for preparing isometric drawings is easy to implement; the lines are all parallel to the isometric axes and the measurements are all made to the same scale.

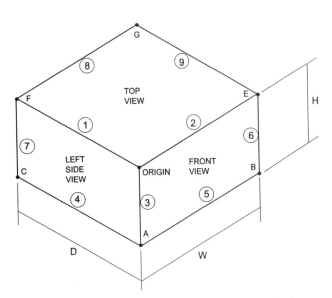

Fig. 10.3 Isometric pictorial of a rectangular block.

The example of a rectangular block was too easy. Let's try a more complex geometry as shown in Fig. 10.4. Examine the three-view drawing of an odd shaped block illustrated in Fig. 10.4. It shows an unusual three-view drawing because it presents the left-side view instead of the more conventional right-side view. The three-view drawing is presented in this manner because the isometric drawing shows the same three-views. For the simple rectangular block, presented in Fig. 10.3, the origin of the isometric axes was placed at the upper left hand corner of the front view. For the more complex geometry shown in Fig. 10.4, the origin is again placed at the same location.

The isometric pictorial, presented in Fig. 10.4, is drawn on isometric paper with only a straight edge to guide the lines. With a steady hand, you may sketch the lines without the straight edge. Isometric paper gives many evenly spaced lines parallel to the isometric axes eliminating the need for the 30° triangle. Begin at point B and draw plane 1 in the region for the left view. Do not draw the entire left-side view, because there is a taper to the block that complicates this view in its isometric representation. Defer completing the left-side view, and move to the top view and draw plane 2. Both planes 1 and 2 are clearly defined in the three-view drawing and are easy to construct on the isometric pictorial. Again the top view is not completed because the step and the taper complicate the geometry. Move next to the front view and add plane 3, which defines the depth of the step. Now return to the top view and draw plane 4 as shown in Fig. 10.4. Because the top view is now complete in the isometric drawing and the height of the step is established, it is easy to draw plane 5. Note plane 5 is on the surface formed by the taper; it does not lie in a plane formed by the isometric axes. This non-isometric plane is located by first drawing the four well defined isometric planes on the pictorial. The location of plane 5 is then clearly established by the position of planes 1 and 4.

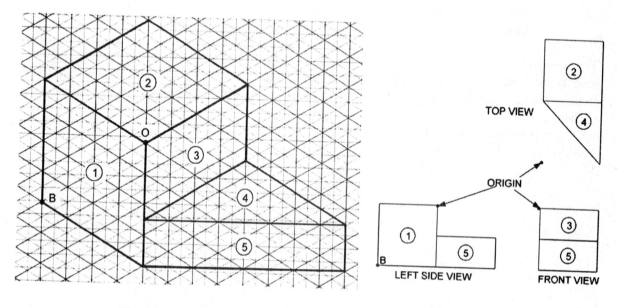

Fig. 10.4 Isometric pictorial of a block with a step and a tapered side.

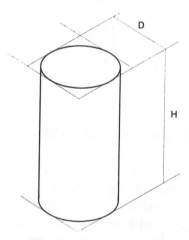

Fig. 10.5 Isometric pictorial of a right circular cylinder of diameter D and height H.

You now understand the procedure for drawing isometric pictorials of simple rectangular blocks and more complex blocks with both a step and a taper. Let's next consider a cylinder of diameter D and height H and illustrate it with an isometric pictorial. To draw the cylinder, lay out two isometric axes located a distance H apart using light construction lines as indicated in Fig. 10.5. On the upper set of axes, draw a square in the top plane with the length of the sides of the square equal to the diameter of the cylinder. Then select an isometric ellipse (35° - 16') from an ellipse template, and draw an isometric ellipse so that it is tangent to each side of the square drawn in the top plane. (If you do not have one of these handy templates, sketch the ellipse in by hand). Note that the ellipse is tangent to the isometric axes at four points as indicated in Fig. 10.5. Move down to the second set of isometric axes, which represent the bottom plane. Take the template and draw an isometric ellipse again. This time draw only the front half of the ellipse, because only that portion is visible when viewing the cylinder from above. The two ellipses are joined with vertical lines at their outer most points to complete the cylinder. The construction lines in Fig. 10.5 remain to aid you in understanding the procedure used in preparing this drawing. To finish the isometric representation, erase the construction lines and shade the cylinder to enhance the visual effect of a three-dimensional object. Shading and shadows will be discussed later in this chapter when this isometric drawing of a cylinder will be referenced again.

10.3 OBLIQUE DRAWINGS

Oblique pictorials and isometric pictorials are similar, because they both utilize parallel lines in constructing the three views. The difference between isometric and oblique pictorials is in the definition of the axes. In oblique drawings, an x, y, and z coordinate system is employed as illustrated in Fig. 10.6. The three coordinate axes divide the sheet into three regions for drawing the front, top and right-side views. With the axes defined as shown in Fig. 10.6, the object is viewed from above and to the right. The z-axis, which is the receding axis in Fig. 10.6, is drawn with a 45° angle relative to the x-axis; however, other angles such as 30° or 60° may be employed to construct the receding axes.

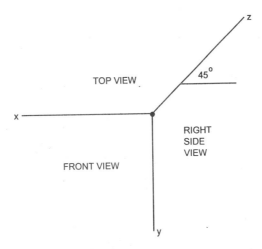

Fig. 10.6 Oblique axes use the x, y, and z coordinate system.

Three different types of oblique drawings are frequently used for pictorials as illustrated in Fig. 10.7:

1. Cavalier oblique is drawn with the receding axis at any angle from about 30° to 60°, but the measurements along all three axes are made with the same scale.
2. Cabinet oblique is drawn with the receding axis at any angle from about 30° to 60°, but the measurements along this axis are half-scale.

3. General oblique is drawn with the receding axis at any angle from about 30° to 60°, but the measurements along this axis varies from half to full-scale.

If you examine the cube represented by the three types of oblique drawings in Fig. 10.7, it is evident that full-scale (true length) measurements are used in the front view in all three types of oblique pictorials. The difference among them is the scale employed along the receding axis. In cavalier oblique, full-scale measurements are made along the receding axis; however, this scale produces a drawing that is out of proportion. The cube does not look like a cube.

In cabinet oblique, half-scale measurements are made along the receding axis. The resulting drawing is in better proportion than the cavalier oblique, but sometimes it appears that the depth dimension along the receding axis is too short. The general oblique where the measurement on the receding axis can be varied from half-scale to full-scale is preferred. The scale is adjusted between these limits to give what appears to the eye to be the correct proportions.

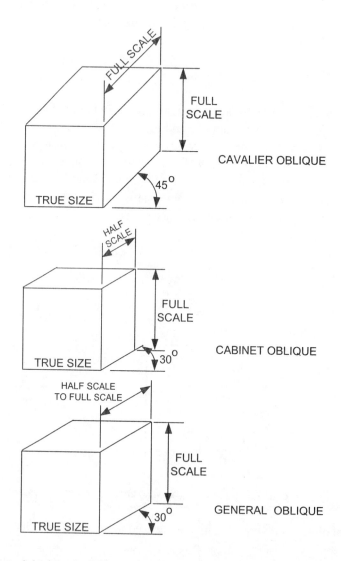

Fig. 10.7 A pictorial of a cube drawn with cavalier, cabinet and general oblique techniques.

To illustrate the procedure followed in drawing an oblique pictorial, examine the three-view drawing of a pair of the connected rectangular blocks as shown in Fig. 10.8. To begin an oblique pictorial, draw the x, y, z-axes using point O, the lower left-hand corner, as the origin. First, draw part of the front view (plane 1) corresponding to the front of the large block. Next, draw part of the top view (plane 2) of the large block. In drawing the front view, use true lengths (on plane 1) to layout the width and height of the large block. On the top view, establish the depth of the large block maintaining proportion by using a scaling factor of ¾.

Note the smaller block obstructs the right-side view of the large block. This obstruction is handled by drawing the small block. Using the information in the three-view drawing, you can place plane 3 on the oblique pictorial. Plane 3 is located by working from the back edge in the top view. The width and height measurements required to locate and size plane 3 in the oblique pictorial are true lengths, but the depth is again scaled by ¾.

After you have drawn plane 3 locating the small block in the pictorial, it is easy to draw plane 4 by referring to the right-side view in the three-view drawing. Complete the drawing of the small block by dropping vertical lines down from the three corners in plane 3 and by closing the sides that form planes 5 and 6.

The resulting oblique pictorial clearly captures the relative proportions and positioning of the two blocks. A comparison of the pictorial with the three-view drawing demonstrates the advantages of the pictorial in visualizing the object—the pictorial is much more effective. The three-view drawing is employed for the precise definition of size and location of all of the features of a component or structure. As such it is dimensioned so that it can be used in the shop for manufacturing. The pictorial is not usually dimensioned because it is not a substitute for a detailed three-view drawing.

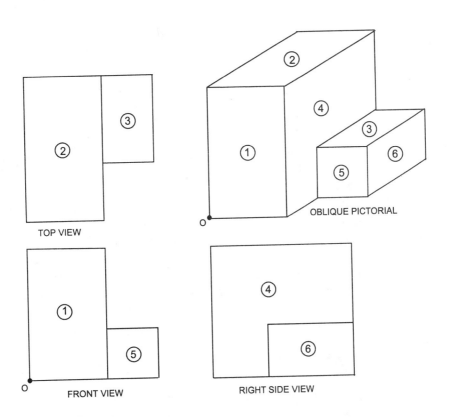

Fig. 10.8 A three-view drawing of a connected pair of rectangular blocks together with an oblique pictorial of the same object.

For two final examples of preparing oblique pictorials, consider the drawing of a cylinder and a block with a circular hole as shown in Fig. 10.9. It is easy to draw a cylinder or a circle on an oblique pictorial providing the required circles are placed on either the front plane or any plane parallel to the front plane. Because both the width and height dimensions are true length and the x-y axes are orthogonal on the front view, the circle is not distorted into an ellipse. You can draw the circle with a compass or a circle template, which is a significant advantage.

To represent the cylinder shown in Fig. 10.9a, draw two circles centered at points A and B. Both points A and B lie on the z (receding) axis, which has been drawn at a 45° angle to the x-axis. The circles are drawn with construction lines using full-scale diameters. The spacing of the two circles along the z-axis is ¾ scale relative to the length of the cylinder. Draw two lines parallel to the z-axis tangent to the circles for the sides of the cylinder. Draw the front circle with a heavy line, and darken the visible portion of the back circle. The resulting pictorial shows a somewhat distorted cylinder. The distortion is due to the fact that the back circle has been drawn with the same diameter as the front circle. When observing an actual right circular cylinder, the back circle would "appear" to be smaller than the front circle. The apparent difference in the size of features located in different planes will be addressed in the next section on perspective drawing.

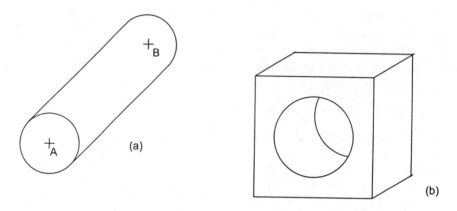

Fig. 10.9 Examples of oblique pictorials.

The final example of oblique pictorial, presented in Fig. 10.9b, shows a simple rectangular block with a central hole. The block has been drawn so that the circle of the hole is observed in its entirety in the front view. The question is how to draw the circle in the back plane? First draw the square locating the back plane in the pictorial with light construction lines. Then find the center of this square and draw the second circle with the same diameter as the front circle. Note, the two circles intersect and only a portion of the circle on the back plane is visible. Darken the construction lines on the back circle over that part of the arc that is visible through the hole. The resulting pictorial is an effective three-dimensional drawing showing the appearance of a hole through a rectangular block. Artists would critique the drawing, because it too distorts the visual image of the object. The scale used for both the back and front planes are the same producing distortion of the image. This distortion is usually acceptable in engineering drawing, but better pictorials are possible with perspective drawings.

10.4 PERSPECTIVE DRAWINGS

Perspective drawings are pictorials that represent observations similar to those made with our eyes or a camera. This method of illustration is critical to the success of an artist making either sketches or paintings. Engineers also use perspective drawings particularly when preparing visuals for those who are not trained to read more conventional three-view drawings. The drawing instruments used are the same as those described previously except for adding a thumbtack and a piece of string to the list.

There is one very significant difference between perspective drawings and isometric or oblique drawings. In both isometric and oblique drawings, the lines defining the edges of an object are parallel to the axes; however, in perspective drawing, the lines defining the receding edges are not parallel. Drawing parallel lines distorts the drawing, because parallel lines appear to converge as they recede in space. The best illustration of this fact is a pair of railroad tracks, shown in Fig. 10.10. Looking down the tracks, you see that the tracks converge to a point. Also the railroad ties and the trees and poles lining the track appear to become shorter in the distant view. Everyone knows that the tracks are parallel. What is going on? To understand what is happening, it is necessary to define four terms used in describing prospective drawing.

1. A picture plane is the surface (i. e. the sheet of paper) of the pictorial. The edges of the paper represent the window through which you "see" a three-dimensional object that is the subject of the perspective drawing.
2. A horizon line divides the sky and the land (or the sea) if you are outdoors. The horizon line is at the level of your eyes and will change with your elevation. In a room the true horizon cannot be located because the walls of the room block the view. In this case, it is assumed that the horizon line is at eye level.
3. A viewing point and direction of view depends on the location of eyes relative to the object. You can look directly at the object, from left to right, right to left, downward, upward, etc. What you see changes markedly depending on these parameters. Look at an object from a window, and change where you stand and the direction of your view. Does the view of the object change?
4. A vanishing point is where parallel lines converge to a point as they recede into the distance. You can clearly identify the vanishing point in Fig. 10.10 where the tracks appear to meet.

Fig. 10.10 Railroad tracks illustrate the visual effect of converging parallel lines
and foreshortening of distant lines.

One Point Perspective Drawings

Depending on the view, an object can be represented with a one, two or three-point perspective. Let's start with the simple one-point perspective and illustrate the approach by drawing a simple rectangular block. In one-point perspective, you place the front view in the picture plane and show its true width and height as illustrated in Fig. 10.11. Then a construction line is drawn to represent the horizon. The location of this line depends on the viewing point and the viewing direction. In Fig. 10.11, you are viewing the block straight-on (not from the right or the left); however, you are above the block. Your eyes look downward and observe the top surface of the block. The elevation of your eyes relative to the top of the block is taken into account by raising the horizon line. Next, locate the vanishing point on the horizon line at the center point behind the front view, because you are looking straight-on at the block. Draw construction lines from the vanishing point to the top corners of the block in the front view as shown in Fig. 10.11. The back edge on the top view is drawn parallel to the front top edge to establish the depth of the block. Note that the back edge is much shorter in length than the front edge. The shortening of the lines on the recessed planes gives an illusion of the third dimension. The edges at the side are darkened to complete the pictorial. Again these edges are converging giving an illusion of depth on the two-dimensional sheet of drawing paper.

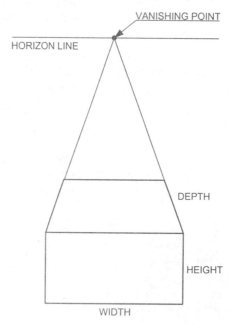

Fig. 10.11 An example of a pictorial drawing of a rectangular block with one-point perspective.

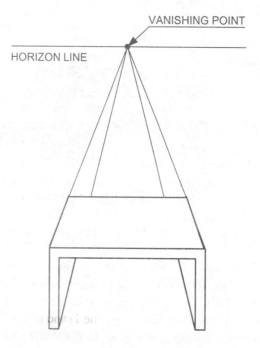

Fig. 10.12 A one-point perspective drawing of a coffee table.

Another example of one-point perspective is the drawing of a coffee table presented in Fig. 10.12. Again the front of the table is drawn to scale in the picture plane. The width and height dimensions are true lengths. Coffee tables are low, so when standing it is necessary to look downward to view the table. In this example, you are centered relative to the table and looking downward at its top surface. To develop a perspective view with this direction of viewing, draw a horizon line at eye level aligning the vanishing point with the center of the table. Next, draw light construction lines from the top outside corners of the table to the vanishing point. Also draw construction lines from the lower inside corners of the table legs as shown in Fig. 10.12. These construction lines form two triangles. The larger of these two triangles is used to provide the length of the back top edge of the table. Next, draw the edges of the tabletop to complete the perspective rendering of the top view. The smaller of the two triangles is used as a guide in drawing the bottom edges of the legs, which are visible under the table.

In a normal perspective drawing, the construction lines, the vanishing point and the horizon line would be erased. These drawing aids have not been erased in Fig. 10.12 because they were used to illustrate the procedure followed in making a one-point perspective drawing.

Two-point Perspective Drawings

The one-point perspective drawing is useful when viewing an object straight-on with its front view shown in the picture plane. However, if the object is rotated so that neither the front or side view is in the picture plane, as illustrated in Fig. 10.13, a two-point perspective is required.

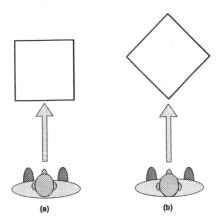

Fig. 10.13 Directions of viewing control the number of vanishing points used in perspective drawing.
a.   Straight-on—one-point perspective.   b.   Angle view—two-point perspective.

Again consider a rectangular block to illustrate the approach followed in drawing a two-point perspective pictorial. From Fig. 10.13, it is evident that only one vertical edge of the block lies in the picture plane. You start with this fact and proceed step by step to illustrate the block with a two-point perspective drawing.

- Draw a vertical line, as indicated in Fig. 10.14, to establish the edge between the left-side view and the front view.
- Draw the horizon line to reflect the fact that you are standing above the block looking downward onto the top surface.
- Place two vanishing points on the horizon line. The vanishing point (VP - R) has been located to the right of the vertical line farther from the vertical line than the vanishing point on the left-side (VP - L), because the block is being viewed at a slight angle from the right toward the left.
- Measure the true length of the vertical line (line 1) and label its ends with the letters A and B.
- Draw four construction lines connecting points A and B with VP - R and VP - L.

- Draw vertical lines (2 and 3) to locate the back edges of the left side and the front of the block. The ends of these lines, located by the construction lines, are labeled (C, D) and (E, F).
- Draw top edge lines (4, 5) by connecting points F, A and D.
- Draw bottom edge lines (6, 7) by connecting points E, B and C.
- Draw construction lines from point F to VP - R and from point D to VP – L. Then locate point G at the intersection of these two lines.
- Draw edge lines (8, 9) by connecting points F, G, and D.

The two-point perspective drawing of the rectangular block is complete as shown in Fig. 10.14. The procedure may seem long and tedious, when it is described in a sequence of steps, but it is not difficult. When you understand the basic concepts of establishing the true length of the vertical line in the picture plane and the vanishing points on the horizon line, the remaining steps are routine.

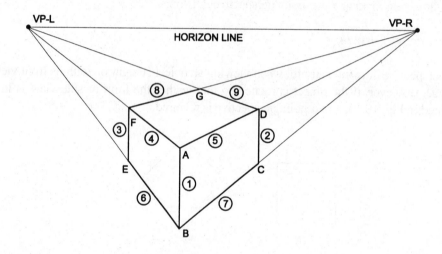

Fig. 10.14 A rectangular block represented with a two-point perspective drawing.

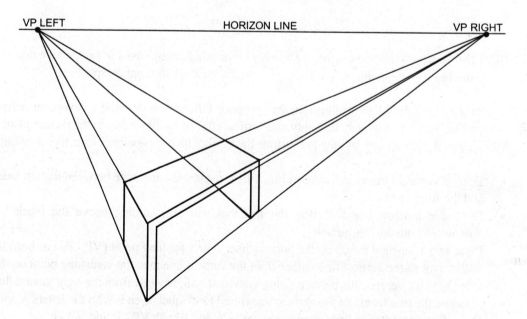

Fig. 10.15 A two-point perspective drawing of a coffee table.

To complete the discussion of the two-point perspective, let's again consider the coffee table. This time, a two-point perspective drawing will be used to illustrate the table. Begin by placing the vertical edge of a table leg in the picture plane as indicated in Fig. 10.15. The edge of the leg is shown in true length on the picture plane. Establish the horizon line and the vanishing points to provide the direction of the view that you are showing. Draw construction lines from the ends of the vertical line to the vanishing points. Vertical lines are then drawn to show the width and the depth of the table. The location of these vertical lines is not established by measurement, because both the width and depth dimensions are not true length in this two-point perspective. Place the vertical lines in a location that maintains—**to the eye**—the correct proportions of the table.

When the vertical lines that bound the left-side view and the front view are drawn, the remainder of the drawing is easy to complete. Construction lines are drawn from the ends of these vertical lines to the vanishing points. The construction lines outline the table. It is only necessary to darken those segments of the construction lines that define the visible edges of the table.

Three-point Perspective Drawings

Three-point perspective drawings are used when the object being illustrated is very tall. Architects drawing a city view with skyscrapers would use three-point perspectives and taper the building as it extends into the sky. Engineers usually deal with smaller objects that usually can be represented in pictorials with either one or two-point perspective drawings. For this reason, the methods used in preparing three-point perspectives will not be described. However, if you are interested in learning much more about perspective drawing, an excellent book by Powell [2] is recommended.

10.5 SHADING AND SHADOWS

Adding shading and shadows to the pictorial drawings makes them appear more realistic. Let's first distinguish between shading and shadowing. If you light an object, some surfaces are exposed to this light and other surfaces are in the shade. Consequently, there is a difference in the intensity of light reflected from these surfaces. Those in the shade are darkened to some degree. When an opaque object blocks the light, a shadow is produced. The shadow occurring on the floor plane is shown as a dark area in the drawing. An example of shading and shadowing of a right circular cylinder is shown in Fig. 10.16. In this example, parallel light rays are illuminating the cylinder from the upper left. (They are included on the drawing only to show the logic for determining the shade and shadow regions). The right half of the cylinder is in the shade and is darkened slightly. A shadow is cast on the plane of the floor upon which the cylinder rests. The depth of the shadow is the same as the depth of the cylinder on the isometric pictorial. The length of the shadow is dependent on the direction of the parallel rays of light. Extending the lines representing the light rays defines the edge of the shadow.

Let's consider another example of shading and shadowing in a two-point perspective drawing of a cube. The drawing, presented in Fig. 10.17, is similar to that shown in Fig. 10.14, so you may assume that the cube can be drawn with a two-point perspective pictorial. In Fig. 10.17, the construction technique is demonstrated for determining the exact size of the shadow when the light is coming from a point source.

Suppose you have completed the two-point perspective drawing of the cube and are ready to shade and shadow the drawing. The process is first to select the location of the light source. It is your choice, but you should place the light source well above the horizon either to the left or the right of the cube. Second, you must select the vanishing point for the shadow. Again it is your choice as long as it is directly below the light source and in the ground (floor) plane. The reasons for these two constraints are evident. The shadow must vanish when the light source is directly overhead, and the shadow must always lie on the ground (floor) plane. In Fig. 10.17, the vanishing point of the shadow was placed on the horizon line; however, it could have been located anywhere on the ground plane under the light source.

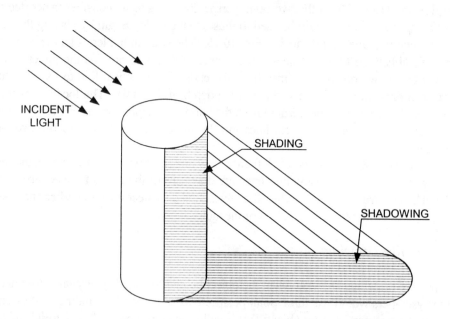

Fig. 10.16 Shading and shadowing of an isometric pictorial of a right circular cylinder.

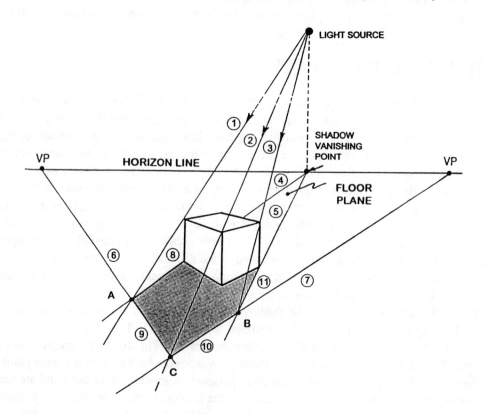

Fig. 10.17 Technique for determining shadow outline on a two-point perspective drawing.

Consider a step-by-step procedure for locating the shadow outline shown in Fig. 10.17:

1.   Draw the light rays (1, 2, and 3) from the source through the three forward corners on the top of the cube.

2. Draw two construction lines (4 and 5) from the vanishing point of the shadow through the two bottom-side corners.
3. Locate points A and B at the intersections of the lines (1 and 4) and (3 and 5).
4. Draw lines 6 and 7 from the left and right vanishing points through points A and B.
5. The intersection of lines (6 and 7) locates point C.
6. Darken segments of the construction lines (8, 9, 10 and 11) to give the shadow outline.
7. Erase all of the construction lines and darken the shadow region within the outline.

The shadow has been added to the drawing, but the front and left-side views of the cube are in the shade. To complete the pictorial, these two sides should be darkened slightly indicating they are in a shadow.

10.6 SUMMARY

Three different pictorial drawings—isometric, oblique, and perspective—have been introduced. Both isometric and oblique pictorials are widely used in engineering to present three-dimension illustrations on a two-dimensional format (a sheet of drawing paper or a computer monitor). The isometric and oblique pictorials are easy to draw because lines are parallel, and most, if not all, of the measurements are to scale. However, exact scaling of dimensions and parallel lines tend to distort these three-dimensional illustrations. The eye or camera "sees" dimensions with different scales, and parallel lines as parallel only if the view is on the picture plane. If the view is on any other plane behind the picture plane, the dimensions are foreshortened. If the parallel lines are on views with receding axes, they tend to converge. These distortions are ignored in engineering drawings when preparing isometric or oblique pictorials.

Perspective drawings provide a more realistic representation of three-dimensional objects. Artists almost always use perspective concepts in their sketches or paintings. In engineering, one-point or two-point perspectives are usually employed to produce realistic three-dimensional drawings that are used in catalogs and maintenance manuals. The techniques of perspective drawing require understanding of four quantities: the picture plane, the horizon line, vanishing point or points, and viewing direction. The techniques for drawing perspectives with converging lines and foreshortened dimensions follow directly from the definitions of these four quantities.

Shading and shadowing is an added technique for making pictorial drawings even more realistic. Methods for determining the outline of shadows either from point light sources or collimated light sources that produce parallel light rays have been described.

REFERENCES

1. Franks, G. Pencil Drawing, Walter Foster Publishing, Laguna Hills, CA 1988.
2. Powell, W. F. Perspective, Walter Foster Publishing, Laguna Hills, CA 1989.
3. Earle, J. H., Engineering Design Graphics, Addison Wesley, Reading, MA 1983.
4. Bertoline, G. R., and E. N. Wiebe, Fundamentals of Graphics Communication, 3rd Ed., McGraw Hill New York, NY, 2002
5. Lockhart, S. D. and C. M. Johnson, Engineering Design Communication, Prentice Hall, New York, NY, 1999.
6. Jensen, C. H. and J. D. Helsel, Engineering Drawing and Design, Glencoe McGraw Hill, New York, NY, 1997.
7. Sorby, S. A., Manner, K. J., Baartmans, B. J. and S. S. Sorby, 3-D Visualization for Engineering Graphics, Prentice Hall, New York, NY, 1998
8. Giesecke, F. E. et al, Engineering Graphics, 7th Ed., Prentice Hall, New York, NY, 2001.

EXERCISES

10.1 What are the three types of pictorial drawings? Why are pictorial drawings used in engineering applications?

10.2 Prepare an isometric pictorial drawing of a rectangular block 35 mm wide by 50 mm deep by 25 mm high. What scale should you use in preparing this drawing? Why?

10.3 Prepare an isometric pictorial drawing of a right circular cylinder 25 mm in diameter by 50 mm long. What scale should you use in preparing this drawing? Why?

10.4 Prepare an oblique pictorial drawing of a rectangular block 20 mm wide by 50 mm deep by 100 mm high. What scale should you use in preparing this drawing? Why?

10.5 Prepare an oblique pictorial drawing of a right circular cylinder 20 mm in diameter by 50 mm long. What scale should you use in preparing this drawing? Why?

10.6 Prepare an isometric pictorial drawing of a rectangular block with a centrally located circular through hole. State the dimensions of the block and the diameter and orientation of the hole.

10.7 Prepare an oblique pictorial drawing of a rectangular block with a centrally located circular through hole. State the dimensions of the block and the diameter and orientation of the hole.

10.8 Prepare an isometric pictorial drawing of a component for the product that your team is developing.

10.9 Prepare an oblique pictorial drawing of a component for the product that your team is developing.

10.10 Prepare a prospective pictorial drawing of the building that houses the Dean's offices. View the building from one of its front corners.

10.11 Draw a pictorial illustration of a sphere with shading and shadowing.

PART IV

SOFTWARE APPLICATIONS

CHAPTER 11

MICROSOFT EXCEL

11.1 INTRODUCTION

Microsoft® EXCEL, a spreadsheet program, is an extremely important tool in engineering because it is useful in many different applications. Techniques for three of these applications will be described including—preparing a parts list, performing calculations, and drawing graphs. There are several different spreadsheet programs on the market, namely EXCEL® by Microsoft, QUATTRO PRO® by Corel, and LOTUS 1 • 2 • 3® by IBM. All of these programs have a similar format and they all have essentially the same capabilities. If you learn to use one program, it is relatively easy to change to a different program with a modest investment in time. EXCEL has been selected because it is the most popular, many universities have adopted it and you will probably have access to it on the computer net during your tenure in college.

The purpose of this chapter is to show you how to navigate on the EXCEL spreadsheet, to make tables, perform calculations, prepare charts and graphs, print the output and save a file. You should recognize that these objectives are limited. The content in this chapter is focused on developing your entry-level skills. Hopefully, you will find the spreadsheet tool important enough to develop a higher skill level with independent study. More complete and detailed treatments are given in the references listed at the end of this chapter.

11.2 THE SCREEN

You load EXCEL from WINDOWS by clicking on the EXCEL icon and the program opens with the spreadsheet display shown in Fig. 11.1. Let's explore this display. At the top you will note six rows, and then below in the center is a big table that is the spreadsheet proper, and finally two more rows at the bottom of the display.

The Six Top Rows or Bars

The top most row confirms that you have opened the Microsoft EXCEL program. The three buttons to the far right side of the title row are common to all Microsoft® programs.

The left hand button with the dash symbol places the program aside (but does not close it) so that you can work in a different program. The middle button is a toggle permitting you to either expand the size of the display on the screen or to shrink it. The right button with the X symbol closes both EXCEL and the file upon which you are working. Do not use the X button if you intend to use EXCEL again during your current session. Instead, use the dash button and set the active program aside. Click on each of these three buttons until you understand their functions.

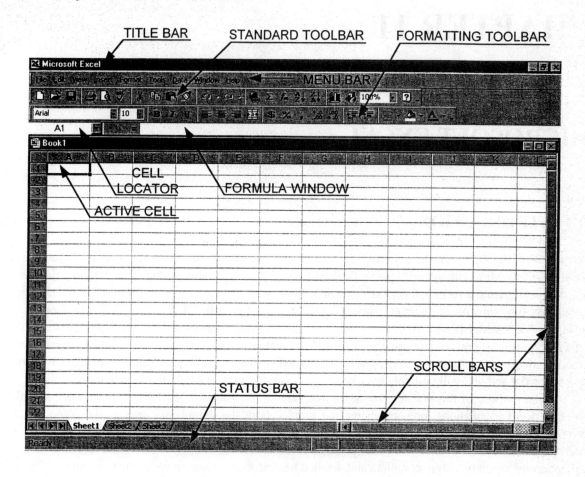

Fig. 11.1 The screen display for Microsoft EXCEL.

The Menu Bar

The second row from the top is called the menu bar; it lists nine different features included in the program beginning with File and ending with Help. The menu bar is similar to that found in other Microsoft® programs. Using your mouse, select one of the menu items and click. A pull-down menu is displayed. The items on each of these pull-down menus will not be listed because you will learn faster by clicking than by reading. The purpose of each of the menu entries is listed below:

1. File is to open or close, save, or print a file or to exit EXCEL.
2. Edit is to change entries that you have already made.
3. View is to change the size of the screen or to add toolbars, a header or footer, etc.
4. Insert is to add cells, rows and columns to the spreadsheet.
5. Format is to change the height of rows and the width of columns.
6. Tools are to check spelling, audit calculations and develop macros.
7. Data is to provide a means for sorting, filtering and forming data entries.
8. Window is to enable you to open a new window.
9. Help is to provide answers to questions that arise as you attempt to use the program.

The Standard Toolbar

The third row from the top exhibits the standard toolbar with 21 icons. The standard toolbar incorporates many icons common to other Microsoft® applications. The pictures on the icons give a clue as to the purpose of each button; however, if you are in doubt, use the mouse and point to a particular button. Do not click. Wait for a moment and a small label appears giving you the function of each button. Again, point and wait until you understand the function of each icon.

The Formatting Toolbar

The fourth row down is the formatting toolbar with 20 buttons. Again many of its icons are common to other Microsoft® applications. Icon labels on each button illustrate the purpose of each button, but if in doubt, you simply point to the button and a label will appear defining the function. You should try using the two buttons to the far left of the format toolbar to change the font type and its size.

The Cell Locator and Formula Toolbar

The cell locator and formula toolbar is found in the fifth row down from the top of the display. It provides information about the active cell location and its contents. The small window to the left defines the active cell location on the spreadsheet (A1 is the active cell when you first open a new book or file). The longer window on the right, called the formula bar, is activated when you make an entry into a cell. The numbers or letters that you have entered into the cell appears in this window. If you wish to edit a cell entry without completely retyping it, activate the cell in question by clicking on it, point with the mouse to place the cursor in the formula block, click and then edit the entry.

Note the = sign just to the left of the formula window. When you click on the = sign, EXCEL interprets the cell entry as a formula. The = sign effectively turns on the computational capabilities of the program.

The Title Bar

The sixth and last row from the top is the title bar. When opening a new file, the title Book 1 is displayed on the left side of this bar. When you save this file and rename it, the new name appears in the title bar replacing Book 1. On the far right side of the tile bar, you will find the three buttons for setting aside, changing the size of the display, and closing the file. These buttons refer to the file; whereas, the three buttons in the topmost row refer to the program.

The top six rows of the EXCEL display have been described, but it is necessary for you to develop an understanding of the menu items and the icons by pointing and clicking. Do not be concerned with the beeps from the computer—that is part of the learning process.

The Spreadsheet

Most of the screen is occupied by the spreadsheet, which is simply a large table with many columns and rows. The columns are labeled across the top with letters A, B, C,.. L and beyond. The rows are labeled down the left side with numbers 1, 2, 3, ….. 22 and beyond. Each small block, called a cell, is identified with its coordinates (e. g. in Fig. 11.1, the cell A1 is outlined with a border indicating it is active). The letter (column location) is placed before the number (row location). The cells are locations in a very large table (spreadsheet) where you enter numbers, text or formulas.

You may move about on the spreadsheet using either the mouse or the keyboard. Since the mouse usually permits the cells to be identified more rapidly, it is the preferred method for placing the cursor over a particular cell. When the pointer is on the spreadsheet, its location is identified with a large plus sign. You may use the mouse to point to any cell on the screen; however, if the cell of interest is not visible on the screen, the scroll buttons or bars located to right or below the spreadsheet are used to bring this cell into the field of view. When the large plus sign is pointed at the correct cell—click; the cell becomes active and is ready to receive the data you enter. You may also move the active cell by using the arrow keys, the tab key, the shift-tab keys, and the control-arrow keys. Try them and note how the active cell moves about the spreadsheet.

The spreadsheet is much larger than it looks on the screen, which shows only columns A through L and rows 1 through 22. (The number of columns and rows visible on the screen will depend on the size of the monitor and the zoom setting. Explore the number of rows on the spreadsheet by holding down the control and the arrow-down key. Note that the number of the last row is 65,536. Depressing the control and the arrow-right keys moves the active cell to the far right of the spreadsheet, and indicates a column heading of IV which is the 256th column. You have a total of (256)(65,536) = 16,777,216 cells on the spreadsheet. Your spreadsheet is huge—let's hope you never have to use all of the available cells!

The Bottom Two Rows

In the next to last row from the bottom of the spreadsheet, you will find three tabs identifying sheet numbers 1, 2 and 3. When opening EXCEL and starting a file, you immediately begin with a workbook for a given project. The program automatically establishes many sheets (pages) in your workbook, although only three are visible. If your project requires more than three worksheets, you may add more by clicking on Insert, found on the menu bar. Then from the drop-down listing you click on Worksheet to add another worksheet. You may enter data and perform calculations on several different worksheets in this file by clicking on the tabs to move from one sheet to another. If you right click on any of the buttons to the left of these tabs, a drop-down menu allows you to select an active sheet number.

The row at the very bottom of the spreadsheet is called the status bar. It provides a brief statement about the function of any button that is being activated and other information about the status of the program. In Fig. 11.1, the status message is "Ready" indicating that the EXCEL program is prepared to accept your data entry in cell A1.

11.3 PREPARING A PARTS LIST

Let's begin to learn techniques for using EXCEL by preparing a parts list for a mailroom scale. Load the program from **WINDOWS** by clicking the Start button, select Programs and then Microsoft Excel. The screen display should resemble that shown in Fig. 11.1, indicating that you are ready to start. It is recommended that you title the file before beginning to develop the parts list. Click on File and select Save As. The Save As dialog box page is displayed as shown in Fig. 11.2. Type in a suitable file name for the project (PARTS LIST) and click the save button. The file name is now displayed on the top row of the spreadsheet.

Begin the parts list by typing its title in cell A1. Since the title should be obvious and easy to read, use bold Arial font and increase its size to 16 point. Skip a row and position the cursor on cell A3 to begin organizing the section headings as shown below:

There are five section headings (columns) in the parts list:

      A. A number which refers to the part number; every unique part has its own part number.
      B. The name of the item.
      C. A description of the item or a drawing number defining the part.
      D. The quantity of parts that will be needed to build a single prototype.
      E. The unit price for the item.

When you type these column headings in cells A3, B3, C3, D3, and E3, it becomes clear that the columns are not the correct width to accommodate the text. The column widths are adjusted by pointing mouse to the line separating A and B on the column headings. When the pointer is positioned correctly, you will observe a short vertical line with arrows pointing to the left and right. Press the left mouse button and drag the column boundary either to the left or the right to adjust the width of column A. Move to the position of the cursor to a position on the column headings between column B and C; repeat the process to size the width of column B. Continue to adjust all of the column widths until they all are the proper width to best show the data included in the parts list.

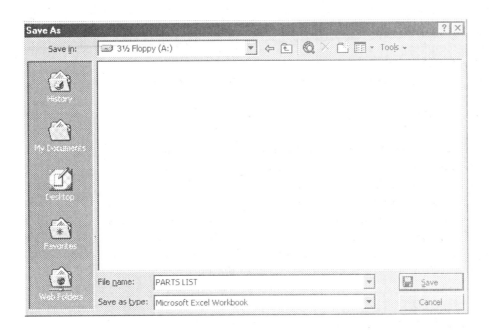

Fig. 11.2 Dialog box for naming a file and saving it.

Table 11.1

Parts list example

PARTS LIST FOR A SOLUTION DISPERSAL SYSTEM				
NUMBER	NAME OF ITEM	DESCRIPTION OR DRAWING NO.	QUANTITY	PRICE
1	STORAGE TANK	PLASTIC WATER BOTTLE 1 LITER SIZE	2	$ 0.10
2	TUBING	TYGON TUBE 1/4 ID, 3/8 OD, 1/16 IN. WALL	2 m	$ 1.25
3	FACE PLATE	PLASTIC SHEET WITH LABEL FOR SCALE	1	$ 1.15
4	SUPPORTS	1/4 X 1 WOOD STRIPS -------DWG. NO. 100-01	12 m	$ 3.25
5	BASE	1/2 X 4 X 4 PLYWOOD BASE -------DWG. NO. 100-02	1	$ 0.50
6	MOTOR	LEGO GEAR MOTOR B775225-017	2	$ 17.00
7	BEAMS	LEGO BEAMS	1	$ 1.00
8	PINION	LEGO 8 TOOTH SPUR GEAR	2	$ 1.00
9	RACK	LEGO RACK GEAR	1	$ 1.00
10	PLATES	LEGO PLATES	1	$ 1.00
11	BRADS	1/32 SHEET METAL --- DWG. 100-08	1 Box	$ 1.45
12	ADHESIVE	GENERAL PURPOSE STRUCTURAL ADHESIVE	1 Tube	$ 1.25

In this example parts list, 12 part numbers are shown. You may have more items because the number of parts depends on the complexity of the design. Type the necessary information for the parts list into the table that is provided by the spreadsheet. Your only concern in this process is that you have activated the correct cell before making an entry. If one of the entries is too long, you can change the column width at any time to accommodate the description. However, it is important to use brief descriptions because the parts list should fit a standard sized sheet of paper.

When you have completed the entries, work on the spreadsheet to improve its appearance. The part numbers, the quantities, and the prices have been centered within their respective columns. Center the entries by marking all of the cells in a given column (point and drag with the mouse and watch a gray background color fill the column). When the gray color covers the column entries you wish to mark, lift the left hand button on the mouse and move the pointer to the formatting toolbar and click on the centering button. On the column for price, the entries were initially made with the default format; however, prices are in terms of dollars. You may change the format of the column by marking it and clicking on the $ button located on the formatting toolbar to show the prices in terms of a currency.

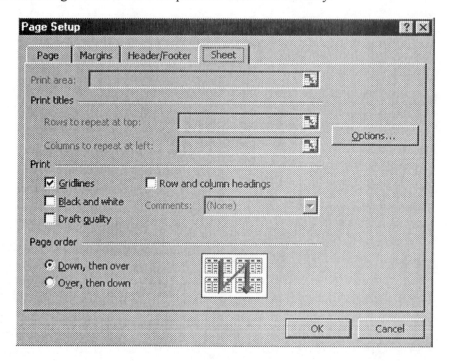

Fig. 11.3 Dialog box for page setup used to adjust margins, page layout and gridlines.

You are ready to print the parts list, but you should check to see if it fits the page, has the correct margins and is readable. Click on File then Print Preview, to examine the appearance of the spreadsheet when it is printed. To change its appearance, click on Setup button that is located on the bar at the top of the page preview screen. The dialog box that appears is presented in Fig. 11.3. From the four tabs available choose the sheet tab. Then point and click on the open square for gridlines so that the cell boundaries (gridlines) are visible when the spreadsheet is printed. Printing the cell gridlines makes the parts list much easier to read. Next, click on the page tab, and select either the portrait or landscape mode. The default setting is the portrait mode because it is more common, but if the spreadsheet is much wider than it is long, the landscape mode is preferred. You can also adjust the size and the quality (dots per inch) of the printing on this page. Finally, click on the margin tab and change the margins so that the spreadsheet is positioned properly on the page.

When you have made all of the necessary modifications, the spreadsheet should look professional (that means excellent). Next, click on the print button, and from the dialog box that appears (see Fig. 11.4) select the number of copies required and print the spreadsheet. Your parts list should compare favorably to the one shown in Table 11.1.

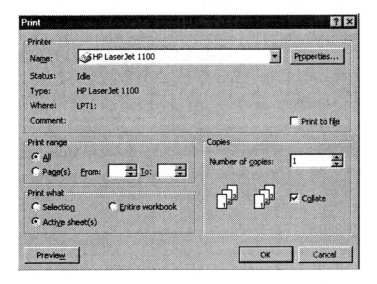

Fig. 11.4 Dialog box used in printing the spreadsheet.

11.4 PERFORMING CALCULATIONS

One of the most important advantages of any spreadsheet program is its capability to perform calculations. EXCEL performs not just one calculation, but a whole sheet full of them. Moreover, EXCEL performs the calculations almost instantaneously and without error. The results can be displayed in two different ways— as an accurate table or a suitable graph. In this section, two simple examples showing the computing power of EXCEL will be demonstrated. The results will be presented in tabular format.

The Rich Uncle

For the first example demonstrating the computational power of EXCEL, consider your rich uncle that was introduced in Chapter 8—Tables and Graphs. Recall that the accumulation of compound interest, which was earned from his initial investment of $5000, was previously described. The result of this analysis, presented in Table 11.2, is a record of the growth of the investment.

Let's recall the relation for the sum of the accumulation and the input data from the previous chapter where:

$$S = P(1 + i)^n = PM \qquad (11.1)$$

Uncle contributed the principal P = $5,000.00, and the interest rate, (i) for the six month compounding period was 5%. (The annual interest rate was 10%). The sum, S accumulated for a total of 80 compounding periods (40 years) is to be determined. Recall, n is the number of compounding periods and M is a multiplier dependent on n.

Load EXCEL and Save As—Compound Interest—to establish the file name. In Table 11.2, you title the spreadsheet and enter text describing the parameters that control the results in the first five rows (A1 to A5). Skip a row and define the headings for Periods, Multiplier and Sum in row 7. Initially, you define only three columns (A, B and C). Place the active cell at location A8 and begin by entering the number of the period starting with zero. It is possible to type all 40 entries into column A, but it is easier to fill in the entries automatically.

To permit EXCEL to do the work, position the cursor on the zero entry in cell A8 and drag the cursor down A column from cell A8 to cell A48 marking the location for the required number of cells for 40 entries. Next, click on Edit, select Fill, and then select Series. A dialog box is displayed as shown in Fig. 11.5 than enables you to select Column indicating that the input data (1 to 40) is filling entries into column A. You also select Linear because the series that you will use to fill column A is a linear series (i.e. 1, 2, 3, etc.). At the bottom of the dialog box, note two small windows—one for Step and the other for Stop. Enter 1 for the Step value and 40 for Stop value and then click on the OK button. The program fills all of the cells from A8 to A48 with numbers starting with 0 in cell A8 and increasing in steps of one until it stops at the number 40 in cell A48. The Fill command listed in the Edit menu item saves time when it is necessary to enter a long series of numbers.

Table 11.2
Accumulation over a forty-year period.

DETERMINING THE RUNNING SUM						
FOR AN INITIAL INVESTMENT OF $5000.00						
ANNUAL INTEREST RATE OF 10%						
COMPOUND INTEREST SEMI-ANNUALLY						
NO TAXES PAID						
Periods	Multiplier	Sum		Periods	Multiplier	Sum
0	1.0000	$ 5,000.00		40	7.0400	$ 35,199.94
1	1.0500	$ 5,250.00		41	7.3920	$ 36,959.94
2	1.1025	$ 5,512.50		42	7.7616	$ 38,807.94
3	1.1576	$ 5,788.13		43	8.1497	$ 40,748.33
4	1.2155	$ 6,077.53		44	8.5572	$ 42,785.75
5	1.2763	$ 6,381.41		45	8.9850	$ 44,925.04
6	1.3401	$ 6,700.48		46	9.4343	$ 47,171.29
7	1.4071	$ 7,035.50		47	9.9060	$ 49,529.86
8	1.4775	$ 7,387.28		48	10.4013	$ 52,006.35
9	1.5513	$ 7,756.64		49	10.9213	$ 54,606.67
10	1.6289	$ 8,144.47		50	11.4674	$ 57,337.00
11	1.7103	$ 8,551.70		51	12.0408	$ 60,203.85
12	1.7959	$ 8,979.28		52	12.6428	$ 63,214.04
13	1.8856	$ 9,428.25		53	13.2749	$ 66,374.74
14	1.9799	$ 9,899.66		54	13.9387	$ 69,693.48
15	2.0789	$ 10,394.64		55	14.6356	$ 73,178.15
16	2.1829	$ 10,914.37		56	15.3674	$ 76,837.06
17	2.2920	$ 11,460.09		57	16.1358	$ 80,678.92
18	2.4066	$ 12,033.10		58	16.9426	$ 84,712.86
19	2.5270	$ 12,634.75		59	17.7897	$ 88,948.50
20	2.6533	$ 13,266.49		60	18.6792	$ 93,395.93
21	2.7860	$ 13,929.81		61	19.6131	$ 98,065.73
22	2.9253	$ 14,626.30		62	20.5938	$ 102,969.01
23	3.0715	$ 15,357.62		63	21.6235	$ 108,117.46
24	3.2251	$ 16,125.50		64	22.7047	$ 113,523.34
25	3.3864	$ 16,931.77		65	23.8399	$ 119,199.50
26	3.5557	$ 17,778.36		66	25.0319	$ 125,159.48
27	3.7335	$ 18,667.28		67	26.2835	$ 131,417.45
28	3.9201	$ 19,600.65		68	27.5977	$ 137,988.32
29	4.1161	$ 20,580.68		69	28.9775	$ 144,887.74
30	4.3219	$ 21,609.71		70	30.4264	$ 152,132.13
31	4.5380	$ 22,690.20		71	31.9477	$ 159,738.73
32	4.7649	$ 23,824.71		72	33.5451	$ 167,725.67
33	5.0032	$ 25,015.94		73	35.2224	$ 176,111.95
34	5.2533	$ 26,266.74		74	36.9835	$ 184,917.55
35	5.5160	$ 27,580.08		75	38.8327	$ 194,163.43
36	5.7918	$ 28,959.08		76	40.7743	$ 203,871.60
37	6.0814	$ 30,407.03		77	42.8130	$ 214,065.18
38	6.3855	$ 31,927.39		78	44.9537	$ 224,768.44
39	6.7048	$ 33,523.76		79	47.2014	$ 236,006.86
40	7.0400	$ 35,199.94		80	49.5614	$ 247,807.21

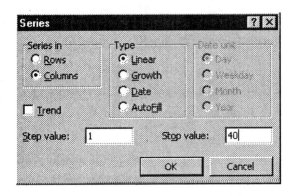

Fig. 11.5 The dialog box for filling a column or row with a series of numbers.

Next, let's locate column B and activate cell B8. You are to calculate the multiplier, M in this column. The multiplier, M is a number that increases with the number of the compounding periods. It is determined from Eq. (11.1) as:

$$M = (1 + i)^n = (1.05)^n \qquad\qquad (11.2)$$

With cell B8 active, you strike the = key, which notifies EXCEL that a formula will be placed in the active cell. Then type the formula in cell B8, which in this case the entry is =(1.05)^A8. You use the cell address A8 because it contains the value of n for the column of calculations being considered. You use the ^ symbol to indicate that you are raising (1.05) to a power. After you have entered and carefully checked the formula strike the enter key and examine cell B8. It should read 1, because the entry =(1.05)^A8 = $(1.05)^n$ = $(1.05)^0 = 1$.

You have successfully employed EXCEL to compute the first multiplier M. Let's determine the other 40 multipliers by copying the first one. Activate B8, and then click the Copy button on the toolbar. You have just stored the formula =(1.05)^A8 in temporary memory; it will be stored there until it is replaced the next time you use the Copy command. Next, mark (select) the B column starting with B9 and continuing to B48 by dragging the cursor down the column. Check that the region is blue where you want to Copy the multiplier term and then click on the Paste button. The cells B8 to B48 show the multiplier M superimposed on the blue background color of the column. Clicking on any cell outside this region eliminates the blue background. In performing the calculations down the B column when using the Copy—Paste commands, EXCEL modified the formula =(1.05)^A8. To check the actions of the program, activate cell B9—note the formula has changed to =(1.05)^A9. EXCEL modifies the formula to accommodate the changes in n as you move down the column. In this instance, it was essential for EXCEL to make this modification. However, in the next example, this type of modification leads to errors. To avoid errors in some extended calculations, the technique necessary to control the way EXCEL modifies the formulas will be shown. Formula control with the Copy—Paste command sequence will be demonstrated more completely in an example presented later in this chapter.

Inspecting the results for the multiplier M in column B, it is noted that the number of decimals is not consistent from cell to cell. Mark all of the cells in column B, click on either one of the two decimal buttons on the formatting toolbar, and observe the number of decimals change. Click on one or the other of these buttons until you display the multiplier with four digits after the decimal point.

To determine the sum, S that is accumulated with time, it is necessary to multiply the multiplier M by the initial investment of P = 5000. You select cell C8 and type =5000*B8. The symbol * is used to indicate multiplication. When you strike the enter key, the entry in cell C8 changes from =5000*B8 to 5000. EXCEL noted that cell B8 contained the number 1, performed the calculation (5000)(1), and displayed the result in cell C8. When you wish to incorporate a cell address into a formula, it can be typed or you can point to the cell with the correct address. After pointing at the cell, click the left mouse button to enter its contents into the formula. In the point and click approach, one of the math symbols (+, −, *, /, ^) **must precede the point and click process.** You activate cell C8 then Copy and Paste this result in the column of cells from C9 to C48. This action computes the sum accumulated at each interval during the 20-year period of the investment (40 interest periods).

Review the quantity in cell C8, and note that the result 5000 is not in the correct format. It is more appropriate to show the results as a currency indicating the sum accumulated from Uncle's investment is in terms of dollars. To change the format:

1. Mark the entire column from C8 to C48 by dragging the cursor down the column.
2. Check that the blue background covers only this region.
3. Point and click on the $ button located on the formatting bar.
4. Clear the blue background by clicking on any cell outside the blue region.

When you converted the C column into a currency format another problem occurred. The results for several cells changed into #######. This symbol indicates that the column width is not sufficient to display the result. To widen column C, you point to the column-heading row between columns C and D. When a line with the double arrows replaces the pointer, depress the left mouse button and drag to the right to widen column C. With a wider column the result in cell C8 reads $5,000.00.

You have a result for the sum S in cell C8. Let's arrange for EXCEL to complete the calculations. Activate cell C8 because it has the required formula, and click on the Copy button on the toolbar. Next mark the cells C9 through C48 to locate where the formula is to be applied. The blue background verifies that the region is marked correctly. Next, click on the Paste button and EXCEL applies the formula to each of the cells marked in the C column calculating the results in each cell. Scroll down the column and check to ascertain of the results are displayed correctly. If you encounter #######, it is necessary to widen the column. The result in cell C48 should read $35,199.94.

You have computed the accumulated sum S for 40 periods, but S was to be determined for 80 periods. Why did we stop at n = 40? The calculation was divided into two parts so that the output, when printed, would fit on one page. It is easy to extend the calculation to n = 80 by using the Copy command.

To continue your work, mark the block with corners at A7 and C8 and click the Copy button. You now have the headings and the initial formula in temporary storage (memory). Move the cursor and activate cell E7. You skip column D to provide space in the table between the two lists of results. After activating cell E7, click on the Paste button and the results from the A7 to C8 block are copied (with modifications for the shift in columns) into the E7 to G8 block. Look at cell E8 and note that the period shown is n = 0. Let's change that value to 40 by editing the number displayed on the formula toolbar. When cell E8 is changed to 40, the results in E9 and G9 are identical with previous results for the period n = 40. This fact implies that the program in EXCEL has adjusted the formulas to account for the fact that you have changed the columns in which the calculations are made. Activate cells F8 and G8 and compare the formulas in them with those in cells B8 and C8. Have you noted the changes automatically made by EXCEL as you copied formulas from one column to another?

To complete the table, activate cell E8 and fill in the entries for n = 40 to 80. You again use the automatic number generator in EXCEL. Click on Edit, select Fill, and then Series. On the dialog box, click on Column, Linear, and indicate a Step value of 1 with a Stop value of 80. Click on OK and then scroll down the spreadsheet to make certain that the correct entries have been generated for column E. Next, mark cells F8 and G8 and Copy them. Mark the F8 – G48 block and Paste the formulas from F8 and G8 into this block. Clear the blue background and scroll down the G column. You will probably note the symbol ####### in many of the cells. Widen column G until all of the results are shown in the currency format. As a check, the result $247,807.21 should appear in cell G48.

You are now ready to print the table showing all of the results. Click on File, then on Print Review. The screen shows the spreadsheet as it will appear when printed. You note that it could be improved with gridlines and a larger left hand margin. To show the gridlines, click on File then the Page Set-up button.

From the dialog box, which appears on the screen, select the sheet tab? Click on the gridlines square and note that a check appears indicating the gridlines will be printed. To adjust the margins, click on the margin tab. Then simply point, click and drag the left margin line until the table is positioned correctly in the sheet shown in the print review screen. Next, click on the OK button to return to the preview of the spreadsheet. Finally, click on the Print button to print the spreadsheet.

You will return to this spreadsheet later when preparing a graph of the results.

Strain on a Simply Supported Beam

In many engineering applications it is necessary to determine the strain developed when a beam is subjected to transverse loads. Beams are long slender members that serve to carry loads applied in a direction perpendicular to their axis. You have seen many beams when driving under bridges that support highway overpasses. When a load is applied in the center of a beam and its two ends are simply supported, the relationship for the maximum strain developed at the load point on the beam is given by the formula shown below:

$$\varepsilon = 3\ W_u\ S/(2bh^2\ E) \tag{11.3}$$

where: W_u is the load (weight) in pound or newton.
S is the span of the beam in inch or meter.
b is the depth of the section of the beam in inch or meter.
h is the height of the section of the beam in inch or meter.
E is the elastic modulus in psi or Pascal.

Let's program this relation in EXCEL to determine the strain induced in the beam at a location under the load point. In examining Eq. (11.3), note that the strain ε depends on five variables—S, b, h, E and W_u. For this example, consider a very small beam and fix three of these variables at the values shown below:

$$S = 4 \text{ in.} \qquad b = \tfrac{1}{4} \text{ in.} \qquad E = 10.6 \times 10^6 \text{ psi} \qquad \text{An aluminum alloy}$$

Let's treat the weight, W_u and the height, h as variables, and explore solution space for the strain, ε. Consider the load, W_u increasing from zero to a maximum of 4 lb in steps of 0.25 lb. Also, consider the height h of the cross-section of the beam varying from 0.01 in. to 0.07 in. in steps of 0.01 in. To determine the strain, display the variable W_u in column A, and the other variable, height, along the 8th row. This arrangement of the spreadsheet is shown in Table 11.3.

The ranges for the load and the height have been selected to cover the region of interest in designing a sensing beam. The sensing beam is for a mailroom scale intended to weigh letters and small packages[1]. The maximum weight in the specifications for this scale is 4 lb. Also, when the prototype of the scale is built, it will be calibrated using a number of small weights to verify the scale's accuracy. The results of this spreadsheet analysis are essential in verifying the calibration.

In the spreadsheet illustrated in Table 11.3, you enter the data for the height of the beam in row 8, columns B to H. In rows 1 to 4, you provide the title of the spreadsheet and information related to the analysis of the strain developed in the beam. In row 6, the equation used in determining the strain is given. Both rows 5 and 7 are blank to provide space above and below the equation. In row 9, you enter the heading for W_u and its units and repeat the data for the height of the beam together with its units. The spreadsheet is now organized with an arrangement that displays the two variables and their units of measure.

---

[1] This product is described in Introduction to Engineering Design: Book 3, Postal Scales, College House Ent., 1998.

Table 11.3
Strain in a simply supported beam

DETERMINE STRAIN							
FOR A SIMPLY SUPPORTED BEAM							
SPAN S =4 in., DEPTH b = 1/4 in., ELASTIC MODULUS E = 10.6 x 10^6							
WEIGHT Wu AND HEIGHT h ARE VARIABLES							
STRAIN = 3WuS/(2bh^2E)							
HEIGHT	0.01	0.02	0.03	0.04	0.05	0.06	0.07
Wu (lb.)	h = 0.01 in.	h = 0.02 in.	h = 0.03 in.	h = 0.04 in.	h = 0.05 in.	h = 0.06 in.	h = 0.07 in.
0.00	0.000000	0.000000	0.000000	0.000000	0.000000	0.000000	0.000000
0.25	0.005660	0.001415	0.000629	0.000354	0.000226	0.000157	0.000116
0.50	0.011321	0.002830	0.001258	0.000708	0.000453	0.000314	0.000231
0.75	0.016981	0.004245	0.001887	0.001061	0.000679	0.000472	0.000347
1.00	0.022642	0.005660	0.002516	0.001415	0.000906	0.000629	0.000462
1.25	0.028302	0.007075	0.003145	0.001769	0.001132	0.000786	0.000578
1.50	0.033962	0.008491	0.003774	0.002123	0.001358	0.000943	0.000693
1.75	0.039623	0.009906	0.004403	0.002476	0.001585	0.001101	0.000809
2.00	0.045283	0.011321	0.005031	0.002830	0.001811	0.001258	0.000924
2.25	0.050943	0.012736	0.005660	0.003184	0.002038	0.001415	0.001040
2.50	0.056604	0.014151	0.006289	0.003538	0.002264	0.001572	0.001155
2.75	0.062264	0.015566	0.006918	0.003892	0.002491	0.001730	0.001271
3.00	0.067925	0.016981	0.007547	0.004245	0.002717	0.001887	0.001386
3.25	0.073585	0.018396	0.008176	0.004599	0.002943	0.002044	0.001502
3.50	0.079245	0.019811	0.008805	0.004953	0.003170	0.002201	0.001617
3.75	0.084906	0.021226	0.009434	0.005307	0.003396	0.002358	0.001733
4.00	0.090566	0.022642	0.010063	0.005660	0.003623	0.002516	0.001848

To begin the programming of the equation for the strain, let's fill column A with information about the weight W_u. Type zero in cell A10 and keep this cell active. Mark column A by clicking and dragging from Cell A10 to cell A26. Then click Edit on the menu bar and select Fill then Series. On the series dialog box, and select Column, Linear, and then enter a Step value of 0.25 and Stop value of 4. Finally, click the OK button and check to determine if the numbers automatically entered into column A are correct.

Next, activate the B10 cell and type the formula for the strain as:

$$=3*A10*4/(2*(1/4)*10.6*10^6*(B8)^2) \qquad (a)$$

This entry is correct for cell B10, but it will cause significant difficulties later when we try to use the Copy and Paste commands to extend the calculation of strain to other cells in the spreadsheet. Let's proceed by copying the formula in cell B10 into cells B11 to B26. To copy the equation into other cells in column B, position the cursor on cell B10 and click on Copy. Then mark the column from B11 to B26. Next, click on Paste. When you examine the results, they are clearly in error. To troubleshoot the problem, activate cell B11 and examine the formula displayed in the formula bar. You observe the following relation:

$$=3*A11*4/(2*(1/4)*10.6*10^6*(B9)^2) \qquad (b)$$

Unfortunately this relation is not correct. In the Copy-Paste operation, EXCEL modified this formula indexing the entries for row locations in both columns A and B by 1. This indexing was correct for the entries in column A, but the indexing produces an error in the entries in column B. Instead of indexing the entries from column B, you must use the data in cell B8 for all the calculations in column B. To prevent unwanted indexing, modify the formula in cell B10 to read:

$$=3*A10*4/(2*(1/4)*10.6*10^6*(B\$8)^2) \qquad (c)$$

The $ symbol before the number 8 locks the value contained in the cell B8 into the equation. When performing the Paste operation and EXCEL indexes down the B column in modifying the formula, the value of the entry in cell B8 is used in all of the calculations. You now Copy cell B10 and Paste down the B column from B11 to B26 to generate the correct results for the strain corresponding to a beam with a height of 0.01 in.

The Copy and Paste operation worked well on column B after the B$8 entry was corrected. Let's try to copy the modified formula in cell B10 into the row of cells C10 to H10. If you Copy, Paste, and examine the formulas in these cells, you will note again that EXCEL has not provided the correct relations. You must modify the entry in cell B10 to properly perform the Copy and Paste operation along row 10.

When you Copy the contents of cell B10 along a **row** to the right, the cell address that defines the weight—A10 changes to B10, C10, and D10 etc. Of course, the content of cells B10, C10, D10, etc. have nothing to do with the weight applied to the beam. Because it is always necessary to multiply by the weight which is listed in column A, it is imperative that the A column designation remain fixed in the Copy-Paste operation. To modify this entry in cell B10, you type:

$$=3*\$A10*4/(2*(1/4)*10.6*10^6*(B\$8)^2) \qquad \text{(d)}$$

The $ symbol before A locks column A into the formula; it remains fixed as you change columns in the Copy-Paste operation.

Now that your entry into cell B10 has been corrected permitting you to Copy and Paste along both the rows and columns, you may proceed. Copy and Paste the formula in cell B10 to cells C10 to H10. Then Copy this row (C10 to H10) and Paste to the block C11 to H26. The program executes these calculations and displays the results as shown in the previous spreadsheet. For the maximum load of 4.00 lb, the strain varies from 0.090566 for a beam 0.01 in. high to 0.001848 for a beam 0.07 in. high.

You should always use your calculator to manually check a few values of the strain to insure that errors were not made in programming the equation into the EXCEL spreadsheet. Finally, the spreadsheet does not indicate the units for the strain. In this case, the absence of units does not present a problem because strain is a dimensionless quantity.

You may also want to change the number of decimals used in reporting the results. Six decimal places are usually employed to represent strain. This is a very large number of decimal places to carry, but strains are usually small and often two or three of these decimal places are expended carrying zeros. You may check the spreadsheet results for the strain to verify this fact. You may change the number of decimals by marking the block containing the results (B10 to H26). Then click on one or the other of the increasing or decreasing decimal keys on the formatting toolbar until the results are presented with six decimal places.

Try these examples and understand and appreciate the power of EXCEL in performing not one but an entire sheet full of calculations. Use EXCEL to solve the homework problems assigned in physics and mathematics courses. You can determine the single answer usually sought in an assigned problem and then easily explore solution space in EXCEL by treating one or two of the quantities in the controlling equation as variables. Try exploring solution space. Understand the importance of determining all reasonable solutions to a well-posed problem in a single analysis with a spreadsheet.

Entering Formulas

In the two previous sections, formulas have been entered by hand using the well known mathematical symbols +, −, * , / and ^ to add, subtract, multiply, divide and raise to a power, respectively. For engineers with a good understanding of the mathematical operations, entering the formulas by hand and making use of these mathematical operators is usually the preferred approach. However, EXCEL has an embedded formula palette to aid in entering and editing formulas in a worksheet. The formula palette, shown in Fig. 11.6, appears when you have activated a cell and clicked on the equal sign located adjacent to the formula window.

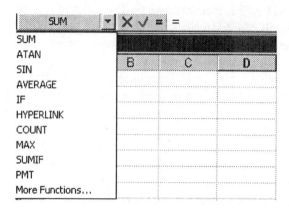

Fig. 11.6 The formula palette.

The formula palette appears on the screen with SUM as the default operation. It will aid you in adding a column or row of numbers and placing the result in the activated cell. Exercises are given at the end of the chapter to provide the opportunity to use the SUM command. Several other commonly employed operators appear, as shown in Fig. 11.6, when you click on the ▼ button. A much more extensive list of operators is available by clicking on the bottom line of the listing—More Functions.

The Paste Function dialog box, which appears on the screen, provides a technique for determining all of the functions that have been programmed into EXCEL. In most engineering courses, the Math and Trig functions are the most important. The Paste Function dialog box, presented in Fig. 11.7, includes a very complete listing of functions beginning with ABS (the absolute value) to TRUNC (an operator which truncates a number with a decimal or fraction to its integer value).

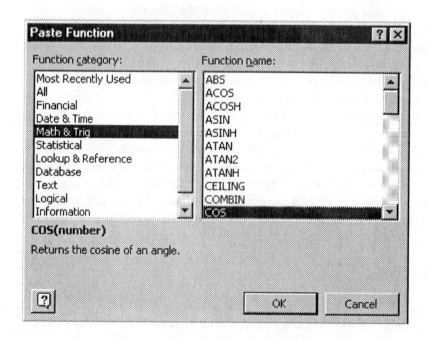

Fig. 11.7 The Paste Function dialog box with the Math and Trig operators activated.

Let's consider using the cosine function to illustrate the technique for using the Paste Function dialog box. Suppose you are to determine the cosine of a 45° angle and to place this result in cell A8. Click on cell A8 and either type = or click on the equal sign on the formula bar. Click the arrow button▼ to the right of the SUM button and then click More Functions to bring up the Paste Function dialog box. Click on the Math & Trig line under the Function category and then select the COS function as shown in Fig. 11.7. The COS dialog box, presented in Fig. 11.8, appears, and you enter the argument of the cosine in the box labeled number. It is important to recognize that the argument of the cosine **must be** given in **radians**. To convert the 45° to radians, divide 45 by 180 and multiply that quantity by π. Unfortunately EXCEL cannot handle π without a fuss. You must type PI() or its decimal equivalent 3.14159265358979. The parentheses () following PI are essential because EXCEL recognizes PI without the parentheses as text—not the number corresponding to π. After the entry for the argument is complete, the result for the cos (45°) is shown at the bottom of the dialog box. With mathematical functions, it is imperative that the argument of the function be enclosed with parentheses as shown in Fig. 11.8

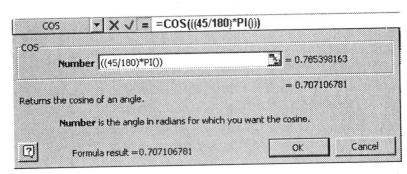

Fig. 11.8 The COS dialog box is typical of the aids available for math and trig functions.

11.5 GRAPHS WITH EXCEL

EXCEL can be employed to produce several types of graphs easily and quickly after you have learned the procedure. To show you the techniques used for producing graphs, several examples are used to demonstrate the capabilities of EXCEL in preparing pie, bar, and X-Y graphs. Let's begin by discussing a pie chart.

Pie Charts

In Chapter 8, Tables and Graphs, a manual technique for preparing a pie chart was demonstrated. If you remember the previous discussion, this chart is used primarily to show the distribution of some quantity. The larger the piece of the pie—the larger one's share of the distribution. To illustrate how to make a pie chart in EXCEL, open a new book and title it—PIE CHART—using the Save As selection under File on the menu bar. Use the top four rows for the title and row six for the column headings. The tabular data, which describe the distribution of time in various assignments for Mechanical Engineers in their initial positions in industry, is entered in cells B7 to B15. These entries are shown in the spreadsheet presented in Table 11.4.

The data in Table 11.4 is interesting, but the information presented does not carry a visual impact. There is no need for great precision in reporting these numbers; the data is reported without using a single decimal place. Clearly, the tabular format is not appropriate. The pie chart presented in Fig 11.9 is a much better method for representing this information. It would be much more effective printed in color.

This pie chart looks very good. How was it generated? After the numerical data on the distribution has been entered in cells B7 to B15, you mark them. Then click on the chart wizard button on the standard toolbar. The chart wizard presents a series of four dialog boxes that guide you through the steps necessary to convert the data displayed on the spreadsheet into a chart or graph of one type or another. The first dialog box, presented in Fig. 11.10 guides you in the selection of the type of chart or graph selected to represent the data entered on the spreadsheet. You will find that EXCEL offers many options from which to choose.

Select the pie chart and note that they can be constructed with six different options. Select the three dimension pie chart, shown in Fig. 11.10 by clicking on its symbol and then click on the Next > button which brings the second dialog box to the screen.

Table 11.4
Listing of data for a pie chart

PIE CHART	
RESPONSIBILITIES OF MECHANICAL ENGINEERS	
FIRST ASSIGNMENT IN PERCENT TIME	
ASSIGNMENT	PERCENT
1. DESIGN ENGINEERING	40
2. PLANT ENGINEERING, OPERATIONS, MAINTENANCE	13
3. QUALITY CONTROL, RELIABILITY, STANDARDS	12
4. PRODUCTION ENGINEERING	12
5. SALES ENGINEERING	5
6. MANAGEMENT	4
7. COMPUTER APPLICATIONS, SYSTEMS ANALYSIS	4
8. BASIC RESEARCH AND DEVELOPMENT	3
9. OTHER ACTIVITIES	7
TOTAL	100

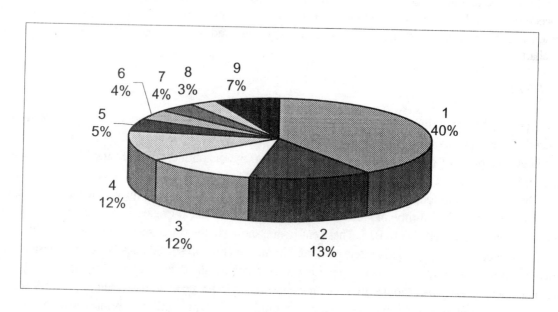

Fig. 11.9 A pie chart showing the responsibilities of Mechanical Engineers in their first assignment.

The second dialog box, titled Chart Source Data, is presented in Fig. 11.11. Its purpose is to identify the range of data to be employed in constructing the pie chart and to provide a preview of the pie chart being developed. In this example, you have already marked the data in column B—cells B7 to B15. You confirm that these cells are correct when clicking on the Next > button.

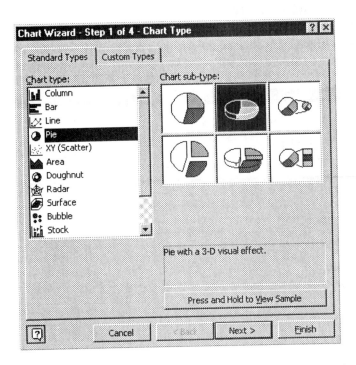

Fig. 11.10 The first of four dialog boxes presented by the chart wizard.

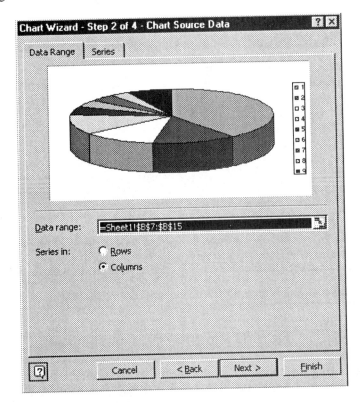

Fig. 11.11 The second dialog box in the chart wizard identifies the range of data to be employed.

The Chart Options dialog box for step 3, illustrated in Fig. 11.12, also previews the pie chart. Because it looks perfect, you do not alter the default settings for the range of data. Also, this dialog box assists you in adding a title to the graph and providing legends to identify the different slices of the pie.

Click on the title tab to add a title. The legend tab yields a dialog box that provides several choices for the placement of the legend. However, in this example, you decide not to show the legend and to leave the legend block blank. Instead you click on the data labels tab and the dialog box shown in Fig 11.13 appears. You identify each slice of the pie with a label and a percent by clicking on the appropriate dot.

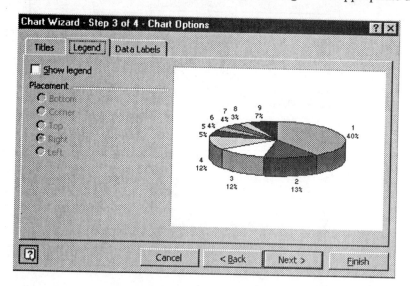

Fig. 11.12 Step three—placement of the legend, title, and data labels and a preview of the chart.

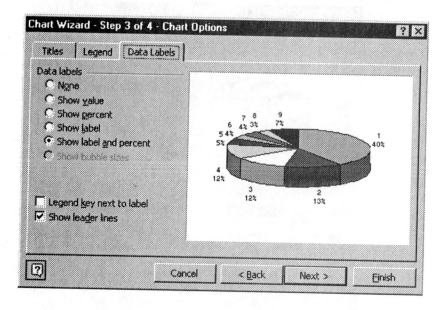

Fig. 11.13 Step three—placement of data labels.

Except for sizing the chart and locating it on the spreadsheet, you have completed the pie chart. You click the Next > button on the Step 3 dialog box and the final page from the chart wizard, presented in Fig. 11.14, provides a choice for the placement of the chart. You decide to show the chart as an object on Sheet 1 of the spreadsheet. Click the finish button and examine the results on the monitor. If the chart is too small or too large, click and drag on a corner of the chart to adjust its size.

When you are satisfied with the appearance and location of the chart, click on File and select Print Preview. You may add gridlines and adjust margins on the Print Preview display. Print the spreadsheet and the pie chart and observe its professional appearance (see Fig. 11.9).

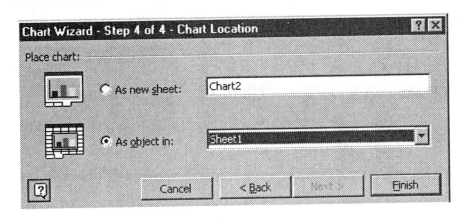

Fig. 11.14 The final step in using the chart wizard involves placing the chart.

Bar Charts

Bar charts are usually employed for showing comparisons. In Chapter 8, Tables and Graphs, an example of a bar chart was introduced that compared the percentage of high school students graduating with three or more years of secondary mathematics and science in 1982 and 1994. It is a very simple comparison, but it indicates the power of visual graphics in carrying a message.

Begin by loading EXCEL, which opens to a new book. Again, you title this book by clicking on File, Save As and then name the file—BAR CHART. The title of the new spreadsheet is entered in the first five rows, and then the data for the graduating high school students is entered in the block A7 to C9, as illustrated in Table 11.5. In the spreadsheet, the column headings are in row 7, and the years used in the comparison are listed in column A. You begin to create the bar chart by marking the region on the spreadsheet where the numerical data exist—B8 to C9. You then click on the Chart Wizard button; the first dialog box shows illustrations for many different chart types. At the top of the list is an option for column charts and another for bar charts. They are nearly the same. The bar charts display comparison categories along the Y-axis and numerical values along the X-axis. The column charts display comparison categories along the X axis and numerical values along the Y-axis. Both the column and the bar chart representations in EXCEL are considered suitable for presenting data comparing two quantities. Suppose you select the three-dimensional bar chart. The first dialog box, presented in Fig. 11.15a, shows the available selections. Near the bottom of this dialog box is a long key—when it is activated a preview of the bar chart is presented. It is important that you employ this feature to determine if the data range has been marked correctly. If the chart is not correct, cancel and revise your selection of the data to be included in the bar chart.

Table 11.5
Comparison of skills of graduating seniors

BAR CHART						
COMPARISON OF GRADUATING HIGH SCHOOL STUDENTS						
COMPLETING THREE OR MORE YEARS OF MATH AND SCIENCE						
FROM 1982 TO 1994						
YEAR	MATH	SCIENCE				
1982	38	33				
1994	60	52				

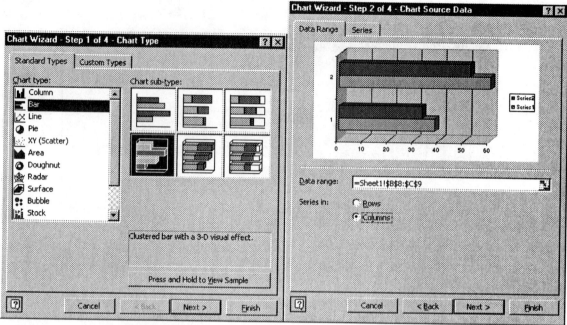

Fig. 11.15a   Fig. 11.15b

The next dialog box (Step 2) provides two tabs (two pages) for data entry—both are extremely important. The first is for the data range, which you marked prior to initiating the chart wizard. However, the series is in a column in the spreadsheet (Table 11.5) and this choice should be made as indicated in Fig. 11.15b. Click on the series tab for the Source Data dialog box and identify the series as MATH and SCIENCE by pointing with the mouse to the headings of the respective columns (B7 and C7). The "category (X) axis labels" is confusing. In EXCEL the (X) axis is the ordinate on a bar chart. Identify the (X) axis labels by giving the cell addresses for the years 1982 and 1994 (A8 and A9). The series page for the source data dialog box is presented in Fig. 11.15c.

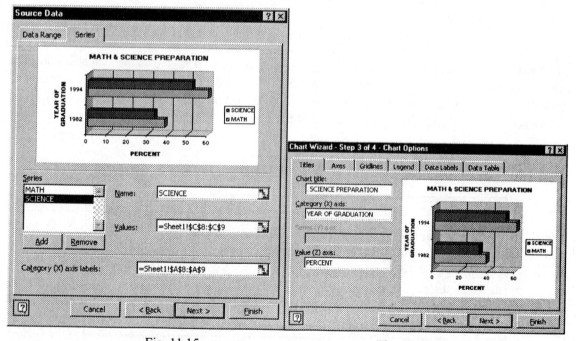

Fig. 11.15c   Fig. 11.15d
Fig. 11.15 Chart wizard dialog boxes showing data entries used to create the bar chart of Fig. 11.16.

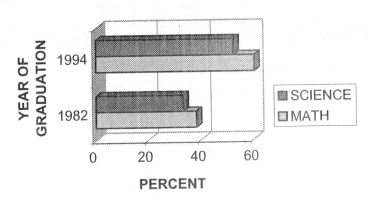

Fig. 11.16 Bar graph presentation of data comparing two quantities at two different times.

The dialog box associated with step three provides several tabs (pages) for entering information on the bar chart concerning the title, defining the axes, providing gridlines, showing legends, labeling data and showing the data table. A preview screen in the dialog box is very helpful in selecting axes and in deciding on the positioning of the gridlines. The page associated with the titles tab is shown in Fig. 11.15d. Click on the various tabs and explore the various techniques for modifying the bar chart.

The final dialog box, Step 4, is identical to that shown in Fig. 11.14; it provides the opportunity for you to select the location of the bar chart. You click on the circle that places the bar chart as an object on a new sheet. The screen of your monitor shows a large bar chart on a sheet labeled CHART 2. When you are satisfied with the appearance of the chart, click on File and Print Preview. Adjust the margins and then print the output. You should try to duplicate the spreadsheet presented in Table 11.5 and the bar chart shown in Fig. 11.16.

X-Y Graphs

The X – Y graph is the visual representation most frequently employed by engineers. It is very effective in visually conveying trends indicated by large amounts of numerical data. Is the trend increasing or decreasing? Is the quantity level or oscillating with respect to time? The X-Y graph quickly and dramatically shows these trends. To demonstrate the method for producing X-Y graphs in EXCEL, consider the strain produced by a centrally-loaded, simply-supported beam as predicted by Eq. (11.3).

You have previously used Eq. (11.3) to demonstrate the technique for performing calculations using EXCEL. You can save some work by recalling the spreadsheet presented in Table 11.3 that gives the results obtained from Eq. (11.3). These calculations provide the numerical data necessary for preparing your X-Y graph. Using the chart wizard, develop an X-Y graph from the results as shown in Fig. 11.17.

Let's discuss the step-by-step procedure for converting the numerical data on a spreadsheet in Table 11.3 to an X-Y graph. Start by marking the data block—A9 to H26. In this case, you have included the headings for the seven columns of Y data that are given in columns B to H of the data block. Remember that the data for the scale and caption along the X-axis are listed in column A. EXCEL will plot the results for the strain automatically along the Y-axis. After the data block is marked (A9 — H26), click the chart wizard button and select the type of chart from the options listed in the first dialog box. **Do not select the line graph** from the available options because it often distorts the data along the X-axis. Instead, **select the X-Y scatter graph** as shown in Fig. 11.18a. Activate and hold the long button on this dialog box to review the X-Y graph. Click on the Next > button, and the Chart Source Data dialog box, associated with Step 2, appears as presented in Fig. 11.18b. The tab for the data range is active, and you check that the data range is correct. You also indicate that the series to be plotted is in columns. Also note that EXCEL has assigned the beam height h in the legend in the preview of the graph. Click on the Series tab, and verify the location of the series of numerical results to be plotted. In this example, a series of numerical results is available for each of the seven beam heights. The cell locations of the values plotted for X and for Y are given in a box for data entry. If these cell locations do not identify the correct parameter they can be changed by clicking on the button on the right side of this box. The correct series designations are indicated in Fig. 11.18c.

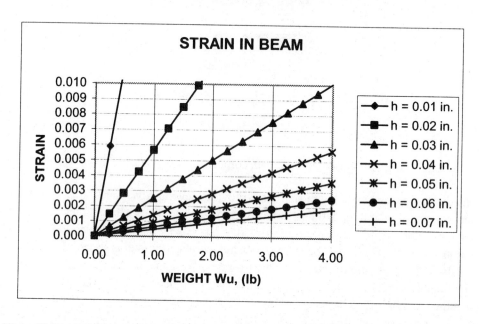

Fig. 11.17 An X-Y graph showing the strain in beams of various heights as a function of applied load.

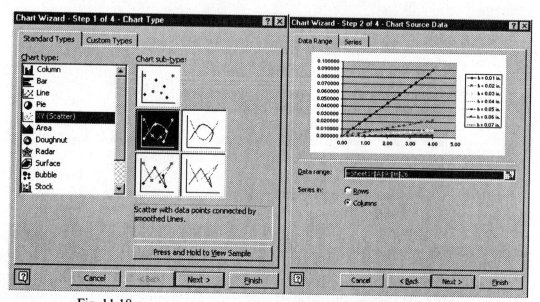

Fig. 11.18a   Fig. 11.18b

Fig. 11.18 Dialog boxes used in constructing X-Y (scatter) graphs.

The Chart Options dialog box, Step 3 in Fig. 11.18d, provides an opportunity to add the title and the captions for the X and Y-axes. You add the title **STRAIN IN BEAM,** and caption the X and Y-axes as **WEIGHT, W$_u$ (LB)** and **STRAIN,** respectively. You may also add or remove grid lines and move the position of the legend by activating the appropriate tab. The draft of the graph is essentially completed when you click the Next > button.

In the fourth and final dialog box, you decide to show the graph on the same worksheet as the numerical data. You click on the graph to activate it, and then drag the corner and/or side markers of the graph to enlarge it. When the graph is the correct size it is positioned with the pointer to a suitable location on the spreadsheet. Before printing the graph, click on File then Print Preview, and check to determine if you have arranged the display so that it conveys the numerical data in tabular form on the spreadsheet as well as properly positioning the X-Y graph.

The combination of the spreadsheet and the graph is an effective technique for presenting the results of theoretical or experimental data. However, in some instances you prefer to show the graph without including the spreadsheet. You may show the graph alone without the spreadsheet by displaying it on another worksheet. Click on the graph to activate it and then click on the Copy icon. Next, click on the tab for Sheet 2 to bring a new spreadsheet onto the screen. Finally, click on the Paste button. The graph from Sheet 1 is copied onto Sheet 2. Click on the graph to activate it, then drag its corners or sides to position and size it on the sheet. Examine the appearance of the graph with Print Review prior to printing. The result is shown in Fig. 11.17.

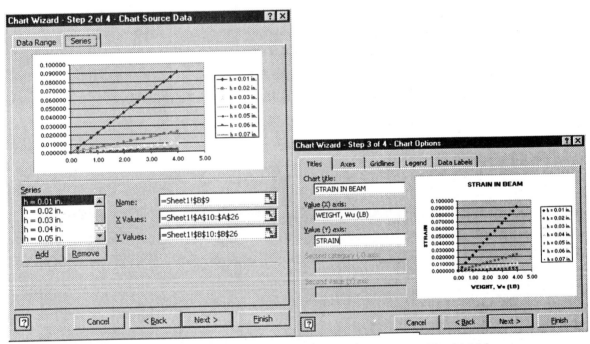

Fig. 11.18c   Fig. 11.18d

Fig. 11.18 Dialog boxes used in constructing X-Y (scatter) graphs.

A close examination of Fig. 11.17 shows that the graph is not exactly the same as that presented in the previews displayed in Fig. 11.18 or the view of the graph shown in Fig. 11.19. The scales on both the abscissa and the ordinate are different and the gridlines for the X-axis are missing in the preview graphs. The choice of scale for the X and Y-axes is very important. The very high values of strain shown in Fig. 11.19 tend to distort the graph. These strains are excessive and indicate that the beams with small heights would fail by plastic bending long before achieving a load of four pounds.

To change the scale on the graph in Fig. 11.19, point to the number 5.00 on the caption for the X-axis and double click the left mouse button. The Format Axis dialog box appears, as shown in Fig. 11.20a, with five different tabs. Select the tab for scale and enter the number 4.00 for the maximum value for the X-axis. To change the scale on the Y-axis, point to the number 0.10 in the caption for the ordinate and repeat the process. Again, select the scale tab and enter the value of 0.01 for the maximum strain to be plotted along the Y-axis. When the new selections are made, the graph is rescaled so that it displays more meaningful values of strain as shown in Fig. 11.17.

A close examination of the graph in Fig. 11.19 indicates that the gridlines for the X-axis are missing. To correct this oversight, click on the graph to activate it; then click on the chart wizard icon. Proceed quickly through the dialog boxes, which are complete because the graph is active. When the third dialog box appears, click on the gridlines tab and select gridlines for the major markings along the X-axis. Click on the finish button to add the grid lines.

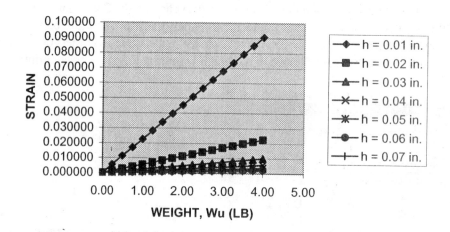

Fig. 11.19 Appearance of the X-Y graph before editing.

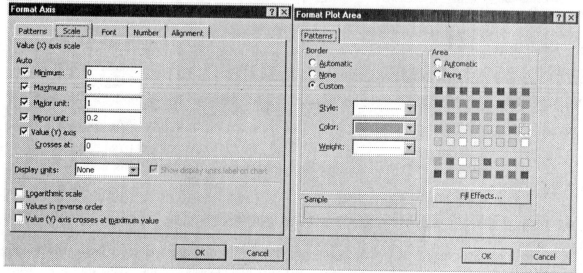

Fig. 11.20a   Fig. 11.20b
Fig. 11.20 Dialog boxes for editing graphs.

Finally, the background color (shading) of the chart area in Fig. 11.19 is too dark. To remove the shading double click on an open area of the graph. The Format Plot Area dialog box appears, as shown in Fig. 11.20b, with selections for the color of the border and the area. Under that part of the dialog box for the area, select white for the background color. Click on the OK button and the background shading vanishes.

11.6 SUMMARY

EXCEL the most popular spreadsheet program in use today has been introduced. This introduction was brief because its purpose was to help you develop limited entry-level skills. You should know how to interpret the screen, the menu and the toolbars. You should also be able to navigate on the spreadsheet with the keyboard or the mouse, and to mark cells or blocks of cells by dragging the cursor with the mouse.

Three examples were described that will be useful in this course and in other courses that you will encounter later in the curriculum. First, a technique for preparing a parts list was illustrated. In this instance, the spreadsheet was utilized to organize a table containing both descriptive and numerical data.

The second example demonstrated techniques for performing calculations using EXCEL to evaluate simple formulas. Methods for performing a single calculation in a cell were demonstrated. Then this calculation was copied to perform a line of calculations by varying one parameter. Finally, a spreadsheet full of calculations was made when two parameters were systematically varied.

The third example involved the preparation of three different charts or graphs. The techniques involved in preparing a pie chart, a bar chart, and an X-Y graph were described. Extensive use of the chart wizard in EXCEL was made. The wizard leads you through the chart construction technique in a four-steps. Techniques for editing X – Y graphs were also discussed.

EXCEL is a tool somewhat like a hammer. With time and practice, you learn to use a hammer— you hit the nail more squarely, with more force and you become more effective. The same is true for any spreadsheet program. You will grow more proficient with time. The benefits provided with a spreadsheet program, during your tenure in college and in your professional and personal life, make the investment of time in learning this program worthwhile.

REFERENCES

1. Harvey, G., Excel 2000 for Windows for Dummies, John Wiley & Sons, New York, NY 1999.
2. Gottfried, B. S., Spreadsheet Tools for Engineers, Mc Graw Hill, New York, NY, 1996.
3. Walkenbach, J., Excel 2000 Formulas, John Wiley & Sons, New York, NY, 2000.
4. Anon, Discovering Microsoft Office 2000: Small Business and Standard, Microsoft Corporation, Redmond, WA, 1999.
5. Warner, N. P. and N. D. Warner, Easy Microsoft Excel 2000, Que, Indianapolis, IN, 1999.
6. Walkenbach, J., Excel 2002 Power Programming with VBA, John Wiley & Sons, New York, NY, 2001.

EXERCISES

11.1 Explore the EXCEL screen and learn the purpose of the icon buttons.

11.2 Explore the EXCEL menu and learn the available options.

11.3 Prepare a parts list for the project your team is developing.

11.4 Suppose that you have a rich grandmother, who has provided an inheritance for you. Unfortunately the money is in a trust, and you cannot place your hands on the loot until you are a mature 50 years of age. If the amount of the inheritance was $10,000.00, and the trust yields 5.0% per annum compounded semi-annually, determine the value of the trust each year until your fiftieth birthday. Also prepare an X-Y graph showing the value of the trust with time. Assume the money in the trust is deposited in a Roth account and not subject to either state or federal income taxes.

11.5 Duplicate the work involved in producing the spreadsheet determining the strain in the beam.

11.6 Prepare a bar chart comparing the percentage of men and women in your high school graduating class with the percentage in the freshman engineering class in 2000. Your high school yearbook will provide information on the gender of your graduating class, and the Office of the Dean of Engineering will provide data on the freshman class.

11.7 Suppose that you decide that the graph in Fig. 11.17 covers too large of a range of height of the beam. Prepare a new graph covering the range in height from 0.04 in. to 0.07 in.

11.8 Suppose that you decide that the graph in Fig. 11.17 covers too small of a range of weight. Prepare a new graph covering the range in weight from zero to 6 pounds.

11.9 Suppose that you decide that the graph in Fig. 11.17 covers too small of a range of height of the beam. Prepare a new graph covering the range in height from 0.01 in. to 0.10 in.

11.10 Suppose that you decide that the graph in Fig. 11.17 covers too large of a range of weight. Prepare a new graph covering the range in weight from zero to two pounds.

11.11 Evaluate the relation $Y = a + bX$ when $a = 2$ and $b = 3$. Let X vary from zero to 15. Prepare an X – Y graph showing the results.

11.12 Evaluate the relation $Y = a + bX$ when $a = 3$ and $b = 5$. Let X vary from zero to 15. Prepare an X – Y graph showing the results.

11.13 Evaluate the relation $Y = a + bX$ when $a = 2$ and $b = 4$. Let X vary from zero to 15. Prepare an X – Y graph showing the results.

11.14 Evaluate the relation $Y = a + bX^2$ when $a = 3$ and $b = 2$. Let X vary from zero to 6. Prepare an X – Y graph showing the results.

11.15 Evaluate the relation $Y = a + bX^2$ when $a = 2$ and $b = 3$. Let X vary from zero to 6. Prepare an X – Y graph showing the results.

11.16 Evaluate the relation $Y = a + bX^2$ when $a = 1$ and $b = 2.5$. Let X vary from zero to 6. Prepare an X – Y graph showing the results.

11.17 Evaluate the relation $Y = a + bX^2 + c X^3$ when $a = 0$, $b = -3$ and $c = 1$. Let X vary from zero to 5. Prepare an X – Y graph showing the results.

11.18 Evaluate the relation $Y = a + bX^2 + c X^3$ when $a = 1$, $b = 2$ and $c = -2$. Let X vary from zero to 5. Prepare an X – Y graph showing the results

11.19 Evaluate the relation $Y = a + bX^2 + c X^3$ when $a = 2$, $b = -1.6$ and $c = 2.4$. Let X vary from zero to 5. Prepare an X – Y graph showing the results

11.20 Evaluate the relation $Y = ae^X$ when $a = 1.5$. Let X vary from zero to 6. Prepare an X – Y graph showing the results.

11.21 Evaluate the relation $Y = ae^X$ when $a = 1.9$. Let X vary from zero to 5. Prepare an X – Y graph showing the results.

11.22 Evaluate the relation $Z = a + b Y + c X$ when $a = 1.32$, $b = 2.8$ and $c = 3.4$. Let $X = 1, 2, 3$ and 4 and $Y = 1, 2, 3,$ and 4. Prepare a graph showing Z as a function of X and Y.

11.23 Evaluate the relation $Z = a + b Y + c X$ when $a = 2.22$, $b = -1.3$ and $c = 0.8$. Let $X = 1, 2, 3$ and 4 and $Y = 1, 2, 3$ and 4. Prepare a graph showing Z as a function of X and Y.

CHAPTER 12

MICROSOFT PowerPoint

12.1 INTRODUCTION

During this course, your team will be responsible for two or three design reviews. You will be required to make a presentation before the class and the instructor. In the design briefing, you will describe the product specification, the important features that your team is incorporating in the design, and illustrate your concepts with engineering drawings. If you want to make a very good presentation, it is important that the audience closely follow your descriptions of the team's activities. To help in this regard, you should use graphics, which enable transmission of information through two senses—audio and visual. This chapter provides instruction for a graphics presentation program (GPP) that facilitates the preparation of the visual aids needed for your presentations.

Microsoft PowerPoint is a graphics presentation program that markedly reduces the time needed for you to prepare professional quality visual aids. It has a wide range of capabilities from simple overhead transparencies to sophisticated on-screen electronic displays. Graphics presentation programs are useful in producing either black and white or colored transparencies, 35 mm slides and full screen projected electronic slides as well as valuable support materials such as hard copy printouts, notes and outlines.

Before beginning to learn about PowerPoint, it is important to understand that preparing a presentation is a five-step process, which includes:

1. Plan the presentation to provide the information necessary for a design review. In planning, determine the amount of time that you have to speak, the size and layout of the room and the type of equipment available for visual aids.
2. Formulate the presentation by preparing an outline that identifies all of the topics that will be described.
3. Prepare the slides needed in your presentation by using a GPP such as PowerPoint to guide you. As you compose the slides, edit them and make certain that there are no misspellings. After you have completed the slides, review them and make the changes necessary for their enhancement. Can you add graphics or bullets to better capture the attention of the audience? Is the font size correct? Have you used the bold, underline or shadow options to the best advantage? If you are using electronic displays, significant latitude exists for enhancing the visuals and controlling the pace of the presentation.
4. Finally, you must rehearse the presentation. It is imperative that you completely understand the material on your slides. Nothing is more deadly than a speaker reading his or her material from the slide to the audience. Rehearse until the presentation is smooth and polished. You should feel confident about the message to be conveyed and your ability to handle the slides and the projector.

12.2 THE PowerPoint AUTO CONTENT WIZARD

Begin by loading PowerPoint. The window that opens, presented in Fig. 12.1, includes the PowerPoint dialog box with options for creating presentations. Select the Auto Content Wizard to help you develop a presentation.

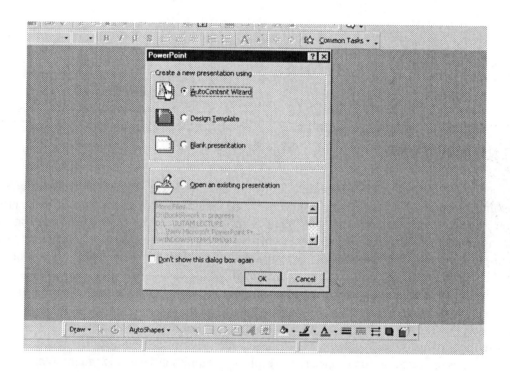

Fig. 12.1 The opening screen in PowerPoint that is used to create a new presentation.

When you click on the OK button at the bottom of the PowerPoint dialog box, a new dialog box for the Auto Content Wizard appears as shown in Fig. 12.2a. This dialog box describes the step-by-step process that PowerPoint follows in structuring the slides for your presentation. Click on the Next > button and the Presentation Type dialog box appears as shown in Fig. 12.2b.

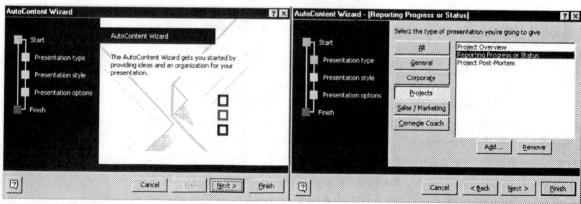

Fig. 12.2a

Fig. 12.2b

Fig. 12.2 The first two dialog boxes associated with the Auto Content Wizard.

From the Presentation Type dialog box, click on the Projects button and note the three options that are available in PowerPoint:

- Project Overview
- Reporting Progress or Status
- Project Post-Mortem

Select Reporting Progress or Status, and click the Next > button. The Presentation Type dialog box presented in Fig. 12.3a appears that prompts you to select the type of visuals to be used. The author has selected black and white overheads since this book is printed using only one color—black. There are much better choices that will be covered later. Click the next > button and the Presentation Options dialog box appears as illustrated in Fig. 12.3b.

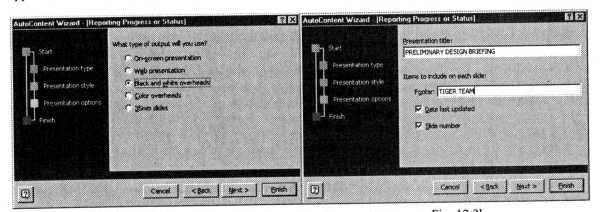

Fig. 12.3a   Fig. 12.3b
Fig. 12.3 Dialog boxes used for selecting presentation style and options.

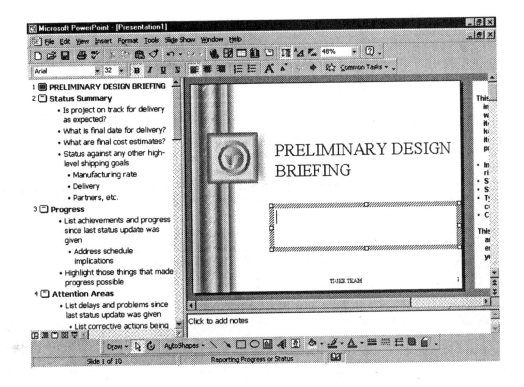

Fig. 12.4 The working screen for preparing a presentation reporting on project progress or status.

Use this dialog box to title the presentation—PRELIMINARY DESIGN BRIEFING—and also to show the team name. Date the overheads and number them by clicking in the appropriate toggle squares. Click on the Next > button and the wizard indicates that you have finished the set up for the presentation. Click on the finish button to obtain the screen illustrated in Fig. 12.4.

12.3 THE PowerPoint WINDOW

The screen shown in Fig. 12.4 is extremely busy. It includes several toolbars and a working area divided into four different regions. Let's first consider the toolbars.

The Toolbars

There are four important toolbars located above the body of the window shown in Fig. 12.4. The top row, called the title bar, displays the file name that you assign to the presentation when you first save your file. In Fig. 12.4, the title is Presentation 1—the default title assigned by PowerPoint prior to saving the file. At the far right side of the title bar, you will find three buttons: the left one sets the program aside, the middle button changes the size of the display and the button marked with the × on the right closes the file.

The second row contains Microsoft's standard menu items. Click on one of these items and a pull down menu is displayed. Click and explore the pull down menu listings until you are comfortable with each item. You should develop a good understanding of what can be accomplished with each of the menu items. The Slide Show menu item is particularly interesting. Some of its advantages will be described later in this chapter.

The third row is the standard toolbar that displays 20 icons and a zoom control symbol ▼ to change the size of the display. Point to any icon and note that its purpose is defined in a small box attached to the pointer. The icons on the left side of this toolbar are common to all of the Microsoft applications; however, those on the right side of the bar are unique to PowerPoint. Several of these icons will be introduced later in an example presentation for a design review.

The fourth row down from the top is the formatting bar. Again, most of the icons are common with other Microsoft applications. With the mouse, point to some of the icons on the right side of the formatting bar and the purpose of each icon is displayed. The Common Tasks button at the far right of this toolbar is particularly important since it enables you to:

- Insert a new slide in the presentation.
- Change the slide layout.
- Select a different design template.

At the bottom of the window, you will find three more rows. The top row in this group is located at the left side of the screen and contains the five icons shown in Fig. 12.5.

Fig. 12.5 The icons used to control the format of the display.

These five icons control the presentation views. From left to right the icons enable the following views:

1. The normal view with the workspace divided into four regions as shown in Fig. 12.4.
2. The outline view with more space provided to show the content of the slides.
3. The slide view enlarges the region for the slides and for the action item notes.

4. The slide sorter enables you to view all of the slides and to rearrange them.
5. The slide show enlarges each slide to fill the screen and clicking the mouse enables you to review the complete presentation.

These icons will be described in more detail later when the preparation of a presentation for a preliminary design review is discussed.

The next to the last row is used for the drawing toolbar. You may add drawings, arrows, clip art, etc. to any slide to enhance its visual impact. The drawing toolbar may be moved to a different location on the screen by clicking on it and dragging it to a preferred location.

The bottom row contains the status bar where the slide number and helpful messages are displayed.

The PowerPoint Workspace

The center area of the window is the workspace where you will prepare the slides. One usually works in the normal view that is illustrated in Fig. 12.4. In the normal view, the workspace is divided into four regions.

The region on the left side of the workspace shows the outline of the presentation. Each slide is numbered and its contents are shown. You can click at any location in the outline and edit the text in any slide. The scroll bar to the right of this outline region enables you to scroll forward or backward in the presentation to locate the slide of interest.

The large region to the right center of the workspace shows a view of the slide in sufficient detail to judge its appearance. Vertical and horizontal scroll bars are also evident to the right and below the slide region. The vertical scroll bar enables you to move either forward or backward in this group of slides. The presentation style and the design template of the slide presented in Fig. 12.4 may not be suitable for your purposes. However, by selecting a more suitable style and design template from the Common Tasks options, you may easily change the appearance of the slide. Changing the style and design template will be discussed later in this chapter.

The region to the right of the slide appears only when the first slide in the presentation is active. This region serves as a note pad to list the action items that occur as a result of discussions occurring during and following the presentation. Following the presentation, these notes are transferred to an action item slide. The technique followed in using the action item pad is shown in Fig. 12.6.

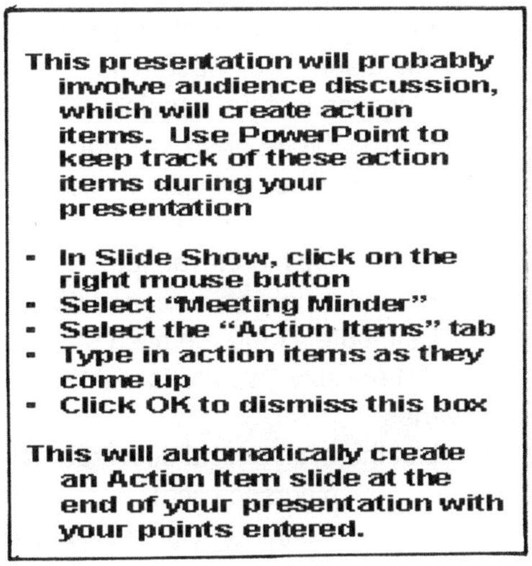

Fig. 12.6 Directions for using the action
item pad in PowerPoint.

This presentation will probably involve audience discussion, which will create action items. Use PowerPoint to keep track of these action items during your presentation

- **In Slide Show, click on the right mouse button**
- **Select "Meeting Minder"**
- **Select the "Action Items" tab**
- **Type in action items as they come up**
- **Click OK to dismiss this box**

This will automatically create an Action Item slide at the end of your presentation with your points entered.

The fourth region is located at the bottom of the workspace just below the slide. It is used to add notes as you prepare the slides. These notes are not visible on the slides projected during the presentation. While the pad for the notes in the normal view of the workspace is very small, it becomes much larger if you shift to the outline view.

The double delta buttons, on the right side of the screen below the vertical scroll bar, are used to move forward or backward in the sequencing of the slides.

12.4 PLANNING AND ORGANIZING WITH PowerPoint

PowerPoint has built-in features to assist you in planning, organizing and timing your presentation. To show these features, the outline embedded in PowerPoint for "Reporting Progress or Status on Projects" has been copied and presented in Table 12.1. This listing provides an excellent guide for you to follow in preparing your design briefings. It includes ten slides that cover the important issues arising in most design projects. Indeed, it may contain many more questions than your team may wish to consider. If this is the case, it is easy to delete them while preparing the presentation.

Table 12.1
PowerPoint outline for reporting progress or status on projects

1. Preliminary Design Briefing
2. Status Summary
   - Is the project on track for delivery as expected?
   - What is the final date for delivery?
   - What are the final cost estimates?
   - Status against any other high-level shipping goals.
     - Manufacturing rate
     - Delivery
     - Partners, etc.
3. Progress
   - List the achievements and progress since the last status update.
   - Address the schedule implications.
   - Highlight the key actions that made progress possible.
4. Attention Areas
   - List delays and problems since the last status update was presented.
   - List the corrective actions being taken.
   - Address the schedule implications.
   - Make sure you understand.
     - Issues that are causing the delays or impeding progress
     - Why the problem was not anticipated
     - If customer will want to discuss any issue with upper management
5. Schedule
   - List the top high-level dates.
   - Keep it simple so audience does not become distracted with details.
   - Distribute a more detailed schedule if appropriate.
     - Make sure you are familiar with the details of schedule so you can answer questions
6. Deliveries
   - List the main critical deliverables.
     - o Yours to client.
     - o Yours to outside services.
     - o Outside services to you.
     - o Other departments to you.
   - Understand your confidence rating to each deliverable.
     - Indicate this confidence level on slides if appropriate.

7. Costs
- List the new projections of the costs.
- Include original estimates.
- Understand the source of differences in these numbers -- be ready for questions.
- Are there cost overruns?
  - Summarize why
  - List the corrective or preventative action you've taken
  - Set realistic expectations for future expenditures

8. Technology
- List the technical problems that have been solved.
- List the outstanding technical issues that need to be solved.
- Summarize their impact on the project.
- List any dubious technological dependencies for project.
- Indicate the source of doubt.
  - Summarize the action being taken or the backup plan

9. Resources
- Summarize the project resources.
  - Dedicated (full-time) resources.
  - Part-time resources.
- If project is constrained by the lack of resources, suggest alternatives.
- Understand that the customers may want to be assured that all possible resources are being used, but in such a way that the costs will be properly managed.

10. Goals for the Next Review
- Date of next status update.
- List the goals for next review.
- Specific items that will be done.
- Issues that will be resolved.
- Make sure everyone involved in project understands the action plan.

Many of these topics are pertinent for your preliminary design briefing, although the answers to many of the questions posed in the outline may be incomplete. In some cases, such as deliveries, the topic may not be applicable. Delete this slide and add another that is more suitable. You may decide to add many slides to completely describe all of the mechanical details involved in the design of the system. However, the preliminary design review, or any presentation for that matter, has a time constraint (usually imposed by the instructor in this class). Your presentation should require 30 minutes or less. Many accomplished speakers believe that 18 minutes is the ideal length for a presentation. This amount of time is sufficient to include substance, but short enough to avoid boring the audience.

The Title Slide

Let's begin with the title slide[1]. Note that you have already provided the title in responding to the questions raised by the Auto Content Wizard. You may want to provide additional information in the subtitle block. To add this material, use the normal view, click on the subtitle block to activate it and you are ready to enter additional material. Keep the title and the subtitle short. You will have the opportunity to expound later in the presentation. Entering information on the slides is the same as typing in a word processing program. Click in the appropriate area and type. An example of a title slide appropriate for a design briefing is shown

---

[1] The anti-icing system for highway bridges will be used as an example in preparing the slides for the design briefing.

in Fig. 12.7. It gives the purpose of the presentation, the title, identifies the team and its members, and gives the date of the presentation.

Examination of Fig. 12.7 shows a marked difference from the view shown in Fig. 12.4. The design template has been changed by clicking on the Common Task button and selecting the "Apply Design Template Option." The dialog box used in selecting a design template is illustrated in Fig. 12.8.

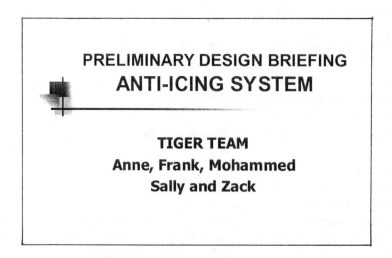

Fig. 12.7 An example of a title slide.

The Blends template was selected because it has a light background color. There are many choices. Click on one of the template options, and examine the preview that appears in the window to the right. Note that many of the presentation templates have a dark background that impairs the visibility of the slide's content. The dark background is particularly detrimental when the light from the projector is insufficient or when the room in which the presentation is made cannot be darkened sufficiently.

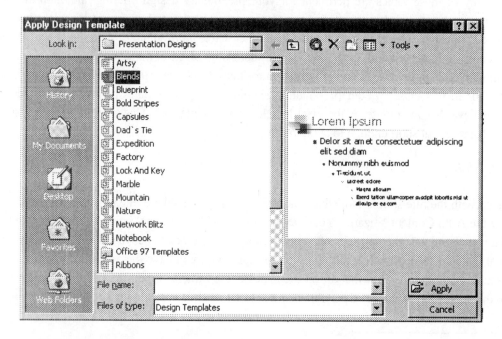

Fig. 12.8 The design template dialog box enables the selection of different presentation templates.

The Status Summary Slide

Let's continue with the preparation for the presentation by considering Slide 2. Shift to the outline view, and respond to the questions posed by PowerPoint for the Status slide. The answers to these questions, adapted for this design briefing, provide the current status of the project. As you make entries in the outline, examine the slide to check the layout of the text. You may wish to change to the normal view when inspecting the appearance of the slide because it enlarges the image of the slide. A typical Status Summary slide is presented in Fig. 12.9.

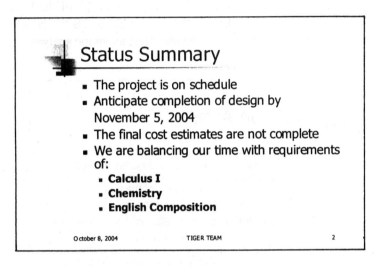

Fig. 12.9 Example of the Status slide.

The Progress-to-Date Slide

Next, consider Slide 3, which describes progress made to date in the development of the project. Again answer the questions posed by PowerPoint by editing the outline. The answers provided, as depicted in Fig. 12.10, are rather general. The various tasks that are complete or nearly complete are cited. Problems that the team has encountered are identified. Details pertaining to design problems, schedule, cost, etc. are described in slides presented later in the design briefing.

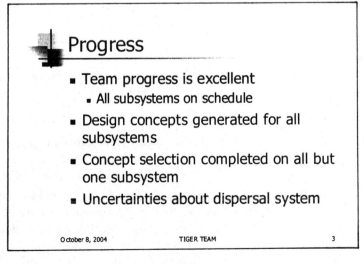

Fig. 12.10 Example of a Progress slide.

The Attention Areas Slide

After you have briefly reviewed the progress made in developing the various subsystems, it is important to identify the problems that the team has encountered. The problems have been identified briefly in the Progress slide. In the Attention Areas slide (#4), the problems are described in more detail. This is your opportunity to raise any issues (questions) that the design team has discovered to date. Do not hide uncertainties; come forward and seek help. The design review is a formal venue for this purpose. The example shown in Fig. 12.11 indicates that the team is encountering problems in developing the system required to disperse the saline solution in a uniform layer over the model of the bridge pavement. This problem has resulted in a significant schedule delay. The team understands the criticality of the issue and has scheduled a meeting with the instructor to resolve the questions about the analysis.

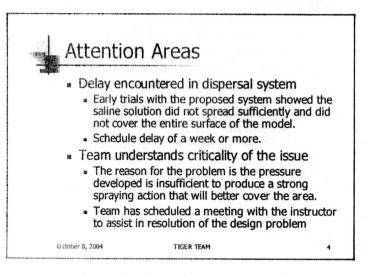

Fig. 12.11 The Attention Areas slide provides an opportunity to seek help in resolving problems.

Goals for the Next Review Slide

The final slide, illustrated in Fig. 12.12, describes your team's plans for action in the near future. Wizard titles this slide as Goals for the Next Review, which is appropriate. Wizard then prompts you to:

- List the goals to be accomplished in the near future.
- Specify detailed task that will be completed.
- Define any issues outstanding that must be resolved.
- Insure the instructor and your peers that all of the team members understand and agree to the action plan.
- Indicate the time of the next design review.

The Wizard's prompts were followed in preparing the concluding slide shown in Fig. 12.12.

You are now ready to print the output. Click on File and select Print. The print page, presented in Fig. 12.13, is adapted for PowerPoint. It contains a selection box labeled—Print What. You have several choices including: slides, handouts with 2, 3 or 6 slides per page, note pages, and an outline view. You can prepare the overheads for your presentation and the handout for the audience. Print your presentation using these options and examine the output to decide which style is most suitable for the handout intended for the audience. The black and white paper versions of the slides can be copied to provide the transparencies necessary for the presentation. If you have your own laser or ink jet printer, the transparencies can be printed directly. Be certain you save the file before exiting the program.

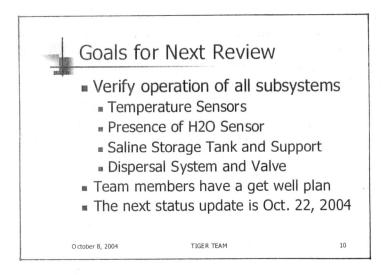

Fig. 12.12 The final slide outlines the actions the team plans to complete before the next briefing.

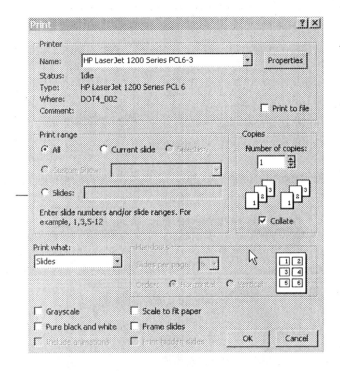

Fig. 12.13 Dialog box for printing the output from PowerPoint.

12.5 ENHANCING THE SLIDES

Suppose you decide to use color slides instead of black and white. It will cost more to use color if you take the job to a commercial copy center, but you may decide that the visual impact of color is worth the added cost. How can you change the black and white slides to color slides? It is easy, and you do not have to retype all of the information into a new version of the presentation.

To change to color slides click on the Gray Scale Preview icon. This button toggles the slide between color and gray scale. To change the design template, Click on the Common Tasks button and select the Apply Design Template option. The Design Template page displayed on your screen has over 30 different options that start with Artsy and end with Sumi Painting. With the mouse, select one of the templates and note that it is displayed for your review in a small preview box. Try several templates before selecting the one you believe is the most appropriate.

If you would like to change the color of the background, go to the menu bar and click on Format, and select Slide Color Scheme. The Color Scheme dialog box that is displayed gives you seven options for the colors of the text, titles and background employed. Click on one of the options and preview the results until you obtain the colors that you believe will be the most effective in your presentation. Be careful and stay with relatively light colors for the background because slides with a lighter background color are more visible to the audience. Dark slides do not project well unless the room is extremely dark. And dark rooms' are an invitation for your audience to sleep.

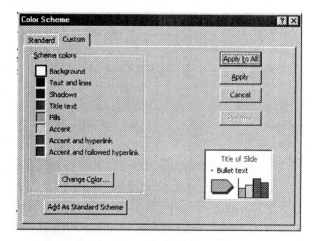

Fig. 12.14 The Color Scheme dialog box showing the colors for each feature on the slides.

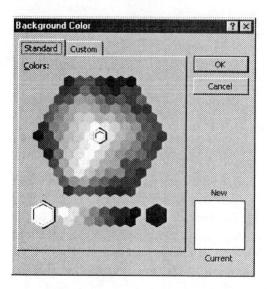

Fig. 12.15 The color hexagon shows all of the available options for customizing your slides.

Suppose that you like some of the colors used, but are unhappy with the color of the text and the bullets. Change them by clicking on Format and select Slide Color Scheme. The Color Scheme page has two tabs—standard and custom. The standard tab illustrates seven choices of color for the titles and bullets. If none of the colors are suitable, click on the custom tab and the Color Scheme dialog box shown in Fig. 12.14 appears on the screen. Prepare to become a painter. You have the opportunity to select the colors that you want to use in the background, shadows, fill, text and title and bullets. To begin changing the colors select a feature on your slide such as the background. Click the Change Color button and the multicolored hexagon, presented in Fig. 12.15, appears with many choices of different shades of the basic colors. Select the color and its shade that you wish to employ for each of the features on the slide. You can have a single design for all of your slides (recommended), or you can have a different color scheme for each slide. Have fun; PowerPoint gives you the means to make great slides.

Some presenters like to use either a header or footer on their slides. A footer is helpful if you give many presentations and want to retain the slides for your files. The information in the footer can include date, occasion for the presentation, location, etc. For example, you may decide to use a footer stating October 10, 2003—Preliminary Design Review, and slide number. How do you incorporate this footer on PowerPoint slides? Click on View on the Menu bar and select Header and Footer. The Header and Footer dialog box is displayed with two tabs—one for the slide and the other for the notes and handouts. Let's select the slide tab and add a footer to your slides. Click on the Date and Time Square, select Fixed, and type the date of the planned presentation. Mark the slide number because it will help you in timing the delivery of the presentation. Click on the footer, and type in a short message to appear in the center region of the footer. Also click on the box indicating that you do not want the footer to appear on the title slide. Delete the footer from the title slide because you want the audience to remain focused on the title in the beginning of the presentation. The Header and Footer page has a small preview window that permits you to review the input prior to printing. The footer developed with this procedure is illustrated in Fig. 12.7. The slide was in beautiful color on the monitor, but the figure is in black and white. The author's apologies, but the use of color in printing in a low-volume, low-cost textbook is prohibitively expensive.

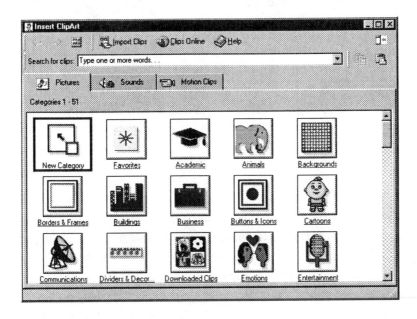

Fig. 12.16 Clip Art dialog box.

Still not happy with the impact that your slides will make? What about adding some clip art? Select a slide that you want to enhance with the addition of some photo or sketch that is available in the Microsoft Clip Art files. You don't know what's available? Not to worry; click on Insert on the menu bar, select Picture and then Clip Art. There is a pause while the clip art files are loaded into the active PowerPoint program. A dialog box entitled Insert Clip Art appears on the screen, as shown in Fig. 12.16, with 51 categories of illustrations. Select the Animal category and 965 illustrations of the selections of clip art available under this category appear in the preview window. To assist in the search, you should type turtle into the line for search topics and hit the enter key. Then select turtle (No. 4) from the eight images available, and click on the insert button.

Now examine the slide you have selected for enhancement. You have a large picture of a turtle— smack in the middle of the slide. To move the picture and change its size, drag it by the one of the little squares either on the corners or the middle of the sides to position the insert appropriately on the slide. The turtle is shown in the lower right hand area of the Status Summary slide in Fig. 12.17. Note, if you push while dragging the object, it becomes smaller, but when the handles are pulled the object becomes larger. Try it, and you will soon manage to both move and size the turtle.

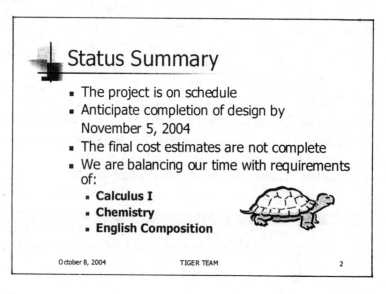

Fig. 12.17 The status slide shown with clip art enhancement.

12.6 REHEARSING

Rehearsing is an essential part of preparing for a design review. Fortunately, PowerPoint has a neat feature that helps in handling the visuals as you rehearse. Go to the first (title) slide and then click on Slide Show, which is located on the menu bar. Select View Show and the screen changes with all of the toolbars disappearing. The slide occupies the entire screen of the monitor and the PowerPoint presentation simulates the screen in the classroom. You can practice your delivery moving from one slide to the next by using the directional arrow keys on the keyboard to forward or reverse the order of presenting the slides.

Timing is an important part of the delivery of a design review. Many presenters make the mistake of delivering material slowly in the beginning and hurrying at the end of the presentation to meet the time constraint. The delivery should be paced to allow more than sufficient time at the end of the presentation. The last slide is the most important in the overall design review. Be certain that there is time near the end of your briefing to present the conclusions properly.

12.7 COMPUTER PROJECTED SLIDES

The ability to project your slides directly from the computer gives you several significant advantages, which are available in PowerPoint.

1. You may store a large number of slides in the computer's memory. If questions arise during your presentation, you can quickly locate any slide necessary to improve your response and/or expand the discussion.
2. You do not need a color printer to make slides projected directly from your computer.
3. You can use special effects incorporated in PowerPoint to give your presentations some real class.

Let's first explore the special effect known as transitions where PowerPoint controls the manner in which you switch from one slide to the next. Go to the title slide in your presentation, click on Slide Show on the menu bar and select Slide Transition. The Slide Transition dialog box that appears is illustrated in Fig. 12.18. Click on the arrow located beside the message window and a list of many different types of transitions is presented. Select one of them and watch the illustration in the small preview window switch from one view to another. Select the type of transition that you like the best. The dissolve transition is an excellent choice. Next, select the speed of the transition—slow, medium or fast. Again you can review your choice in the preview window. Finally, select the advance method. Until you are a real pro at delivery of design briefings, it is suggested that you advance the slides by clicking the mouse. Click the Apply button when you have completed the set up of the transition of the title slide.

Click on the Slide Sorter View button and the first six slides of your presentation are displayed. The first slide has a small slide transition icon visible near its lower left corner. A new toolbar also appears at the top of the window replacing the formatting toolbar. You are now ready to select the type of transition for the remaining slides. Select slide 2, and click on the transition arrow button on the new toolbar. You can select the type of transition from the long list for slide 2.

Proceed to mark any slide that you wish to transition in the Slide Sorter View, and to select the type of transition that you want to use when presenting that slide. After you have made your choice, it is confirmed by the presence of a small transition icon just below the slide when you have the Slide Sorter View active as shown in Fig. 12.19. If you employ the same transition for all of the slides, mark all of them (Edit/Select All), and select the type of transition that you believe is the most appropriate for your presentation.

Fig 12.18 Dialog box for Slide Transition in PowerPoint.

PowerPoint also permits you to build your slides progressively for the audience. It is a bit like watching the construction of a building from the foundation to the roof. For slides, start with the title and add one line at a time to the text in the block below the title. To modify your slide to incorporate this feature, repeat the slide transition procedure described in the preceding paragraph. The new toolbar that is evident just above the window has a box that is titled No Build Effect. Click on the arrow beside that box to reveal a wide choice of techniques for building your slides. Pick an option; fly from the left is very common. When you select an option for building the slide, a second small icon shows under the view of that slide in the slide sorter view as shown in Fig. 12.19. You can proceed to build your slides, one by one, or you can use the same style on all of them. Select all of the slides on the slide sorter view by pressing the Control + A keys.

When you have completed the transitions and the building effects, you are ready to review the dynamic nature of your electronic presentation. This is real high tech. Go to slide 1 and click on the Slide Show icon button on the horizontal scroll bar. As you advance from slide to slide with arrow keys on the keyboard or by clicking the mouse button, you will see the transitions and the lines of text for each bullet item being moved onto the screen.

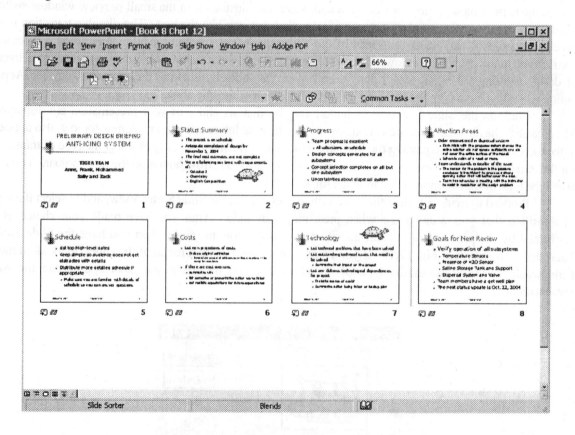

Fig. 12.19 Slide sorter view with transition and building effects icons below the individual slides.

If you have computer and a computer controlled projector available in the classroom, electronic presentations should be mandatory. Creating electronic slides and using them in a professional presentation are skills that are important for you to master. The transitions keep the audience on their toes and the building effects permits you to pace the flow of information. It takes less than an hour to learn how to prepare a very professional, dynamic presentation using PowerPoint. Take this opportunity to develop your computer graphic presentation skills.

12.8 SUMMARY

You have been introduced to PowerPoint, a graphics presentation program that is an extremely useful tool in preparing for design reviews or any other type of presentation. It is suggested that you take a few hours to acquire PowerPoint skills. At the very least, you will master the skill necessary to produce professional quality slides or overheads. If appropriate equipment is available in the classroom, you will also develop the ability to make presentations with computer-projected slides.

The screen for PowerPoint has been briefly described. As with all Windows based programs produced by Microsoft, many of the icons on the toolbars have a common purpose. The coverage given here is brief because learning is more rapid if you click, study and understand. Hopefully, this limited coverage with the examples described will be helpful.

It is recommended that you begin learning to use PowerPoint by preparing your preliminary design review using the Auto Content Wizard. The wizard organizes your presentation, arranges the sequencing of your slides and provides excellent suggestions for content in each slide.

After you have completed the standard presentation with the Auto Content Wizard, you should enhance the slides to improve their appearance. Procedures have been described for changing the design template, the colors of the background and the text, and by adding a footer and clip art. Tasteful and appropriate clip art is particularly impressive and easy to implement.

Finally, techniques for preparing electronic presentations were introduced. An electronic presentation depends upon the availability of a computer-controlled projector for the classroom. If the equipment is available, the use of a computer-controlled projector in your presentation has significant advantages. Employing transitions, as you move from one slide to the other, maintains audience interest. Utilizing build effects, where the text is presented one line at a time, has the significant advantage of controlling the flow of information to the audience. Also procedures for preparing and rehearsing a computer-aided presentation are described in this chapter.

EXERCISES

12.1 Prepare a title slide suitable for a final design briefing using PowerPoint.
12.2 Prepare a title slide suitable for a preliminary design briefing using PowerPoint.
12.3 Prepare a status slide that provides an overview for a preliminary design briefing using PowerPoint.
12.4 Prepare a progress slide using the slide outline format in PowerPoint. After completing the slide in the outline format, edit the slide, and print a copy of it.
12.5 Prepare a pair of slides describing your progress in the development of one of the subsystems for the project that you are developing.
12.6 Prepare a slide that describes the attention areas facing your team develops its project.
12.7 Prepare a slide showing your plans for action in the near future.
12.8 Using clip art, enhance a slide by inserting a caricature of an eagle in its lower right hand corner.
12.9 Prepare a transition from the title slide to the second slide. What type of a transition did you select?
12.10 Prepare a dissolve transition from the status summary slide to the progress slide.
12.11 Prepare a slide using the building effect where single lines of text are added to the slide one by one as you click the mouse button.
12.12 Add sound (chimes) to the transition from one slide to another during the sequencing of your slides.

PART V

COMMUNICATION

CHAPTER 13

LIBRARY RESEARCH SKILLS

By Margaret M. Cunningham

13.1 INTRODUCTION

This chapter is designed to guide you through the basic steps in library research. University libraries are complex and challenging places. Even those of you who were comfortable using your high school or public library can be intimidated by the size and complexity of a university library system, and its variety of online, print and nonprint resources. It can be challenging at first to understand how the Library of Congress classification scheme works, how the book and periodical stacks are arranged, how a microfilm reader operates, how government documents are organized, how to place an interlibrary loan request or locate materials placed on reserve, in addition to learning your responsibilities as a user of an academic library.

A main goal of this chapter is to help you work smart by being effective and efficient in your use of libraries. The library research skills you learn can be applied to any project regardless of the topic. Before you go any further, read over these time-saving tips:

- **Be proactive.** Take the time to learn the basic steps in library research. You will save valuable time in the end if you know *how the library works*.

- **Start early.** Library research *always* takes longer than you think. If you think you will need to spend an hour in the library, you will probably need two hours or more. Give yourself enough time to take new directions or unexpected turns. Depending on the size of your campus, you may be in competition with many other students and faculty for the same materials. Items are not always on the shelf when you need them.

- **Be realistic.** Plan to use a variety of resources to construct your assignment. A well-researched project will lead to academic success.

- **Seek help and ask lots of questions.** Librarians understand that it is not always easy for students to know what is the most appropriate resource to use, or how to locate an item in the library. Ask, ask, ask. It can save a great deal of your valuable time.

- **Evaluate your findings.** Critically evaluate the relevancy of the information you intend to use in your research, especially if you find the information freely on the web.

- **Copy, paste and cite all work not original to you.** When possible, e-mail the results of your searches to yourself so you can cut and paste bibliographic references and parts of documents into your reports. You must remember that academic honesty is vital and the penalties for plagiarism are severe. Properly cite borrowed ideas, terms, sentences, charts, images or paragraphs, found in print or on the web, with accurate footnotes and a complete bibliography. Unsure whether a source you used should be cited? Never guess. Ask your instructor or a librarian for guidance.

- **Carry your student ID card with you.** You will need it if you want to borrow a book or access a library resource, such as a periodical database or electronic journal that requires authentication.

- **Consider bringing these extra essentials with you to the library.** Carry a pen or pencil, paper or notebook, for taking notes and recording accurate bibliographic citations, a computer disk, some money or a copy card to make photocopies or to print out search results. Most libraries charge for printing.

13.2 OVERVIEW OF ACADEMIC LIBRARIES

Academic libraries collect the kinds of materials needed to support the courses being taught and research being conducted by faculty and graduate students. These materials include books, periodicals (newspapers, journals, magazines, newsletters, trade publications), government documents, compact disks, DVDs, technical reports, video and sound recordings, and online resources such as periodical databases, reference works, e-journals, e-books, and more.

Some campuses have such huge collections of library materials that it is necessary to divide it into smaller, more manageable collections. For example, on the University of Maryland, College Park campus there are seven libraries; they are: **Art, Architecture, Performing Arts, Chemistry, McKeldin Library** (main library), the **Engineering and Physical Sciences Library** and special collections in **Hornbake Library**.

13.3 TEAM OF FRESHMEN RESEARCHERS

For this introduction to library research, the basic steps (and some missteps) will be illustrated by the experiences of a team of four student researchers: Amir, Kim, Valentina and Mark. They are freshmen taking their first university engineering class, ENES 100 – Introduction to Engineering Design. Their instructor has divided the class into teams and every team is assigned the same design project. To gather background information for their project, they need to use the library. Let's join them on their library research expedition.

13.4 INITIAL STEPS – WHERE DO YOU BEGIN?

Amir, Kim, Valentina and Mark have been assigned by their instructor a semester-long project to design, build and test a **weighing machine**. In addition to building one, they must write a detailed report, including a bibliography that they will give to the instructor at the end of the semester. Just like these four students, you may have been assigned a project, or you might have a topic of your own in mind. Save time by clarifying the specific parameters of the assignment by meeting with your instructor or teaching assistant. After you understand the scope of the project, establish what is it that you need to create, assemble and test. You will soon begin a search for materials that describe *weighing machines*. The library is the source for these materials, and it is your task to find them.

Many research materials exist in print, and many others are available online from periodical, technical reports and patent databases, and the free web. A library's web site is a good starting point from which to find materials and to use the library's catalog to determine what books and journals are in the collection, and available on the shelves.

13.5 INTRODUCING LIBRARY RESOURCES

Identifying the most appropriate resources to use is critical to success. There is not one perfect book, article, web site, encyclopedia, handbook, manual or periodical database that has all the information you need. Plan to use a variety of resources to conduct your research, support your findings and project design.

The four freshmen have never designed and built a *weighing machine[1]*. They agree that they need to learn what *scales* look like and how they work, and to gather other background information. They ask the librarian at the Engineering and Physical Sciences Library for help in getting them started by directing them to the best resources to use. Amir, Kim, Valentina and Mark decide to divide up the work among them, then reconvene and share what they have learned and gathered.

Reference Materials

Reference materials, in print, are usually kept close to the information or reference desk of any library. Reference books normally do not circulate, that is, you cannot borrow them; you must use them in the library. There are many types of reference books, including encyclopedias, handbooks, dictionaries, almanacs, directories, manuals, telephone books, maps and more. Some of these may be available online, usually linked to a library's home page.

Kim and Valentina set out to find some background information on *weighing machines*. They decide to ask a librarian for help—this will almost always save time. One of the librarians on duty that day, Jim, suggests they begin by using an encyclopedia. Encyclopedias contain short articles that present a broad overview of topics and list a few references useful for further information. Jim suggests that Kim and Valentina consider using one or more of the following encyclopedias:

Access Science (McGraw Hill's encyclopedia of science and technology on the web)
Encyclopaedia Britannica (available on the web as *Britannica Online*)
Encyclopedia of Physical Science and Technology
Harper's Encyclopedia of Science
International Encyclopedia of Science and Technology
Van Nostrand's Scientific Encyclopedia.

Using the online version of *Encyclopedia Britannica,* called *Britannica Online*, Kim and Valentina do a search on the word:

scale

Although their search does not find specific articles on the topic, it does locate a good general introduction and overview of the term *balance*, and related weighing devices. Kim and Valentina begin recording this information in a notebook. Valentina asks the librarian to show her how to properly cite this online encyclopedia article.

Kim and Valentina decide to look in another online encyclopedia available from the library's web site: *Access Science*. Learning that the term *balance* might be better search term to use than

scale, Kim searches *Access Science* by entering the term:

balance

The first several records listed as search results appear to be on target. There are descriptions with illustrations on different kinds of mechanical and electronic balances; this information will provide the

---

[1] The project used in this chapter to illustrate library search methods is the design of a weighing machine.

team with valuable ideas on how to build a working model of a *scale*. Valentina asks Jim to show her how the library's print system operates so that she can obtain a copy of the encyclopedia article. Kim and Valentina ask the librarian how to properly cite this online source. At the end of the article is a bibliography on sources cited. Jim shows Kim and Valentina how to copy and paste some of the referenced book titles into the library's catalog search screen. They find several books available in the Engineering and Physical Sciences Library that can be checked out or borrowed.

Jim mentions to Kim that handbooks or manuals differ from encyclopedias in that they contain concise information, often of a technical nature about particular subjects. The students consult the following:

CRC Handbook of Mechanical Engineering   *Mechanical Engineer's Handbook*
Handbook of Mechanical Engineering   *Scale Handbook*
Kempe's Engineers Yearbook   *Scale Manual*

Kim and Valentina find detailed background information and clear descriptions of how *scales* function, with diagrams and formulas.

"What else can we use?" asks Valentina. The librarian suggests that they consider consulting other reference works such as thesauri, glossaries, or almanacs that contain synonyms, definitions, facts, figures and statistics. He pulls off the shelf a dictionary of engineering terms and says, "Use this dictionary if a word is unfamiliar. Consider using new terms to enhance your search strategies."

Another useful source is *Grainger General Catalog*, which is a product catalog of maintenance, repair and operating supplies and related information. Jim also suggests they consult *Thomas Register of American Manufacturers*, which is a directory of engineering companies. It can be used to find descriptions of the work performed by specific engineering companies, and their contact information (address, FAX, telephone number, e-mail address). Having helped students in previous semesters taking ENES 100, Jim knows they will need to contact companies that design and build scales and include this information in their final report that accompanies their design project.

Finding Books Using the Online Library Catalog

While Kim and Valentina are looking through reference works to gather information, Amir and Mark run into a mutual friend. Eun-Ju is a sophomore and has already taken the introductory engineering design class. He tells them that books are good sources for acquiring in-depth information on *weighing machines*. The only problem they might encounter looking for books, Eun-Ju warns, is that "they might already be borrowed by someone else, or may not contain the most recent information." Unsure of how to find books in the library, Amir asks another librarian, Julie, to show him how to find books.

"The online library catalog is the resource you need to use to locate books on your topic. It lists all the holdings of the library," says Julie. Amir, Mark and the librarian walk over to a computer terminal in the reference area. The computer monitor displays the library's home page. "As you can see," Julie says, "there is a direct link from the library's home page to the library's catalog. Just click on the link that says *Catalog*. A library catalog may be searched several different ways—by key word or words anywhere, by subject, title, or author or call number. Since you have a specific topic in mind—*weighing machine*—executing a *words anywhere* or *key word* search would be the first step to find books on that topic." Amir enters the words:

weighing machine

Julie says, "the catalog provides you with a list of records that contain the words you entered. Depending on the library catalog system you use, consider doing a phrase search by using quotation marks around your terms. For example a search on "weighing machine" will find catalog records that contain this phrase, where words are linked together in this exact word order. Also, consider truncating the word *machine*. Truncation enables you to locate the variant spelling of terms. In this catalog system, the truncation symbol is the asterisk (*). Adding it to the word *machine** will retrieve records that contain the words *machine, machines* and *machinery*. Examine your results and click on a book title that interests you. The catalog record will state the availability, call number and item status." Mark asks, "what is meant by availability and item status." "*Availability* indicates the library location of the book (e.g., Engineering and Physical Sciences Library, book STACKS). *Item status* tells whether the book is available on the library shelf or has been borrowed by someone," says Julie.

Amir and Mark search the library's catalog, using a variety of search terms and find a few books about *weighing instruments*. All the books are located in the Engineering and Physical Sciences Library. They write down the call numbers, examine a floor map of the library and go into the book stacks to retrieve them. A short time later they return to the reference area with two books between them. However, they could not find one of the most promising titles that had been published just this year. When they check the catalog again, they notice that the status has a "due date." They ask Julie about it and she suggests that they *request* the book from the person who has it checked out. It is a simple procedure using the online catalog. Since Amir does not have his student ID card with him, Mark performs the online request. He will be notified—probably in two to three weeks that the book is available for him. Julie explains that when you place a request through the catalog, the library sends a notice to the person who has the book checked out. The person does not know your name nor do you know their name. Even though the library has long loan periods, everybody, including faculty, are subject to returning items when requested. If Mark gets notification that the book he wants has been returned, he will need to act on it quickly. Julie mentions that the procedure for requesting or recalling books can vary between library systems.

Periodical Databases

Meanwhile, Valentina and Kim have found a table in the library where they can work. They have pulled out from their class notes a description of the research requirements for their design project. Amir and Mark walk over to meet them. "Hey, we found some books on *weighing instruments*," says Mark. "Great!" Kim says, "now, according to our assignment we have to locate some professional or scholarly journal articles on this topic. It says we need to use periodical databases—what are they and where do we find them?"

They walk over to the reference desk and ask Bob, another librarian, who says, "Periodical databases are also called journal databases or indexes or sometimes are simply referred to as databases. They are used to locate articles published each year in magazines, scholarly journals, newspapers, and newsletters. Databases can contain information on a variety of subjects, such as science, business, literature, or may contain subject specific information, such as civil engineering. Some databases provide you with the entire full text article, but many others provide only citation information (title of the article, author, and title of the journal, volume, issue, date and pages) as well as an article summary or abstract. Academic libraries subscribe to periodical databases primarily in online form, making them available from links off the library's home page."

"This library subscribes to a variety of engineering databases that contain professional literature or scholarly articles. Some are more appropriate to use than others, depending on the topic. What are you researching?" asks Bob. "*Weighing machines* or *scales*," says Kim. "We are all taking ENES 100." "Oh, I have helped many students over the past few years with this assignment," says Bob, "so I suggest you begin by doing some preliminary searches in these two databases: *Applied Science and Technology Abstracts* and *NTIS* (an acronym for the *National Technical Information Service*). *NTIS* will help you find technical reports on your topic."

The team members consult with each other and discover that none of them has ever searched a periodical database before. "Okay, let's use *Applied Science and Technology Abstracts*." They gather around a computer terminal in the reference area and connect to the database.

Bob asks the students, "Do you know about phrase searching, and using AND and OR, also called Boolean connectors, to search periodical databases?" Amir tells him that Julie, the other librarian, instructed them on how to do a phrase search, using quotation marks, and how to truncate terms, using the asterisk, to search the library's catalog. "Can we apply these search options when searching a periodical database," asks Mark. "Yes and no," says Bob, "some databases allow you to search for exact phrases using quotation marks, while others do not. The truncation symbol can vary from database to database; they can include: * ? ! + $. To save time, check the HELP or SEARCH TIPS section of any database; it will give you information on which commands are used in a given database." "About Boolean connectors," says Valentina, "I think AND is used to connect different words or different concepts together, but I can't remember what OR does." "Yes," says the librarian, "AND is used to connect words or phrases that are different or unrelated in meaning, when you want the search to find both words in the same record. AND *narrows* your search." In *Applied Science and Technology Abstracts*, you can use quotation marks around words to form phrases, and use the asterisk (*) as the truncation symbol. Here are some examples using AND:

> weighing instruments **and** design
> "weighing instrument*" **and** design*

Bob continues to explain, "OR is used to connect concepts that are similar in meaning, when you want the database to find any or all of the words or phrases in the same record. OR *broadens* your search." Examples include:

> balance **or** weighing machines **or** scales
> balance **or** "weighing machine*" **or** scale*

Bob offers another tip: "You can also combine phrase searching, truncation and AND and OR in the same search sentence. Make sure to use place parentheses around terms connected using OR. Here are some search examples

> design* and (balance or "weighing machine" or scale)
> "weighing instrument*" and (design or construction)
> "weighing instrument*" and (design or construct*)
> (design or construct) and (balance or scale)

"Let's try some searches and see what happens," says Amir. "Kim, how about starting us off?" Kim enters the search:

> (scale* or balance*) and (design or construct or build)

The database responds by listing an enormous number or results, many of which appear irrelevant. Bob suggests, "perhaps your search is too broad. Modify your search to find information on specific kinds of *scales* or *balances,* such as *digital* or *electronic scales* or *electronic weighing scales.*" "Okay," says Kim. She modifies the search as follows:

> "digital scale*" or "electronic scale*" and (build* or construct*)

Bob indicates that the next step for them is to evaluate their results. "Some of the records retrieved will contain the kind of information you are looking for. Others will contain your key words, but will have no relevancy to the kind of information you have in mind. These records are referred to as *false hits*." "We cannot be sure which records are relevant and which are not from the title without reading the abstract first," says Amir, "so, let's take the time to read through them carefully."

"Also, depending on the kind of search you do, you might find too many results. If that happens, you may narrow your search by adding more key words connected using AND. On the other hand, if your search retrieves too few results, expand your search by thinking of additional synonyms or related words and connect them using OR," suggests Bob. Mark says, "I'm glad we decided to start our research today because it takes longer than you think."

Amir, Mark, Kim and Valentina read through and examine their search results, selecting the ones that seem most relevant. They quickly determine that this database does not always contain the full text of articles, but sometimes contains only abstracts or summaries--as the title of the database suggests. Jim says, "If you don't want to print the list of article citations you are interested in reading, you can e-mail it to yourself, or print a copy. After you do this, I will show you how to locate these articles in the library."

"What's a citation or article citation?" asks Amir. "An article citation indicates the author's name, the title of the article, the title of the journal the article was published in, the volume and issue number, the date of publication and page numbers," says Bob. "Also," says Kim, "the information we use from any article in our report must be accurately recorded in our list of references or bibliography." Valentina confirms this and points to a record on the screen. She shows the others what information is included in an article citation:

Title: Scale reads and writes barcodes.
Source: Design News   vol. 43 (Jan 5 1987), p. 26.

"Don't some databases have full text articles?" asks Mark. "Yes," says the librarian, "some producers of science databases provide not only citation and abstract information but provide the full text of articles. For records that pertain to your topic, look for those that have links to the complete or full text document, usually indicated by a hyperlink or perhaps a full-text icon. For the rest you will need to record the complete citation, then connect to the library's catalog and search to see if the library subscribes to the journal title you need."

"Okay, let's begin searching this periodical database," says Mark. The team works hard, trying various key word search combinations. They experiment by using the truncation symbol to find variant spelling of terms, and refine or limit their search by a specific time period (e.g., 1993-2004). They spend time reading and printing the abstracts of articles that potentially look useful.

Amir suggests that they not limit their findings to only one source and try another database. The students try the *NTIS* database next. "What kind of information does *NTIS* have?" asks Valentina. "*NTIS* contains technical reports of U.S. government-sponsored research. These reports are issued by federal, state, and local governmental agencies, by universities, and by private companies. *NTIS* even includes some international technical reports. Therefore, *NTIS* has different kinds of information on *balances* and *scales* than what you found searching *Applied Science & Technology Abstracts*," says Bob.

The students learn that they can use AND and OR in *NTIS*, just like in *Applied Science and Technology Abstracts*. Instead of writing down journal citations, the students write down the report numbers (e.g., PB93855195); technical reports often have more than one report number under which they might be filed. Working with Bob to locate these reports they learn that the library owns many, but not all, of the reports listed in the *NTIS* database. The reports that the library does have are available on microfiche, filed by report number, in the Technical Reports Center of the library.

Technical reports that the library does not own may be special-ordered from the National Technical Information Service, for free. They may take about two weeks to receive.

Using the Online Catalog to Locate Journals

To find the full text of some articles found while searching the *Applied Science and Technology Abstracts* database, the students recorded a list of citations and now will need search the library's online catalog. "I thought you couldn't search the catalog to find journal articles," says Mark. Bob, "Yes, that's right. You do not use the library catalog as a search tool to find articles on your topic; that is why you searched the *Applied Science* and *NTIS* databases to find articles and technical reports. However, if the full text article is not available in the database, you do need to use the catalog to determine if the library owns the journal and volume you need."

"Instead of doing a *key word* or *words anywhere* search in the catalog, do a *title* or *title beginning with* search on the title of the journal. For example, you would do a title search on *Design News*. If the library has the journal, it usually appears on the first page of results. Click on the title then click on *Holdings* or *Availability* to determine if the library has the volume you need. If *yes*, record the library location and call number. If the name of the journal you need does not appear in your list of results, the library probably does not own the journal. But check your spelling to be sure," says Bob.

Amir types the title of one of the journals on his list, but it does not show up in the catalog. The librarian advises, "It could be a journal our library does not own. No library, no matter how big their collection is, is able to subscribe to all the periodicals ever published. Libraries across the city, state, nation, and around the world share materials needed by their patrons. If this library does not have the journal and/or volume you need, you can request that the Interlibrary Loan office obtain a copy of the article for you. They will procure a copy for you at no cost. You simply fill out an online request form, linked from the Interlibrary Loan web page, and the library does the rest on your behalf."

Amir and Mark have to go to class, so the team stops their research for the day.

Patents

The next morning, Kim, Valentina, Amir and Mark are back in the library, this time looking for patents on their topic. By now they feel very comfortable talking to the librarians, so they find Julie at the reference desk and ask for his help with patent searching.

"Patents can help with your design project because they have detailed descriptions and drawings of specific devices," says Julie. "You will need to look at a number of patents, because they are often written in 'patentese,' the highly technical and legal language that is necessary to define inventors' legally protected ideas." Julie and the students walk over to a computer terminal in the reference area. Julie suggests that they search the *United States Patent and Trademark Office* web site:

http://www.uspto.gov

Julie points out the features on the home page of the *U.S. Patent and Trademark Office*. "Kim," says the librarian, "how about getting your team started by clicking on *Patents*, then click on *SEARCH patents*. From the list of search options made available, I suggest that you begin by doing a *Quick Search*." Julie goes on to mention that it is possible to use quotation marks to form a phrase and use the truncation symbol $ to find variations on word endings. Kim enters the following search:

"digital scale$"

This search results in 281 patents. "I heard our Teaching Assistant talking to some students about *programmable scales*. Let's add another word to our search," says Amir. Mark refines the search as follows:

"digital scale$" and programmable

This search results in 61 patents since 1976. "Let's look at this one **6,538,215 Programmable digital scale**," as Kim clicks on that number to obtain the full text. "You can use recent patents to find older ones," the librarian advises, "scrolling further down the screen, after the sections for *Abstract* and *Reference Cited*, is a list of earlier *U.S. Patent Documents*. We see that this current patent for a *programmable scale* has cited an earlier one (no. 763,967,690) from July 1996."

Scrolling down the page, they see the *Current U.S. Class* is **177/25.16** as well as having four other classes: 177/25.19, 600/547, 702/173 and 128/921   "You can copy and paste any one of these class numbers (e.g., 177/25.16) into a search," Julie explains, "but you need to add **ccl/** before the number. Keep in mind that the U.S. patent database is searchable by key word, inventor name, and company name, back to 1976, but patents dated earlier than 1975 all the way back to 1790 can be searched *only* by knowing the patent number and by the subject classification number. This means that you must find classification numbers to search pre-1976 patents by subject." Kim copies and pastes the current class number (e.g., 177/25.16) and clicks back in the patent database to the *Refine Search* screen. "Once you know the class," Julie says, "you can search specifically on that type of patent." Kim enters the search:

ccl/177/25.16

This search results in 19 patents; all have searchable text and images linked to their patent descriptions. "Keep in mind that outside of the library," says Julie. "you need a special *tiff* viewer to view the patent images." "There's a lot to remember!" exclaims Valentina. "There are many helpful pages on the web," says Julie, "such as Patent and Trademark Resources on the Internet:"

http://www.library.ucsb.edu/subj/patents.html

"However, the primary web site to remember is the U.S. Patent and Trademark Office page:"

http://www.uspto.gov

"It has a lot of basic and introductory information, but it links you to highly technical and legal resources, also." The students seem somewhat overwhelmed, as they prepare to head back to class. Julie reassures them, "Patent searching is a lot to learn, but you all catch on very well."

Beyond the Library – Web Search Engines

The freshmen researchers, Valentina, Mark, Kim and Amir, have worked hard to gather information on the topic of *weighing machines*. "What other resources can we use?" asks Kim. Valentina says, "A friend of mine surfs the net all the time to find information, and by the net I don't mean the library's web site. Maybe we should do that." Kim says, "I believe that everything we need for our project is on the web." "That's what I think too," says Amir, "we should have searched the web from the start." "I'm not so sure about that. In my experience," says Mark, "I find a lot of good stuff on the web that matches my personal interests. But when I used it to do research for my English class, I found it difficult to come up with a few good pieces of information, and I found them only after spending a lot of time looking through lots of junky sites." "But we used a non-library site—the U.S. Patent and Trademark Office web site to find authoritative patent information on scales, so there are sites out there we can use," says Kim. "Yes," says Mark, "but how do you locate these important web sites? I don't easily come across that many useful sites when I surf online." "Let's ask Julie for some advice," says Valentina. "We really can't afford to waste time."

The students go over to the reference desk. Julie is just getting off the telephone, "Hello," she says. "We are having a debate about surfing the web to find information on our ENES 100 project. Some of us find it easy and others difficult to locate useful information. What do you suggest?" "I am sure we would all agree that the web is a wonderful invention. We wouldn't be able to access all the library's online resources without it. However, there is a lot of misinformation about the web," says Julie. "People believe that *everything you ever wanted* is available on the web, that everything on the web is *free*, that everything published on the web is *authoritative* and should be trusted, and that everything on the web is *up-to-date*. This belief is just not so."

"Many students start off using a search engine to surf the web and waste a lot of valuable time weeding though hundreds of useless results to find one or two useful sites. I commend you all because your team has gone about starting your research the right way. Since your aim was to find credible, reliable information, you used some of the library web resources such as an online encyclopedia to find an overview of *scales* or *balances*. You also used the library's online catalog to find books, and periodical databases to locate articles. Then you searched the U.S. Patent and Trademark Office web page to find patents on scales."

"So there are different kinds of web sites?" asks Kim. "Yes," says Julie. "For example, the library's web site is created by librarians who are experts in the field of information access. People who click on the links from this page, and use its resources, can trust that they have been edited and checked for accuracy by scholarly organizations and publishers before being purchased by the library, and then carefully evaluated by librarians before being added to the library's collection."

"Beyond this site there are lots of useful web sites produced by other academic institutions, by government agencies, by non-profit organizations, by companies and corporations, and by individuals. However, keep in mind that non-experts create the vast majority of web sites. Anyone with a little expertise using a web editor and access to a server can create a web page. But, can you trust the information? You need to take the time to evaluate what seems reasonable and discard what seems unreliable or questionable," Julie says, "the web is a great research tool so long as you know how to use it effectively and spend the time to evaluate your results."

"Let's take a minute and talk about searching. Which search engines do you most often use?" asks Julie. "Google," says Amir, Kim and Valentina. "I use Google and Yahoo!" says Mark, "I want to be able to search the web as efficiently as we learned to search the library's periodical databases." "Using Yahoo! I type in my key words, and then look through my results, which are sometimes accurate but most times not really on target. I seem to waste a lot of time."

"Let me suggest a few search tips," says Julie, "How about one of you leading us through some searches in Google?" "I'll do it," says Amir as he turns the keyboard and proceeds to type in the URL for Google. "How about we do a search on the word *scale*?" Amir enters the word:

scale

Julie says, "Most people believe that a search engine searches the entire web and brings back to the searcher all the possible results or web pages that contain information on their topic. This is false. A search engine searches only a portion of the web; it searches the web pages indexed in its database of web sites. That is why if you do the same search using another search engine you obtain different results."

Google responds with a result list of millions of hits on the word *scale*. "Let's change our search to *digital scale*," suggest Kim. Amir adds a word to perform a search on:

digital scale

"The search engine responds with almost 2.5 million hits – a lot less than the millions listed for the initial search. The more words you add the narrower or more focused your search becomes. The search engine tries to find the words *digital* and *scale* together in the same web page," says Julie. Google will automatically put an AND (the Boolean connector) between the words, so Amir's search was actually:

digital **and** scale

Not all search engines use AND as the default connector. Amir tries a third search using OR:

digital scale **or** electronic scale

The search engine finds web pages that have the word *scale* as well as the words *digital* or *electronic* (or both). Now the search results are even fewer. "I know that you can put quotation marks around words to make a phrase," says Amir. He modifies the search as follows:

"digital scale" **or** "electronic scale"

This last search is more specific, since the students are looking for web pages that contain the specific phrases *"digital scale"* or *"electronic scale"* in the records retrieved. Adding a plus + sign in front of a word or phrase will tell the search engine that this term *must* appear in the list of results. Amir tries another search:

"digital scale" **or** "electronic scale" +design

In this search, web pages with the phrase *"digital scale"* or *"electronic scale"* will also include the word *design*. Keep in mind that web pages containing your key words many more times are listed first or at the top of the results list, and are more relevant to the topic, as compared to web pages listed towards the bottom of the list of results since they contain fewer of your key words and are probably less relevant. "But remember," Julie says, "you have used only one search engine. When you compare the results found using Google to another search engine, the results will be quite different. Also, each search engine works differently from each other. It is best to read the HELP or SEARCH TIP screens to determine what commands or features work in that search engine, learn how the Advanced Search features function, and how to customize settings." Julie says, "Consider adding one more word to our search statement." Amir types:

"digital scale" **or** "electronic scale" +design +build

This search results in fewer than 100 hits. "Okay, I'll leave you to search on your own," says Julie. Amir scrolls through the first 10 results. The students examine them. "You can tell by part of the URL (or Web address) called the domain," says Mark, "the group that created the web site." The most common domain names are:

.com - a company or commercial organization **.net** - a network of computers
.edu - an educational or academic institution **.org** - a not-for-profit organization
.gov - a U.S. government agency, department **.mil** - a U.S. military department

Valentina is looking over Amir's shoulder and sees a lot of web sites that have **.com** or **.org** in the address. The team gathers around the workstation and they try several new searches. A short time later, Julie returns and asks, "How is it going?" "It's a lot tougher that you would think to find some basic information on how to build and design a *digital scale*," says Valentina. "We found lots of commercial sites that give brief descriptions of *digital* or *electronic scales*, but they want you to place an order, or fill out a request to obtain their online product catalog," says Mark. Kim adds, "Other company sites ask you to send them your design specifications so that they can send you an estimate of what it would cost to build the *scale* you want." Mark adds, "It's amazing how a simple search on this topic can lead you to find so many things." "Well," says Valentina, "we did find a couple of sites we believe have some credible information, and we also wrote down a list of company web addresses that we can contact if we need to consult them about building materials."

"When we search *Applied Science and Technology Abstracts*, all the articles we found were published in scholarly publications and were authoritative. How do you know we can trust the information we find on the net that are non-library sites?" asks Valentina. "Let me show you," says Julie. "The library has designed a web page to help people understand how to evaluate web sites." The students click through the various links on the library's home page to find the *How to Evaluate Web Sites* page. Amir writes down the web address for the group. Listed here are some criteria you can apply to web sites to determine:

Accuracy
Anyone can create a web page. It is important to find out who is the author and what are the author's qualifications or expertise; this will help you determine the credibility and reliability of the information.

Purpose
Determine the purpose of the web site by looking closely at the content of the information. Some sites provide links to *About Us* or *Who Are We* or a *mission statement*, detailing its purpose. The purpose of other sites might not be obvious at first. Take the time to thoroughly explore a web site to determine if the information is subjective (biased or opinionated) or objective (factual), or mixed.

Currency
The regularity of updating information is vital for some types of information and not so important for others. For example, sites that provide historical information (such as the presidential papers of Thomas Jefferson) do not have to be updated as often as sites that provide news or stock market information.

Design

Design, organization and searchability are important considerations. web sites can be useful sources of information; but if they are slow to load, difficult to navigate, search or read, their contribution or usefulness will be lost.

"Great," says Mark, "let's apply these criteria to the web pages we found and are considering using in our report."

Web Sites of Professional Organizations

Several days later, Mark and Valentina are wondering whether they have overlooked anything. They try searching the library's web site for other resources they can use. They come across links to: Professional Organizations, Standards and Careers.

Professional organizations, associations and societies are dedicated to exploring, developing and promoting all aspects of the discipline they represent. Their Web pages usually have links to their membership, publications, news releases, calendar of events, special interest groups, awards of excellence, training, professional development and more. Here are a few:

Professional Organizations

ASM International Foundation (http://www.asm-intl.org)
Aerospace Industries Association (http://www.aia-aerospace.org)
American Chemical Society (http://www.acs.org)
American Institute of Aeronautics and Astronautics (http://www.aiaa.org)
American Institute of Chemical Engineers (http://www.aiche.org)
American Nuclear Society (http://www.ans.org)
American Society of Civil Engineers (http://www.asce.org)
American Society for Metals & Metallurgy (http://www.asm-intl.org)
American Society of Mechanical Engineers (http://www.asme.org)
Institute of Electrical and Electronics Engineers (http://www.ieeeusa.org)
Institute of Industrial Engineers (http://www.iienet.org)
American Institute of Mining, Metallurgical and Petroleum Engineers (http://www.tms.org)
National Society of Professional Engineers (http://www.nspe.org)
Society for Mining, Metallurgy, and Exploration (http://www.smenet.org)
Society of Petroleum Engineers (http://www.spe.org)

Standards

Organizations or societies often develop design and specification standards for technologies such as *balances* or *scales*. Here are a few:

American Society for Testing Materials (http://www.astm.org)
International Organization for Standardization (http://www.iso.ch)
National Institute for Standards and Technology (http://www.nist.gov)

Valentina says, "Let's search the web using the following search strategy:"

"professional society" +engineering

"This will help us find other scholarly associations in the field of engineering," says Valentina. "Great," Mark says, "I learned the other day that besides professional societies, we can access descriptions of careers and occupations using the *Occupational Outlook Handbook Online,* a searchable database produced by the U.S. Bureau of Labor Statistics, found at:"

http://stats.bls.gov/oco/

Mark clicks on OOHSearch/A-Z Index and does a search on the following occupation:

civil engineer

This handbook allows users to locate information on specific engineering related occupations. Professional societies for each occupation are linked at the end of each article. For this profession, he was able to locate information such as: *nature of the work of civil engineers, employment, job outlook, earnings, and sources of additional information.*

13.6 SUMMARY

The steps taken by Amir, Kim, Mark, and Valentina in the library and on the web illustrate the basic research process for finding good sources of information that will help in completing an assignment, project, or paper. The steps can be time consuming, even with guidance from librarians, so it is important not to wait until just before an assignment is due to begin the process.

The students began their research with encyclopedias, in order to obtain descriptive background information on their topic. They also looked in other reference books, such as directories, handbooks, glossaries and product catalogs. Then they searched the library's online catalog to find a few books on their topic.

Their next step was to use an online database to search for periodical articles. In the field of engineering, they discovered that the *Applied Science and Technology Abstracts* has relevant citations and some full text articles to journal articles, and the *NTIS* database is very important for locating technical reports. They learned how to use Boolean connectors and truncation to improve their search results.

When they found a citation to an article in a journal, they looked up the title of the journal (*not* the title of the article) in the library's catalog to see whether the library subscribes to that journal. They discovered that valuable sources for detailed descriptions and models are patent databases, which are searchable on the web. There is a lot to learn about patent searching, but librarians are eager to help.

The students also learned that a lot of valuable information can be found on the web – providing that one narrows the search using quotation marks around phrases and the + sign to indicate very important concepts. Also, it is critical to evaluate the quality and credibility of information found on the Web, since so many non-experts or organizations and individuals with biases can create a presence on the web.

Lastly, the students discovered some useful web sites for locating information on careers, professional organization and societies, and standards.

The amount of information the team uncovered in a few days of research was overwhelming. They could not possibly use all of it for this project, or for any assignment. They would have to be very selective in picking and choosing among all these resources. With their improved understanding of how to evaluate these information resources for relevance to their need and for scholarly credibility, however, they would be able to make good choices.

REFERENCES

1. Hanrahan, T., "The Best Way to Search Online," Wall Street Journal, December 6, 1999, p. R-25.
2. Bolner, M. S. The Research Process: Books and Beyond, Kendall Hunt Pub. Co., Dubuque, IA, 1997.
3. Jones, S, Doing Internet Research: Critical Issues and Methods for Examining the Net, editor, Sage Publications, Thousand Oaks, CA, 1999.
4. MacDonald, R. M., Successful Keyword Searching: Initiating Research on Popular Topics Using Electronic Databases, Greenwood Press, Westport, CT, 2001.
5. Schlein, A. M. Find It Online: The Complete Guide to Online Research, Facts on Demand Press, Tempe, AZ, 2000.

6. <u>Navigating the Sea of Information</u> [video recording] produced for The Provost Advisory Committee for Video Production; produced by the staff of the Center for Instructional Development and Distance Education, University of Pittsburgh, Media Inc, Media, PA, 1997.

EXERCISES

13.1 Using the library's online catalog, prepare a list of four or five books published since 1990 on the product your team is designing.

13.2 Other than the periodical databases listed in this chapter (*Applied Science and Technology Abstracts* and *NTIS*), list two additional engineering or science periodical databases that you could use to research the product your team is designing.

13.3 Using a periodical database, prepare a list of five article citations, published since 1990, on the product your team is designing. A complete citation includes: the author (s), title of the article, title of the journal, the volume, issue, date and page numbers.

13.4 Referring to your list of five journal article citations (see exercise 13.4), choose one. Locate the full text article and in your own words briefly summarize it.

13.5 Generate a list of the basic steps needed to perform a patent search using the *U.S. Patent and Trademark Office* Web site: http://www.uspto.gov.

13.6 Prepare a list of web addresses of five different search engines, other than those mentioned in this chapter (*Google, Yahoo!*).

13.7 Evaluate this web site http://www.scalesgalore.com using the criteria listed in this chapter, such as: accuracy, purpose, currency, design and organization.

13.8 Search the web to find the home page for the *American Society of Mechanical Engineers*. What is their web address? What is their purpose or mission? List five features listed on their home page.

13.9 Using the print or online version (http://www.bls.gov/oco) of the *Occupational Outlook Handbook*, do a search on *engineers*. List the difference in median annual earnings between civil engineers and electronic engineers.

13.10 Using the web, develop a one-page history of the Ford Motor Company.

13.11 Using the web, develop a one-page history of the General Motors Corporation.

13.12 Using the web, develop a one-page history of the largest manufacturing company in the State in which you reside.

13.13 Go to this address http://collegehousebooks.com and write a brief history of this company.

13.14 Search a suitable periodical database to find an article by the author of this textbook (see Exercise 13.14).

13.15 What is the name of the Dean of the college of engineering at your institution? Search a suitable periodical database to find an article written by this person.

13.16 If you could give one tip to your best friend about library research, what would it be?

13.17 Cite three patents that are related to the product your team is designing. Write a brief description of each patent and list the claims allowed by the examiner.

13.18 Perform a product search and identify products that are similar to the one your team is designing.

13.19 What is the URL for your library's home page?

13.20 List at least three things that you learned after reading this chapter that you did not already know.

CHAPTER 14

TECHNICAL REPORTS

14.1 INTRODUCTION

It is important that you enjoy writing, because engineers often have to prepare several hundred pages of reports, theoretical analyses, memos, technical briefs, and letters during a typical year on the job. It is essential that you learn how to effectively communicate—by writing, speaking, listening, and employing superb graphics. Communication, particularly good writing, is an extremely important skill. Advancement in your career will depend on your ability to write well. It is recognized that you will be taking several courses in the English Department and other departments in Social Sciences, Arts and the Humanities, which will require many writing assignments. These courses should help you immensely with the structure of your composition and the development of good writing skills. Most of the assignments will be essays, term papers, or studies of selected works of literature. However, there are several differences between writing for an engineering company and writing to satisfy the requirements of courses such as English 101 or History 102.

In college, you write for a single reader, namely the instructor. He or she must read the paper to grade it, subsequently determining how well you are doing in class. In industry, many people, within and outside the company, and with different backgrounds and experience, may read your report. In class, the teacher is the expert, but in industry the writer of the report is supposed to be the one with the knowledge. Many of those working in industry do not want to read your report because they are busy: their phone is ringing continuously, meetings are scheduled back to back, and many important tasks need to be completed before the end of the day. They read—if not skim—the report only because they must be aware of the information that it contains. They want to know the key issues, why these issues are important and who is going to take the actions necessary to resolve them. In writing reports in industry, you cut to the chase. There is no sense in writing a 200-word essay when the facts can be provided in a 40 to 50-word paragraph that is brief and cogent. An elegant writing style, so valued by instructors in the arts and humanities, is usually avoided in engineering documentation. In writing an engineering document, do not be subtle; instead be obvious, direct and factual.

In the following sections, some of the key elements of technical writing, including an overall approach, report organization, audience awareness, and objective writing techniques will be explained. Then a process for technical writing, which includes four phases: composing, revising, editing and proofreading, is described.

14.2 APPROACH AND ORGANIZATION

The first step in writing a technical report is to be humble. Realize that only a very few of the many people who may receive a copy of your report will read it in its entirety. Busy managers and even your peers read selectively. To adapt to this attitude, organize your report into short, stand-alone sections that attract the selective reader. Three very important sections—the title page, summary and introduction are located at the front of the report so they are easy to find. At the front of your report, they attract attention and are more likely to be read. In college, you call the page summarizing your essay an abstract, but in industry, it is often called an executive summary. After all, if it is prepared for an executive, perhaps a manager will consider it sufficiently important to take time to read it. Follow the executive summary with the introduction and then the body of the report. A common outline to follow in organizing your report is provided below:

- Title page
- Executive summary
- Table of contents, list of figures and tables
- Nomenclature—only if necessary
- Introduction
- Technical issue sections
- Conclusions
- Appendices

The title page, the executive summary and the introduction are the most widely read parts of your report. Allow more time polishing these three parts, as they offer the best opportunity to convey the most important results of your investigation.

Title Page

The title page provides a concise title of the report. Keep the title short—usually less than ten words; you will have an opportunity to provide more detail in the body of the report. The authors list their names, affiliations, addresses, and often their telephone and fax numbers and e-mail address. The reader, who may be anywhere in the world, should be provided with a means to contact the authors to ask questions. For large firms or government agencies, a report number is formally assigned and is listed with the date of release of the report on the title page. A title page of a report on a computer code for designing combustors is shown in Fig. 14.1. This report was prepared by several authors from different organizations and was sponsored by NASA (National Aeronautics and Space Administration) and ARL (Army Research Laboratory).

Executive Summary

The executive summary should, with rare exceptions, be less than a page in length (about 200 words). As the name implies, the executive summary provides, in a concise and cogent style, all the information that the busy executive needs to know about the content in the report. What does the big boss want to know? Three paragraphs usually satisfy the chief. First, briefly describe the objective of the study and the problems or issues that it addresses. Do not include the history leading to these issues. Although there is always a history, the introduction is a much more suitable section for historical development and other background information related to the problem. In the second paragraph, describe your solution and/or the resolution of the issues. Give a very brief statement of the approach that you employed, but reserve the details for the body of the report. If actions are to be taken by others to resolve the issues, list the actions, those responsible, and the dates of implementation. The final paragraph indicates the importance of the study to the business. Cost savings can be cited, as can improvements in the quality of the product, gains in the market share, enhanced reliability, etc. You want to convince the executive that your work was worth its cost, valuable to the corporation and that you performed admirably.

NASA
Technical Memorandum 107204

Army Research Laboratory
Memorandum Report ARL–MR–307

ALLSPD-3D
Version 1.0a

K.-H. Chen
Ohio Aerospace Institute
Brook Park, Ohio

B. Duncan
NYMA, Inc.
Brook Park, Ohio

D. Fricker
Vehicle Propulsion Directorate
U.S. Army Research Laboratory
Lewis Research Center
Cleveland, Ohio

J. Lee and A. Quealy
NYMA, Inc.
Brook Park, Ohio

April 1996

National Aeronautics and
Space Administration

U.S. ARMY

RESEARCH LABORATORY

Fig. 14.1 Illustration of a title page of a professional report.

Introduction

In the introduction, you restate the problem or issues. Although the problem was already defined in the executive summary, **redundancy is permitted in technical reports**. Recognize that you have many readers, so important information, such as a clear definition of the problem, is placed in different sections of the report. In the introduction, the problem statement can be much more complete by expanding the amount of information describing the problem. A paragraph or two on the history of the problem is in order. People reading the introduction are more interested than those reading only the executive summary; therefore, more detail is appropriate.

Following the problem statement, establish the importance of the problem to both the company and the industry. Again, this is a repeat of what was included in the executive summary, but your arguments are expanded. You can cite statistics, briefly describe the analyses that lead to the alleged cost savings, give a sketch of customer comments indicating improved quality, etc. In the executive summary, you simply stated why the study was important. In the introduction, however, arguments are presented to convince the reader that the investigation was important and worth the time, effort and cost.

In the next segment of the introduction, you briefly describe the approach followed in addressing the problem. You begin with a literature search, followed by interviews with select customers, a study of the products that failed in service, interviews with manufacturing engineers, and the design of a modification to decrease the failure rate and improve the reliability of the product. In other words, tell the reader what actions were required to solve the problem.

The final paragraph in the introduction outlines the remaining sections of the report. True, that information is already covered in the table of contents, but again, you repeat to accommodate the selective reader. Also, placing the report contents in the introduction gives you the opportunity to add a sentence or two describing the content of each section. Perhaps you can attract additional reader interest in one or more sections and convince a few people to study the report more completely.

Organization

An organization for the report was suggested by the bullet list shown at the beginning of this section. However, you should recognize that an organization exists for every section and every paragraph in the report. In writing the opening paragraph of the section, you must convey the reason for including this section in the report and the importance of its content. You certainly have a reason for including this section in the report, or you would not have wasted the time required to write it. By sharing this reason with your readers, you convince them that the section is important.

A paragraph is written to convey an idea, a thought or a concept. The first sentence in the paragraph must convey that idea, clearly and concisely. The second sentence should describe the significance or importance of the idea. The following sentences in the paragraph support the idea, expand its scope, and give cogent arguments for its importance. State the idea in the first sentence and then add more substance in the remaining sentences. If the report is written properly, a speed-reader should be able to read the first two sentences of every paragraph and glean 80 to 90% of the important information conveyed in the report.

As you add detail to a section or a paragraph, develop a pattern of presentation that the reader will find easy to read and understand. First, tell the reader what he or she will be told, then provide the details of the story and finally summarize with the action to be taken to implement the solution. Do not try this approach in the theme that you have to write for English class, because the instructor will clean your clock for being redundant. Technical writing is not the same as theme writing. In technical writing, you will often reiterate facts to insure that the message is conveyed to the readers. Often one or more of the readers must take corrective actions for the benefit of the company and the industry.

You will frequently include analysis sections in engineering reports. In these sections, the problem is stated and then the solution is described. After the problem statement and the solution, you introduce the details. The reader is better prepared to follow the details (the tough part of the analysis) after they know the solution to the problem.

14.3 KNOW YOUR READERS AND OBJECTIVE

Another important aspect to technical writing is to know something about the readers of your report. The language that you use in writing the report depends on the knowledge of the audience. For example, I am writing this book for engineering students. I know your math and verbal skills from the SAT scores required for admission. I recognize that you have very good language skills and even better math skills. You are bright and articulate. I believe that the content in this book closely matches your current abilities. However, there are a few sections in the book that deal with topics usually found in the business school.

Profits and costs are important in product development, and although they may not be as interesting as a computer program, they need to be clearly understood.

When you write a report for a class assignment, you understand that the instructor is the only reader with which you need to communicate. You also know that he or she is knowledgeable. In that sense, the writing assignments in a typical college class are not realistic. For example, suppose your manager asks you to write an assembly routing (a step by step set of instructions for the assembly of a certain product) for the production of your new widget in a typical factory in the U. S. What language would you use as you write this routing? If the plant is in Florida or the Southwest for example, Spanish may be the prevalent language. You should also be aware that a significant fraction of the factory workers in the U. S. are functional illiterates and many others detest reading. In this case, it might be a good approach to use fewer words and many cartoons, photographs and drawings.

Know the audience **before** you begin to write, and adapt your language to their characteristics. Consider four different categories of readers:

1. Specialists with language skills comparable to yours.
2. Technical readers with mixed disciplines.
3. Skilled readers, but not technically oriented.
4. Poorly prepared readers who may be functional illiterates.

Category 1 is the audience that is the least difficult to address in your writings; the audience in the last category is the most difficult. Indeed, with poorly prepared readers, it is probably better to address them with visual presentations conveyed with television monitors and videotapes.

A final topic, to be considered in planning, is the classification of the technical document that you intend to write. What is the purpose of the task? Several classifications of technical writing are listed below:

- Reports: trip, progress, design, research, status, etc.
- Instructions: assembly manuals, training manuals and safety procedures.
- Proposals: new equipment, research funding, construction and development budget.
- Documents: engineering specifications, test procedures and laboratory results.

Each classification of writing has a different objective and requires a different writing style. If you are writing a proposal for development funding, you need persuasive arguments to justify the costs of the proposed program. On the other hand, if you are writing an instruction manual to assemble a product, arguments and reasons for funding are not an issue. Instead, you would prepare complete and simple descriptions with many illustrations to precisely explain how to accomplish a sequence of tasks involved in the assembly.

14.4 THE TECHNICAL WRITING PROCESS

Whether you are writing a report as part of your engineering responsibilities in industry or as a student in college, you will face a deadline. In college, the deadlines are imposed well in advance and you have a reasonable amount of time to prepare your report. In industry, you will have less time, and an allowance must be made for the delay in obtaining approvals from management before you can release the report. In both situations, you have some limited period of time to write the report. The idea is to start as soon as possible. Waiting until the day before the deadline is a recipe for disaster.

Many professionals do not like to write; consequently, they suffer from writer's block. They sit at their keyboard hoping for ideas to occur. Clearly, you do not want to join this group, and there is no need for you to do so. A technique is described in this section that should help you to avoid writer's block.

Define Your Task

Start your report by initially following the procedures described in the previous section. Understand the task at hand, and classify the type of technical document that you have to write. Define your audience, before starting to write to establish the level of detail and the language to employ. Prepare a skeletal outline of the report. Start with the outline presented previously and add the section headings for the body of the report. This outline is very brief, but it provides the structure for the report. This structure is important because it has divided the big task (writing the entire report) into several smaller tasks (writing one section of the report at a time).

Gather Information

It is impossible to write a report without information. You must generate the information to be presented in the report. You can use a variety of sources: interviews with peers, instructors or other knowledgeable persons who are willing to help; complete literature searches at the library; read the most suitable references; conduct additional Internet searches for information. Take notes as you read, gleaning the information applicable for your report. Be careful not to plagiarize. While you can use material from published works, you cannot copy the exact wording; the statements must be in your own words. If the report has an analysis, go to work and prepare a statement of the problem, execute its solution and make notes about an interpretation of the solution. Successful methods for gathering information were described previously in Chapter 13.

Organize Data

When you have collected most of the information that is to be discussed in your report, organize your notes into different topics that correspond to the section headings. Then incorporate your topics in an initial outline, which will grow from a fraction of a page to several pages as you continue to incorporate notes in an organized format.

Compose Document

You are now ready to write. Writing is a tough task that requires a great deal of discipline and concentration. It is suggested that you schedule several blocks of time and reserve these exclusively for writing. The number of hours that should be scheduled will depend on the rate at which you compose, revise, edit and proofread. Some authors can compose about a page an hour, but most students need more time.

In scheduling a block of time for composing, you should be aware of your productive interval. Most writers take about a half an hour to come up to speed, and then they compose well for an hour or two before their attention and/or concentration begins to deteriorate. The quality of the composition begins to suffer at this time, and it is advisable to discontinue writing if you want to maintain the quality of your text. This fact alone should convince you to start your writing assignments well before the deadline date.

While you are writing, avoid distractions. Writing requires deep concentration—you must remember your message, the supporting arguments, the paragraph and sentence structure, grammar, vocabulary and spelling. Find a quiet, comfortable place and focus your entire concentration on the message and the manner in which you will present it in your report.

Naturally, some sections of a report are easier to write than others. The easiest are the appendices, because they carry factual details that are nearly effortless to report. The interior sections of the body of the report carry the technical details, which are also easy to prepare, because describing detail is less concise and less cogent. Do not become careless on these sections because they are important; however, each sentence does not have to carry a knockout punch.

The most difficult sections are the introduction and the executive summary. It is advisable to write these two sections after the remainder of the first draft of the report is completed. Usually the introduction is written before the executive summary. Although the introduction contains much of the same information as the executive summary, it is more expansive. The introduction can be used as a guide in preparing the executive summary.

14.5 REVISING, EDITING AND PROOFREADING

Writing is a difficult assignment. Do not expect to be perfect in the beginning. Practice will help, but for most of us, it takes a very long time to improve our skills because writing is such a complex task. Expect to prepare several drafts of a paper or report, before it is ready to be released. In industry, several drafts are essential because you often will seek peer review and manager reviews are mandatory. In college, you have fewer formal requirements for multiple drafts, but preparing several drafts is a good idea if you want to improve the report and your grade.

First and Subsequent Drafts

The first draft is focused on composition, and the second is devoted to revising the initial composition. Use the first draft for expressing your ideas in reasonable form and in the correct sections of the report. In the second draft, focus on revising the composition. Defer the editing process, and concentrate on the ideas and their organization. Make sure the message is in the report and that it is clear to all of the readers. Polish the message later in the process.

Several hours should elapse between the first and second drafts of a given section of the report. If you read a section over and over again, you soon lose your ability to judge its quality. You need a fresh, rested brain for a critical review. In preparing this textbook, the author composes on one day and revises on the next day. Revising is always scheduled for an early morning block of time. When you are rested, concentration is usually at its highest level.

Let's make a clear distinction between composing, revising, editing and proofreading. Composition is writing the first draft where you formulate your ideas and organize the report into sections, subsections and paragraphs. Unless you are a super talented writer, your first draft is far from perfect. The second draft is for revisions where you focus on improving the composition. The third draft is for editing where errors in grammar, spelling, style, and usage are corrected. The fourth draft is for proofreading where typographical errors are eliminated.

Revising

When you revise your initial composition, be concerned with the ideas and the organization of the report. Is the report organized so that the reader will quickly ascertain the principle conclusions? Are the section headings descriptive? Sections and their headings are helpful to the reader because they help organize his or her thoughts. Additionally, they aid the writer in subdividing the task and keeping the subject of the section in focus. Are the sections the correct length? Sections that are too long tend to be ignored or the reader becomes tired and loses concentration before completing them.

Question the premise of every paragraph. Are the key ideas presented together with their importance before the details are included? Is enough detail given or have you included too many trivial items? Does the paragraph contain a single idea or have you tried to include two or even three ideas in the same paragraph? It is better to use a paragraph for every idea even if the paragraphs are short.

Have you added transition sentences or transitional phrases? The transition sentences, usually placed at the end of a paragraph, are designed to lead the reader from one idea to the next. Transitional phrases, embedded in the paragraph, are to help the reader place the supporting facts in proper perspective. You contrast one fact with another, using words like **however** and **although**. You indicate additional facts with words such as **also** and **moreover**.

Editing

When the ideas flow smoothly and you are convinced that the reader will follow your concepts and agree with your arguments, begin to edit. Run the spell checker, and eliminate most of the typographical errors and the misspelled words in the report. Search for additional misspelled words, because the spell checker does not detect the difference between certain words like **grate** or **great**, or say **like** and **lime** and between **from** and **form**.

Look for excessively long sentences. When sentences become 30 to 40 words in length, they begin to tax the reader. It is better to use shorter sentences where the subject and the verb are close together. Make sure that the sentences are actually sentences with a subject and a verb of the same tense. Have you used any comma splices (attaching two sentences together with a comma)? Examine each sentence and eliminate unnecessary words or phrases. Find the subject and the verb, and attempt to strengthen them. Look for redundant words in a sentence, and substitute different words with similar meaning to eliminate redundancy. Be certain to employ the grammar check incorporated in the word processing program. It is not perfect, but it does locate many grammatical errors.

Proofreading

The final step is proofreading the paper to eliminate errors. Start by running the spell checker for the final time. Then print out a clean, hard copy to use for proofreading. Check all the numbers and equations in the text, tables and figures for accuracy. Then read the text for correctness. Most of us have trouble reading for accuracy because we read for content. We have been trained since first grade to read for content, but we rarely read for correctness.

To proofread, read each word separately. You are not trying to glean the idea from the sentence, so do not read the sentence as a whole. Instead, read the words as individual entities. If it is possible, arrange for some help from a friend—one person reading aloud to the other with both having a copy of the manuscript. The listener concentrates on the appearance of each word and then checks the text against the spoken word to verify its correctness.

14.6 WORD PROCESSING SOFTWARE

Word processing software is a great tool to use in preparing your technical documents. The significant advantage of using word processing software is the ability to revise, edit and make the necessary changes without excessive retyping. You can mark, delete, cut, copy and move text. You can easily insert words, phrases, sentences and/or paragraphs.

While working with word processing software, it is difficult to revise, edit, and proofread on the screen because only a portion of the page is visible. Better results are obtained if you print a hard copy and view the entire page. With the entire page, you can see several paragraphs at once and can check that they are in the correct sequence. Use double spacing when printing the first few drafts to give adequate space between the lines for your modifications.

After you have completed the revisions on the hard copy, make the required modifications to the text using word processing software and save the results to your floppy disk. It is recommended that you keep only the most recent version (draft) of the document. If you save several versions of the document, it is necessary to keep a logbook of the changes to each version. If the writing takes place over several weeks, it is easy to lose track of what changes you made and which version of the hard copy goes with which electronic copy. You will find it easier to keep one electronic file (on your floppy disk) of the most recent draft with a hard (paper) copy to serve as a back up.

Word processing software has several features that are helpful in editing. The spellchecker finds most of your typographical errors and provides suggestions to correct many misspelled words. The search or find command permits you to systematically examine the entire document so you may replace a specific word with a better substitute. A thesaurus is available to help you with word selection, but you must be careful when using it. Make sure you understand the meaning of the word that you select; do not try to impress the reader by using long or unusual words. Short words that are easily understood by the reader are preferred in technical writing.

Formatting the report is another significant advantage of word processing software. You can easily produce a document with a professional appearance. The formatting bar permits you to select the type of font, the point size (the height of the characters) and emphasis such as bold, underline or italic.

You can also format the page with four different types of line justification, which are commonly available. I am using word processing software to prepare this textbook, and applying "justify" alignment for both the right and left margins. The word processing software automatically added the spaces required in each line and aligned both margins. In designing your page, use generous margins. One-inch margins all around are standard unless the document is to be bound; then the left margin is usually increased from 1.00 to 1.25 inch.

Tables and graphs can be inserted in the text. Take advantage of the ability to introduce clip art into the text or to transfer spreadsheets and drawings produced in other software programs into your report. Position the tables and figures as close as possible to the location in the text where they are introduced. Try to avoid splitting a table between two pages. If the table is longer than a page or two, consider placing it in an appendix. Identify figures and photographs with a suitable caption. The caption, placed directly beneath the figure, is to reinforce the message conveyed by the illustration.

14.7 SUMMARY

Writing is a difficult skill to master; most engineers experience significant problems when they prepare reports early in their careers. Unfortunately, the writing experiences in college do not correspond well with the writing requirements in industry. In college, you write for a knowledgeable instructor. In industry, you write for a wide range of people with different reading abilities. Moreover, the audience often varies from assignment to assignment. In both college and industry, you write to meet deadlines imposed by others. While writing may not always seem to be fun, there are many techniques that you can employ to make writing much easier and more enjoyable.

The first technique is organization; an outline for a typical report was suggested. Gather information for the report from a wide variety of sources, and generate notes that you can use to refresh your memory as you write. Sort the notes and transpose the information to expand the outline for your report.

When you organize the outline, but before you begin to write, determine as much as possible about your audience. They may be technically knowledgeable regarding the subject or they may be functionally illiterate. The language that you use in the report will depend on the reader's ability to understand. Also understand the objective of your document. Is it a report, an extended memo, a proposal or an instructional manual? Styles differ depending on the objective, and you must be prepared to change accordingly.

There is a process to facilitate the preparation of any document. It begins with starting early and working systematically to produce a very professional document. Divide the report into sections and write the least difficult sections (appendices and technical detail portions) first. Defer the more difficult sections, such as the executive summary and the introduction until the other sections have been completed.

The actual writing is divided into four different tasks—composing, revising, editing and proofreading. Keep these tasks separate:

- Compose before revising
- Revise before editing
- Edit before proofreading
- Proofread with great care.

Multiple drafts are necessary with this approach, but the results are worth the effort.

Word processing software does not substitute for clear thinking, but it is extremely helpful in preparing professional documents. Word processing saves enormous amounts of time in a systematic editing process. It enables a mix of art, graphics and text neatly integrated into a single document. Word processing software has a thesaurus and word search features that are helpful in editing. It essentially turns a computer and printer into a print shop so that you have wide latitude in the style and appearance of your professional documents.

REFERENCES

1. Eisenberg, A., Effective Technical Communication, 2nd Edition, McGraw Hill, New York, NY, 1992.
2. Goldberg, D. E., Life Skills and Leadership for Engineers, McGraw Hill, New York, NY, 1995.
3. Elbow, P. Writing with Power, 2nd Edition Oxford University Press, New York, NY, 1998.
4. Alreb, G. T., C. T. Brusaw and W. E. Oliu, The Handbook of Technical Writing, 7th Edition, St. Martins Press, New York, NY, 2003
5. Struck, W. C. Osgood and R. Angell, Elements of Style, 4th Edition, Allyn & Bacon, Needham Heights, MA, 2000.
6. Pickett, N. A. et al, Technical English: Writing, Reading and Speaking, Longman, Reading MA, 2000.

EXERCISES

14.1 Prepare a brief outline of the organization for the final report describing the development of the product that your team is designing.
14.2 Prepare an extended outline of the final report for the development of the product that your team is designing.
14.3 Write a section describing one of the subsystems for your team's project.
14.4 Write a section covering the testing of a subsystem of the product that your team is designing.
14.5 Write the Introduction for the final report on the product that your team is developing.
14.6 Write the Executive Summary for the final report on the product that your team is developing.
14.7 Revise the Introduction that another team member wrote.
14.8 Edit the Introduction after it has been composed and revised by others.
14.9 Proofread the Introduction after it has been composed, revised, and edited by others.
14.10 Edit the final report after other team members have completed it.
14.11 Proofread the final report after other team members have completed it.

CHAPTER 15

DESIGN BRIEFINGS

15.1 INTRODUCTION

You communicate by writing, speaking and with a number of different visual methods. All these modes of communication play a role as you attempt to convey your thoughts, ideas and concerns to others, ands they all are important. Let's focus your attention on speaking in this chapter. You use speech almost continuously in your daily life—why do you need to study about design briefings? There are several reasons. You usually speak with your friends and family in an informal style. You know them and feel comfortable with them. They know you. They are concerned with your well being, genuinely like you and are usually interested in what you have to say. A professional presentation is different. It is a formal event that is usually scheduled well in advance. The audience may include a few friends, but usually it consists of peers (coworkers) and strangers. The time you have in which to convey your message is limited, and the audience may not be interested in your message. A person or two in the audience may be managers who control your advancement in the company. There are many reasons for tension headaches as you prepare for a professional presentation.

The design briefing is extremely important to both the product development process and to your career. Information must be effectively transmitted to your peers, management, and any external parties involved in the project. Clear messages accurately defining problems that the development team is effectively addressing are imperative. On the other hand, ambiguous messages are often misunderstood; hinder the definition of the problem and lead to delays in implementing timely solutions. Clearly, such messages should be avoided. The design briefing provides an opportunity to review the status of a specific product development. It also permits peers to share their ideas with you and affords management an open forum for assessing the quality of your work and the progress made by your development team. Since the professional presentation is critically important, you should become knowledgeable about some effective techniques for accurately delivering your messages to a mixed group of strangers, peers and acquaintances.

15.2 SPEECHES, PRESENTATIONS AND DISCUSSIONS

To begin, let's distinguish between three types of formal methods of oral communication—speeches, presentations and group discussions.

Speeches

The speech is the most formal of the three modes of communication. In the year of a presidential election, there are always many examples of speeches. Everyone is besieged with political addresses that clearly illustrate the key features of speeches. These speeches are given to large audiences in huge rooms, stadiums or coliseums. The setting is usually not appropriate for visual aids although the speaker usually uses a teleprompter. The audience is diverse with a wide range of concerns. Those listening to the speech are of widely different ages with different interests and persuasions. The speech is carefully scripted with the speaker usually reading or closely following the script. Ad-libbing is carefully avoided. Communication is one way—from the speaker to the audience. Questions are usually not appropriate and generally not appreciated. Time is strictly enforced unless you are the President of the United States giving the State of the Union address. Fortunately, engineers are rarely called upon to make speeches; hence, this topic will not be developed further in this book.

Presentations

Professional presentations differ from speeches. Presentations are made to smaller audiences in smaller rooms that are usually equipped with electric power, light controls and projection devices. You depend on visual aids, demonstrations, simulations and other props to help convey your message during a typical presentation. The audience is knowledgeable (about your topic) and usually has many common characteristics. The presentation is very carefully prepared because of its importance. It has order and structure, but it is not scripted. The flow of information is largely from the speaker to the audience; however, questions are permitted and often encouraged. The speaker is considered the expert, but discussion, comments and questions give the audience an opportunity to share their knowledge of the topic. Time is carefully controlled and is often insufficient from the speaker's viewpoint. The professional presentation is a vitally important mode of oral communication that you must quickly learn to master.

Group Discussions

Group discussions are also very important to the design engineer. The audience is smaller with much more in common. The group may all be members of a development team. The topic being discussed is narrowly focused. The speaker (central person) serves as a moderator, and he or she is an expert on the topic being considered. However, members of the discussion group (audience) may be as knowledgeable as the speaker (moderator). The moderator works in two modes. He or she may act as a presenter giving brief background information to frame an issue that prompts any member of the group to discuss the topic. The moderator may also direct questions to a group member known to be the most knowledgeable on the issue being addressed. The speaker (moderator) controls the flow of the information, but the flow is clearly multi-directional—from speaker to the group and from one group member to another. When a group (team) discusses a design issue, time is difficult to control and the content and range of coverage are strongly dependent on the skill of the moderator. Group discussion is very important in industry because the leader of a development team will often use this method of communication to identify problems and to initiate efforts directed toward their solution. Group discussion methods will not be described in this textbook; however, it is recommended that you watch *Washington Week in Review* on the PBS television network to gain some insights. While the participants are journalists, you can adopt many of moderator's clever techniques to guide engineering discussions.

Design Briefing

A design briefing is a type of professional presentation. It is the method you will use in speaking before the class and will be an educational topic emphasized in this course. Indeed, you probably will be required to make presentations describing your product development on at least two occasions—the preliminary and the final design reviews. In these two design reviews, you will employ professional visual aids to assist in delivering the information required to describe the status of your project with conciseness and clarity.

15.3 PREPARING FOR THE BRIEFING

A design briefing is too important to take casually. You should prepare very carefully to insure that you will accurately report the status of your team's development and to identify problems or unresolved issues. There are three aspects that you should consider in your preparations:
- Identify the audience
- Plan the organization of the presentation and your coverage of each topic
- Prepare interesting and attractive visual aids.

Identifying the Audience

In a typical college classroom, it is much too easy to identify the audience. You have your peers, the instructor and perhaps a visitor from industry. In an industrial setting, the audience will be more difficult to identify because it will be much more diverse. The size and diversity of the audience depends on the magnitude of the development project. A small briefing will have 10 to 20 people in attendance with most participants from within the company. A large briefing may include 50 to 100 representatives who are both internal and external to the company. The characteristics of the audience are important because you must adapt the content of the presentation to the audience. Classify your audience with regard to their status, interest and knowledge before you begin to plan the style and content for your presentation.

The status of the audience refers to their position in the various organizations they represent. Are they peers, managers or executives? If they are a mix, which is likely, who is the most important? If you are preparing a design briefing for high-level management, it must be concise, cogent and void of minor detail. Executives are busy, stressed and always short on time. High-level executives and managers are impatient and rarely interested in the technical details that engineers love to discuss in their presentations. Recognize these characteristics and adjust the content and the timing of your presentation accordingly. The executives are interested in costs, schedule, market factors, performance, risk or any other critical issue that will delay the development or increase projected costs of either the development or the product. Usually you will have only 5 to 10 minutes to convey this information at an executive briefing.

First and even second level managers have more time and are more interested in you and the designs that you are creating. If you prepare the presentation for this lower level of management, you can safely plan for more time (10 to 20 minutes). These managers are likely to be engineers, and they will share your knowledge of the subject. In fact, they probably will be more experienced, expert and knowledgeable than you. They will want to know about the schedule and costs because they share responsibilities with the higher-level management for meeting these goals. However, they are also interested in the important details, and you can engage them in a discussion of subsystem performance. The first and second level managers usually control the resources for the development team. If you need help, make sure that they receive the message in enough detail to provide the assistance required. Do not hide the problems that your team is encountering; managers do not like surprises. If you have a problem, make sure they understand it and are able to participate in its solution. If you hide a problem from management and it causes a delay or escalation in costs, the manager will take the heat and you will be in a very cold doghouse.

Peer Reviews

Design briefings for peers are often less tense, and you will have more time (20 to 30 minutes). You will address schedule and cost issues because everyone needs to know if you are meeting the milestones on the Gantt charts. However, the main thrust of your presentation will pertain to design details. If you are working on a subsystem that interfaces with other subsystems, you will cover your subsystem in sufficient detail for all to understand its geometry, interfaces, performance, etc. It is particularly important that the interface issue be fully addressed in the presentation. If four subsystems are to be integrated in a given product, many details about all four subsystems must be addressed to ensure that the integration goes smoothly. Suppose, for example, that your team is developing a power tool with a motor that draws a current of 10 amps, and some team member intends to employ a switch, rated at 6 amps, to turn the power on/off. The switch will function satisfactorily in the short-term tests with the prototype, but it will malfunction in the field with extended usage sometime after the product has been released to the market. Clearly, the two persons responsible for the interface between the motor and the switch did not communicate properly. The integration of the two subsystems failed.

The most important purpose of design briefings with peers is to make certain that all of the details have been addressed. You strive to integrate the subsystems without problems arising when the prototype is assembled and tested.

Peer reviews in the absence of management are very beneficial in the development process. The knowledge of most of the team members is about the same although some members are more experienced than others. The topic is usually a detailed review of one subsystem or another. The audience is fresh and capable of providing a critical assessment of the technology employed. Your peers may check the accuracy of your analysis, comment on the choice of materials, discuss methods to use for manufacturing and assembly, and relate their experiences with similar designs on previous products. Peer reviews afford the opportunity for the synergism that makes good products better. They also provide a forum for passing on important lessons from the more experienced designers to the beginning designers in a friendly, stress-free environment.

The subject of the presentation is self-evident in a design review. You are developing a product, and the content of the design briefing will deal with one or more issues regarding this development. There is a choice of topics and considerable latitude regarding the content. Although advice has been given in previous paragraphs about matching content with the interests of your audience, you must understand the importance of knowing who will be in your audience. Executive reviews serve a different purpose than peer reviews, and the content and the time allotted for the presentation is adjusted accordingly.

For every type of review, there are two **absolute** rules that you must follow. First, **you must know the material absolutely cold**. It only takes an audience about two seconds to understand that you are faking it. Also, some presenters fail to rehearse adequately, and the audience often misinterprets this lack of preparation for insufficient knowledge. In either event, when they realize that you are not the expert you allege to be, your presentation fails. You have **lost** the opportunity to effectively communicate your message. Second, **be enthusiastic**. It is all right to be calm and cool, but do not be dead on arrival. You must command the attention of your audience or they will quickly turn off. You control the attention of the audience only if you are enthusiastic and knowledgeable in presenting your material (story).

15.4 PRESENTATION STRUCTURE

There is a well-accepted structure for professional presentations. If you are familiar with PowerPoint®, a graphics presentation program (GPP) marketed by Microsoft, you are already aware of the templates included in this program for various types of presentations. An excellent structure for a design briefing is shown in the outline for a professional presentation presented below:

- Title
- Overview
- Status
- Introduction
- Technical Topics
  - First Topic
    - Background
    - Status
  - Second Topic
    - Background
    - Status

- Final Topic
  - Background
  - Status
- Summary
- Action Items

Let's examine each of these topics individually and discuss the content that should be included on the visual aids that accompany your presentation. Recognize that the visual aids (35 mm slides, overhead transparencies or electronic projection) control the flow and the information that you will present. For this reason, the topics listed above in the context of information included on the presentation slides will be addressed.

Title Slide

The title slide[1], illustrated in Fig. 15.1, obviously carries the title of the presentation. However, it also states your affiliation and the names of the team members that contributed to the work. Remember, since you may be reporting on the work of the entire team, it is necessary for you to acknowledge their contributions on the title slide and in your opening remarks. A brief title—ten words or less—is a good rule to follow in drafting the title for your presentation. It is also a good idea to use descriptive words in the title. For example, you could use **PRELIMINARY DESIGN BRIEFING: ANTI-ICING SYSTEM**. Five words and you have informed the audience of the topic (a design briefing or review), the type of briefing (preliminary), and the product being developed (an anti-icing system). A significant amount of information is successfully conveyed in five words

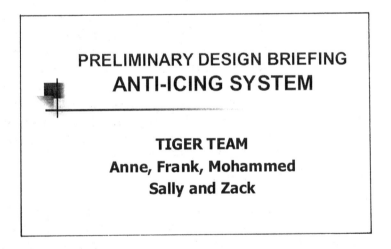

Fig. 15.1 Illustration of an informative and descriptive title slide.

Below the title, the team name together with the name of each team member involved in the development is listed. It is essential that you share the ownership of the material covered in your presentation. Often a single member of the team covers the work of several people in a design briefing. It is not ethical to make the presentation without acknowledging the contributions of others. Students typically use first names in identifying team members, but in industry the complete names of the team members are used. If necessary, division and department affiliations are also stated.

---

[1] Reference will be made to slides in this discussion; however, the information provided is also applicable for overhead transparencies.

It is also a good idea to date the title slide and to cite the occasion for the presentation. If you make many presentations, this information is useful when referring to your files, allowing you to easily reuse some slides after selective editing and/or updating.

Overview Slide

The second slide presents an overview of the design briefing. The overview for a presentation is like a table of contents for a book. You list the main topics that you intend to cover in the presentation. In other words, you tell the audience what you plan to tell them in the next 10 to 20 minutes. The number of topics should be limited. In a focused presentation, you can cover about a half dozen topics. If you press and try to cover ten or more topics, you will encounter difficulty maintaining the attention of the audience throughout the presentation. Constrain the tendency to tell all—focus on the important issues. In a design review, the topics usually are organized to correspond with the subsystems involved in the product under development. For a component redesign, you concentrate on the issues guiding the component modification being considered. An example, of an overview slide for a preliminary design review for an anti-icing system for highway bridges, is illustrated in Fig. 15.2.

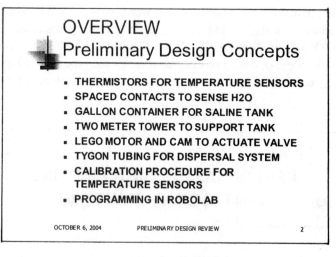

Fig. 15.2 The slide with the overview indicates the content in your presentation.

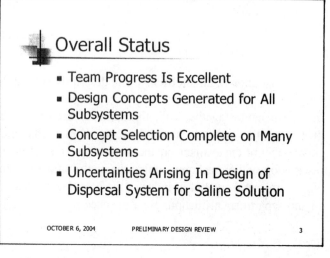

Fig. 15.3 Status slide indicates overall progress and accomplishments while signaling potential problems.

Status Slide

The third slide is used to present the product development status. You should report on the progress made by the team to date. It is a good idea to incorporate a Gantt chart showing the development schedule either on the status slide or a separate slide. The progress of the team on each task is defined on the chart and is immediately conveyed to the audience. If there are problems or uncertainties, this is the time to air the difficulties that the team is encountering. An example slide, shown in Fig. 15.3, illustrates uncertainties that the team has with regard to the design of the dispersal system for the saline solution.

Introductory Slide

The fourth slide covers the introductory material such as background, history or previous issues. It is important that the audience understand the product, the main objectives of the development, the role the team has in that development and the most pressing of the current issues. This introductory slide permits you to set the stage for the body of the presentation that follows. The audience responds better to your presentation when they know in advance the background and the topics that you will cover. They are better prepared mentally to receive your message.

Technical Topics

The body of the presentation usually deals with five or six technical topics. In most instances, the topics selected correspond to the more critical subsystems involved in the design of the product. In the redesign of a component, the topics deal with the issues arising when considering new design concepts. Avoid trying to cover a large number of topics; it is better to report on a limited number of topics thoroughly than to rush through a dozen topics with incomplete coverage. It is suggested that you use two slides for each topic. The first slide describes the progress made by the team in generating design concepts and indicates the criteria employed in selecting the best concept. The second slide covers the status of developing the selected concept. The type of information reported on this slide includes accomplishments, outstanding issues, lessons learned, etc. An example of the status of the early stages of the design of a mailroom scale is presented in Fig. 15.4.

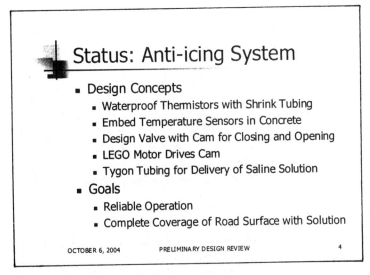

Fig. 15.4 Information presented in describing the early design of an anti-icing system.

Concluding Slides

After the technical topics have been covered, you must conclude the presentation with a decisive pair of slides. Suggestions for two concluding slides are presented in Figs. 15.5a and 15.5b. The next to last slide in the presentation, shown in Fig. 15.5a, is titled "Key Issues." This is your opportunity to identify very important issues (questions) that your team has discovered. Do not hesitate. This is the time to introduce the uncertainties and to come forward and seek help. Design reviews are not competitive. In a design review, you seek help regardless of your status or role as a member of the audience. Managers will arrange support for your team if it is required. Peers will make suggestions and introduce fresh approaches to unresolved problems that the team may find useful. The design review is a formal process in the product development in which everyone participates. Take advantage of this review to identify uncertainties and to seek help whenever your team needs assistance.

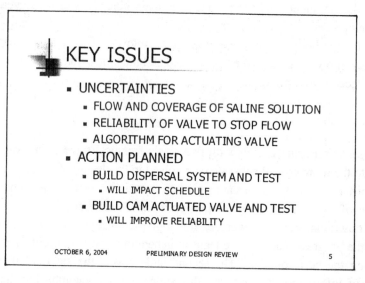

Fig. 15.5a The "key issues" focuses the team and management on your most significant problems.

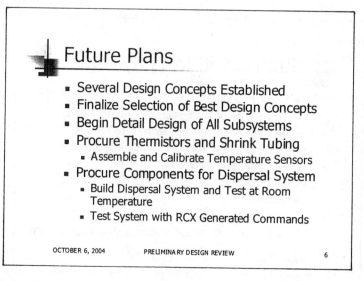

Fig. 15.5b The "future plans" slide gives you the opportunity to describe your approach for solving problems that the team is encountering.

The final slide presented in Fig. 15.5 is to present your plans for the future. Management wants to know how you intend to solve the issues that you have disclosed and the amount of time and money required for the solutions. Your peers will be interested in the technical approaches that you intend to follow. You should also define any requirement that you have of anyone in the audience. If you need help from your peers or more resources from the manager, be certain this assistance is clearly indicated on this final slide. You should distinctly identify those individuals who have agreed to provide the requested support. If action items are listed, clearly identify the responsible individual and estimate the time to completion. The planning incorporated in the final slide is very important. It is the prescription for your team's get-well program.

15.5 TYPES OF VISUAL AIDS

Because visual aids control the content and the flow of the information conveyed to the audience, they are essential for a technical presentation. Slides provide visual information that reinforces the verbal information. With visual aids you communicate using two of a person's five senses. Take time to carefully select the type of visual aid that will most effectively convey your message.

In selecting visual aids, you are usually limited—first, in your ability to produce the visual aids within the time and financial constrains and second with regard to the availability of equipment and the room in which the presentation will be made. Is the room equipped with a 35 mm slide projector, an overhead projector or a computer-controlled projector? Another important consideration is the control of the light intensity in the room. Most rooms have light switches used to control the overhead lights. However, some rooms do not have blinds and bright sunshine will cause problems with the visibility of 35 mm slides when they are projected.

Computer Projected Slides

Computer projected slides are an excellent choice if the room can be darkened. Computer projection of slides has several advantages:

- The presentation is easy to prepare using PowerPoint.
- The slides are in color are more effective than black and white illustrations.
- Advanced visual effects (slide transition and slide building feature) enhance the impact of the presentation.
- It is also easy for you to copy your presentation onto a floppy disk.

A computer is attached to a digital projection system that stores your presentation material. You employ a mouse to click through your slide line by line. Finally, low-cost handouts with four or six slides per page can be prepared, which help the audience follow your presentation. The only disadvantage of computer-projected slides is that sometimes a relatively dark room is required for projection. With a darkened room, eye contact with the audience or even reading their body language is not possible. You lose your visual contact with the audience that is so important in reading their response to your presentation style.

Overhead Transparencies

Projecting overhead transparencies is probably the most common method of presenting visual material. Overhead projectors are relatively inexpensive (a few hundred dollars compared to a couple of thousand dollars for the computer controlled projectors). Hence, overhead projectors are available in most rooms. Overhead projectors are also the best choice if the room cannot be darkened. The projectors are sufficiently bright to provide quality images without extinguishing the lights. Bright rooms have an advantage because they help you keep the audience alert. Overhead transparencies are also easy to make on a copy machine or a laser printer. If you are willing to sacrifice the benefits of color, utilize inexpensive black and white transparencies. If you want to use color in the presentation because it is more effective, the costs increase significantly when using a color laser printer or a color copier to prepare the overheads. However, color ink-jet printers can be used to produce good quality color transparencies at a reasonable cost.

35-mm Slides

The 35 mm slide presentation is a third option for visual aids; however, the equipment required is not always available. Projectors for 35-mm slides are common, but in an engineering college, they usually are much more difficult to find than the overhead projector. A slide presentation is the preferred option if you have many photographs to project because the quality of colored 35-mm slides is excellent. The slides are inexpensive and easy to prepare if you have a camera and the time necessary for the film processing. It is possible to prepare colored slides from your PowerPoint files, but conversion from the digital format to the 35-mm format requires special equipment. Again, a significant disadvantage is that the room must be darkened to view the image from a 35-mm projector. Also, the relatively long distance required from the typical projector to the screen precludes the use of very small conference rooms for presentations.

Although ineffective, sometimes speakers will employ two or three different types of projectors. Such a tactic should be avoided as the switch from one projector to the other is disruptive. If possible, use only one type of projector. If you must use more than one type of projector to properly present your message, try to minimize the number of times you switch from one device to the other.

Other Types of Visual Aids

There are several other visual aids sometimes used in professional presentations. Videotapes are common; unfortunately, the smaller size of TV monitors often makes viewing difficult for a large audience. However, if you want to show motion or group dynamics, video clips are clearly the best approach. Video cameras are readily available and after practice you can become a reasonably good video producer.

Motion picture films, particularly of older material with historical interest, are very effective in developing a long-term prospective. However, finding an 8-mm or 16-mm motion picture projector may be difficult—allow time in your schedule for making the necessary arrangements.

Hardware, or other materials used in the product development, is sometimes passed around the audience during a presentation. The hands-on opportunity for the audience is a nice touch, but you pay for it with the loss of attention by some members of the audience during the inspection period. It is recommended that you defer passing materials to the audience during your presentation. Instead, after the presentation, invite the audience to inspect exhibits that are placed on a strategically located table. This approach maintains the attention of the audience throughout the presentation while giving those interested in the hardware time for a much more thorough examination.

15.6 DELIVERY

Excellent presentations require well-prepared slides and a smooth, well-paced delivery. There are many aspects to the delivery part of the presentation process including: dress, body language, audience control, voice control and timing.

Attire

Let's start with dress. Broadly speaking, there are four levels of dress. The highest level, black tie for men and formal gowns for women, is a rare event and is never appropriate for design briefings. The next level, business attire for men and women, is sometimes appropriate for design briefings. A conservative Eastern company may have a dress code requiring business attire; whereas, a less formal Western company would encourage more casual attire. Casual attire should not be confused with sloppy attire. Casual attire is neat and tasteful, but without suits, ties and white shirts for men and suits for women. Although becoming more common in the workplace, sloppy attire (old jeans and a sweatshirt) is strictly taboo. You should select tasteful business or casual attire when you dress for the presentation. While it is recognized that most students these days prefer a relaxed, collegiate style of dress, you should resist the impulse to dress too casually. If you look like a homeless person, who will believe your message?

Body Language

Posture is another important element in the presentation. In the words of former President Reagan, "stand tall." Your body language signals your attitude to the audience—you are the presenter and in control. Make sure you are calm, cool and collected. Nervous gestures with your hands, rocking on the balls of your feet, scratching your head, pulling on your ear, etc. should be avoided.

Audience Control

Before you begin your presentation, take control of the audience. One approach is to pause a moment before projecting the title slide and immediately make eye contact with the entire audience. How do you look at everyone in the house? You scan the audience from left to right and then back looking slightly over their heads. Occasionally drop your eyes and make eye contact for a second or two with one individual and then another. Pause long enough for the audience to become silent (10 to 20 seconds). If someone is rude and continues to talk, walk toward them and politely ask for their attention. When you have everyone's attention, project the title slide and begin your delivery.

If you have rehearsed, you will not need notes. The slides carry enough information to trigger your memory. If they do not, you have not rehearsed long enough to remember the issues. Scripting the presentation is not recommended. People who script will eventually start to read their comments, which is a deadly practice. Rehearse until you are confident. Make notes to help you rehearse, but do not use them during the presentation.

Voice Control

When you begin to speak, be certain everyone can hear you. If you are not sure of how far your voice carries, ask those in the back of the room if they can hear. If you use a microphone with an amplifier and speakers, be careful; the tendency is to speak too loudly. Try to control the loudness of your voice within the first minute of the presentation when you introduce the topic with the title slide. The audience will read that slide with anticipation, and they will bear with you as you adjust the volume of your voice.

Do not mumble. Speak slowly, clearly and carefully enunciate each word. It is better to present your material too slowly than too rapidly. If you can improve your enunciation, you will be able to increase the rate of your delivery without losing the audience.

Don't run out of breath while speaking. Learn to complete a sentence, pause for breath (without a gasp) and then continue with the next sentence. There is nothing wrong with an occasional pause in the presentation, provided the pause is short.

Timing

If you forget some detail, do not worry. Skip it, and move to the next point that you are trying to make. If the detail is critical, count on one of your team members to raise the issue in the question and answer period. Most speakers forget some of the material that they intended to present and introduce some additional items on an ad-hoc basis. Usually, the audience will never be aware of your omissions or additions.

Studies have clearly shown that people can listen to conversation faster that the presenter can speak. The trouble with listening to someone who speaks very rapidly is not their rate of speech, but their enunciation. Many people who speak rapidly tend to slur their words, and the audience has trouble understanding poorly pronounced words. The trouble you have in listening to a person speaking too slowly is to stay with his or her message; while you are waiting for the next word, your mind goes off on a mental tangent. You anticipate that your mind will return to the speaker in time for the next word, but unfortunately the mind is sometimes tardy. You miss keywords or even complete sentences before your attention returns to the speaker.

Listening is a skill, and many people have not developed it properly. As a speaker, it is your responsibility to keep the listener, even the poor ones, on the topic. You may use several techniques to maintain the attention of the audience. Employing the screen is probably the most effective tool for this purpose. As you project the slide, walk to the screen and point to the line on the slide that corresponds to the topic that you are addressing. If a few members in the audience are coming back from their mental tangents, you reset their attention on the current topic.

When referring to the image projected on the screen—and good speakers use this technique—position yourself correctly as shown in Fig. 15.6. Stand to the left side of the screen being careful not to block anyone's view. Face the audience and maintain eye contact. When you need to look at the slide, turn your head to the side and read over your shoulder. The 90° turn of your head and the glance at the slide should be quick because you want to continue to face the audience. Under no circumstance should you stand with your back to the audience and begin reading from the slide as if it were a script.

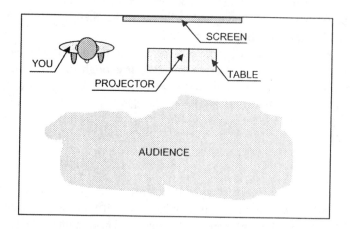

Fig. 15.6 Typical room arrangement for a presentation. Position yourself to the left of the screen and face the audience.

As you develop the discussion, point to key phrases to keep the audience focused on the topic. Engage both of their senses. Deliver your message through their eyes and ears simultaneously. When you point to a line on the slide, point to the left side of the line. People read from left to right—so start them from the initial point on the left side.

As you speak, modulate your tone and volume. Avoid both a monotone and a singsong delivery. A continuous tone of voice tends to put the audience asleep. A singsong delivery is annoying to many listeners. If you have a high-pitched voice, try to lower the pitch because a high-pitch tone is bothersome to many people.

If a question is asked, answer it promptly unless you intend to cover material later in the presentation that addresses the question. The answer should be brief, and you should try to avoid an extended dialog with some member of the audience. If a member of the audience is persistent with several related questions, simply indicate that you will speak with him or her off-line immediately after the presentation. The timing of your presentation can be destroyed by too many questions. Some questions are helpful because they permit audience participation, but excessive questions cause the speaker to lose control of the topic, the flow and the timing of the presentation. Too many questions cause the format of the communication to change from a presentation to a group discussion. Group discussions have a purpose; however, they are different than a design briefing.

15.7 SUMMARY

The design briefing is extremely important both to successful product development and to your career. Information must be effectively transmitted to your peers, management and to external parties involved in the project. Clear messages, which accurately define the problems that the entire development team can effectively address, are imperative. The design briefing provides the opportunity to inform management and peers of your progress and problems.

There are three types of formal communication: speeches, professional presentations, and group discussions. The characteristics of these three types of communication are described. For the engineer, the professional presentation is the most important and the formal speech is the least important. The main emphasis of this chapter is on the professional presentation with a focus on the design briefing.

Preparation for a presentation is essential. Accomplished speakers often spend the better part of a day preparing for a 20-minute briefing. There are three important elements in the preparation. First, know the characteristics of your audience, and be certain that the presentation's content corresponds to their interests. Second, organize the presentation in a manner acceptable to the audience. Insure that the organization is efficient so that you use the allotted time effectively. Prepare high-quality visual aids (slides) that will help you convey your message and rehearse until you command the subject material.

A structure for the presentation is recommended that is commonly employed in design briefings. This outline includes four initial slides—title, overview, status and introduction. The body of the presentation is devoted to technical details using about eight to twelve slides. Two closure slides are employed for a summary and a listing of action items.

The type of visual aid selected is important because the visual transmission of information will markedly affect the outcome of the presentation. Computer projected slides have some significant advantages, but the availability of equipment often precludes their use. Overhead transparencies are the most common medium. However, if you have a presentation that includes many photographs and a large room in which to deliver the presentation, 35 mm slides are the most suitable medium.

The delivery makes or breaks a presentation. You have been provided with two pages of details about how to do this—and not that. However, the best way to learn to deliver a design briefing is by practicing your presentation. On your first two or three attempts, have a friend video tape the event. Then review your delivery style, technique and appearance during a trail presentation. You will identify many problems of which you were not aware. Another suggestion is to use one of your free electives to take a speech course. The experience gained in a typical communications course will not help much in crafting the content to include in a design briefing, but it will help you to develop very important delivery skills.

REFERENCES

1. Wilder, L. Talk Your Way to Success, Simon and Schuster, New York, NY, 1986.
1. Eisenberg, A., Effective Technical Communication, 2nd ed., McGraw Hill, New York, NY, 1992.
2. Goldberg, D. E., Life Skills and Leadership for Engineers, McGraw Hill, New York, NY, 1995.
3. Elbow, P. Writing with Power, 2nd Edition Oxford University Press, New York, NY, 1998.
4. Struck, W. C. Osgood and R. Angell, Elements of Style, 4th Edition, Allyn & Bacon, Needham Heights, MA, 2000.
5. Pickett, N. A. et al, Technical English: Writing, Reading and Speaking, Longman, Reading MA, 2000.

EXERCISES

15.1   Write a brief description of the characteristics describing your classmates in this course. Focus on the characteristics that will influence the language and content in your design briefings.

15.2   Prepare an outline for a preliminary design review. Include in the outline the titles of all of the slides that you intend to use.

15.3   Prepare the title slide for your preliminary design review.

15.4   Prepare an overview slide for your preliminary design review.

15.5   Prepare a status slide for your preliminary design review.

15.6   Prepare the key issues slide for your preliminary design review.

15.7   Beg, borrow, but do not steal a video camera. Video tape a practice presentation by a teammate. Critique his or her presentation with good taste.

15.8   Locate all of the projectors that you are permitted to borrow for the presentations to be made in this class.

15.9   Practice the delivery of your presentation employing the most suitable projector that you can borrow.

15.10   Measure and record the time required to present your design briefing. As you rehearse, continue to measure the time. As the presentation becomes more polished, the time required for the presentation should decrease. Does it?

15.11   Prepare a group of slides for a design briefing that utilizes the slide transition and slide building features of PowerPoint.

PART VI

ENGINEERING AND SOCIETY

CHAPTER 16

ENGINEERING AND SOCIETY

16.1 INTRODUCTION

It should be made clear from the beginning of this chapter that engineering and society are closely joined. Engineering is an integral part of society, and consequently, engineering must be treated as a social enterprise as we continue to develop sociotechnical systems. When engineers develop a new product or improve an existing system, we change society. The magnitude of the change differs from one product to another, and the effect on society may be trivial or enormous, but no matter what, there is an impact. For example, when a few companies introduced a random orbit sander a few years ago, the time, effort, and cost of producing flat, smooth, scratch-free surfaces was reduced. These sanders helped several hundred thousand workers in the finishing business to produce high-quality coatings on furniture, cabinets, counters and floors. The results also benefited many millions of customers who purchase furniture each year or remodel kitchens. This new product had a modest, but beneficial, impact on society. There was, however, a small negative effect in that some emissions to the environment occurred during manufacturing of the product and virgin resources were consumed. Overall though, society has clearly gained by the introduction of this product.

When transistors were first developed from new ultra-pure semiconductor crystals, the impact was revolutionary. While the first transistor was developed in the mid 1950s, the integrated circuit was introduced in 1959, and the microprocessor was released in 1971, the technological revolution resulting from the introduction of these three products in the microelectronics industry is still underway. Micro-electronic based products are markedly affecting the way in which society lives today. The media refers to the current period as the "information age". New products, which are introduced almost daily, are driven by technology. Products based on new advances in microelectronics will continue to change the way everyone works and plays for the foreseeable future.

You often hear about new products and changing life styles. Television, radio, Web sites and even newspaper advertising keep everyone well informed of the availability and prices of both old and new products. However, you rarely hear much about the actual technology involved even when the product is new and revolutionary. The media and the public usually are not interested in learning about the technology employed in a new product; this is unfortunate because it would be easier to introduce new or modified systems if the public were more technologically literate.

Engineers are, in part, responsible for this apparent wall between the public and the understanding of technology. Engineers often speak in such technical terms that those few individuals interested in our work become confused. As engineers, we must clearly recognize the need to effectively communicate to society. We can begin by learning how to explain why and how products work in terms more easily understood by the general public. We must also be more enterprising in understanding society and their changing views relative to technology. Lastly, we need to think on a more global scale. Engineers develop sociotechnical systems, which are often large and complex. These systems rely on hundreds of products and many different types of services to function efficiently. Even though, as an engineer, you may be dedicated to developing a single component, subsystem or product that is only a small part in a much larger system, you should understand the entire system. Those development teams that have a thorough understanding of all aspects of the complete sociotechnical system develop the superior designs.

Let's consider an example of a sociotechnical system—the commercial air transport system. Do we have technical products involved in this system? **Yes**, literally hundreds of advanced state-of-the-art products. Do we provide technical services for the airlines? **Yes**. Does an engineer working for General Electric in the development of a new jet engine need to understand what Boeing is doing in the design of a new aircraft or modifications of an existing aircraft? Does that same engineer need to know about the sound levels that the public will accept as the plane lands and takes off? Is the engineer working for Boeing required to understand the causes of each and every airline accident? Again, **yes**.

From these simple questions, it should be evident that sociotechnical systems are extremely complex. It will require significant effort for you to understand the detail of the interaction of the many products incorporated in these systems with the public in general. Nevertheless, engineers that understand and appreciate the improvements, which both the customer and society seek, will develop the most successful products.

Finally, as an engineer, you must understand that the systems developed are adaptive. They are dynamic and change with time in response to several societal and economic forces. First, society changes. The requirements of society and the customer in the 1950s are much different than either will accept today. The customers, a segment of society (more on this fact later), typically expect that the new model of the product will be much improved over the older model and be lower in price. Others in society, not necessarily the customers, will expect (perhaps demand) that you manufacture the product without degrading any aspect of the environment.

An example is useful to distinguish between the customer and society. Let's consider the automobile as the product. The customer of an automobile demands style, comfort, mobility, convenience, affordability, reliability, performance and value. Society requires (through federal regulation) safety, fuel efficiency and reduced emissions and pollution. The fact that the requirements from the customer and society often conflict should be recognized. Conflicting requirements challenge engineers to create improved designs that advance the state of the art in all areas of technology.

The governments (Federal, State, County and City) are also involved in the relationship between engineers and society. Various government agencies frequently issue regulations that affect both the sociotechnical systems and the product. Consider the commercial air transport system again. Obviously, safety is a very serious issue. The Federal Aviation Administration (FAA) is responsible for issuing regulations to insure safe operations of commercial aircraft. However, 1996 was not a good year for the FAA or the commercial airlines because several very serious accidents occurred with significant loss of life. As a consequence, of the fire in the cargo bay, which caused the crash of a Boeing 737 aircraft in the Florida Everglades and the loss of all aboard, the FAA is now requiring the airlines to retrofit all planes with a fire suppressant system. Regulations from government agencies drive a portion of the changes to any sociotechnical system.

This understanding of the relationship between engineering and society is relatively new. If you examine the five-volume tome, *A History of Technology*, [1] published by Oxford University Press over the period from 1954-58, you will find a chronological, deterministic catalog of technological development. However, the editors almost completely ignore the relation between engineering and society. In his review of this tome, Hughes [2] states that the editors belatedly conclude their five-volume treatise with an afterthought essay on technology and the social consequences. Today, the relationship is more clearly recognized although educators have not yet managed to include adequate treatment of this very complex relationship in a typical engineering curriculum. A few examples of the relationships between engineering and society will be described in this chapter. For a more complete treatment, read reference [3].

16.2 ENGINEERING IN EARLY WESTERN HISTORY

Technology has impacted the way that we live since the beginning of recorded history. Long before the first engineering college was founded, intelligent, hard-working men[1] worked as engineers developing new products and processes. You may study history and cite many developments that changed, and for the most part, benefited society.

Prior to about 6000 BC, there was little evidence of technology. Men and women were hunters and gatherers. They scrounged for food and lived in caves if they were lucky and shacks or huts if they were not. The significant developments in the period from 6000 to 3000 BC were in agriculture with the domestication of animals and the cultivation of grains. With time, it was possible to insure a relatively stable, if limited, food supply and people began to congregate together in villages and small cities.

The first technology involved the building of roads, bridges and boats for local transportation. People clustered in villages and small cities wanted to transport food, water and other materials over short distances. The geographic regions where technology was evident were extremely local. Early Western history involved only Mesopotamia, Egypt, Greece and then Rome. In a few cities, there were aqueducts to transport water and some sewerage. In most regions, only the ruling class lived well. The great monuments of Egypt and Greece were built with slave labor. In fact, slaves (people who were on the losing side in a war or those born into slavery) provided most of the power required for building, mining and agriculture for many thousand more years. The rise of religion, after the coming of Christ, is often credited with the reduction in the prevalence of slavery. However, the steam engine developed by Thomas Savery, Thomas Newcomen and James Watt in the early 18th century signaled the beginning of the end of the need for power provided by humans (slaves and other unfortunate souls in bondage). As society enters the 21st century, the development of the low-cost, high-performance computers and a wide variety of specialized software programs portends another beginning of the end for the need for many employees performing repetitive tasks or clerical work.

The Egyptians

The Egyptians built many huge monuments and were masters in moving large blocks of stone from distant quarries to a construction site and then elevating these blocks to their respective positions in gigantic structures. Some estimates indicate that 20,000 to 50,000 men were required to drag the 20 to 50 ton blocks of stone used in the construction of these monuments. In spite of the majestic grandeur and durability of the monuments, the new innovations introduced by the Egyptian engineers were not outstanding. They constructed temples in a simple fashion using only uprights (columns) and cross pieces (short, deep beams). While the arch was known during this period, it was constructed with mud bricks and never adapted to stone construction. Classical treatments of the history of technology tell us very little about the impact of these developments on society. However, it is apparent that the construction of a pyramid or a temple required a significant segment of the working population for many decades. This effort probably did little to improve the welfare of even the **free** segment of the population. It is also evident that the effort required of the slaves to move the countless large, heavy stones must have been brutal.

[1] Historical accounts of engineering achievements indicate that the profession was male dominated. The author is sensitive to the gender issue in engineering education, but it must be recognized that the presence of women in the profession is very recent when considered relative to a historical span of about 8000 years.

The Greeks

Although they had very capable engineers, the Greeks are better known for their contributions to science. Thales, Euclid and Aristotle are recognized for their contributions to the development of the basics of geometry and physics. This new knowledge of geometric relations was applied almost immediately in architectural engineering. However, scientific understanding during this period was so limited that nearly 2,500 years elapsed before this knowledge was expanded sufficiently to be useful to the engineering community.

Greek engineers built cities, roads, water (hydraulic) systems and some machinery. Archimedes and Hero were innovators who used the pulley, screw, lever, and hydraulic pressure to develop cranes, catapults, pumps, and several hydraulic devices. The Hellenistic period brought better buildings and local roads, which improved living conditions for the upper class. However, the construction of the city infrastructure was still performed largely by slaves because the use of other forms of power was nonexistent. The Hellenistic world was divided into extremely independent city-states. Politics of the period did not foster cooperation between these governmental entities; as a consequence, very few long roads were built. Transportation was largely by sea in boats equipped with oars powered by a galley of slaves. Some of the boats were fitted with a square sail, but these were so poorly designed that they were effective only on down-wind tacks.

The Romans

The Roman engineers followed the Greeks, and while they were considered by some historians to be less inventive, they were masters of detail. As the Roman Empire spread from the Middle East to Scotland, the Roman engineers built long roads connecting the distant cities; roads so well constructed that they were in service for many centuries. The large cities constructed in the conquered lands had paved streets and heated homes with water supplies and sewer systems. The Roman engineers had little or no theoretical knowledge, but they had developed practical methods of construction that served their purposes well. During this period, the subjects of statics and mechanics of materials were not understood. The theory used to design beams would not be developed for another 1,700 years. The Roman engineers based their construction on experience, and they used very large safety factors. These empirical methods, while crude by today's standards, sufficed to produce many bridges, roads, aqueducts and cities. A great bridge over the Tagus River in Spain supported a road nearly 200 m long with six spans 60 m above the river. Considering it was completed at about 100 AD, this bridge was a remarkable engineering feat. The durability of the Roman construction is well documented by their structures some of which are still in use today—2,000 years after they were placed in service.

The Roman engineers deviated from the Egyptian and the Greeks in their construction techniques by using much smaller building stones and bricks. They were able to produce massive structures with the smaller blocks by utilizing mortar and cement. The use of smaller stones and bricks and cement clearly impacted society. Buildings could be constructed with much less effort because thousands of slaves were not necessary to move a single block of material. Life for the slaves clearly improved although the construction work was still very demanding. Also, the upper classes benefited from the more rapid construction of buildings and houses for their cities.

While the Roman engineers were outstanding in the construction of civil infrastructure, they did very little to advance the use of power in driving the few machines in use during that period. Humans were still used in most instances to power pumps, mills and cranes. In rare cases, water wheels were used at power mills to grind grain, but these were of the undershoot design and much less effective than the overshoot design that was developed later. While references on the history of technology are quiet on the issue, the life of a worker powering a pump or a mill must have been very difficult.

Eventually, the vast Roman Empire crumbled, and the world entered a period that the historians classify as medieval civilization from 325 to 1300 AD. The fact that barbarians stripped and destroyed the cities does not imply that progress stopped, but it slowed progress to a snails pace. When Rome collapsed, the Byzantine Empire was the next civilization to follow. Significant engineering achievements were made in this period in developing dome structures and curved dams. The author has omitted discussion of these developments to keep this historical treatment brief.

Early Agriculture and the Use of Animals

In spite of the slow development of engineering and the lack of the construction of monuments during the Middle Ages, an early agricultural revolution occurred. White [4] has described the changes that occurred as agriculture moved from the dry sandy soil of Italy into the heavy alluvial and wet soil of northern Europe. The plows that were effective in the light sandy soils would not turn the sod in the heavy, moisture-laden soil of the north. A new, heavy-wheel-plow was invented that would cut through the grass sod and turn over the soil. However, this plow required eight yoked oxen to provide the large force necessary to pull it through the heavy sodded soil. Very few farmers of the day owned eight oxen. They had to combine their land, share their oxen and cooperate in plowing and planting. The combination of farmers and their land holdings eventually led to a manorial society.

It may be difficult to imagine that the invention of a new plow made such a major change on the way people lived. Today farmers are able to produce all of the food needed in the U.S., while exporting excesses in significant quantities, with less than 2% of the population. However, before the advent of engines and motors (stream, gasoline and electrical), men and women struggled to grow enough food to keep from starving. In the Middle Ages, 90% of the population worked in agriculture, and they had much less to eat than we do today.

Oxen were used as a source of power for plowing and transport on the farms. However, oxen are very slow and they consume large quantities of food. At that time, horses were of limited use because of inadequate harnesses and saddles and the frequent problem of splitting hoofs. (Early harnesses fitted about the horse's neck tended to strangle the animal.) These problems were alleviated when a newly designed horse collar, which transferred the load to the shoulders of the horse, was introduced and when iron horse shoes were fitted to their feet. Horses gradually replaced the oxen because they were 50% faster, worked longer and required less food.

In early warfare, the horse was not very effective in frontal attacks because stirrups for the rider did not exist. Consequently, horse mounted warriors were of limited value because they could not adequately thrust their lance while staying mounted on the horse. With the advent of the stirrup, the horsemen could stand-up, lean forward and thrust the lance through the defenders' shields without losing their seat. The simple addition of stirrups to saddles had a profound influence on society. Feudalism evolved with an aristocracy based on warriors conquering and defending landholdings. A new battle strategy with armored knights mounted on large strong horses was dominant for several centuries [4].

In early history, small technological improvements produced major changes in the form of government, the type of society and the way people of all classes lived. It is remarkable that for about 5,000 years, humans were the prime source of power for most activity. Progress in agriculture and engineering proceeded together to provide more food, better housing, roads and bridges and eventually horse- and oxen-powered plows and wagons.

16.3 ENGINEERING AND THE INDUSTRIAL REVOLUTION

Civilization has always been power limited; even today, space travel is severely constrained by the inadequacies of rocket power. Until the last three centuries, animals, windmills and waterwheels provided the only power available for agriculture, construction, milling, mining, and transportation. Progress was often slow because power was not available when and where it was needed. Life, in the power starved 17th century, was rather bleak for the general population. They worked from dawn to dark as farmers or craftsmen. Only those endowed with landed estates and/or money lived what you would consider a comfortable life.

The Development of Power

The first of several breakthroughs in developing new sources of power occurred in the 18th century. In Great Britain, Savery, Newcomen and Watt developed stationary steam engines. While these early steam engines were woefully inefficient (only 0.5%), they were produced in relatively large numbers (500 existed in 1800) and were employed to pump water and power mills (textile and flour).

Because the steam engines were replacing horses as power sources, James Watt conducted experiments to measure the power that a "brewery horse" could provide. His measurements indicated this horse generated 32,400 (foot-pounds/minute) of power. The horse's output was rounded to 33,000 (foot-pounds/minute) and this value is still used today as the conversion factor for one horsepower (HP).

Most of the 18th century was devoted to innovations directed toward improving several different models of steam engines. Applications were largely limited to stationary power sources because engines, of that day, were large, heavy and inefficient. However, with improved metallurgy and machining methods, it was possible in the 19th century to reduce the size and the weight of the engines while maintaining their output. These improvements permitted the steam engine to be adapted to power riverboats, ocean-crossing ships, rail locomotives and even a few steam-powered automobiles.

The development of the steam engine and its adaptation to several modes of transportation formed the foundations of the industrial revolution. The life style of almost everyone was markedly changed with the emergence of steam-powered factories and much more effective transportation. Fewer people were employed in agriculture, but many more were working under appalling conditions in factories and mines. Twelve-hour days, seven days a week, with only Christmas as a holiday during an entire year was the norm. Manufactured products, such as clothing and housewares, were available to a larger share of the population at reduced prices, but the margin between earnings and the money necessary for a factory worker, miner, or farmer to stay alive remained very narrow. Steam power had relieved many thousands of men and horses from brutally hard labor, but they were released from one dull job for another. On a more positive note, the progress in agriculture and industry enabled the population to triple during this period (1660 to 1820).

16.4 ENGINEERING IN THE 19TH AND 20TH CENTURIES

For the eight thousand years of recorded history prior to about 1800, the advance of technology was extremely slow. Agriculture dominated society with most of the population growing food, which was consumed locally. Factories were being developed and simple products needed on the farm, in the home or by the military were produced. Transportation of goods was usually limited to sailing ships on the seas or horse drawn barges on rivers or canals.

During the 19th and 20th centuries, technology literally exploded. The many scientific discoveries of the preceding millennium expanded knowledge sufficiently to allow countless engineering applications. To illustrate this explosion, let's define modern technology to consist of the following five elements:[2]

1. Abundant available power.
2. Available food in sufficient quality and quantity to insure a nutritious diet for everyone.
3. Transportation by air, land and water with associated infrastructure to provide safe, rapid, convenient and inexpensive access.
4. Communication between everyone, anywhere, anytime at a reasonable cost.
5. Information (knowledge) storage, retrieval, and processing in seconds at any location, anytime at negligible cost.

---

[2] It is assumed in this listing that housing, food and housewares are available to all from the technological-based industrial complex already existing in developed countries. A significantly different listing would be necessary for third-world countries.

While the advances in technology have been astonishing, particularly in the last 50 years, there are still significant deficiencies. Rocket power for launching space vehicles currently constrains space travel. Traffic jams in large cities extend the working day for many millions of people commuting to their jobs. Today nearly 45,000 people die in accidents on the highways every year. Travel on airlines is becoming unreliable, difficult and often unpleasant. Communication in many parts of the world is still not available to the public. Twelve years ago the author vacationed near a small fishing village in Mexico and learned that the radiophone at the tourist hotel was the only phone in the entire area. However, the recent introduction of the cell phone has greatly improved communication in underdeveloped countries. The cost of cell phone towers is much less than the expense of installing land lines required for conventional phone service. Many people are just beginning to explore the different ways to use the massive amount of information that has recently become available to those fortunate enough to own a personal computer and a modem.

Several important inventions with approximate dates are listed below. Think about these inventions, and how they have affected your life.

- Batteries 1800.
- Steam locomotive 1825.
- Electric generator 1831.
- Steamship 1835.
- Electric motor 1835.
- Telegraph 1837
- Baltimore to Wheeling railway 1850.
- Telephone 1876.
- Internal combustion engine 1876.
- Incandescent lamps 1880.
- Central electric power generation 1882.
- Steam turbines 1889.
- Alternators (ac generators) 1893.
- Radio 1896.

- Automobile 1900.
- Airplane 1903.
- Vacuum tubes 1906.
- Television 1935.
- Jet engines 1940.
- Atomic bomb 1945.
- Nuclear bomb 1950.
- Digital computer 1950.
- Transistor 1953.
- Integrated circuit (IC) 1959.
- Nuclear power 1960.
- Communication satellite 1970.
- Microprocessor 1971.
- Cellular phones 1982.
- Internet 1990.
- Electronic commerce 1996

Obviously, this list is not complete—many more important innovations could have been included. The idea is to show you that the period from 1800 to 2000 was packed full of new technology. This technology markedly impacted the way that everyone lives, works and plays.

You may debate the exact year of the invention. However, the exact date is rarely important since an invention usually evolves into a commercially successful product after several iterations with improvements or modifications by several designers. You should not worry about the exact dates. Instead, you should recognize the continuous progress and observe the shift in emphasis from one period to another. In power, technology evolved from animals, to steam, to internal combustion (gasoline) and to nuclear energy. Power, from gasoline fueled internal combustion engines, became mobile and electricity was widely distributed. In transportation, technology evolved from the horse and buggy and sailing ships to automobiles, airliners, nuclear powered submarines and space shuttles. In communications, technology progressed from the pony express to the telegraph, telephone, fax machines, cellular telephones, pagers and the Internet. In information, technology is currently evolving from a hard copy library system to a digital multi-media information bank available at any time, to everyone, at nearly zero cost.

Have these advances in technology changed society? It is believed that they have had a profound beneficial effect. Farmers produce an overabundance of food with less than 2% of the population working in agriculture. Factory workers manufacture most of our consumable products less than 20% of the work force. The workweek is shorter, and we have many more affordable products. Most families own at least two cars and home ownership is at an all time high. Concerns are expressed regarding the improvements in the standard of living over the past 20 years, particularly for those people at the lower end of the income spectrum. Technology has probably not helped this group as much as others. Global trade has permitted many of the semi-skilled tasks to be moved to third world countries, where the cost of labor is a small fraction of the costs in the developed countries. Consequently, many of the previously well-paid factory jobs in manufacturing in the U.S. have been moved offshore and this process has negatively affected those without marketable skills.

16.5 GREATEST ENGINEERING ACHIEVEMENTS—20TH CENTURY

Recently the National Academy of Engineering[3] (NAE) working with 27 professional engineering societies complied a listing of the 20 greatest achievements in engineering in the 20th century. What made these achievements great? They are notable because of the very significant benefits of each accomplishment to all members of society. An excellent publication outlining the engineering achievements that transformed our lives is presented in Reference [6].

1. Electrification—in the 20th century, widespread electrification gave us power for cities, factories, farms, and homes and forever changed our lives. Thousands of engineers made it happen, with innovative work in fuel sources, power generating techniques and transmission grids. From streetlights to supercomputers, electric power makes our lives safer, healthier, and more convenient.
2. Automobile—perhaps the ultimate symbol of personal freedom, the automobile is a showcase of engineering ingenuity and the world's major transporter of people and goods.
3. Airplane—with its speed and efficiency, air travel makes personal and cultural exchange possible on a global scale.
4. Water Supply and Distribution—water is our most precious resource, and improvements in its treatment, supply and distribution have changed our lives profoundly, virtually eliminating waterborne diseases in developed countries and providing clean and abundant water for communities, farms and industries.
5. Electronics—electronics provide the basis for countless innovations including CD players, TVs, and computers to name only a few. Electronic products improve the quality and convenience of modern life.
6. Radio and Television—radio and television were major agents of social change in the twentieth century, entertaining millions and opening windows to other lives, to remote areas of the world and to history in the making.
7. Agricultural Mechanization—tractors, combines and hundreds of other agricultural machines increase the productivity of farms, reduce the need for manual labor and improve our ability to feed the world.
8. Computers—perhaps the defining symbol of 20th century technology, the computer has transformed businesses and lives around the world, increased our productivity and opened up access to vast amounts of knowledge.
9. Telephone—a cornerstone of modern life, the telephone brings the human family together and enables the communications that enhance our lives, industries and economies.
10. Air Conditioning and Refrigeration—air conditioning and refrigeration make it possible to transport and store fresh foods and adapt the environment to human needs. Once luxuries, they are now common necessities which greatly enhance our quality of life.
11. Highways—vast networks of roads, bridges and tunnels connect our communities, enable goods and services to reach remote areas, and encourage economic growth.

---

[3] More details for the Greatest Engineering Achievements of the 20th Century are provided at the Web site for the National Academy of Engineers—www.greatachievements.org

12. Spacecraft—the human expansion into space is perhaps the most awe-inspiring engineering achievement of the 20th century. The development of spacecraft has thrilled the world, broadened our knowledge base and led to thousands of new inventions and products.

13. Internet—in a few short years, the Internet has become a vital instrument of social change, transforming business practices, educational pursuits and personal communications. By providing global access to news, commerce, and vast stores of information, the Internet brings people together and adds convenience and efficiency to our lives.

14. Imaging—From tiny atoms to distance galaxies, modern imaging technologies have expanded the reach of human vision with tools such as X-rays, telescopes, radar, and ultrasound.

15. Household Appliances—household appliances have greatly reduced the labor involved in everyday tasks, giving us more free time and enabling more people to work outside the home.

16. Health Technologies—medical imaging devices, artificial organs, replacement joints, and biomaterials are a few of the advanced health technologies that improve the quality of life for millions.

17. Petroleum and Petrochemical Technologies—petroleum and petrochemicals provide fuel for cars, homes and industries and the basic ingredients for products ranging from aspirin to zippers.

18. Laser and Fiber Optics—lasers are used in supermarket scanners, surgical devices, satellites and other products. Today fiber optic fibers provide the infrastructure to carry huge amounts of information via laser light, spurring a growing revolution in telecommunications.

19. Nuclear Technologies—the harnessing of the atom led to dramatic changes in the 20th century by providing a new source of electrical power, improving techniques for medical diagnosis and treatment and transforming the nature of war forever.

20. High-performance materials—from the building blocks of iron and steel to the latest advances in polymers, ceramics, and composites, engineering materials are used in thousands of important applications.

16.6 NEW UNDERSTANDINGS

In this discussion of the historical development of technology, we have been optimistic about the very positive effects of technology on society. In the past the public accepted, without serious questions, all forms of technology and the risks that were involved. However, the situation has changed to a remarkable extent in the past 50 years. Our leadership (engineers included) has, to a large degree, lost the public trust. An excellent example of this loss pertains to the technology employed in the generation of electricity using nuclear power. Since the Three Mile Island incident that occurred in Pennsylvania in 1979, the general public in the U.S. has opposed nuclear power. There is a valid public perception that the technological risks associated with nuclear power were significantly understated. John O'Leary, a deputy director for licensing at the Atomic Energy Commission (AEC), stated that "the frequency of serious and potentially catastrophic nuclear incidents supports the conclusion that sooner or later a major disaster will occur at a major generating facility" [7]. This statement was made several years prior to the accident at Three Mile Island. In spite of this warning of the likelihood of a serious accident, federal, state or local governments have not planned for orderly emergency evacuation of nearby residents or for containment measures to limit the possible contamination of many of power reactors.

The Chernobyl accident that occurred in the former Soviet Union in 1986 proved that O'Leary was correct. The No. 4 reactor at the Chernobyl facility exploded sending a radioactive plume in the air so high that an alert nuclear plant operator in Sweden detected it. Many died in this radioactive explosion and many more have died since then from radiation poisoning. High radiation levels affected countries as far away from Chernobyl as Italy.

The nuclear industry and the government agencies regulating it, world wide, have not been candid with the public. Nor has the public been very interested in intelligent debate—resorting instead to large, ugly demonstrations to express their displeasure. The public needs an accurate and honest assessment of the risks of any nuclear power project by the industry, the regulators and experts. The risks must be weighed against the benefits to society. Sometime in the future fossil fuels will be come unavailable, scarce or costly. Indeed, we recently witnessed a sharp increase in the price of crude oil at the beginning of the Iraqi War. When a serious fuel crisis occurs, we may find that nuclear power is the only viable alternative for power generation from carbon burning plants. Regulations pertaining to carbon emissions may also limit burning of coal or oil to generate electrical power. Natural gas, favored by both industry and the public to generate energy while minimizing environmental impact, is also in short supply.

The purpose here is not here to argue the case for or against nuclear power. It is used to illustrate the fact that understating or poorly assessing risk brings a severe penalty—loss of public trust followed by a ban on technical development in the subject area.

As engineers, we must significantly broaden our perception regarding technological risk. Today, many engineers dismiss failure due to human, operator or pilot error because they are not machine errors. This attitude is **absolutely wrong**. The **human operator and the machine** constitute a system and affect public safety accordingly. The machine and its operator must be included in system design since they both constitute an interacting set of weaknesses and capabilities [8].

Another new understanding pertains to geography. Prior to the Second World War (WW–II), international trade was very limited. As a country we produced what we consumed, mined our own minerals and pumped our own oil. Geography was very important because it constrained trade by law and by distance. After the war, laws were changed and trade agreements (GATT, NAFTA, etc.) that lowered or eliminate tariffs were arranged. International trade was encouraged so much so that the U.S. now consistently imports more goods and materials than it exports. The annual deficit in the trade balance for the U. S. is expected to exceed 500 billion dollars in 2004. This deficit represents the equivalent to more than 12,000,000 high-income jobs paying $40,000 a year.

Previously, the cost of shipping limited trade among countries separated by large distances. However, following WW-II many technological improvements were implemented in the shipping business: the super tanker was developed—greatly improving the efficiency of shipping huge quantities of oil. Also very large containerized ocean vessels were developed reducing the cost and time of loading and unloading cargo. Large efficient diesel engines were installed on the ships permitting increased speed and significant reductions in shipping time and costs.

With the time and costs of shipping greatly reduced and the legal barriers to trade removed, the country's commerce is often conducted in a global marketplace. Engineers compete worldwide to develop world-class products or services. Products designed in the U.S. may be produced anywhere in the world, or products designed in Europe or Asia may be produced in the U.S. Today, design and production activities are located to minimize costs and to maximize benefits to the customers and/or to the corporate entities.

The third new understanding is in the role of the governments (federal, state, county and city). Prior to WW-II, governments were small and taxes were relatively low. The primary role of the federal government was to insure national security. The state governments worried mostly about transportation infrastructure, and the local governments concerned themselves with education. Today, the situation is markedly different. Governmental agencies, at all levels and in response to their interpretation of the law (either old or new), issue regulations which pertain to topics of concern to society including the protection of the environment, safety, health and energy conservation.

You can debate the wisdom of many of the regulations, but that is not the point. The regulations currently exist and new ones will continue to be issued with an alarmingly high frequency. As an engineer, you must be prepared to serve society-at-large as reflected by these governmental regulations. If you question the wisdom of some of the regulations, the best approach is to become involved in the political process so that you may have the opportunity to influence them. Engineers tend to resist constraints because they limit the freedom of their designs and add to the cost of the product. You may think of regulations as government imposed constraints, but they are really barriers imposed by society-at-large. Society controls the regulatory process because it has the voting power to change the decisions of the politicians drafting the laws and of the bureaucrats formulating the regulations.

16.7 BUSINESS, CONSUMERS AND SOCIETY

Engineers serve three constituencies—business, the customer and society-at-large. The business leaders (management) look to engineers to develop products, services and processes that meet global competitive challenges. Consumers seek more convenient, reliable, enhanced and value-laden products at reduced prices. Society-at-large, through elected politicians, trail lawyers and public interest groups, demand action leading to solutions of problems involving safety, health, energy conservation and preserving or improving the environment. It is an overlapping set of constituencies, as shown in Fig. 16.1, because society encompasses business, governments, and consumers.

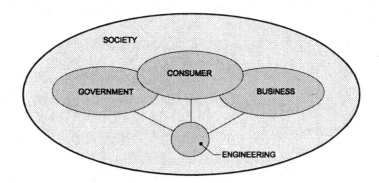

Fig. 16.1 Schematic illustration of the overlapping constituencies served by engineers.

To discuss the issues raised by serving three constituencies, consider the U.S. automotive industry as an example. Marina Whitman has described many aspects regarding the three-way demands. Since this discussion is based on her excellent contribution, you may want to study reference [9] in depth.

Business demands are simple although they are often difficult to meet. Any business must be profitable to continue to exist. Losses can occur over the short term, but if the red ink is prolonged the company is forced by its creditors into bankruptcy proceedings. Making a well-defined profit every quarter is the usual goal of management. For many years following WWII, it was relatively easy for U.S. businesses to make money. Bombing and shelling during the war had destroyed most of the manufacturing capability and a significant part of the infrastructure in every developed country in the world except in the U.S. It was easy to be a leader when the competition was without factories adequate infrastructure. For the automobile industry, the golden years began to erode in the late 1960s, and survival became a life or death battle in the 1980s. Problems for the auto industry came from global competition (mostly Japan). The Japanese firms provided much more reliable automobiles at lower costs than those produced in the U.S.

To show the extent of the market penetration by imported automobiles, recall that the imports accounted for only 4% of the market in 1962, but they controlled 30% of the market by 1990. This devastating loss of market share was not limited to the auto industry; several other industries including optics, consumer electronics, ship building, steel, etc. suffered even more serious losses.

Engineers and management in the U.S. firms had lost the competitive edge after decades of modest competition from the Europeans. The Japanese automobile companies (Toyota, Honda, Nissan, etc.) provided more affordable, reliable, appealing, fuel efficient and value-laden cars than either the American or European companies. In recent years, the loss market share has been reversed by improved management and technical methods (concurrent design and systems engineering[4]), but regaining the market share of the 1960s will be a very long and difficult process.

---

[4]Systems engineering means many different things to different groups. It is defined here as a cost and time efficient method to specify, design, develop and integrate all of the individual subsystems so that the total system for the product meets the requirements of customers, business and society-at-large.

The second constituency, the consumers, has been well recognized in the past decade. The importance of keeping the customer pleased is foremost in the minds of both management and engineering. Systematic methods for establishing the consumers' needs have been developed and placed into practice [10]. Many companies in the U. S. currently follow a product development process that begins with the customer interviews and proceeds through all of the phases of product development. You do not want to underestimate the importance of this second constituency; however, engineers have usually kept the customer in mind during the design of new products. The newer methods [10, 11] tend to insure that the customers' requirements are more accurately defined by the product specification before beginning a development.

The third constituency, society-at-large, generates the most challenging requirements. Society-at-large is represented by government agencies and a variety of public action groups. Officials of the government agencies (bureaucrats) produce regulations that define requirements for automobiles that may affect safety, fuel conservation or the environment. The public action groups create pressure and influence politicians in setting public policy that eventually generates new requirements.

Engineers often have some difficulties with regulations and/or proposals from the public action groups. Lawyers write the laws and public servants write the regulations. Because lawyers and public servants are often technically illiterate, they sometimes seek technical support prior to drafting binding regulations. However, government bureaucrats, prior to a thorough study by knowledgeable representatives of the public and engineering communities, sometimes place regulations into effect. The difficulty with this procedure is that the demands of the customer, including style, comfort, affordability, reliability, performance, convenience, etc., may be in serious conflict with societal imposed regulations on safety, fuel efficiency, emissions, etc.

A very recent example illustrates what many believe to be an ill-advised regulation [12]. The National Highway Transportation Safety Administration (NHTSA) mandated requirement for dual airbags in every vehicle licensed in the U.S. Most people consider air bags a valuable safety feature because several thousand lives have been saved in serious frontal accidents since they have been mandated. However, many children and small adults (mostly women) have been killed by rapidly inflating airbags. As of January 2004, 230 people have died by rapidly deploying air bags in low severity crashes since the use of this safety feature was mandated [13]. More than 90 percent of the passenger airbag fatalities have been children and infants. They account for about 60 percent of all airbag related deaths. Small women are also at risk particularly if they cannot position themselves far enough away from the airbags. The statistics cited above are confirmed deaths due to rapidly deploying airbags. In addition, there were many unconfirmed air bag related fatalities where NHTSA did not send an investigative team to the accident scene to confirm the cause of death.

In a recent ruling, that modifies the existing standard, NHTSA reduced the impact speed of the crash test from 30 to 25 MPH [13]. NHTSA will also require vehicles be tested using a family of dummies—an average size man, a small woman and 1, 3 and 6 year-old children. This new standard will be phased in over a three-year period beginning in 2004. Hopefully these changes will enable the protection of large adults while eliminating the dangers of air-bags to small women and children.

The killing of women and children by a safety device is a **very serious concern**. Moreover, the regulation prohibits anyone other than the car's owner to disable even the passenger side air bag[5]. If you are a concerned parent, you can disable it yourself, but few owners have the required skills. Also, if you ruin

---

[5] A recent ruling from the NHTSA allows auto dealers to install an on-off switch to control the passenger side air bag; however, NHTSA requires an extremely complex process for drivers to obtain a waiver, which is necessary before the dealer will perform the necessary modifications to the deployment system. First, you have to obtain a NHTSA request form. Then you must swear—under threat of criminal penalty—that you are too short to be safe with an air bag deployment, or cannot always put the children in the back seat, or have one of five NHTSA approved medical conditions. If you option for the medical conditions you must have a letter from a medical doctor confirming the condition. To date, relatively few people have fought the bureaucratic battle with NHTSA and obtained waivers. As of October 30, 1999, 57,183 requests have been authorized, but only 11,195 switches have been installed.

the mechanism and/or the bag, the cost of replacement may approach $1000. Because of their high costs, air bags have replaced stereo systems as the theft of choice if your car is robbed.

The issue is not whether you should have air bags or not. They have proven to be a very effective safety device for most adults either driving or riding as passengers. The issue is with the rigidity of the federal mandates and the failure of the regulators to:

1. Update the regulations to accommodate different behavior of drivers.
2. Provide more design flexibility in the requirements.
3. Provide a means of permitting the driver to activate or deactivate the passenger side bag.

The air bag regulation was written in 1977 when only 14% of the population even bothered to use their seat belts. By 1994, when the seat belt mandate was fully effective, 49 of 50 states were enforcing seat belt laws and usage had risen to 68%. While seat belts do not eliminate the need for air bags, the belts increase the time allowed for the deployment of the air bag. The NHTSA did not change the air bag regulation to accommodate the increased usage of seat belts until early in 1997. Responding to the controversy about the risk due to air bags, NHTSA issued a new regulation permitting automakers to install airbags 25% to 30% less powerful than the current bags. The new regulation gives the automakers two years to implement the change. In the meantime on many cars manufactured before 1999, air bags continued to be deployed at a speed of 200 MPH to protect an unbelted adult male. This is great for the big guys; however, for kids or smaller women, a 200 MPH kick in the face is extremely dangerous. When seat belts are used, more time is available to deploy the air bag, and the speed of deployment may be reduced from the dangerous value of 200 MPH.

Industry representatives recognized these problems and warned the NHTSA of the dangers to children. Rather than change the regulation to provide more flexibility in the design of air bag systems, NHTSA drafted a warning label and industry added instructions in the owner's manuals explicitly cautioning against placing a front facing child safety seat in the car's front seat. The manuals indicate that it is safe to place a child in the passenger seat if the child is buckled-in with a seat belt, **but** the seat must be in its full rear position. The manual also states that children should never lean over with their faces near the air bag cover while the car is in motion. The warning labels and instructions in the owner's manual are a result of poor design that was forced by an inflexible regulation. Currently safety devices (air bags) are mandated for automobiles that are dangerous to small women and children. Warning labels and instructions in manuals reduce the consequences of litigation (there is another law indicating that the user must be warned of possible dangers), but they are totally inadequate to protect children who are often too young to read. The solution is simple—young children should not ride in the passenger seat.

It would be relatively easy to modify the design of the control system to include an on/off switch with a warning light on the control panel of the automobile. With this feature the driver could deactivate the passenger side air bag whenever he or she wished to do so. Another design approach entails a modification of the pneumatic system to reduce the explosive force with which air bags deploy. This modification would probably have taken place several years ago when seat belt usage finally became widespread. However, the federal regulation did not permit modification then, and only recently has it allowed modification after the death of 168 (and counting) men, women and children.

Another problem with air bags that is not understood by the public is related to sodium azide, which is often used as the propellant for inflating the bag. Sodium azide is very toxic. When the air bag is deployed, you may be contaminated; the usual practice following an air bag deployment is to take the person or persons involved to the hospital for decontamination.

The public is concerned, if not angry. The NHTSA has received many letters from owners requesting the waivers required to have trained mechanics at the dealers to disconnect the dangerous air bags. Many mechanics refuse to disconnect the airbags for fear of future litigation. A recent survey of 700 repair shops indicated that two-thirds refused to install the on-off switches. There appears to be a significant amount of confusion about the remedy. Recently the NHTSA has provided a listing of auto dealers willing and certified to install switches on their Web site.

The current disarray is alarming. The interaction of the government agencies representing the public-at-large, business and the consumer is woefully inadequate. The public and business needs solutions often in a relatively short time. Yet, the regulatory system is too cumbersome to provide rapid solutions. The awkwardness comes from inflexibility on the part of the government and inadequate cooperation between the automakers. Competition on the part of business to increase market share is understandable, but safety features should be excluded from the normal competitive secrecy. The government agencies must become much more flexible and recognize that they cannot decree away all danger.

The difficulties between the government and business are not new. An interesting summary of the regulatory relationship between government and business was recently published by Jasanoff [14]. She is quoted below:

> Studies of public health, safety and environmental regulation published in the 1980s reveal striking differences between American and European practices for managing technological risks. These studies show that U.S. regulators on the whole were quicker to respond to new risks, more aggressive in pursuing old ones and more concerned with producing technical justifications for their actions than their European counterparts. Regulatory styles, too, diverged sharply somewhere over the Atlantic Ocean. The U.S. processes for making risk decisions impressed all observers as costly, confrontational, litigious, formal and unusually open to participation. European decision making, despite important differences within and among countries, seemed by comparison almost uniformly cooperative and consensual; informal, cost conscious and for the most part closed to the public.

The assessment by Sheila Jasanoff gives us clear direction. It is important to reduce the adversarial relationship between business and government. As engineers working for both government and business, we should be an important element in the future to produce a regulatory system with more flexibility, cooperation, cost consciousness and consensus.

16.8 CONCLUSIONS

The standard of living in the U.S. will be determined by the interplay of three powerful influences:

- New and rapid technological advances.
- Business (management) response to global consumer demands.
- Social demands as evidenced by legislative and regulatory requirements.

Engineers have always provided the leadership in technological advances, and they are expected to continue to provide business and the public-at-large with cutting-edge, world-class technology.

Business requires capital, technology and astute management to remain competitive. In addition to providing the technological base for a company, engineers frequently serve in management. If management career path appeals to you, plan on extending your education in a business school. The combination of a Bachelor of Science degree in engineering and a Master of Science in business provides a very solid foundation for a career path in a technically oriented business.

In the past, engineers usually have not been actively engaged in societal issues. They lack patience in dealing with the public and become irritated by the public's lack of technical literacy. They fume at the inefficiencies of government agencies and their lack of flexibility. They become outraged at the arrogance of bureaucrats and at their lack of concern for time and costs. They withdraw from public forums and focus on new developments and products. In the meantime, the public-at-large grows wary of many important sociotechnical systems and the government resorts to litigious processes, which impedes real progress.

It is unfortunate that engineers have not been a significant force in dealing with societal issues. When society, business and the customer create conflicting demands, engineering expertise is essential in crafting well-balanced solutions. The conflicting demands provide new opportunities and complex challenges for engineers. To take advantage of these opportunities, the engineers of tomorrow will require enhanced communication skill, greater disciplinary flexibility, a better understanding of the mechanisms of regulatory agencies and a much broader perspective relative to societal demands.

REFERENCES

1. Singer, C. E. J. Holmyard, A. Hall, and T. Williams, Eds. A History of Technology, Oxford University Press, New York, NY, five volumes, 1954-1958.
2. Hughes, T. P., "From Deterministic Dynamos to Seamless-Web Systems," Engineering as a Social Enterprise, ed. Sladovich, H. E., National Academy Press, Washington, D. C. 1991, pp. 7-25.
3. Sladovich, H. E., ed. Engineering as a Social Enterprise, National Academy Press, Washington, D. C. 1991, pp. 7-25.
4. White, L., Jr., Medieval Technology and Social Change, Oxford: Clarendon Press, NY, 1962
5. Kirby, R. S., S. Withington, A. B. Darling, and F. G. Kilgour, Engineering in History, Dover Publications, New York, NY, 1990.
6. Constable, G. and B. Somerville, A Century of Innovation: Twenty Engineering Achievements that Transformed Our Lives, Joseph Henry Press, Washington, D. C., 2003.
7. Ford, D. F., Three Mile Island: Thirty Minutes to Meltdown, Viking, New York, NY, 1982
8. Adams, R. McC. "Cultural and Sociotechnical Values," Engineering as a Social Enterprise, ed. Sladovich, H. E., National Academy Press, Washington, D. C. 1991, pp. 26-38.
9. Whitman, M. v. N., "Business, Consumers, and Society-at-Large: New Demands and Expectations," Engineering as a Social Enterprise ed. Sladovich, H. E., National Academy Press, Washington, D. C. 1991, pp. 41-57.
10. Clausing, D. Total Quality Development: World-Class Concurrent Engineering, ASME Press, New York, NY, 1994.
11. Schmidt, L., et al, Product Engineering and Manufacturing, 2nd Edition, College House Enterprises, Knoxville, TN, 2002
12. Payne, H. "Misguided Mandate," Scripps Howard News Service, Knoxville News-Sentinel, January 12, 1997, p. F-1.
13. Anon, Q & A: Airbags, Insurance Institute for Highway Safety, http://www.hwysafety.org/safety_facts.qanda/airbags, 2004.
14. Jasanoff, S. "American Exceptionalism and the Political Acknowledgment of Risk," Daedalus, Vol. 119, No. 4, 1990, pp. 61-81.

EXERCISES

16.1 Recall a product that you or your family has purchased in the past month or so. Write a brief paper describing both the positive and negative impacts of that product on society.

16.2 The Federal Aviation Administration enacted a regulation affecting the mailing of packages weighing more than 16 ounces. The regulation requires you to present the package to a postal clerk for mailing and prohibits the mailing of the package from a postbox. Write a paper covering the following issues:
- Describe the regulation in more detail.
- Why is the FAA writing a regulation affecting the U.S. Post Office?
- Do you believe that the regulation will be effective for its intended purpose? Please give arguments supporting your viewpoint.
- What actions will the post offices (40,000 of them) have to take to make the regulation effective? What actions do the post offices actually take with regard to the regulation?
- Will these actions be costly? Estimate the costs to both the public and the individual. Assume that a postal clerk is paid $15.00/hour and that an individual considers his or her free time worth $15.00/hour.
- Give your assessment of the cost to benefit ratio for this regulation?

16.3   Suppose that you were a slave in the time of Rameses II and were assigned to the task of constructing his statue. For those not up to date on Egyptian statues, this one weighs 1,000 tons and was 56 feet tall. Write a paper describing your daily tasks.

16.4   Horses were not used extensively to relieve man from brutal work or to provide a significant advantage in military actions until nearly 1000 AD. Write a brief paper describing both the social and technical reasons for the very long time required to effectively employ the horse in either military or commercial enterprise.

16.5   Suppose that you were living on a manor in England in about 1300 AD. Describe your lifestyle and indicate how technology affects your work and play. Select in which of the two classes of society that you existed.

16.6   Describe where power is available in the U.S. today. Define the type of power in your response. Also, include an example when you personally suffered because of a lack of power.

16.7   Why was California subjected to rolling blackouts in the summer of 2002?

16.8   Write a paper comparing the lifestyles, as you imagine them, for a man or woman living today and living in the year 1000. Did technology make a difference in the quality of life?

16.9   Find the Web site for the National Academy of Engineering and click on Greatest Achievements for the 20th Century. Select one of the 20 achievements and prepare a paper describing the history of that particular achievement.

16.10   Find the Web site for the National Academy of Engineering and click on Greatest Achievements for the 20th Century. Select one of the 20 achievements and prepare a paper describing the time line for that particular achievement.

16.11   Why is the public-at-large opposed to power generation with nuclear energy? Is the public-at-large correct in their collective assessment? Explain why you formed this opinion.

16.12   Why does a global marketplace exist today? Does global trade improve our standard of living? What does our current trade deficit have to do with wages for factory workers? Does technology help or hinder the balance of trade deficit?

16.13   If you were the director of the National Highway Transportation Safety Administration (NHTSA), what action would you take regarding the public concern about air bags?
   • Explain the reasons for your actions.
   • Explain how you would convince the major automakers to agree with you.
   • Prepare an outline you would follow in the press conference announcing this action.
   • Describe how you would handle a public interest group disagreeing with your ideas.
   • Describe how you would handle the public interest groups that agree with you.

16.14   Prepare a paper describing the information provided by NHTSA on air bags. A significant amount of information is given on its Web site at http://www.nhtsa.dot.gov/airbags.

16.15   From the NHTSA Web site, find the Web site for the Insurance Institute for Highway Safety and examine the page on Airbag Statistics. Determine the ratio of deaths caused by air bags to the estimated number of lives saved by air bags. Write a position paper about actions that should be taken in the future regarding air bags to insure public safety.

16.16   Using the NHTSA Web site, find the name and location of a dealer near you that is willing and authorized to install on/off switches for air bags.

16.17   Recently side airbags have been installed in some automobiles. Are these side air bags a threat to small children seated near a door? What has the NHTSA done to mitigate this threat?

16.18   What social science courses should you take during your undergraduate program to broaden your prospective and aid you in dealing with sociotechnical issues?

16.19   Is there a technical elective in your program of study that deals with sociotechnical issues?

16.20   Prepare a paper discussing the issue of air quality regulations and the design of automobiles.

CHAPTER 17

SAFETY, RISK AND PERFORMANCE

17.1 LEVELS OF RISK

When engineers design a new product or modify an existing one, an element of risk is often introduced in society-at-large. Sometimes the risk is minimal, and the resulting damage from a malfunction or failure is small. However, in some instances the risk may be large and the damage may be catastrophic.

Two simple examples are in order to illustrate the concept of risk with attendant benefits to society. Suppose you design a new model of an electric, pistol-grip drill that is powered from the standard 120 volt, 60 cycle single-phase power supply from a local utility company. The drill operator is at risk from an electric shock, although the probability for injury is very small. It is your responsibility as an engineer to minimize this risk while maintaining the advantage in performance of a new improved model of an electric-powered drill.

In the late 1950s a new aircraft—the Lockheed L-188 Electra powered by turbine driven propellers—was introduced and placed in service by several airlines. Three of these aircraft crashed, due to structural failure in flight, before the Federal Aviation Administration (FAA) grounded the fleet of Electras. An investigation revealed that a torsional resonance occurred under certain operating conditions causing the engines to vibrate violently and break away from the wing. This was a very serious engineering error that caused the death of over a hundred innocent passengers and significant losses of property. Clearly, this example illustrates a case where the risk far outweighs the benefits of performance of a new aircraft design. It also illustrates the seriousness of design errors and inadequate testing of prototypes of products that pose danger to the public.

Acceptable Risk

How do you protect an operator from electric shock? The operator is holding an electric motor with 120 volts across the armature coils and the field coils in his or her hands. The answer in this instance is to build the case for the drill from a tough, durable plastic that is structurally strong and an excellent electrical insulator. The case keeps the operator from touching any part of the electrical circuit powering the motor. In fact, Black & Decker® a major manufacturer of small power tools, uses double insulation to provide two independent insulation barriers to prevent the operator from making contact with any part of the electrical circuit. Operators may still manage to receive an occasional shock, but it will not be easy. Accidental contact of the operator's hands with a live electrical circuit is a rare event. In fact, the risk of electrical shock is so small that a worker routinely picks up a power tool and employs it to perform some task without even a passing thought about the possibility of receiving an electrical shock. Operating a power tool does involve risk, but it is acceptable because the operator is not consciously concerned about his or her safety while using the tool.

Voluntary Risk

Let's move up the risk ladder and consider flying in a commercial airliner from point A to point B, some 1,200 miles distant. Is it safe to make this trip? Most people appreciate that it is reasonably safe to fly commercial airliners, but they realize that there is a small probability of a crash. The statistics show the probability of a fatality p_k in an airline accident is about one in a billion passenger miles flown ($p_k = 1/10^9$). Hence, if you make the round trip from point A to B, your probability P_k of being killed due to an airline accident is:

$$P_k = s\, p_k = (2)(1200)/10^9 = 2.4 \times 10^{-6} = 0.00024\%$$

where s is the number of miles in a round trip.

This (2.4 chances in a million) is a very low probability for a fatal accident; consequently, most people do not worry much about the possibility of dying in a crash when they board an airliner. However, if you are a sales engineer flying 100,000 miles a year, every year for 20 years, your total mileage accumulates to 2×10^6 miles. Your probability of being killed in a crash increases to:

$$P_k = 2 \times 10^6/10^9 = 2 \times 10^{-3} = 0.2\%$$

The probability has increased significantly if you accumulate the miles flown by a traveling professional over a 20-year period. Knowing there is one chance in 500 that you will die in an airline crash, would you still want to be a sales engineer flying weekly to meet with customers? Could you tolerate this level of risk? Would you be apprehensive?

Everyone has some tolerance for risk. People weigh the speed, convenience, cost and risk of flying against that of traveling by train or by car. In almost all instances, travelers select the plane for long distances and the car for short distances. Travelers select the train only in those rare instances when it combines relatively good service (high speed with frequent trains) to convenient (downtown) locations. Most people do not contemplate the risks involved in travel because, except for driving a car, fatal accidents are rare events.

Unfortunately, fatal accidents in automobiles are not rare events. Preliminary estimates by the National Highway Traffic Safety Administration (NHTSA) showed that 42,850 occur on U.S. highways in 2002. In addition 2,914,000 serious injuries occur in these accidents. Death on the highway is the leading killer of Americans 1 to 35 years old. It is clearly more dangerous to drive than to fly when you compare the probability of fatalities and/or injuries on a per mile basis. How do you handle the higher risk associated with driving? Rationalization—I am an excellent driver, and it will not happen to me. Not necessarily a true statement, but the rationalization of one's superior driving skills alleviates the worry and concern about a fatality or injury producing accident. The real risk still remains.

When driving, you are an active factor in determining the risk. Driving habits (high-speed, reckless steering, tailgating, etc.) and driving skills affect—to a large degree—the level of risk involved. When flying, you are a passive participant. You have voluntarily decided to fly and sometimes even have a choice of airlines. You may pick the airline with the best safety record. (Southwest Airlines is the best with no accidents in their entire 30-year history). However, even knowing the facts, you are placed at risk and have little choice in the matter.

Involuntary Risk

Suppose you live 40 miles northeast of a nuclear power plant. Is 40 miles sufficiently far away to avoid significant fallout from a serious explosion involving the nuclear reactor at the power plant? Remember the winds are usually from the southwest, so that the fall-out plume will be pointed in your direction in the event of an accidental release of gasses containing radioactive particles.

The risk associated with radioactive fall-out, air pollution, toxic chemicals, polluted ground water, etc., is a serious concern to almost everyone; people dislike being exposed involuntarily. They expect governmental regulators to control the environment and to reduce the level of the risks of accidental exposure. Unfortunately, governmental agencies are not always effective in protecting the population from these exposures. Accidents have occurred in the control of reactors, and citations for safety violations by inspectors of the Nuclear Regulatory Commission are commonplace. From a public perception, the risk of a serious malfunction of a nuclear reactor became unacceptably high in the U.S. and many other European countries. After the accident at the Three Mile Island facility in Pennsylvania in 1979, the commercial nuclear industry died in this country. Some plants that were under construction were completed and others were converted to fossil fuels, but no new nuclear plants have been started since that time. With the more recent and much more serious accident at Chernobyl in 1986, new plans for power generation using nuclear energy in the U.S. have been placed on hold for the foreseeable future.

There is a very large difference between the acceptable level of risk depending on whether the risk-producing activity is voluntary or involuntary. The degree to which one personally controls the risk is also extremely important. Activities like skiing, scuba diving, horseback riding, hang-gliding, mountain climbing, dirt biking, etc. carry extremely high levels of risk. Yet, intelligent people swarm to the ski resorts and pay large amounts of money to potentially break their bones. Why? They have voluntarily decided that the thrill or other pleasures derived from the activity is worth the risk. Also, they control the level of risk. They can choose the "bunny" slope and minimize the risk or the "black diamond" slope to maximize the thrill.

Engineers must recognize that some level of risk is involved in almost all the products produced. It is imperative that this risk be minimized while maintaining an acceptable level of performance. Also, it is essential that this risk be acceptable to customers and to society-at-large. An appropriate balance between risk and performance must be achieved[1]. It is important that engineers, business and governmental regulatory agencies cooperate to provide a realistic assessment of the risk level. Finally, the public should be made aware of the risks involved and should not be astonished by news releases describing the gruesome details of victims involved in these relatively rare accidents.

17.2 MINIMIZING THE RISK

An engineer minimizes risk by preventing failure of each and every component of the system. This task is not easy as there are several different ways in which components can fail. Parts fail by breaking and aging, by corrosion or fatigue, by overload and burning, etc. In some instances, you may anticipate these failures and replace the parts before they malfunction. This controlled replacement of finite life parts is known as scheduled maintenance. In most cases, engineers try to design each component so that failure will not occur during the anticipated life of the product.

Safety in Design

A complete treatment of methods to avoid failure would require more coverage than a single chapter or even an entire textbook. However, an analytical procedure for designing tension members will be introduced to provide an illustration of a conservative design that insures safety. These tension members will be sized (made sufficiently large) so that they exhibit an acceptable safety factor. In other words, they will accommodate a service load higher than anticipated in service over their entire life cycle.

Let's begin by introducing a tension member as shown in Fig. 17.1.

---

[1] Cost is an issue when balancing safety, risk and performance. As a guide to cost, the Environmental Protection Agency (EPA) uses a worth of $6.1 million for what economist call a "value of statistical life" in their cost benefit analyses for new regulations.

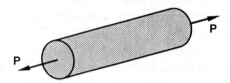

Fig. 17.1 A tension member is a long thin rod subjected to axial load P.

When the tension rod is subjected to an axial load P, an axial stress σ develops that is uniformly distributed over the cross section of the rod. This uniformly distributed stress acting on the rod is illustrated in Fig. 17.2.

The magnitude of this stress is determined from the formula:

$$\sigma = P/A \qquad (17.1)$$

where σ is the stress given in units of pound/square inch (psi) or newton/square meter (Pa).
P is the load expressed in units of pound (lb) or Newton (N).
A is the cross sectional area of the rod specified in square inch (in^2) or square meter (m^2).

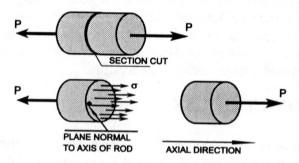

Fig. 17.2 The stress σ is distributed uniformly over the cross-sectional area of the tension rod.

Two examples demonstrating the technique for computing stresses in axially loaded tension members are presented below:

EXAMPLE 17.1

If the axial load P applied to a circular tension rod is 10,000 lb, determine the axial stress σ that is developed if the rod is ½ in. in diameter.

Solution

The cross sectional area is given by:

$$A = \pi r^2 = \pi(0.25)^2 = 0.1963 \text{ in}^2$$

From Eq. (17.1), the axial stress σ is:

$$\sigma = P/A = 10,000/0.1963 = 50,930 \text{ psi}$$

where psi is the abbreviation for lb/in^2.

At this stage of the analysis, it is not possible to interpret the significance of the answer $\sigma = 50,930$ psi. More information is needed pertaining to the strength of the material from which the tension member is fabricated. When the strength of the rod's material is known, it is possible to make a comparison of the magnitude of the stress relative to this strength and remark on its significance.

In this first example, U.S. Customary Units were used to express the axial load in pounds, the cross sectional area in square inches, and the stress in psi or pounds per square inch. In the next example, the International System of Units (SI) will be employed where the load is expressed in Newton, the cross sectional area in meters, and the stress in Pascal.

EXAMPLE 17.2

If the axial load P imposed on a circular rod is 50,000 Newton (N), and the radius of this rod is 5 mm, determine the axial stress σ.

Solution

First, determine the cross sectional area A as:

$$A = \pi r^2 = \pi (5 \times 10^{-3})^2 = 78.54 \times 10^{-6} \text{ m}^2$$

Then, from Eq. (17.1), the axial stress is given by:

$$\sigma = P/A = (50 \times 10^3 \text{ N})/(78.54 \times 10^{-6} \text{ m}^2) = 636.6 \times 10^6 \text{ Pa} = 636.6 \text{ MPa}$$

where the abbreviation Pa stands for Pascal—the unit used for stress in the SI system.

A Pascal is equal to a newton per square meter (N/m^2). When calculating stresses in the SI system, it is common to obtain very large numbers for the stress when σ is expressed in Pascal. For this reason, MPa (Mega Pascal) is often employed as the unit for σ where $\text{MPa} = 10^6 \text{ Pa} = \text{N/(mm}^2)$.

The stress on a tension member can be accurately determined if you can ascertain its size and the load that it must resist. The calculated stress is a number with associated units—either psi or MPa. As a number, it is not very informative until the stress is compared to the strength of the material from which the tension member is fabricated.

Let's suppose that the tension rod used in Example 17.1 is machined from very strong alloy steel exhibiting a yield strength of 100,000 psi. Do you believe the tension rod is safe? Will it retain its shape under load? Will it fail by yielding (stretching under load and not recovering completely when the load is removed)? These questions may be answered by making a simple comparison of the applied stress σ to the strength S of the material. In Example 17.1, the axial tensile stress $\sigma = 50,930$ psi and the strength of the material from which the rod was machined is S = 100,000 psi. The comparison shows that the yield **strength** of the rod is **greater** than the applied **stress**; therefore, the rod **will not fail** by yielding or breaking. But how safe is the rod? You know it will not fail; however, some measure of the safety of the rod should be established. Is the safety of the rod marginal, or is the rod too safe? Why would any structural element be too safe?

Safety Factor

To respond to the safety issue for the rod, determine the ratio of the strength to the stress, and define this ratio as the safety factor SF:

$$SF = S/\sigma \qquad (17.2)$$

For the stress imposed on the rod described in Example 17.1, the safety factor is determined from Eq. (17.2) as:

$$SF = 100,000/50,930 = 1.96$$

The safety factor is unitless. The value of SF = 1.96 indicates that the tension rod is almost twice as strong as it must be to resist yielding under the applied load. You have a reasonable safety factor and can be confident that the rod will perform safely in service. Your only concerns are the accuracy with which the load has been predicted and the quality control for the material used in manufacturing the tension rods. If the factors listed below are satisfied:

- You have extensive experience with the application.
- You are certain of the magnitude of the load.
- You have confidence in the manufacturing division in tracking their materials.
- You have verified the strength of the materials received from suppliers.

Then you usually can be satisfied that a safety factor of about two is adequate. However, if you are not certain about the load and/or the materials, a safety factor of two is not sufficient. You should increase the safety factor to accommodate your ignorance of either the applied load or your lack of control over the material employed. Safety factors ranging from two to four are commonly employed in design. Safety factors of less than two require considerable care, expense and expertise. The use of relatively low safety factors is justified only for very high-performance applications or for components that are produced in very high volume. In these special situations, the engineering analysis, prototype testing, quality control inspections, maintenance inspections and documentation must be extensive.

Margin of Safety

The margin of safety MS is sometimes used to describe the degree of safety incorporated into the design of a structural component. The margin of safety should not be confused with the safety factor as they are different quantities. The margin of safety is defined as:

$$MS = (S - \sigma)/\sigma = SF - 1 \qquad (17.3)$$

Using the results from Example 17.1 and Eq. (17.3), the margin of safety is given by:

$$MS = 1.96 - 1 = 0.96 \text{ or } 96\%$$

The tension rod has a margin of safety of 96%, which implies that the strength of the material exceeds the applied tensile stress by 96%. If the load on a tension member can be determined, it is easy to size (adjust the cross sectional area) of the rod to provide any margin of safety that is deemed necessary to satisfy management's concerns and to meet professional obligations to society-at-large.

17.3 FAILURE RATE

Sometimes components fail by burning out, aging or wearing out—not necessarily by breaking or yielding (excessive permanent deformation). When conducting a failure analysis for wear or aging, the safety factor is not a relevant parameter. You must cope with wear, aging, or burn out by using other methods. To begin, let's recognize that all components do not wear out at the same time. Some people drive their car 25,000 miles before the brakes wear out and others may drive 60,000 miles or more. Some people can use their personal computers for several years before a component fails; yet others have problems after only a few months in service. To analyze these differences in service before failure, a mortality curve for both mechanical and electronic components is introduced in Fig. 17.3a.

To determine the failure rate FR, records of the performance with time-in-service of a large number of components are maintained. You begin your record of performance with N_0 components that are placed in service at time $t = 0$. After some arbitrary time t, some number N_f will have failed and the remainder N_s will have survived. At any arbitrary time, you may write:

$$N_f + N_s = N_0 \qquad\qquad (17.4)$$

With time, N_f increases and N_s decreases, but the sum remains constant and equal to N_0. The failure rate FR is defined as:

$$FR = N_f / (N_0 \times t) \qquad \text{when } t > 0 \qquad (17.5)$$

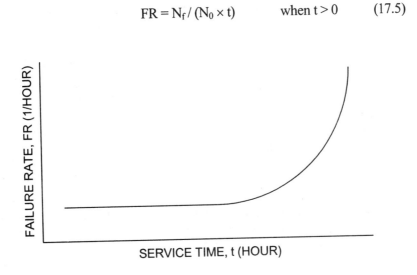

Fig. 17.3a Mortality curve for mechanical components showing failure rates with time in service.

When the results of Eq. (17.5) are plotted with respect to time, the component mortality curve presented in Fig. 17.3a is obtained. For mechanical components, where testing is performed on the assembly line, the failure rate is small when the components are relatively new. However, the failure rate increases non-linearly with time in service because parts begin to wear, corrode, or they are abused.

The failure rate for electronic components is usually characterized by the well-known bathtub mortality curve presented in Fig. 17.3b. This curve exhibits three different regions of interest—each due to a different cause. The high failure rates in the first region, $t < t_1$, are due to components manufactured with minute flaws that were not discovered during inspection and testing at the factory. In the center region of the curve, for $t_1 < t < t_2$, the failure rate FR_0 is small and nearly constant with time. Clearly, this is the best region to operate if you are to maximize the reliability of the product. Near the end of the service life, when $t > t_2$, the failure rate increases sharply with time as the components begin to fail due to the effects of aging.

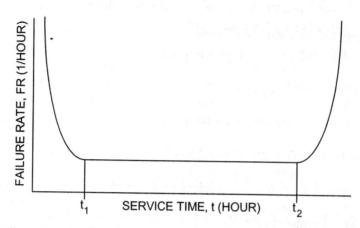

Fig. 17.3b Mortality curve showing the failure rate for electronic components with time in service.

For high-reliability systems, it is important to eliminate the inherently defective components associated with early failures. For mechanical systems, inspecting and testing the system's performance to eliminate flawed components produced in manufacturing accomplishes this objective. This task is not easy for the electronic component manufacturers because the feature sizes on the components are so small that inspection requires very high magnification. Even with elaborate inspection schemes, not all of the flaws can be detected. A better procedure to eliminate the few inherently flawed components is to conduct what is known as burn-in testing prior to incorporating the components into an electronic product. With burn-in testing, all of the components that are produced are operated for a time t_1 (as defined in Fig. 17.3b). The flawed components burn out (fail) and are eliminated. The surviving components that will function with a much lower failure rate FR_0 are used in fabricating high reliability products.

Clearly, it is important to design components with low failure rates. If the failure rates are low, then the time between failures will be high. In fact, manufacturers sometimes cite what is known as the mean time between failures (MTBF) in their product specifications. The MTBF and the failure rate are related by:

$$MTBF = 1/FR \qquad (17.6)$$

You should understand the concept of component failure rates and how they vary over the life of the component. How is this information about the failure rate used to establish the reliability of a specific component to perform safely over the anticipated life of a product?

17.4 RELIABILITY

Component Reliability

To answer the question of component and system reliability, you must introduce the concept of probability or chance of occurrence. In considering probability, let's return to the test conducted to determine the failure rate and use the data collected for N_f, N_s, and N_0. The probability of survival P_s, which varies with time t, may be determined from:

$$P_s(t) = N_s(t)/N_0 \qquad (17.7)$$

And the probability of failure P_f is given by:

$$P_f(t) = N_f(t)/N_0 \qquad (17.8)$$

Both the probability of survival and the probability of failure are functions of time. With increased time in service, the probability of survival decreases and the probability of failure increases.

From Eqs. (17.7) and (17.8), it is evident that:

$$P_s(t) + P_f(t) = 1 \qquad (17.9)$$

Substituting Eq. (17.8) into Eq. (17.9) yields:

$$P_s(t) = 1 - [N_f(t)/N_0] \qquad (17.10)$$

If Eq. (17.10) is differentiated with respect to time t and the result is rearranged, it is possible to write:

$$[N_0/N_s(t)][dP_s(t)/dt] = [-1/N_s(t)][dN_f(t)/dt] = - FR(t)dt \qquad (17.11)$$

Consider the term $[1/N_s(t)][dN_f(t)/dt]$ and note it is the instantaneous failure rate $FR(t)$ associated with a sample size N_s at time t. Next, integrate Eq. (17.11) and assume that $FR(t) = FR_0$, is a constant. The relation obtained gives the probability of survival as a function of the failure rate.

$$P_s(t) = e^{-(FR_0 t)} \qquad (7.12)$$

where e = 2.71828 is the exponential number.

Let's consider an example to demonstrate the method for determining the reliability of a mechanical component.

EXAMPLE 17.3

Suppose a company maintenance records exist that show the number of failures of a certain component over an extended period of time. Analysis of this data indicates that the failure rate is essentially constant with 1 failure per 10,000 hours of service. Does that sound good to you? Will the reliability be adequate? Let's determine the probability of survival (the reliability) of this mechanical component with time.

Solution

Substituting $FR_0 = 1/10,000$ h into Eq. (17.12), yields the results for $P_s(t)$ shown in Table 17.1.

Table 17.1
Reliability $P_s(t)$ of a mechanical component
with increasing service life ($FR_0 = 1/10,000$ h)

Time (1000 h)	Time* (years)	$FR_0 \times t$ (unitless)	Reliability $P_s(t)$
1	0.5	0.1	0.905
2	1	0.2	0.819
5	2.5	0.5	0.606
10	5	1.0	0.367
20	10	2.0	0.135
50	25	5.0	6.738×10^{-3}
100	50	10.0	4.540×10^{-5}
200	100	20.0	2.061×10^{-9}

*The conversion from hours to years of service life is based on 2,000 hours/year.

Reliability varies markedly with service life. For a short service life, say a year, the reliability is reasonable with 81.9% chance of surviving. However, for long life, say 10 years, the reliability drops to only 13.5%. For very long life 50 to 100 years, failure is almost certain. While the failure rate of 1 in 10,000 hours appears satisfactory in an initial assessment, the reliability resulting from this failure rate is disappointing. You cannot be certain of a service life of 10,000 hours prior to failure. In fact, there is only a probability of 36.7% to survive for 10,000 hours. If a reliability of 90% is required for a service life of 10,000 hours, the failure rate FR_0 must decrease to about 1 failure in 100,000 hours.

System Reliability

The failure of a component may or may not cause failure of a system. When a system is comprised of several components, its reliability will depend on the probability of failure of the individual components and their arrangement. There are two possible arrangements—series or parallel. Consider your automobile to illustrate both the series and parallel arrangement of components. For lighting the highway during the night, an auto is equipped with two headlights. This is a parallel arrangement because the design incorporates two identical components (the headlights), which essentially perform the same function. If one headlight burns-out, you can still drive. Your visibility is impaired to some degree; the system has been compromised; however, it has not failed.

On the other hand, suppose you intend to start your engine. Consider the components involved in this action:

- Ignition switch
- Battery
- Wiring
- Solenoid relay
- Starter motor
- Starter motor clutch and gear
- Engine ignition components (spark plugs and points)
- Fuel availability
- Fuel pump
- Fuel lines

If any of these components should fail, the engine will not start when you turn the key. A complete series of successful components is required for the system to function. To start your engine, every component must properly function since the failure of a single component will result in a failure of the entire system.

Reliability of Series Connected Systems

Let's consider the reliability of a system involving three components that are in a series arrangement, as shown in Fig. 17.4.

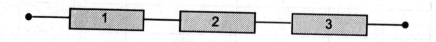

Fig. 17.4 A system comprised of a series arrangement of three components.

Because all three of the components in this series arrangement must operate successfully for the system to perform correctly, the probability of survival of the system (P_s^s) is a product function given by:

$$P_s^s = P_{s1} \, P_{s2} \, P_{s3} \qquad\qquad (17.13)$$

where the superscript s refers to the entire system of components.

If you combine Eqs. (17.12) and (17.13), it is possible to express the system reliability in terms of the failure rate of the individual components as:

$$P_s^s = e^{-(FR_1 + FR_2 + FR_3)t} \qquad\qquad (17.14)$$

When the number of components in a system is increased with a series arrangement, the system reliability is decreased markedly. Also, the system reliability continues to decrease with time in service.

EXAMPLE 17.4

Consider the case where n components are arranged in a series where n increases from 1 to 1000. Let's also assume that the reliability of each of the n components is the same ($P_{s1} = P_{s2} = \ldots\ldots= P_{sn}$) at some time during the operating life of the system. Evaluate the system reliability for three different values of the component reliability—$P_s = 0.999, 0.99,$ and 0.90.

Solution

The results obtained from Eq. (17.13) are shown in Table 17.2.

Table 17.2
Reliability of a system with n series arranged components
as a function of component reliability P_s

Number of Components, n	$P_s = 0.999$	$P_s = 0.990$	$P_s = 0.900$
1	0.999	0.990	0.900
2	0.998	0.980	0.810
5	0.995	0.950	0.590
10	0.990	0.904	0.349
20	0.980	0.818	0.122
50	0.950	0.605	*
100	0.905	0.366	*
200	0.819	0.134	*
500	0.606	*	*
1000	0.368	*	*

* System reliability of less than 1%.

Examination of the results presented in Table 17.2 clearly shows the detrimental effect of placing many components in a series arrangement. For components with a high reliability ($P_s = 99.9\%$), the system reliability P_s^s drops to about 90% with 100 series arranged components. However, with components with a lower reliability ($P_s = 90\%$), the system degrades to a reliability of less than 1% when the number of components approaches 50. The lesson here is clear—if you connect a large number of components together in a series arrangement to develop a system, then the individual component reliability must remain extremely high over the entire service life of the system.

Reliability of Systems with Parallel Connected Components

Recall the description of a highway lighting system on an automobile that includes a pair of headlights. This system is an example of a parallel arrangement because both headlights must fail before the system fails. It is possible to drive with one light, although the State Troopers might warn the driver of the dangers of doing so. The human body has several parallel systems; you have two hands, two eyes, two legs, two feet and two ears, etc. that increase a person's ability to function in the event of a mishap.

A system with a parallel arrangement of three identical components is represented with the model shown in Fig. 17.5.

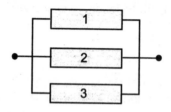

Fig. 17.5 A system with a parallel arrangement of three identical components.

These parallel systems are considered to be redundant because all of the components must fail before the system fails. Accordingly, the relation for the probability of system failure is written as:

$$P_f^s = P_{f1}\, P_{f2}\, P_{f3} \qquad (17.15)$$

Substituting Eq. (17.9) into Eq. (17.15) gives:

$$1 - P_s^s = (1 - P_{s1})(1 - P_{s2})(1 - P_{s3}) \qquad (17.16)$$

Expanding Eq. (17.16) and reducing the resulting expression gives:

$$P_s^s = P_{s1} + P_{s2} + P_{s3} - P_{s1}\, P_{s2} - P_{s2}\, P_{s3} - P_{s1}\, P_{s3} + P_{s1}\, P_{s2}\, P_{s3} \qquad (17.17)$$

Let's examine Eq. (17.17) to determine if redundancy (the number of parallel components) improves reliability.

EXAMPLE 17.5

Suppose the probability of success P_s for all the components in a parallel system is the same. However, P_s is considered as a variable increasing from 0.1 to 1.0. Show the effect of component redundancy by considering a system comprised of one, two and three components in parallel.

Solution

The results obtained from Eq. (17.17) are shown in Table 17.3.

Table 17.3
Effect of the degree of redundancy on reliability
$$P_{s1} = P_{s2} = P_{s3}$$

Component Reliability	Single Component	Two Components	Three Components
0.10	0.10	0.19	0.271
0.20	0.20	0.36	0.488
0.30	0.30	0.51	0.657
0.40	0.40	0.64	0.784
0.50	0.50	0.75	0.875
0.60	0.60	0.84	0.936
0.70	0.70	0.90	0.973
0.80	0.80	0.96	0.992
0.90	0.90	0.99	0.999
1.00	1.00	1.00	1.000

An examination of the results presented in Table 17.3 shows that the degree of redundancy improves the reliability of a system with a parallel arrangement of components. If you have relatively poor component reliability, say 60%, and employ two of these components in a parallel arrangement, the system reliability improves to 84%. Add a third component to the parallel arrangement and the reliability increases to 93.6%. Adding more components in parallel will continue to improve the reliability, although it is a case of diminishing returns.

The improvement in reliability by employing parallel (redundant) components in building a system is a distinct advantage. However, one rarely if ever, enjoys a free lunch. The corresponding disadvantages are the increased costs and the added power, weight and size of the system. These are significant disadvantages, and redundant design is used only when component reliability is too low for satisfactory system performance or when a failure produces very serious consequences.

For a more complete discussion of component and system reliability, see reference [1].

17.5 EVALUATING THE RISK

When the space shuttle Challenger exploded during launch on January 28, 1986, a Presidential Commission [2] was established to:

1. Review the circumstances surrounding the accident and determine the probable cause or causes for the explosion.
2. Develop recommendations for corrective or other actions, based on the Commission's findings and determinations.

Richard Feynman, a Nobel Prize winning physicist from the California Institute of Technology, was appointed to the Commission. Dr. Feynman, near the end of his career and his life, took the appointment very seriously and devoted his entire time and energies to the investigation. One of his many contributions during the review was to determine the probability of failure of the space shuttle system.

During the investigation, Dr. Feynman asked three NASA engineers and their manager to assess the probability of failure P_f of a mission due to the failure of one component—the shuttle's main rocket engine. Secret ballots by the engineers indicated two estimates with $P_f = 1/200$ and one with $P_f = 1/300$. Under pressure, the program manager finally estimated the risk at $P_f = 1/100,000$. There was such a large difference between the manager's and the engineer's assessment that Dr. Feynman concluded that "NASA exaggerates the reliability of its products to the point of fantasy" [3].

The safety officer for the firing range at Kennedy Space Center, who had been under considerable pressure to remove the remotely controlled destruction charges from the shuttle, did not believe the reliability figures cited by NASA. He had collected data for all of the 2900 previous launches using solid rocket boosters. Of this total, 121 had failed. This data provide a very crude estimate of $P_f = (121)/(2900) = 0.042 = 4.2\%$ or about one chance for failure in every 24 launches. The safety officer considered this high risk of failure to be an upper bound, because improvements made since the early launches had improved the reliability of the solid booster motors. Also, the pre-launch inspections on the shuttle were more thorough. His estimate of risk accounting for taking these improvements was $P_f = 1/100$. Dr. Feynman [3], after extensive interviews with engineers, reliability experts, managers from NASA and many subcontractors, concluded that the shuttle "flies in a relatively unsafe condition with a chance of failure of the order of 1%."

After the Challenger accident NASA began a series of comprehensive assessments of risk of flying the shuttle fleet in space. In 1995 the probability of a catastrophic failure was assessed at 1one in every 145 flights. After a series of safety related upgrades to the shuttle system, a new assessment in 1998 placed the probability at one in 245 flights. In October of 2002 NASA most recent assessment placed the probability at one in 265 flights or about 0.39%. How accurate are these assessments? The disintegration of the shuttle Columbia over Texas as it reentered space on February 1, 2003 was the second shuttle failure in 112 flights. This relatively high probability of failure $(2/112) = 0.01786 = 1.786\%$ raises questions about the validity of NASA's safety assessments [4, 5].

Clearly, risk assessment is a difficult task. Each component in a complete system must be evaluated to ascertain its failure rate. Then the components are placed in either a series or parallel arrangement, or some combination of the two, to model the system prior to determining its reliability. The system reliability may be lower or higher than the component reliability depending on the number and the arrangement of the components.

Testing to determine component reliability is possible in some instances when the components are relatively inexpensive. However, establishing reliability estimates by testing requires a very large number of trials that often destroy the component. If you want to show a probability of failure less than $P_f = 1/1000$, you must test considerably more than 1000 components to failure. Obviously, it would not be possible to test more than 1000 booster rockets when the cost of a single test is several million dollars.

Often the probability of failure of a component must be estimated based on previous experience with similar applications. In some instances, it is possible to compute the probability of failure; however, these calculations require considerable knowledge of the spectrum of loading and the ability of the material to resist fracture. When analytical methods are inadequate and engineering judgment is required to assess the probability of failure, the estimate should be made by a senior engineer with considerable experience and expertise. Even then, the estimate should be pessimistic rather than optimistic. A frank and honest estimate of P_f, based on all of the data and knowledge available, is much better than unrealistic appraisals that give an unwarranted feeling of safety.

17.6 HAZARDS

When designing a product, there are usually risks involved in either the production of the product or in its use in the marketplace. The public-at-large is exposed. What can you do to minimize the risk? In previous sections, the concept of safety factor, component and system reliability and evaluating the risk were briefly discussed. There is another approach to minimize risk. Namely, developing an ability to recognize the hazards involved. You must clearly recognize potential hazards before taking the necessary precautions in your designs to minimize the risks associated with them.

A Listing of Hazards

Mowrer describes a complete list of hazards and provides an extended discussion of each in reference [6]. You are encouraged to read his chapter in this reference and to use the extensive checklists incorporated in his coverage. A very brief excerpt from Mower's reference is provided in this section to assist you in recognizing the many hazards that should be considered when designing a product. A list of hazards includes:

1. Dangerous chemicals and chemical reactions
2. Exposure to electrical circuits
3. Exposure to high forces or accelerations
4. Explosives and explosive mixtures
5. Fires and excessive temperature
6. Pressure
7. Mechanical hazards
8. Radiation
9. Noise

Dangerous Chemicals

Chemicals can be nasty and you must appreciate the extreme dangers of exposures to certain toxic substances. Have you read about the leak that developed in a storage tank in Bhopal, India, in 1984? The storage tank contained the toxic chemical methyl isocyanate used in the manufacture of pesticides. It was a significant leak with about 80,000 pounds of the chemical released to the environment. Three thousand nearby inhabitants were killed, 10,000 permanently disabled and another 100,000 injured. The cause of the failure was not in the design of the tank, but in the training of the individuals responsible for the plant maintenance and operation. Nevertheless, a catastrophic accident occurred because several workers and managers, in positions of responsibility, did not adequately understand the dangers of this very toxic chemical.

Exposure to Voltage and Current

What about exposure to electrical circuits? Almost everyone has been shocked by the standard 120 volt, 60 cycle electrical power supplied by the local utility company. Why worry? You should worry because electrical shocks are **dangerous. Yes!** Even the 120-volt supply can cause major problems. Your body acts like a resistor and limits the current flowing from the voltage source through your hands, arms, legs, etc. The problem is that the resistance of your body is variable. It depends on the moisture on your hands, the type of soles on your shoes and even the moisture on the ground. If your hands are dry, your shoes have rubber soles and you are standing on a dry floor when you touch one wire of the circuit with only one hand, then you will probably not be harmed because you have arranged a very high resistance path to ground. Consequently, the current flow through your body will be very small. However, if your hands are wet and you touch both of the wires (the white and the black) from the electrical supply, one with the left hand and the other with the right hand, you have placed 120 volts across your heart. You can be electrocuted with only 120 volts under these conditions. The moisture on your hands greatly reduces the effective resistance of the body.

    Do not take chances with electricity. Insulate the operator from the circuits preferably with two independent layers of insulation. High voltages are even more serious than low voltages because the currents flowing through one's body increase dramatically. When the current flow through the human body increases to about 10 to 100 mA (milli-amperes), there are very serious consequences to the respiratory muscles. Higher currents of 75-300 mA produce problems in a person's heart function.

The electrical current I flowing through the body can increase in two ways: first, by increasing the applied voltage V, and second, by decreasing the resistance R offered by the body. Ohm's law gives the relation among the voltage V, current I, and the resistance R as:

$$I = V/R. \tag{17.18}$$

Using adequate electrical insulation in the design of products with electrical power, markedly increases the resistance R and decreases the current I to negligible amounts.

Some people think that low voltages (5 or 10 volts) are safe because they barely feel a tingle when touching a low voltage circuit. However, some low voltage circuits particularly on high performance computers carry substantial (100 or more ampere) currents. If you short a circuit, with high current flow, an arc occurs generating significant amounts of heat. Also, the flash of the arc can damage a person's eyes and the heat may produce serious burns. In your designs, insulate and/or shield even low voltage circuits.

The Effects of High Forces and Accelerations

High forces and high accelerations (or decelerations) go hand in hand. Newton's second law requires the connection ($\mathbf{F} = \mathbf{ma}$). If the brakes are suddenly applied in an automobile, the car decelerates quickly and the passenger (without seat belt or air bags) is thrown into the windshield. You must always be concerned with acceleration or deceleration because of their effects on the human body. Military pilots are trained to withstand high accelerations (high Gs). A good pilot with a well-designed flight suit can pull 7 or perhaps 8 Gs before losing consciousness. However, a civilian will become irritated at less than 2 Gs. If you want to feel an acceleration thrill, go to an amusement park and ride the roller coaster. There is no need to incorporate high accelerations into the design of most new products. A reasonably hot sports car that accelerates to 60 MPH in six seconds requires an acceleration of only 0.46 Gs. You should be careful to keep both acceleration and decelerations low when you design products that move and change velocity.

Explosives and Explosive Mixtures

With the bombing of the U. S. S. Cole and the disaster at the Federal building in Okalahoma City, most people have become aware of the disastrous effects of large explosions. The gas pressures generated by the blast destroy very substantial buildings, break glass in a very large region and kill or injure many people. The population is generally aware of the characteristics of common explosives like dynamite and ANFO (ammonia nitrate and fuel oil). In most designs, you do not encounter a need to accommodate these traditional explosives. What you must recognize are other less apparent agents that act like explosives under special circumstances. Fuels such as natural gas, propane, butane, etc. can leak and combine with air to produce an explosion when ignited. Boating accidents are common when gasoline leaks in an engine compartment. The resulting mixture of gasoline fumes and air explode when subjected to a small spark, destroying the boat and killing or injuring the passengers. Still another unusual source of fuel for an explosion is dust. When handling large quantities of a combustible solid, dust (fine particles) is generated. If these particles are suspended in air, the resulting mixture will certainly explode when ignited. A grain elevator exploded in Westwego, Louisiana, in 1977 killing 35 people when an explosive mixture of combustible particles (dust) from the grain and air was ignited. More recently on June 8, 1998, the De Bruce grain elevator in Haysville, Kansas exploded killing several workers.

Fires

Over a million fires occur in the U.S. every year. Some are vehicle fires (400,000) and others are structural fires (650,000). Many people die in the fires (4,700), and many more are injured (28,700). Clearly, fire is a serious problem. People are killed, injured or traumatized and property is lost (eight or nine billion dollars worth per year). Some of these fires are, simply put, stupid. For example, he or she who smokes in bed may some night fall asleep and set the house on fire. A surprisingly large number of fires (about 100,000 per year) are deliberately set—apparently to collect the fire insurance, to take revenge or as a sick kind of diversion.

Engineers have a responsibility to decrease the number of fires resulting from the products that they design. Did these fires start with an appliance or a motor that overheated for some reason? Determine the reason for the overheating and redesign the product so that excessive heat will not be generated. Why did the automobile burst into flames? Did a fitting on the gasoline line leak? Redesign to eliminate the fittings or specify a fitting that will not fail under the prevailing conditions. There are many solutions to the problem of fires in America. As a nation, we are far too casual about fires. Carelessness in personal practices, poor design in products intended for the home and business and fraud (arson) to collect fire insurance reimbursement for lost property are tolerated.

Pressure and Pressure Vessels

Pressurized fluids are used for many good reasons, and in most cases, pressure vessels (the containers that hold the fluids) perform very well. It was not always the case. In Boston during the winter of 1919, a huge tank about 90 feet in diameter and 50 feet high fractured. It contained two million gallons of molasses, which flooded the local area. Twelve people died and another 40 were injured in this accident.

It is relatively easy to design a pressure vessel today. In fact, the American Society for Mechanical Engineers (ASME) has developed a code that engineers follow in their design to produce pressure vessels that are certified as safe for service. However, on occasion, tanks fail in service. The difficulty is usually with the steel plates that are welded together to manufacture the tank. The steel employed in both the plates and the welds have to exhibit high fracture toughness at low temperatures, and the welds must be free of large flaws. If you have the responsibility of designing a pressure vessel, follow the ASME code, make certain the welding procedures followed in manufacturing produce crack-free welds, and specify steel for the plate and welding rods, which is sufficiently strong and fracture resistant. If you intend to become a mechanical engineer, you will have the opportunity to learn how to design pressure vessels, select suitable materials for their construction and specify welding procedures in courses presented later in the curriculum.

Mechanical Hazards

Mechanical hazards are features, which exist on products that may cause injury to someone nearby. Consider as an example an electric fan used to cool a room. Is the fan blade adequately guarded, or is it possible for someone to stick his or her finger into the rotating blades? Suppose the cabinet you recently designed to hold a special tool has a sharp corner at hip level. Can someone walk into the cabinet and break skin or bruise a hip on that corner? You have designed a center post crane to lift material and move it over a 25-foot diameter area. Will the center post be stable under all conditions of loading or will it collapse? You have designed a new pizza machine that rolls the dough into sheets exactly 0.120 inch thick. Have you provided protection for the entrance to the rolls that prevent the operator from inserting his or her fingers in the machine? You have designed a wonderful guard that prevents a person from exposing their hands and arms in operating a punch press. However, the guard is attached to the machine with two small screws. Have you used locknuts and/or safety wires to insure that the screws will not loosen during the operation of the press? Are the screws large enough to resist failure by shear?

There are many mechanical hazards encountered in designing equipment and products. Always examine each component, and look for sharp points, cutting edges, pinch points, rotating parts, etc. Do not count on peoples' good judgment. If it is possible to insert a finger into the machinery, even if it is a foolish act, you can be certain that sooner or later someone will do so.

Radiation Hazards

Radiation hazards are due to exposure to electromagnetic waves. The damage produced depends on many different factors such as:

- The strength of the source.
- The degree to which the emitting radiation is focused.
- The distance from the source.
- The shielding between the source and the object being radiated.
- The time of exposure to the radiation.

The radiation spectrum is divided into several different regions—very short wave length, visible light, infrared, and microwave radiation. The short wave length radiation is the most dangerous (x-rays, gamma rays, neutrons, etc.) with serious health risks (cancer) for overexposures. The hazards due to UV, visible light and IR are usually due to excessive exposure where serious burns to the skin occur. For very intense radiation, even short exposures are detrimental to the eyes. Microwave radiation is absorbed into the body and may result in the heating of one's internal organs. Because the means used by organs to dissipate this heat is not know nor the effect of the localized increase in temperature, it is prudent to avoid exposure to microwaves. In America, the Offices of Safety and Health Administration (OSHA) has issued a regulation limiting the power density of microwave energy to 10 mW/cm^2 for an exposure of 6 minutes or more. In designing products, where operators may be exposed to microwave radiation, shielding should be employed to reduce the power density well below the regulatory levels.

Noise

Noise levels that occur in the environment may be damaging to your hearing, may interfere with your work or play and may degrade the quality of your life-style. Most noise is man-made, although occasionally a storm and Mother Nature provides the sounds of thunder and wind. If you listen occasionally to a band playing rock and roll music, there is a temporary shift in the threshold of audibility to a higher level of pressure. However, if you play in the rock and roll band almost every night for an extended period, the shift in the threshold of audibility becomes permanent and your hearing becomes impaired.

Noise is a pressure disturbance that propagates through some medium such as air. The velocity of propagation through air is 344 m/s at room temperature. The pressure disturbance is oscillatory usually with many different frequencies present. The frequency content of the pressure waves depends on the source of the sound. A note from a violin will have much higher frequencies that a note from a tuba. The intensity of the noise is measured with a sound level meter that consists of a microphone, amplifier and a display meter that provides a reading in decibels (dB).

As a general guideline, the threshold for audibility is less than 25 dB[2] before a person is considered handicapped. In addition, there is a threshold for feeling noise-generated pressure at about 120 dB and another threshold for pain between 135 and 140 dB. When designing a product, the noise level is a serious consideration. Clearly, the feeling and pain levels of noise intensity must be avoided, but what levels are satisfactory? The U.S. Environmental Protection Agency (EPA) has established standards, which provide guidance to engineers in the design of products. For example, the results presented in Fig. 17.6 show the relation among the sound pressure level, the communicating distance and the degree of speech intelligibility.

---

[2] What constitutes normal hearing differs from one authority to another. A reference by the American Academy of Ophthalmology and Otolaryngology was used in writing this section.

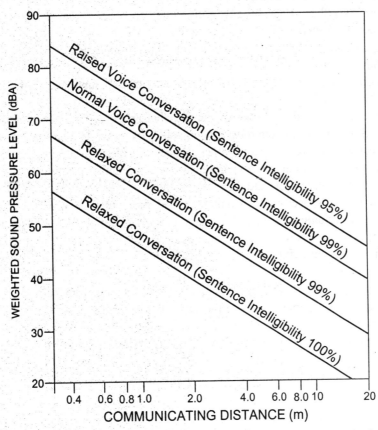

Fig. 17.6 Outdoor distances for intelligible conversation with a background of steady noise [8].

17.7 SUMMARY

When a product is introduced in the marketplace, there is usually some risk involved to the workers making the product, customers and society-at-large. Hopefully, the risks will be small and acceptable to all concerned because of the significant benefits produced for the customers and/or the sociotechnical system. In determining if the risk is acceptable, the public needs an accurate and honest assessment of the consequences of a failure, and the probability of the occurrence of a serious accident. It is the responsibility of the engineering profession to assist business administrators and governmental agencies in these assessments.

While the public must accept risk, engineers must make every attempt to minimize the probability of harm within reasonable constraints on cost and performance. There are many excellent engineering methods for ensuring safe design. The concepts of stress, strength, safety factor and margin of safety illustrate an approach to minimizing risk due to failure by fracture. Recommendations for a range of commonly accepted safety factors have been given. Issues to consider in selecting the safety factor to employ when sizing components have been described.

Sometimes it is understood that failures will occur in service. In these cases, it is necessary to determine the probability of the failure event. To illustrate methods for determining probability of failure, the concepts of failure rate and mean time between failures were introduced. It was shown that the failure rate could be determined simply by keeping records of the failures of components as a function of time after they were placed in service. These concepts are important, but they should not be confused with the probability of failure or survival.

Component reliability was introduced, and a method for computing reliability from the failure rate was shown and demonstrated. In developing this reliability equation, Eq. (17.12), the failure rate was assumed to be constant over the service life. It is important to observe that the probability of survival decreases as the service life is increased and the probability of failure increases with time in service. For reliable service for a long period of time, the failure rate must be extremely low.

System reliability is different than component reliability. A system is usually composed of many components. If the components are arranged in series, they must all function for the system to operate correctly. The reliability of a series connected system is determined from Eq. 17.13. It is evident from the data presented in Table 17.2 that the number of components arranged in a series markedly lowers the reliability of a system. Sometimes a system must function all of the time. If this is the case, a system failure cannot be allowed, because of its very negative consequences. In these instances, engineers design with a number of redundant components arranged in a parallel system. Redundancy improves system reliability, as shown in Table 17.3, but at increased cost and the requirement for more power, weight and size.

The explosion of the space shuttle Challenger and more recently Columbia was described to illustrate the importance of evaluating the risk. It is often a very difficult problem and an analytical solution for the risk is often not possible. Nevertheless, it is a professional responsibility to prepare an intelligent, accurate, honest and frank estimate of the risk, and to insure that all of the principals involved are aware of the consequences of a failure.

Finally, a number of hazards that cause injury or death has been introduced. Unfortunately the listing is relatively long. It is important that you recognize the hazards and be vigilant in your designs to avoid them. Avoidance often does not require extensive calculation from elaborate formulae, but rather a detailed assessment of each component in the system and a good measure of common sense.

REFERENCES

1. Dally, J. W., Packaging of Electronic Systems: A Mechanical Engineering Approach, McGraw Hill, New York, NY, 1990, pp. 288-296.
2. Lewis, R. S. Challenger: The Final Voyage, Columbia University Press, New York, NY, 1988.
3. Feynman, R. "Personal Observations on the Reliability of the Shuttle," Appendix to the Presidential Commission's Report, Ayer Co., Salem, 1986.
4. Mowrer, F. W., Introduction to Engineering Design: ENES 100, McGraw Hill, New York, NY, 1996, pp. 149-172.
5. Whoriskey, P., "Shuttle Failures Raise a Big Question," Washington Post, February 10, 2003, p. A-9.
6. Gugliotta, G and R. Weiss, "Dangers of Gauging Space Safety," The Washington Post, February 17, 2003, p. A-14.
7. Cunniff, P. F., Environmental Noise Pollution, Wiley, New York, NY, 1977, pp. 101-115.
8. "Information on Levels of Environmental Noise Requisite to Protect health and Public Welfare with an Adequate Margin of Safety," EPA Report, March 1974.
9. Powers D. G. and J. Proctor, Editors, Lockheed L-188 Electra, World Transport Press, April 1999.

EXERCISES

17.1 Have you been involved in one or more automobile accidents since you began to drive? Were you or anyone else injured or worse yet killed? Estimate the total mileage you have driven over this period and calculate your accident rate (number/mile). What were the reasons for the accident or accidents? Comment on your driving behavior and its influence on the accident rate.

17.2 Are the benefits of driving worth the risks of injury or death? Determine your probability of being killed in a fatal accident while driving this year. Hint: Statistics on fatal accidents are always listed in the World Almanac and on the web site for the NHTSA. Make any assumptions necessary for your analysis, but justify each of them.

17.3 Your co-worker designs a tie rod (a tension member) from a ¼ diameter inch steel bar with a strength of 62,000 psi. If the team leader has indicated that he or she wants to maintain a safety factor of 2.6, determine the maximum load that can be applied to the tie rod.

17.4 If a component is designed with a safety factor of 2.4, what is its margin of safety?

17.5 If engineers are so smart and have all of these great computer programs to determine stresses, why is it necessary to specify a margin of safety or a safety factor when designing a structural member?

17.6 NASA's space shuttle utilizes two booster rockets fueled with solid propellant to provide the thrust required for launch. Is this a redundant system? State your reason for this conclusion.

17.7 If the probability of failure of one of the solid propellant, booster rockets on the space shuttle is 1/1000, determine the reliability of the solid booster rocket system.

17.8 You are designing a very large computer system to contain the database for a world-wide reservation system. It is estimated that at any instant 500 operators will be accessing the database, and another 4000 operators will soon be ready with their requests for computer availability. The mainframe computers that you plan to employ each have a MTBF of 8,000 hours. Present a design of the computer system that will insure that all 500 operators have a 99.9 % probability of being served. In your design show all calculations and carefully list your assumptions. Justify the costs involved if each mainframe computer employed in the system is valued at $200,000.

17.9 Janet is an engineer in the transmission department of the Fink Motor Corp. Her job is to record data on the mileage prior to failure of the new lightweight transmission that has been placed in 300,000 of the new 2003 model of the Clunker. She records the following data:

Mileage (1000 miles)	No. Failures, N_f
0 - 5	65
5 - 10	38
10 - 20	20
20 - 30	18
30 - 40	22
40 - 50	34
50 - 60	90
60 - 70	521
70 - 80	1,364
80 - 90	6,677
90 - 100	15,592

Determine the failure rate as a function of service life expressed in terms of mileage, and prepare a graph showing your results.

17.10 Using the data from the table in Exercise 17.9, determine the probability of survival and the probability of failure of the transmissions as a function of service life measured in terms of mileage. Prepare a graph of your results.

17.11 If the transmissions in the Clunker were under warranty for 100,000 miles, what would be the consequences for the Fink Motor Corporation?

17.12 Prepare a listing showing the ratio of fatal events per flights for the airlines in the U. S. and Canada since 1980. The data needed to prepare this listing is at http://airsafe.com/airline.htm.

17.13 Based on the results of Exercise 17.12 cite the three airlines with the best safety records. Cite the three airlines with the worst safety records.

17.14 Will the results of Exercise 17.13 affect your planning for air travel in the future?

17.15 Prepare a paper describing the top ten fatal air transport accidents. The web site listed in Exercise 17.12 contains the data necessary for you to write this paper.

17.16 Prepare a paper discussing your reaction to the risk assessment of the space shuttle system by NASA. Give arguments for and against NASA's position and practices regarding safety of the shuttle and its crew both before and after the Challenger and Columbia accidents.

17.17 Describe an incident when you received an electrical shock. Something obviously went wrong to cause this incident. Please indicate the problem. Was there a design deficiency? What could be done to prevent the incident from reoccurring?

17.18 You are designing a cabinet-mounted, self-cleaning oven that requires very high power levels to heat its interior surfaces until they are free of all the splattered, burnt-on grease. As the lead engineer on this design team, what precautions should you take to insure that the oven would not be the source of a fire during its anticipated 20-year life?

17.19 Automobiles are the most commonly used mode of transportation in the U. S. and many other developed countries. Describe a least one-design flaw pertaining to safety of many sports utility vehicles. What government agency is charged with the responsibility of insuring safety of automobiles in America?

17.20 There are many fatal accidents each year involving crashes with large trucks. The Federal Motor Carrier Safety Administration (FMCSA) is the government agency responsible for regulations involving trucks. Write a paper describing possible rules the agency could impose on the trucking industry to reduce the number of fatal truck crashes. Cite reasons for the new rules and predict the benefits that would result if they were implemented. Discuss the economic consequences for your suggestions.

CHAPTER 18

ETHICS, CHARACTER AND ENGINEERING

By James W. Dally and Sheryl H. Ehrman

18.1 INTRODUCTION

This chapter on ethics was written shortly after the 1996 presidential election, and it has been revised every year since then. During this period the newspapers and TV news and talk shows covered many ethical issues emanating from both Congress and the White House. Ethical lapses occurred on nearly a weekly basis, keeping the news services busy reporting on questionable behavior of our elected officials from both political parties. These officials are the people the voters have entrusted and empowered to write and approve the laws governing the country. The news media has pointed out numerous instances where political leaders have skirted the truth if not the law.

During the period from the initial writing of this chapter to its revision in May of 2001, the ethical situation has deteriorated. Ethical issues continue to be reported by the news media with alarming frequency. For example[1], U. S. District Judge Royce Lambert cited the former Secretary of the Interior, Mr. Bruce Babbitt, and the former Secretary of the Treasury, Robert Rubin, for contempt of court for withholding evidence regarding the Indian trust funds. (The government cannot account for $2.4 billion of these funds). Judge Lambert concluded that the employees of the Interior and Treasury Departments had "engaged in a shocking pattern of deception of the court. I have never seen more egregious conduct by the federal government."

A few years ago President Clinton admitted to having an affair with an intern about half his age. He initially claimed on national television that he did not have sex with **that** woman. However, when it was clear from the results of tests with that woman's dress that he had had sex, he admitted it. Previously, the president, in still another legal proceeding, denied the existence of an intimate relationship under oath. When the facts of the president's behavior became apparent, the House of Representatives impeached Mr. Clinton for lying and obstructing justice, and a federal judge held him in contempt of court for lying.

The lax ethical practices by many of our elected officials have led to a very serious skepticism about the merits of many of the government institutions and agencies. The fact that only about 50% of the eligible voters took time to cast their ballots in the closely contested 2000 elections is testimony to the deep skepticism toward the government and our elected officials.

As engineers, we want to preserve, if not enhance, the confidence of the public. To gain public confidence, it is essential that we all behave in an ethical manner every day of every year. Whether we act as individuals or as professionals, consistent ethical behavior is mandatory. Our primary employer, business, must also act in an ethical manner. Since engineering developments are sponsored, financed and controlled by businesses, corporate and engineering behaviors are inseparable. Our actions and words always must be above reproach.

---

[1] See the article entitled, "Shredding 162 Boxes," Wall Street Journal, December 9, 1999.

18.2 CONFUSION ABOUT ETHICAL BEHAVIOR

Many people are confused about ethical behavior. The results of several polls indicate that Americans are less ethical today than in previous generations. Sixty-four percent of a sample of 5000 individuals admits to lying if it does not cause real damage. An even larger percentage of those sampled (74%) will steal providing the person or business that is being ripped off does not miss it [1]. Many students (75% high school and 50% college) admit to cheating on an important exam [2]. Most students have not developed a moral code to use as a guide for their behavior. Poor behavior, when the author was in high school, included making too much noise, chewing gum in class, talking, littering, running in the halls, etc. Today, these relatively minor transgressions have been replaced with more serious problems involving mass murder, drug and alcohol abuse, pregnancy, suicide, rape, robbery and assault. Extremely negative changes. Why?

There are many reasons for the degradation of ethical behavior over the past 50 years. Let's discuss two key reasons. First, institutions (family, church and schools) that once taught ethics are much weaker today than 50 years ago. Many families have been torn apart by an extremely high divorce rate and single parent homes are common. The concept of "love, honor and cherish, in sickness and in health" is often ignored.

The influence of our religious organizations has deteriorated due to serious declines in membership. Unfortunately, their finances often limit their activities to current membership. Many social problems are with people who are not well known to a church, synagogue or mosque. The public has also become much more secular with no religious affiliation. Our public school administrators are so concerned with possible litigation that they have largely removed the teaching of moral values from their curriculum. Given the current constrains on discipline in the public school system, it is often difficult for the administrators and teachers to maintain orderly classrooms. Consequently many students graduating from high school have not been given an adequate opportunity to develop a personal code of ethics. The result of the failure of our institutions to teach ethics is that many of our young people are not aware of what is right and what is wrong. All actions are not gray. Many are plainly right, and many others are obviously wrong.

The second reason for the deterioration in ethics on the part of our younger population is the current state of ethical standards in society. Everyone is exposed to cynical behavior and selfish attitudes on TV, magazines, and newspapers every day. The "bad guy" often walks away from severe crimes with minor penalties. Minor infractions are frequently ignored. Killers and child sex offenders are paroled after remarkably short periods of incarceration. Popular TV programs frequently show examples of lax ethics and self-indulgences to generate a popular but sick kind of humor. The result is to create confusion in our youth as they try to develop some system for judging right from wrong. Our youth mirrors the behavior of society leading to cynicism, selfish attitudes, and dishonesty and irresponsible conduct.

18.3 RIGHT, WRONG OR GRAY

Christina Sommers [2] develops an interesting viewpoint regarding what is considered right, wrong and controversial by society-at-large. She argues that there are some ethical issues that are not clearly defined, and that society has not reached a consensus regarding the correctness of these issues. You can develop cogent arguments for and against a specified viewpoint for controversial issues such as abortion, affirmative action, capital punishment, etc. Sommers refers to these controversial issues as "dilemma" ethics, which should be distinguished from "basic" ethics. With "dilemma" ethics, society-at-large is not certain about what is right, wrong or gray. There are several attitudes prevalent in society—sometimes the issue is so important to segments of our population that demonstrations are organized to elevate one viewpoint or another.

No attempt will be made in this textbook to resolve any of the "dilemma" issues. These issues have been unresolved for decades, and they will require much more time and understanding before our society is willing to reach the consensus necessary for resolution. It is much more productive to consider "basic" ethics where right and wrong or white and black are much more clearly recognized by the majority of society.

Right Versus Wrong

It is easy to define right and wrong. In fact, the understanding of right and wrong dates back to Aristotle [3] and fundamental doctrine has not changed much since then. Let's start with the four classic virtues:

- Prudence
- Justice
- Fortitude
- Temperance

Prudence refers to careful forethought, good judgment and discretion. In other words, think before you act and exercise care and wisdom. If you act, will your action cause problems, now or later, for you or anyone else? Are you careful about your remarks concerning others? One of my rules is not to speak ill of others under any circumstances. Is the action that you are about to undertake in your best interest? Will that action be appreciated by your family, friends, coworkers, managers and peers?

Justice involves fairness, honor, keeping your word, honesty, truthfulness, etc. It is clearly wrong to lie, cheat, steal and break promises. It is also wrong to tolerate those about you who do so[2]. The author's behavior with regard to justice is governed by the golden rule. Do unto others as you would have them do unto you. It is a very simple rule, and it is effective.

Fortitude is about courage and persistence. One usually relates fortitude with warfare and/or battle. Warriors stand their ground and exhibit a very special brand of courage when their lives are at extreme risk. However, there are other brands of courage that must be exhibited in more typical circumstances. Will you stay with an idea, even if it is not popular, if you know that it is the "right" approach? Showing determination and resolution is sometimes difficult when the risks of failure are high or when peers encourage you to abandon your idea.

Temperance, of course, refers to moderation in drinking alcoholic beverages. However, temperance has much wider implications with regard to ethics. Temperance can also mean control of human passions such as anger, lust, hostility, exasperation, and lechery. With regard to food and beverages, it implies self-control and moderation in consumption. Clearly, much is to be gained by avoiding overindulgence in food, drink, and in restraining your emotional extremes.

There are other virtues such as loyalty and obedience that are less commonly discussed these days. With the restructuring and massive downsizing that has taken place in the business world in the past two decades, the concept of company loyalty to the employee and vice versa has been destroyed. Hopefully, loyalty to the family is an invariant, at least in those situations where a family in the true sense exists. The concept of obedience was seriously damaged—if not destroyed—in this country by the Vietnam conflict. The government ordering young people to fight in a war of questionable value was too much for many to endure. Many young men refused to obey the orders to report for duty and fled the country to avoid persecution.

Because of these events (many others could also be added), the virtues of loyalty and obedience have shifted from the "basic" to the "dilemma" category, and arguments can be advanced on both sides of these questions. What do you think of being loyal to an employer who will eliminate your position during a period of **increasing** profits? How do you feel about completing your tax return for the Internal Revenue Service? Talk about straining one's patience!!!

---

[2] Many people are willing to tolerate about them those who cheat, steal and break promises. They do not believe in imposing ethical constraints on friends or associates. This no tolerance concept is difficult to establish in many honor codes, and the author is probably in the minority on this issue.

Theological Virtues

In addition to the four basic virtues, prudence, justice, fortitude and temperance, there are three religious (theological) virtues, which include:

- Faith
- Hope
- Charity

The theological virtues are well known, so it is not necessary to elaborate on their importance. However, it is stressed that their merits are definite and irrefutable. It is good to be faithful, hopeful and charitable. If you want to be even more complete, add other ethical constants such as humility and respect to the list.

While the emphasis in most college courses on ethics is with social dilemmas where arguments and counter arguments are to be developed, most individual behavior can be judged by very clear and well-understood virtues that are part of "basic" ethics. There is no need to be confused, nor is there any reason to impair your "basic" code of ethics with issues raised in the study of moral relativism.

18.4 LAWS AND ETHICS

The governments (federal, state and local) play a role in ethical behavior, because they generate laws and set public policy. Laws (mostly at the state level) are written to aid people in their pursuit of a morally correct and safe life. The purpose of these laws is to prohibit a well-defined set of vices. For example, it is illegal to sell drugs, rob banks, commit burglary, kill or abuse another human, engage in prostitution or pursue deviant sexual practices (rape, among other offenses).

These laws, and many more like them, serve to create a moral ecology in which society exists [4]. The laws require visible or outward conformity to a publicly accepted moral code of behavior, but do nothing to inhibit illegal behavior unless it is observed, reported and prosecuted. An unscrupulous person can sell drugs if he or she is not observed in the act. Even if they are observed, they can avoid penalty if they are not arrested and fully prosecuted for one reason or another. Even if they are observed, detained, prosecuted, and convicted they may receive a suspended sentence. The law does not insure that infractions **will not** occur; the law simply indicates that certain behavior **may** not be tolerated.

In some instances, laws are not a suitable approach for creating a moral ecology. An example of an unsuitable law in this country was prohibition, which was enacted by a constitutional amendment that was ratified in 1919. When it became evident that the law was causing more problems than it solved, it was repealed in 1933. The government wanted to make it illegal to consume alcoholic beverages, but society-at-large wanted to drink. Bootlegging, racketeers, and speakeasies followed bringing crime and violence to the neighborhoods.

It soon became apparent that the law was not a suitable approach to solve a temperance problem. In such cases, the government uses public policy to discourage people from pursuing a vice. Policies are adapted that limit (but do not preclude) the consumption of alcoholic beverages. Hours of sale are restricted, licenses are required, taxes are imposed, the number of outlets is constrained, etc. Public policies are conveyed by means of rules and regulations—not laws. The policies should be crafted to strengthen families, communities and churches by discouraging, but not prohibiting, irresponsible conduct.

18.5 ETHICAL BUSINESS PRACTICES

Corporations are entities that under law are treated as persons. A corporation, acting like a person, has the right to conduct business and the obligation to pay taxes. Just as individuals are judged by their ethical behavior, corporations are judged by their ethical conduct [5]. The chief executive officer (CEO) for a corporation sets the moral and ethical tone for the organization. Any large business is organized with several layers of management. If the CEO operates the business with the highest ethical standards, the middle and lower level managers will follow the example set at the highest level in the corporation. However, if lax ethical standards are permitted, a pattern is set and questionable ethics will become the corporate style and standard.

In an excellent paper, Baker [5] has developed a list of issues that occur daily in a typical corporation. The manner in which a corporation deals with these issues establishes its character. Baker's list is shown below:

1. How people, employees, applicants, customers, shareholders, and the families who live near our facilities are treated?
2. Is everyone free of systemic or individual practices of discrimination?
3. How do we spend our shareholders' money on our expense accounts as an institution and as individuals?
4. What is the level of quality that goes into our products? Do we meet our customers' expectations for quality? Do we meet our own standards for quality?
5. What is our concern for safety, not only for our employees, but also for our customers and our neighbors in communities in which we operate?
6. How "ethically" do we compete?
7. How well do we adhere to the laws of the locales, regions, and nations in which we do business?
8. What is acceptable gift giving and gift taking?
9. How honest are our communications to our employees and our public advertising?
10. What are our corporate and personal positions on public policy issues, and how do we promote those positions?

This is a long list, but we certainly could add more items to Baker's catalog of concerns—sustainability and the environment for instance. Adding more items is not as important as recognizing that corporate decisions are necessary on a daily basis by its managers and employees. These day-to-day decisions determine the corporate character. The ethical judgments that serve as the basis for most of the decisions will depend on individual personal values (virtues) and experience. The quality and consistency of the ethics applied in these decisions will markedly affect the success of the corporation, its management and its employees.

18.6 HONOR CODES

We will not lie, steal or cheat, nor tolerate among us anyone who does [6]. This is a fourteen word honor code that cadets learn almost as soon as they step foot onto a military academy or college campus that has strong ties to the military (Virginia Military Institute and the Citadel). The military academies (Navy, Army, Air Force and Coast Guard) are educational institutions that serve to train career officers for the services. Honor and integrity is a very large and important element in their educational program.

Honor codes are much more widespread than in military institutions. There is a growing trend to introduce honor codes on campuses where they are absent and to strengthen them on campuses where they already exist [7]. An honor code typically forbids lying, cheating and stealing. When a peer notes an infraction of the code, the code requires that student to bring the case forward. A student committee hears the case. Although faculty members may testify if called upon, they have no control over the proceedings. The penalty imposed on those found guilty of breaking the honor code is determined by the student committees—university administrators serve only to implement the student's decisions.

The honor codes are very well respected by the student body and fully appreciated by the faculty. Students generally support the honor codes because they also are concerned with cheating by their peers. They want to play the game fairly and on a level playing field. When the honor system is established, the students handle the responsibility for academic integrity with careful consideration. Punishments are handed out only after complete investigations involving representation of both sides of the case. In some instances, the individual being judged by the student committee retains lawyers.

The faculty is also very much in favor of the honor system. They never have appreciated proctoring exams. With the increase in cheating over the past 25 years, proctoring has evolved into policing, which is very distasteful. The advantage of an honor code to the faculty is that the unproctored examination transfers the responsibility for policing to the students. With an effective honor system, students often are permitted to take an exam at a time and place of their choosing.

Of course honor codes are not perfect. Some students cheat regardless of the code. The military academies, where the code is the strictest and the most rigorously enforced, have suffered the most notorious lapses in standards. Scandals at the U.S. Naval Academy have been the most prevalent and widespread in recent years with as many as 133 midshipmen involved in cheating on an electrical engineering exam several years ago. To make matters worse, the administrators at the academy appear to have interfered with the system, and in doing so, lost the trust of the midshipmen.

The lesson from the academies with regard to the effectiveness of the honor codes is that rigid codes and rules reduce cheating. However, if the students see a way of beating this very rigid system, they will occasionally take advantage of the opportunity. Honor codes are most effective when they produce an ambience of trust between the students, faculty and the administration. Cheating, lying, and stealing will generally be reduced. For honest students, the scoring on exams will be fairer. For faculty the distasteful task of policing is eliminated. The feeling of trust between the faculty and the students is enabling. With time and prolonged success, the honor code and the trust that it engenders becomes an essential part of the educational process.

18.7 CHARACTER

The author had the privilege of teaching for a year (1995-96) at the U.S. Air Force Academy. It was a wonderful experience for me for many reasons—both personal and professional. One of the most important reasons was the opportunity to participate in a well-planned approach to incorporate character building into the educational process. Character was a key educational outcome in every class that was taught at the Academy.

It is not possible to transplant character into an individual, and it is not possible to accomplish much in efforts to explicitly "teach" character. Character comes implicitly when individuals develop a personal set of ethical responsibilities. Let's consider some of these attributes that form the foundation for a person's character.

- **Be committed to excellence**. This attribute simply means to do the best that you can do in both your personal and professional endeavors. To make this commitment, you have to assess your capabilities. How good are your skills? How determined are you in achieving your goals? How much time can you commit without endangering your health? If you know yourself, it is possible to measure your achievements on a realistic scale. Engage to the full measure of that scale.
- **Respect the dignity of everyone**. Respect is the foundation for all achievements. We work, live and play in a diverse world. You must appreciate everyone that you encounter in life. Race, gender, ethnicity and religion are not criteria for judgment. Recognize the potential of those with whom you work and study. Support and encourage them and carefully avoid demeaning criticism. Teamwork, so essential in today's workplace, requires that you accept the differences inevitable in a diverse population and that you fully embrace fellow team members.
- **Integrity**. A single word describes a vitally important attribute of character. Integrity means that you decide to do the right thing solely because it is the right thing to do. You often face decisions that test your integrity. Do you instinctively make the right (honorable) decision? For example, as you parked your car last night, your bumper scraped the fender of the adjacent car. No one observed your accident. Do you leave a note with your name, address and phone number? Or did

you split and drive to another nearby parking lot? How many parking lot scrapes can you find on your car? Did you find a note from the party inflicting the damage claiming responsibility? A person with integrity will consistently make the correct decision, not because it is right for their personal benefit, but because it is clearly the "right" course of action.

- **Be decisive**. When you encounter a problem in engineering or your personal life, there is an information-gathering period followed by a decision. Some folks do not want to make that decision. They prolong the information-gathering period; they procrastinate; they hem and haw; they pass the buck. When you establish the facts, evaluate them and make a timely decision. It will be your decision if you made it in isolation. If you made it within the framework of a team, it should represent a consensus of the members of the team.

- **Take full responsibility**. Decisions are made and actions follow. Often your decisions produce a winner, but sometimes they result in a loser. If the outcome was a loser and you participated in the decision process, it is your responsibility. Step up, accept the responsibility, and lead the effort to fix the problem. Never, under any circumstances, begin to participate in a **finger pointing** exercise. Finger pointing does not resolve problems; it exacerbates them.

- **Be temperate**. Self-discipline is an essential part of character. Eat, drink and be merry in moderation. Overindulgence is disgusting. No one appreciates a drunk or a glutton. Self-discipline insures control of the human passions, sensual pleasures, anger, rage and frustration. With self-control, it is possible to attain consistently high levels of achievement.

- **Exhibit fortitude**. On many occasions, you will encounter significant difficulties in completing the task at hand. It is easy to give up and divert your attention to more pleasurable undertakings. You can quickly forget the missing assignment, the incomplete solution, the late review or the unfinished drawing. Who is to know? Stamina, mental toughness and discipline are attributes that keep us on task. Stay with the job, and not only complete the work, but do it well and with dispatch.

- **Understand the significance of spiritual values**. Many, but not all, of us have a faith. Most of us endorse a set of theological beliefs. Your beliefs and your church may be different than mine. Nevertheless, it is essential that I respect your convictions and you should respect mine. We all need to be sensitive to the important role that religion occupies in the mental comfort of many people. We must accommodate a diversity of beliefs by supporting the right of an individual to choose his or her faith and to pursue the ceremonies offered by the church representing this faith.

18.8 ETHICS OF ENGINEERS

Most of this chapter has been devoted to a discussion of issues pertaining to individual behavior, ethics, virtues, morals and character. A personal code of ethics is the cornerstone to achieving a meaningful sense of moral values. Professional ethics are also important, but they must begin only after a person has developed a well-understood personal code of ethics. Hopefully, the preceding sections of this chapter will be helpful in any attempt that you make to establish a sense of right and wrong.

There are several codes of ethics for engineers. Most of the founding societies of engineering have developed and distributed codes and guidelines for professional conduct and faith statements. The codes of ethics for the different engineering societies are all similar. The fact that each society has their own code, is more to insure a complete distribution of the code than to pursue unique ethical issues.

Let's consider the Code of Ethics of Engineers, presented in Fig. 18.1, which is sponsored by the Accreditation Board for Engineering and Technology (ABET). The code is divided into two parts. The first part deals with four fundamental principles, and the second part contains seven fundamental canons. The Engineers' Council for Professional Development first advanced this code in 1977. The fact that the code remains without modification for more than 25 years indicates that professional ethical values are as constant as personal ethical values.

*Accreditation Board for Engineering and Technology**

CODE OF ETHICS OF ENGINEERS

THE FUNDAMENTAL PRINCIPLES

Engineers uphold and advance the integrity, honor and dignity of the engineering profession by:

I.   using their knowledge and skill for the enhancement of human welfare;

II. being honest and impartial, and serving with fidelity the public, their employers and clients;

III. striving to increase the competence and prestige of the engineering profession; and

IV. supporting the professional and technical societies of their disciplines.

THE FUNDAMENTAL CANONS

1. Engineers shall hold paramount the safety, health and welfare of the public in the performance of their professional duties.

2. Engineers shall perform services only in the area of their competence.

3. Engineers shall issue public statements only in an objective and truthful manner.

4. Engineers shall act in professional matters for each employer or client as faithful agents or trustees, and shall avoid conflicts of interest.

5. Engineers shall build their professional reputation on the merit of their services and shall not compete unfairly with others.

6. Engineers shall act in such a manner as to uphold and enhance the honor, integrity and dignity of the profession.

7. Engineers shall continue their professional development throughout their careers and shall provide opportunities for the professional development of those engineers under their supervision.

111 Market Place, Suite 1050, Baltimore, MD 21202-4012

*Formerly Engineers Council for Professional Development. (Approved by the ECPD Board of Directors, October 5, 1977)

AB-54 2/85

Fig. 18.1 ABET's code of ethics.

The fundamental principles seek to insure that the engineer will uphold and advance the integrity, honor and dignity of the profession. These goals are common to your personal goals that were discussed previously; however, the approach to achieve the professional goals is different as indicated below:

1. We are selective in the use of our knowledge and skills so as to ensure that our work is of benefit to society.
2. We are honest, impartial and serve our constituents with fidelity.
3. We work hard to improve the profession of our discipline.
4. We support the professional organizations in our engineering discipline.

The seven fundamental canons in the ABET Code of Ethics of Engineers, presented in Fig. 18.1, are explained in considerable detail in guidelines that are used to expand and clarify the relatively brief statements that represent the "regulations" that shape our professional behavior. The detailed guidelines may be obtained from ABET[3].

18.9 ETHICS IN LARGE ENGINEERING SYSTEMS

Engineers design and build many different products each year that are included in large and complex sociotechnical systems. Usually these products and the systems are conservatively designed with adequate safety factors, carefully tested, and perform well in service for extended periods of time. They provide a much needed service without endangering either individual or public safety. However, from time to time mistakes are made in the initial design. These mistakes are usually detected in prototype testing and eliminated prior to releasing the product to the market place. In very rare circumstances, a mistake or several mistakes are overlooked, for a variety of reasons, and the system is released and placed in service with an unacceptably high probability for failure. An undetected mistake often leads to a catastrophic accident. The space shuttle system clearly falls into this category. If you are a space supporter, you may argue that the risks are worth the benefits. But if this is the case, unprepared and untrained high school teachers should not be invited to ride along to enhance the image of the space program. It is one thing to order a career astronaut into peril, but totally a different proposition to invite an uninformed civilian to participate in a very dangerous project.

In Chapter 17, the probability of failure (or an accident) has been discussed in considerable detail. You must understand that there is always some risk of failure when designing high performance systems. It is important to learn to accept a trade-off between risk, safety and performance—it is an inherent part of the process. In most complex sociotechnical systems (air transportation for example), there is a small but finite risk for failure with a subsequent loss of life and property. The public must know this risk. It is the responsibility of the engineering community, industry and the government to alert the potential customers to the dangers involved. Knowing the risk, the customers can decide whether or not they want to use the system.

Some people worry more than others and place a very high value on safety. They require a probability of failure of nearly zero. Do you know someone who will not travel by flying? (John Madden, the popular professional football announcer, travels from game to game each week on a special bus). Other folks love the thrill of a risk, and they are willing to accept a much higher probability of an accident. They think hang-gliding or skiing on black diamond slopes is great sport.

18.10 THE CHALLENGER ACCIDENT—A CASE STUDY

With this background on risk, safety and performance, let's begin a discussion of the Challenger accident [9]. The Challenger was one of the original four orbiters built by the National Air and Space Administration (NASA) to serve the space shuttle system. The rocket fuel (liquid hydrogen) on the Challenger 51-L mission exploded 73 seconds after launch on Tuesday, January 28, 1986. The crew of six and a civilian passenger were killed, and the space shuttle was lost.

The Challenger accident has been selected as a case study because it illustrates several ethical issues in the engineering and management of large, complex and inherently dangerous systems: Some issues that will be raised are:

[3] The Accrediting Board for Engineering and Technology (ABET) is located at 111 Market Place, Suite 1050, Baltimore, MD 21202. Visit their web site at http://www.abet.org.

1. The design of the shuttle and the selection of the contractors involved many political considerations [9].
2. The lack of communications between key people and organizations was a significant factor in the accident.
3. The interface between the upper-level administrators (business managers) and the engineers was an important element in the decision to launch on that disastrous morning.
4. Public attention and opinion, not safety, markedly affected the decision-making process.
5. The risk potential was not known by the public and not appreciated by top administrators and politicians directly involved in the launch decision.
6. Christa McAuliffe, a high school teacher and mother of two children, was killed in the accident. Why?

Background Information

To set the stage for the accident, you have to go back to the early 1970s. NASA had been very successful with the Apollo missions (even with the problems of Apollo 13), and looked forward to larger and more aggressive space endeavors. They proposed an integrated space system that would include a space station, space shuttle, space tug and manned bases on both Mars and the moon. This agenda sounded wonderful until the public examined the price tag. The public did a quick look and wanted no part of it. A poll indicated that the public believed that Apollo had been too costly. The politicians, always driven by the polls, took note and reduced NASA's budget. NASA recognized the need for a new, cost-effective project to follow Apollo that the public (and politicians) would buy. Responding to these political pressures, NASA proposed the space shuttle that would serve the military, the scientific community and the rapidly growing commercial business of placing satellites in orbit. The space shuttle system, illustrated in Fig. 18.2, was marketed as a relatively routine space transport system. Even the name "shuttle" connected with the airline shuttle services that routinely fly every hour from one large city to another.

To make the space shuttle system a commercial success, NASA proposed a fleet of four orbiters [10], which would eventually fly on a weekly basis (NASA initially set a goal of 160 hours for the turnaround time for an orbiter). They planned on nearly 600 flights in the period from 1980 to 1991. The early estimate of the cost of a launch was $28 million with a payload delivery cost of $100 to $270 per pound. After some operational experience, the cost estimates proved to be completely unrealistic. Launches actually cost on the order of $280 million, and the cost to place a pound of payload in orbit on the shuttle was in excess of $5,200 [11]. Early experience with the space shuttle system indicated that NASA could not hold their schedules and their costs were running more than a factor of 10 higher than the original estimates. Because of its poor performance, NASA was struggling to improve its image as 1985 ended.

Prior to the launch of the Challenger on January 28, 1986, twenty-four shuttle flights had been made. These flights were all successful in that they returned to earth with all crewmembers safe. They also showed the operational capabilities of the shuttle system in placing commercial satellites into orbit, repairing satellites in space and salvaging malfunctioning satellites. However, these flights also indicated that the shuttle system had many serious problems. The three main liquid rocket engines were too fragile with many critical components. Some of the tiles in the heat shield needed repair and/or replacement after every flight. The computers and the inertial navigation systems experienced occasional failures. The brakes and landing gear were stressed to the limit when the orbiter (an 80 ton dead stick glider) landed at speeds ranging from 195 to 240 MPH. Finally, the seals in the solid fuel booster rockets showed distress (sometimes extensive) in 12 of the 24 previous launches.

The record of the shuttle during the 1981-1985 period showed a consecutive series of successful launches, but with many prolonged delays to repair failing components in a very large, highly stressed system. The maintenance records showed so many problems that NASA estimated it took three man-years of work preparing for a launch for every minute of mission flight time. Clearly, the word shuttle to describe such a transportation system is a misnomer.

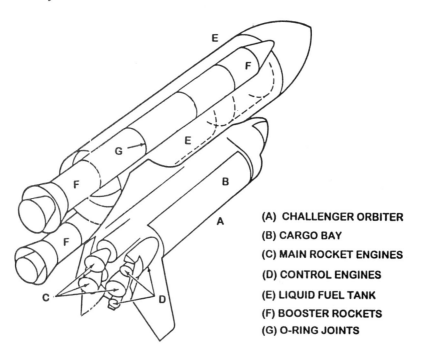

(A) CHALLENGER ORBITER
(B) CARGO BAY
(C) MAIN ROCKET ENGINES
(D) CONTROL ENGINES
(E) LIQUID FUEL TANK
(F) BOOSTER ROCKETS
(G) O-RING JOINTS

Fig. 18.2 Illustration showing the main components on the NASA space shuttle.

The Solid Propellant Booster Rockets

The explosion of the main fuel (liquid hydrogen) tank on the Challenger was due to the failure of the O-ring seals on the solid fuel boosters that were adjacent to the hydrogen fuel tank. To understand the seals and their purpose, it is essential that you appreciate the size and function of the two booster rockets used to provide much of the thrust necessary for the launch. They are enormous cylinders—12 feet in diameter and 149 feet tall. Each cylinder is filled with 500 tons of a solid propellant consisting of a rubber mixture filled with aluminum powder and the oxidizer—ammonium perchlorate. When ignited, the propellant burns to produce an internal pressure in the motor case of about 450-psi (lbs/in^2) at a temperature of about 6000 °F. The expanding gasses exit the rocket motor case through a nozzle, to produce about 2.6 million pounds of thrust from each booster. These booster rockets are essentially very **big** Roman candles!

Big was the nub of the problem. The solid rocket boosters were made by Morton Thiokol in Utah, but launched from the Kennedy Space Center in Florida. They were much too long to ship across the country as a single cylinder. To circumvent this problem, the motor casing was fabricated in segments— each 27 feet long. The segments were filled with propellant and shipped by train to the launch site in Florida. They were then assembled in a special facility near the launch pad. This procedure solved the shipping problem, but it created different problem. The rocket casing is a pressure vessel that must contain very hot gasses at a pressure of about 450 psi. The joints, where the cylindrical segments of the motor case were fitted together, must be sealed so that these hot gasses will not leak and cause damage to adjacent components of the launch vehicle.

The seal was made with a pair of rubber O-rings fitted over one finger of a clevis type of joint as shown in Fig. 18.3. The O-rings, 0.280 inch in diameter, were compressed between the two surfaces affecting a seal that prevents the pressurized fluid from leaking past the joint. A putty like compound was used to prevent the hot gases from eroding the O-rings. The pins locked the segments together (axial constraint only), but did not clamp the clevis fingers about the center finger.

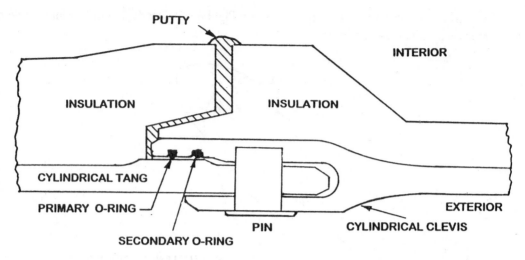

Fig. 18.3 Design of the O-ring seal used at the circumferential joints of the solid rocket motor cases.

The sealing of the segmented cylinders with the O-rings was not a new concept. A single O-ring had been employed previously on the Titan III rocket effectively sealing the circumferential joints on a smaller diameter motor case. Morton Thiokol, the contractor building the boosters, introduced the second O-ring to provide a margin of safety in the event the first O-ring failed. It all sounded good, and the initial test firings of the booster in Utah were apparently successful.

Unfortunately, one test does not validate the safety of even a simple system. To insure high reliability of a component, many tests are required. Moreover, this test firing had little or no bearing on problems associated with the reuse of the solid rocket boosters (up to 20 times). After splash down following a launch and recovery from the Atlantic Ocean, the cylinders were shipped back and forth between Florida and Utah.

Failure of the O-ring Seal

Motion photographs taken at the final launch of the Challenger indicated that the O-ring seals failed almost immediately after ignition of the booster. The hot gasses cut through the O-rings and the joint of the booster. A hole was formed in the wall of the booster at the joint, and a flaming jet of white-hot gasses escaping from the booster rocket cut through the wall of the adjacent tank containing the liquid hydrogen. The launch was effectively over. While many things happened in the last five seconds before the devastating explosion—all bad—the penetration of the adjacent tank, which contained liquid oxygen and liquid hydrogen, spelled the disastrous end of the mission.

It was a terrible day, and we as a nation were in shock while we mourned the loss of the crew and the schoolteacher/mother. It was sometime later, when the facts were brought to the public about the space shuttle program that we learned the shuttle should not have been permitted to fly that day. The on-site engineers understood the very high probability of failure of the O-ring seals. Moreover, knowledgeable engineers at Morton Thiokol tried without success to prevent the launch.

The story containing all of the facts about the failure of the O-ring seal is too long to be covered here, but you are encouraged to read references [9 – 14], where very complete and well-written accounts are given. The essential elements leading to the catastrophic failure are listed below:

1. The circumferential joint changed shape when the motor case was pressurized, and the gap, which the O-rings filled, increased markedly in size. A schematic illustration of the new gap geometry is presented in Fig. 18.4.

2. The new gap opening was so large that the back-up O-ring probably could not seal the joint. When the booster case was pressurized, the seal depended on a single O-ring—not two.

3. The hot exhaust gases had eroded the O-rings on 12 of the previous 24 launches indicating some leakage about half of the time, and on a few occasions, significant erosion of the rings occurred. There was also clear evidence that the putty failed to keep the hot gasses from attacking the rubber O-rings.

4. The joints moved during the early launch sequence as the orbiter engines and then the booster engines were ignited. This motion required the O-ring seals to be flexible and to reseat and reseal continuously during these movements.

5. The temperature the night before the launch dropped to 22°F and had only increased to about 28°F by the time of the launch.

6. The O-ring seals were not certified to operate below a temperature of 53°F by the contractor responsible for the solid rocket boosters.

7. Tests conducted by the contractor Morton Thiokol in 1985, six months before the accident, indicated that the O-ring seal was not effective at low temperatures. (At 50°F the O-rings would not expand and follow the movement in the joint during the period of operation of the booster.) In fact, at room temperature 75°F, it required 2.4 seconds for the O-rings to reseat and seal pressurized gasses after a gap in the joint was opened.

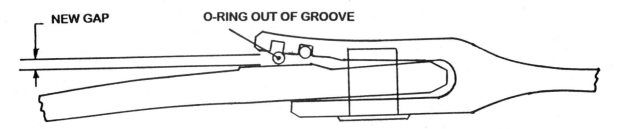

NEW GAP   **O-RING OUT OF GROOVE**

Fig. 18.4 Rotation of the joint due to cylinder pressurization opens the gap in the seal region.

The seven facts listed above show very clear evidence of two problems. First, the design of the seal in the circumferential joint was marginal and should have been fixed much earlier in the program. You do not inspect a piece of burnt and eroded O-ring and walk away from the problem.

Second, the launch should have been postponed due to cold weather for several different reasons. The very low temperatures were well below the limits to which the system had been certified. The rubber in the O-rings becomes very stiff at these low temperatures and cannot respond quickly enough to accommodate joint movement. The O-rings were almost guaranteed to leak. Also, there was considerable ice on the rocket motors and the fuel tank. Pieces of this ice might damaged the rocket motors and the tiles on the heat shield of the orbiter when it separates and falls during the violent vibrations that occur just after ignition but before lift-off.

Ignoring the Problem

Problems usually do not go away when they are ignored. They persist and sooner or later failure will occur. Although the seals had failed on 12 of the previous 24 flights, the failures were not catastrophic because the leaks were small and at locations which did not endanger the adjacent components. The boosters on the shuttle operated for only a few minutes before they were cut loose to fall into the Atlantic Ocean for recovery at a later time. So those concerned with the launch schedule tolerated the small leaks for short periods of time. The trouble with the leak on the Challenger was that it was large, and the jet of hot gases issuing from the leak was pointed directly at the liquid hydrogen fuel tank. The jet of hot gases acted exactly like a cutting torch, and the huge thin-walled fuel tank was penetrated within seconds.

The O-ring seals were inadequate and the joint needed to be redesigned and modified. In fact, during the development of the boosters in 1977–79, several memos were written by engineers at the Marshall Space Flight Center indicating their concern over the design of the joint. They recommended that the joint be redesigned because "the adequacy of the clevis joint was completely unacceptable." The suppliers of the O-rings stated, "The O-ring was being required to perform beyond its intended design and that a different type of seal should be considered."

Apparently there was a major breakdown in communication, and none of these memos and reports from the Marshall Space Flight Center was forwarded to Morton Thiokol, the designers and builders of the boosters. A problem, properly detected by engineering at the NASA Center in charge of monitoring the technical aspects of the development of the boosters, was not pursued to its logical conclusion. Instead of insisting on a redesign of the joint, Marshall Space Flight Center approved the boosters in September of 1980. The pressures of cost overruns and repeated schedule slippage sometimes make managers (and engineers) accept flawed designs. It is a very shortsighted practice and not in the best interest of safe design to do so.

NASA accepted a booster rocket with inadequate seals, but the bird flew. On the first launch, with Columbia, there was no reported damage to the seals. The O-ring seals failed on the second flight. One of the O-rings was burnt with about 20% of the thickness of the ring vaporized. The ring failure did not affect the mission, but it was clear that the putty was not protecting the rings from the hot gasses. Marshall Space Flight Center reacted to the field experience by reclassifying the joint to a "Criticality 1 category." This was the official recognition on the part of NASA that a seal failure could result in "loss of mission, vehicle and crew due to metal erosion, burn through and probable case burst resulting in fire and deflagration." The failure of the O-rings by erosion continued with high frequency (50% of the missions), but NASA decided to accept the situation and did not recommend remedial action.

It appears that NASA decided to live with the seal problem at least until a new lighter booster motor could be designed with improved joints. These new boosters were to be ready about six months after the Challenger exploded. A very real lesson—in too-little-too-late.

Recognizing the Influence of Temperature

From the initial development phase of the boosters, the O-ring seals had been recognized as a problem. A problem that NASA classified as very serious, but one that they must have erroneously believed did not elevate the risk to unacceptable limits. In January of 1985, a year before the accident, the Challenger was launched on a cold day at a temperature of 51°F. An examination of the joints after the recovery of the boosters showed that a number of O-rings were severely damaged and evidence of extensive gas leakage was observed. The contractor, Morton Thiokol, and the monitor, Marshall Space Flight Center, reviewed the O-ring seal problems. During this review, Morton Thiokol noted that low temperatures were a contributing factor to the inability of the O-rings to seal properly. The condition was considered undesirable, but was acceptable. However, the O-ring problems persisted and NASA eventually placed a launch constraint on the space shuttle.

The launch constraint would prohibit launches until the problem had been resolved or reviewed in detail prior to each launch. Unfortunately, NASA routinely provided launch waivers for each subsequent flight to avoid delays in an unrealistic launching schedule. The problem of O-ring erosion was common; one or more joints on 50% of the launches exhibited significant erosion. Fortunately, the erosion and attendant leaks had not caused any serious difficulty until the Challenger accident. The problem was expected and accepted as routine.

A Management Decision

The very cold weather on the night before the fatal launch was no surprise. The weatherman (or woman) was on the mark in predicting the temperature and the ice, which formed on structures that night. The engineers at Morton Thiokol, in Utah, were very concerned about the effects of the very low temperatures on the ability of the O-rings to function properly. Teleconferences took place between the Morton Thiokol engineers and the NASA program managers. The engineers argued to postpone the launch until the temperature increased to at least 53°F.

The NASA program managers were very unhappy about the engineer's concerns. They did not want the extended delay required to wait for warmer weather. Senior vice presidents from Morton Thiokol sensed the displeasure from the customer and took over the decision process. Senior managers decided to keep the customer happy and reversed the decision of the Vice President of Engineering not to launch at these very low temperatures. Senior management at Morton Thiokol signed off on the launch ignoring the fact that the temperature at launch time was expected to be 25°F lower than the lowest temperature which engineering believed the seals would be effective (53°F).

The senior management at Morton Thiokol caved in to client pressure. The engineers were unanimous in their opposition to launch. Senior management asked the engineers to prove that the O-ring seals would fail. Of course, they could not state with 100% certainty that the rings would fail. Failure analysis is performed in terms of probabilities. The risk had elevated as the temperature decreased, but the engineers could not prove conclusively that the seals would fail in a manner that would detrimentally affect the mission. Senior managers at Morton Thiokol and program managers at NASA concluded erroneously that the risks were low enough to proceed with the launch.

Approvals at the Top

NASA has a four-level approval procedure to control the launch of the shuttle. This sounds good, but for a multilevel approval process to be of any value, information must flow freely from the bottom of the organization to the top of the chain of command. The technical discussions concerning the ability of the O-rings to function at the very low temperatures took place between NASA administrators and engineers at level 4 and Morton Thiokol. Program managers at Marshall Space Flight Center, level 3 administrators, were also included in these discussions. While the discussion was extended, with clear polarization between engineering and management, not a word of these concerns was conveyed to the top two levels of management at NASA.

Approvals for the launch were given by the level 2 administrator for the National STS Program in Houston, TX, and the level 1 administrator at NASA's Headquarters in Washington D.C. These approvals were essentially automatic because the administrators in charge had been isolated. They were not privy to the management decision to fly with very cold O-ring seals. They were not informed of the engineering recommendation to postpone the launch and to wait until the temperature increased to at least 53°F.

The lesson here is clear—approvals by the very high level managers are worthwhile only if they are informed decisions. If complete information is not presented to the executives for their evaluation, then their approval is meaningless. These administrators are not knowledgeable, and they add no value in an informed decision making process.

18.11 THE COLUMBIA FAILURE

On February 1, 2003 the Columbia shuttle disintegrated as it reentered the atmosphere at the end of its 16-day mission. In Mission Control, Columbia's re-entry appeared normal until 8:54:24 a.m.[4], when engineers monitoring the sensors informed the flight director that four hydraulic sensors in the left wing were indicating—off-scale low—a reading that falls below the minimum capability of the sensor. At 8:55:00 a.m. nearly 11 minutes after Columbia had re-entered the atmosphere, its wing leading edge temperatures reached their normal level of nearly 3,000 °F. At 8:55:32 a.m. Columbia crossed from Nevada into Utah while traveling at Mach 21.8 at an altitude of 223,400 ft. At 8:58:20 a.m. as Columbia crossed from New Mexico into Texas, a thermal protection system tile was lost. At 8:59:15 a.m. the engineers monitoring the sensors informed the flight director that the sensors monitoring the pressure on both left main landing gear tires were lost. The flight director then informed the Columbia's crew that Mission Control was evaluating these sensor readings, and added that the crew's last transmission was not clear. At 8:59:32 a.m. a broken response from Columbia was recorded but it was cut off in mid-word. Videos made by observers on the ground less than a minute later, presented in Fig. 18.5, revealed that Columbia was disintegrating.

Fig. 18.5 A video photograph of Columbia disintegrating as it is passing over Texas while traveling at about Mach 19.5 at an altitude of about 210,000 ft.

Columbia's History

Columbia was the first space-rated Orbiter, and it made the Space Shuttle Program's first four orbital test flights. Because it was the first of its kind, Columbia differed slightly from the other four Orbiters—Challenger, Discovery, Atlantis, and Endeavour. Built to earlier engineering specifications, Columbia was slightly heavier, and it was not able to carry sufficient cargo to conduct cost effective missions to the International Space Station. For this reason, Columbia was not equipped with a Space Station docking system, which gave it more space in its cargo bay for longer apparatus. Consequently, Columbia usually flew science missions and serviced the Hubble Space Telescope.

The final flight of Columbia was designated STS-107. This was the Space Shuttle Program's 113th flight and Columbia's 28th. The flight was nearly trouble-free. Unfortunately, there were no clear indications to either the crew onboard Columbia or to personnel in Mission Control that the flight was in trouble as a result an impact by at least one large piece that occurred shortly after launch. Mission management failed to note a few clues revealed during launch that the Orbiter was in trouble and take corrective action.

STS-107 was an intense science mission that required the seven-member crew to form two teams, enabling round-the-clock shifts. Because the extensive science cargo and its extra power sources required additional checkout time, the launch sequence and countdown were about 24 hours longer than normal. Nevertheless, the countdown proceeded as planned, and Columbia was launched

---

[4] Times are given as Eastern Standard Time (EST.).

on January 16, 2003. At 81.7 seconds after launch, when the Shuttle was at an altitude of about 65,600 feet and traveling at Mach 2.46 (1,650 mph), a large piece of insulating foam came off an area where the Orbiter attaches to the external tank. This piece of foam impacted the leading edge of Columbia's left wing at a velocity of about 500 MPH. The foam impact was not detected by the flight crew or observed by ground personnel. However, the next day, during detailed reviews of launch camera photography revealed a large piece of foam striking the Orbiter. This foam impact and resulting damage to the left wing of Columbia had no effect on the daily activities performed by the astronauts during the 16-day mission. This mission had met all its objectives prior to reentry.

The Accident Investigation

Immediately following the failure of the Orbiter, NASA established a Columbia Accident Investigation Board to identify the chain of events that caused the Columbia accident. Evidence the Board considered included:

- Film and video during launch.
- Radar images of Columbia in orbit.
- Amateur video of debris shedding during reentry.
- Onboard sensor data from the onboard recorder recovered after the accident.
- Analysis of the debris recovered.
- Computer modeling.
- Impact and wind tunnel tests.

Fig. 18.6 The upper circle shows the foam covered bipod on the external tank and the lower circle shows the impact point on the leading edge of the Columbia's left wing.

The reason for the loss of Columbia and its crew was a breach in the thermal protection system on the leading edge of the left wing. The breach was initiated by a piece of insulating foam that separated from the left bipod ramp of the external tank and struck the wing in the vicinity of the lower half of a reinforced carbon-carbon panel 81.9 seconds after launch. During re-entry, this breach in the thermal protection system allowed superheated air to penetrate the leading-edge insulation. This hot air (plasma) heated and softened the aluminum structure of the left wing and weakened the structure until aerodynamic forces caused loss of control, collapse of the wing, and subsequent disintegration of the Orbiter. A photograph of Columbia prior to launch, showing the area of the foam released from the external tank and the location of its impact on the left wing of the Orbiter is presented in Fig 18.6.

The external tank is the largest component of the Space Shuttle. It serves as the main structural component during assembly, launch and supports both the solid rocket boosters and the Orbiter. It also serves as the cryogenic, propellant tank for the space shuttle main engines. It contains 143,351 gallons of liquid oxygen at – 297° F in its upper tank and 385,265 gallons of liquid hydrogen at – 423 ° F in its lower tank.

The Orbiter is attached to the external tank by two umbilical fittings at the bottom and by a "bipod" at the top. The bipod is attached to the external tank by fittings to the right and left of the external tank centerline. The bipod fittings, which are bolted to the external tank, are located above the flange joint at the inter tank.

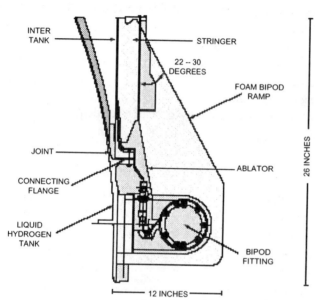

Fig. 18.7 Drawing of the bipod showing the foam insulation that covered the bipod fitting.

The external tank is coated with two materials, which serve as its thermal protection system. The inner coating is a dense composite ablator for dissipating heat, and the outer coating is low-density closed-cell foam for high insulation efficiency. The external tank thermal protection system is designed to:

- Reduce heat flow into the cryogenic tanks to minimize the vaporization of the oxygen and hydrogen prior to launch.
- Maintain the temperature of the external surfaces sufficiently high to prevent the formation of frost or ice.

A combination of factors led to the loss of the left bipod foam ramp (see Fig. 18.7) during the ascent of Columbia shortly after launch. NASA personnel believe that pre-existing defects in the foam were a major factor and that these defects were necessary to induce the foam to fail. However, analysis indicated that pre-existing defects were not the only factor responsible for foam loss. Other factors are involved such as the technique for producing and applying the foam, particularly in the bipod region. The foam used to insulate the external tank consists of two chemical components that must be mixed in an exact ratio and is then sprayed according to strict specifications. The foam is applied to the bipod fitting in a manual operation to form the ramp shown in Fig. 18.7. This process is probably the primary source of defects in the foam. Dissection of these foam ramps, conducted after the accident, revealed defects such as voids, pockets, and debris. These defects are due to a lack of control of several parameters in the manual spraying used to apply the foam. The quality control of the foam is exacerbated by the complexity of the underlying hardware configuration. The dissection studies showed that the defects usually occurred at the boundaries between each layer as the ramp is formed by the repeated application of thin layers.

It is important to recognize that the failure of the foam insulation on previous shuttle missions was not a rare event. Foam loss has occurred on more than 80 percent of the 79 previous Shuttle missions for which imagery is available. Foam was lost from the left bipod ramp on nearly 10 percent of missions analyzed where the left bipod ramp was visible following external tank separation. For many of the missions, it was not possible to determine if foam was lost because the launches were at night or the external tank bipod ramp areas were not in view at the time when the photographs of the external tank were taken.

It is believed that the impact of a large piece of foam damaged the leading edge of Columbia's left wing. The leading edge of an Orbiter's wing is comprised of 22 panels fabricated from a reinforced carbon-carbon (RCC) composite. To prevent oxidation, the outer layers of the carbon substrate are converted into a 0.02 to 0.04 inch thick layer of silicon carbide. The components of the Orbiter's wing leading edge provide the aerodynamic load bearing, structural, and thermal control capability with temperatures exceeding 2,300 °F. The design requirements for the leading edge of the wing included 100 missions with minimal rework, limiting the temperature of the aluminum wing structure to less than 350 °F, and withstanding a kinetic energy impact of 0.006 foot-pounds. The engineering specification clearly that the wing leading edge would not be required to withstand impact from debris or ice, because these objects would not pose a threat during the launch sequence.

The risk of micrometeoroid or debris damage to the RCC panels has been evaluated several times. Hypervelocity impact testing, using a variety of projectiles, as well as low-velocity impact testing with projectiles of ice and other materials, resulted in a design change that improved the resistance of the leading edges to impact damage. Analysis of this design change predicted that an Orbiter could survive re-entry with a ¼ to 1-inch diameter hole in the lower surfaces of the RCC panels depending on panel location.

Analysis of maintenance reports indicated that the RCC leading edge panels from all of the Orbiters had been struck by objects throughout their operational life. However, none of the panels had been completely penetrated. A sampling of 21 post-flight reports noted 43 hypervelocity impacts. The largest damage zone was 0.2 inch in diameter. The most significant low-velocity impact was to one of Atlantis' panels. The damaged area was 1.9 inches by 1.6 inches on the exterior surface and 0.5 inches by 0.1 inches in the interior surface. The substrate of the panel was exposed and oxidized. After inspection the severely damaged panel was replaced. A study concluded that the damage was caused by a strike by a man-made object, possibly during ascent. This panel damage to Atlantis was a clear warning of an impending disaster.

Upon completion of its exhaustive study, the Columbia Accident Investigation Board concluded that the cause of the accident was a breach in the thermal protection system on the leading edge of the left wing. The breach was initiated by a piece of foam that separated from the left bipod ramp of the external tank and impacted the left wing in the vicinity of the lower half of RCC panel 8. The conclusion that foam separated from the external tank bipod ramp and struck the wing in the vicinity of panel 8 is clearly documented by photographic evidence. Sensor data and the aerodynamic and thermodynamic analyses confirmed that the breach was in the vicinity of panel 8. This data and subsequent analysis also indicated subsequent melting of the supporting structure, the spar, and the wiring behind the spar. The detailed examination of the debris also pointed to panel 8 as the breach site. Impact tests established that a large piece of foam could breach an RCC leading edge panel. These tests and also convinced those who were discounting of the analytic evidence presented during the investigation.

Hard Lessons to Learn

Dr. Diane Vaughan, an authority on risk assessment, testified before the Columbia Accident Investigation Board and stated: "What we find out from a comparison between Columbia and Challenger is that NASA as an organization did not learn from its previous mistakes and it did not properly address all of the factors that the presidential commission identified."

Organizational failures always occur usually with higher frequency in larger organizations with several layers of management. Failures happen regardless of the safeguards and systems an organization employs. Failures are even more common in high-risk organizations like NASA— missions fail and astronauts are killed. NASA uses several risk-avoidance systems that are designed to insure that the instruments and astronauts sent into space complete their missions and return safely. However, NASA has failed in three cases to achieve this objective when sending astronauts in space or when preparing space systems to do so. The Space Shuttles Challenger and Columbia tragedies as well as the Apollo launch pad fire in 1967 are examples NASA's failures that resulted in the deaths of 17 astronauts. Organizational failures played an important role in all three cases.

As was the case with O-ring damage in the years before the Challenger launch[5], the accident to Columbia did not represent the first time that foam had detached from the external tank and caused damage to an Orbiter. There had been many reports of impacts on previous flights by pieces of foam that were much smaller than that one which severely damaged Columbia. Orbiters would often return with many damage sites due to both hypervelocity and low velocity impact. Some sites were larger than 1 inch in diameter. There are numerous reports in the Problem Reporting and Corrective Action (PRACA) system, which show that some of the thermal barrier damage was probably due to foam failure during launch and assent. Not only was foam impacting not considered in the design of the thermal protection system, but also the engineering specification for the Orbiter clearly states that nothing should impact the shuttle during launch.

However, foam had been failing and striking the Orbiter with high frequency and a team of chemists and engineers were working on improving the foam and the techniques used for applying it. The studies of impacts with foam concluded that they did not pose a "flight safety" risk because experience showed that the small pieces of foam that had been coming off the external tank only caused small pits and craters on the Orbiters. Unfortunately, the engineers did not consider a large piece of foam traveling at high velocity[6] when they declared foam failure not to be a "flight safety" issue. The impact of such a large piece of foam on a vulnerable area of the Orbiter was unprecedented. (So was the three days of abnormal cold before the Challenger launch.)

Foam failure that regularly damages the thermal protection system on a large fraction of the flights was not considered to be dangerous because the orbiters were returning safely with damage that could be repaired or ignored. What the engineers or management did not considered was the possibility of a larger or heavier piece of foam hitting the Shuttle in a particularly vulnerable area like the leading wing edge. This is a classic example of Vaughan's "normalization of deviance" [16] where an unpredicted anomaly becomes routine.

Normalization of Deviance or Accepting Problems Instead of Solving Them

Vaughan [16] has developed the concept of **normalization of deviance** to explain the way technical flaws escape the scrutiny of the various safety boards within a large organization such as NASA. Frequently these boards observe unanticipated problems, which continue to occur on a regular basis without leading to an accident. People begin to believe the problems are benign and are lead to the

---

[5] See Section 18.10 for a complete description of the Challenger launch.
[6] Photographic analysis revealed that one large piece and at least two smaller pieces of foam had separated from the bipod ramp area of the external tank. The large piece of foam was about 24 in. long and 15 inches wide. It was rotating at 18 RPS and moving with a velocity of about 500 MPH when it impacted the Columbia's wing.

pragmatic notion of "acceptable" deviance. The leadership at NASA found it very expensive and time consuming to establish the cause of some problems and to incorporate added inspections in the regular maintenance cycle of the Shuttle system without strong evidence of "flight safety " risk. Under the pressures of the Shuttle's flight schedule and limited budget, NASA did not commit significant resources on problems that its management did not classify as "flight safety" risks that could cause loss of an Orbiter. This practice resulted in disincentives for the engineers to determine the source of problems, even though scenarios could be envisioned that would cause significant "flight safety" risks. NASA frequently cleared flights as operational based on previously successful flights that had completed their missions while exhibiting clear evidence of a design problem. This reasoning prompted Richard Feynman [13] in his analysis of the Challenger accident to remark— "When playing Russian roulette the fact that the first shot got off safely is little comfort for the next."

The Role of NASA's Management Structure in Columbia's Accident

NASA's organizational and physical infrastructure was developed in the days of Apollo program when it had a large budget and a sharply focused mission—to land a man on the moon before the Russians. A layered bureaucratic group of middle and upper level managers functioned well and the Apollo program accomplished its goals with only one serious mishap in 1967 and a near catastrophic mission on Apollo 13. After more than 30 years, NASA's organizational structure is nearly the same although its missions have changed markedly. There is reason to question if their organizational structure is sufficiently flexible considering its plans for the future and its ongoing projects.

For the Shuttle program, there should be a mechanism for engineers to bypass the bureaucracy and hierarchy, especially in the pre-launch and launch processes. Suppose the engineers had succeeded in calling off the launch of the Challenger due to the effect of cold weather on the O-ring seals. The flight would have been cancelled, Challenger would have been taken off of the launch pad and the solid rocket booster disassembled to replace the damaged O-rings. This action would have been expensive but not nearly as costly as the loss of crew and the Orbiter. Similarly, if an engineer needs a certain type of data, there should be a way to bypass the formal bureaucratic procedures and to obtain the data with dispatch. Engineers have many intuitions and hunches that take time and resources to translate into analysis and data. These intuitions need to be respected, given credence, explored and welcomed by upper management [17].

Probability of Failure

Richard Feynman served on the Presidential commission that investigated the Challenger accident, which occurred on January 28, 1986. It was near the end of his life, and the famous physicist devoted significant amounts of time to the investigation. After the investigation, he prepared a paper describing his personal observations on the reliability of the Shuttle [18]. This paper is available on the website given in Reference 18. We suggest you read it, because we are only reviewing a small portion of his paper. The following paragraphs are direct quotes from Feynman's personal observations.

1. It appears that there are enormous differences of opinion as to the probability of a failure with loss of vehicle and of human life. The estimates range from roughly 1 in 100 to 1 in 100,000. The higher figures come from the working engineers, and the very low figures from management. What are the causes and consequences of this lack of agreement? Since 1 part in 100,000 would imply that one could put a Shuttle up each day for 300 years expecting to lose only one, we could properly ask—"What is the cause of management's fantastic faith in the machinery?"
2. If a reasonable launch schedule is to be maintained, engineering often cannot be done fast enough to keep up with the expectations of originally conservative certification criteria designed to guarantee a very safe vehicle. In these situations, subtly, and often with apparently

logical arguments, the criteria are altered so that flights may still be certified in time. They therefore fly in a relatively unsafe condition, with a chance of failure of the order of a percent (it is difficult to be more accurate).

3. Official management, on the other hand, claims to believe the probability of failure is a thousand times less. One reason for this may be an attempt to assure the government of NASA perfection and success in order to ensure the supply of funds. The other may be that they sincerely believed it to be true, demonstrating an almost incredible lack of communication between themselves and their working engineers.

4. In any event this has had very unfortunate consequences, the most serious of which is to encourage ordinary citizens to fly in such a dangerous machine, as if it had attained the safety of an ordinary airliner. The astronauts, like test pilots, should know their risks, and we honor them for their courage. Who can doubt that McAuliffe was equally a person of great courage, who was closer to an awareness of the true risk than NASA management would have us believe?

5. Let us make recommendations to ensure that NASA officials deal in a world of reality in understanding technological weaknesses and imperfections well enough to be actively trying to eliminate them. They must live in reality in comparing the costs and utility of the Shuttle to other methods of entering space. And they must be realistic in making contracts, in estimating costs, and the difficulty of the projects. Only realistic flight schedules should be proposed, schedules that have a reasonable chance of being met. If in this way the government would not support them, then so be it. NASA owes it to the citizens from whom it asks support to be frank, honest, and informative, so that these citizens can make the wisest decisions for the use of their limited resources.

6. For a successful technology, reality must take precedence over public relations, for nature cannot be fooled.

As one studies the Columbia's fate, we appreciate the significant value of the personal observations made by Richard Feynman nearly 30 years ago. It appears that his message was lost on NASA as they continue to ignore significant operational problems in their efforts to met schedule with a constrained budget. Dr. Feynman estimated that the Shuttle flies in a relatively unsafe condition, with a chance of failure of the order of a percent. With the additional Shuttle flights since the Challenger accident it is possible to test his estimate with the additional flight experience. The Columbia accident was the second in 113 flights; thus the estimate of the probability of failure P_f of future Shuttle flights is:

$$P_f = N_f / N = 2/121 = 1.65\%$$

This estimate is the same order as Dr. Feynman predicted after the Challenger accident and 1,700 times much higher than the probability of failure of 0.001% (1/100,000) predicted by NASA managers.

18.12 BRIDGESTONE/FIRESTONE TIRE RECALL—A CASE STUDY

The Challenger accident described previously is a tragic example of high-risk decision-making on the part of senior management at Morton Thiokol and NASA. In their attempt to launch on schedule, they ignored a well-known, serious problem with a rubber O-ring seal. The actions of senior managers resulted in the deaths of the astronauts on board and a serious setback to the United States space program. During the late summer and fall of 2000, rubber again made the news, but this time it concerned tires, and many more lives were affected.

Background Information and Timeline

On February 7, 2000, a Houston television station (KHOU) aired a segment prepared by reporter Anna Werner, and producer David Raziq drawing attention to a large number of vehicle crashes that appeared to be caused by tread separation in the Radial ATX model tire, made by Bridgestone/Firestone. In May of 2000, the National Highway Traffic Safety Administration (NHTSA) initiated an investigation into the accidents. Over the next several months, an increasing number of crashes were attributed to tread separation. Finally, in response to mounting pressure from consumers and retailers, on August 9, 2000 Bridgestone/Firestone announced a recall of 6.5 million tires, most of which had been installed as original equipment on Ford Explorer Sport Utility Vehicles. At the time of the recall, the tires had been implicated in 46 deaths in automobile crashes in the United States, a number that would later climb to nearly 150 by December 2000.

How Tires Are Currently Manufactured

The manufacture of tires requires a multi-step process involving many raw materials. Tires themselves are quite complicated. A schematic showing a cutaway of a radial tire is shown in Fig. 18.8.

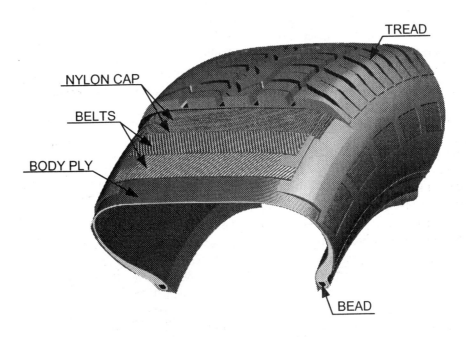

Fig. 18.8 Cutaway view showing construction details of an automobile tire.

In a typical tire factory, rubber arrives in bales. This rubber is placed into a machine called a Banbury mixer, and mixed with ingredients such as oils, carbon black, pigments, antioxidants and other ingredients that have been chosen to give a final rubber mixture with specified properties. The tread and sidewalls are made of different compositions. Rubber is also used in the tire for coating the ply and bead. In the next step, the rubber mixtures are processed in a breakdown mill, where the rubber mixtures are passed between pairs of large rollers over and over again to further mix the ingredients. The rubber mixtures are then extruded into continuous strips that are used for the sidewalls and tread. The rubber strips are sent through a coating apparatus that applies adhesive, and, after cooling, the strips for the tread/sidewalls are cut into the lengths needed to make each tire.

The ply, or body of the tire, is made by coating rubber onto a belt of fabric made from tough polyester or nylon cords. The bead of the tire forms its backbone. The bead is made by arranging steel wires into a ribbon, coating the ribbon with rubber for adhesion and then forming the ribbon into hoops. Additional layers in a tire include the belts, which may be made of steel, polyester or nylon. These layers serve to reinforce the ply layers and hold the tread flat on the road. Belts also improve tread life because they resist damage from impacts and penetration. These components are assembled in a tire-building machine that pre-shapes the tire. The final processing step is to take the so-called 'green tire' and place it in the curing press, similar to a waffle iron, where it is heated to 300 °F. The final tread and sidewall patterns are embossed onto the tire in this step.

After curing the tires are inspected. Because several potential flaws are not visible, sample tires are pulled from the line at random and dissected or x-rayed to examine the internal parts of the tire for defects [19]. More details about the manufacturing process are provided in an excellent description found at:

http://www.goodyeartires.com/about/diversity/how_built.html

Manufacturing Questions

The tires that were subject to the recall were made at a various manufacturing facilities, but one Firestone factory, located in Decatur, Illinois, attracted suspicion when some of its former workers testified about the lack of quality control at the plant, and about the incentives that the workers had to increase production at the expense of quality. The workers reported that they were ordered to use an awl to puncture blisters that developed in the tire after the curing press. Steel belt cords were reportedly exposed to humid air in some sections of the plant, leading to condensation onto the belt cords. This condensation could lead to corrosion of the metal, possibly resulting in poor adhesion within the layers of the tire. Workers said that sometimes pockets of solvent were left in the tire; during curing, the solvents vaporize often creating debonded areas between the tire layers. Disturbingly, workers also testified that inspections of finished tires were often skipped in order to meet production quotas [20].

The Ford Motor Company released statistics that showed that the Wilderness tires made at the Decatur plant during the period from 1996 to 1999 failed at a rate that was 10 times greater than the failure rate for Wilderness tires made at all of Firestones other plants combined [21].

Usage Questions

Some of the problems with the tires may have arisen because of the way the tires were used by the consumers. The failing tires that were recalled were standard equipment on Ford Explorers. Ford Motor Co. had recommended that the tires be inflated to a lower air pressure, 26 pounds per square inch (psi), which was lower than the 30-psi recommended by Firestone. The reason Ford lowered the specified tire pressure was to reduce the risk of rollover, which is a problem for some sport utility vehicles because of their higher center of gravity. Reduced tire pressure would also improve the vehicle's ride. At a congressional hearing in September of 2000, Firestone's position was that a tire pressure of only 26 psi resulted in "a very low safety margin". Adding to the confusion, a Ford official stated that Firestone had agreed to the 26-psi recommendation for a decade. Following this hearing, Ford changed its official recommendation to 30-psi [22].

Design Questions

If you were aware beforehand that the tires were going to be used at pressures of 26 psi rather than 30 psi, do you think a different tire design would have helped to improve the tire's safety? If you consider the product development process described earlier, the first step is to identify the customer needs; the second step is to establish the product specification, based in large part, upon those needs. The end product, ideally, is a tire that is safe under a range of operating conditions that exhibits a long trouble free life. If the specifications are based upon operating conditions that are not consistent with conditions found in practice, this specification is flawed and can result in a design that may not be adequate.

At the same congressional hearings in September 2000, [23], Ford Motor Company and Bridgestone/Firestone executives continued to battle with each other. Firestone's executive vice president, John Lampe, testified that in addition to the possibility of manufacturing problems, the design of the tires might have been responsible for the failures. However, with this admission that the tire design may have been to blame, Firestone then shifted some of the responsibility to Ford by declaring that they had been "exploring a theory that the design specifications (provided by Ford) were too intolerant for small variations in manufacturing".

In manufacturing, some variation from piece to piece and in the final product is to be expected. Quality control is used to insure that final products that do not meet specifications do not leave the plant. Earlier in this book, we discussed how performance, appearance, manufacturing, assembly, maintenance, environment and sustainability and safety must be considered in the design of a system, subsystem or component. These are to be considered along with the needs of the customer. Considering manufacturing, if you set tolerances that are unreasonably narrow, you might find that a large percentage of the final product does not pass the final quality control step because of unavoidable variations in manufacturing each component and in assembling the final product. This practice would result in a much higher cost per product unit, because of the large waste of labor and resources used to make the rejected products. If you set tolerances that are too wide, the final product may be deficient in other ways, such as assembly if the components do not fit together well, and it may not have the necessary margin of safety. There are tradeoffs to be made at every step of the design process as all of these factors are taken into account.

The use of nylon caps between the tread and the steel belts has been shown to prevent tread separation; nylon caps are incorporated into tires sold in other countries, but experts believe that the caps are not necessary in the U.S. because speed limits are lower here than in other countries [24]. What is your opinion? How much extra would you be willing to spend on tires that had a much lower risk of failure?

Government Oversight and Testing Problems

Why did it take a television news story to initiate the chain of events that lead to the recall? Why were these defects in the tires not discovered using government mandated safety tests? The answer to this question may lie in the tests themselves.

The current government approval processes that tires must pass in order to be declared safe include: tests for endurance, strength and whether or not the tire stays on the rim when it fails. The tires being recalled passed these tests easily in 1990 and again in 1997. However, these tests were developed 30 years ago. Tire technology has changed significantly in 30 years, and so have driving habits and vehicle design. The NHTSA is now working to update their standards and to design new tests that will replicate conditions that lead to the tire failures, so that future defects can be detected before the tires are sold to the public [25].

What about the Firestone's product test program? According to House Representative W.J. Tauzin of Louisiana, Chair of the House Consumer Protection Subcommittee, Firestone conducted test samples of these tires in 1996, by running high-speed durability tests in a laboratory. The tests resulted in 31 tire failures in 229 trials, with 20 of the tire failures resulting from tread separation. Firestone made design modifications to the tires in 1998, but no recall was initiated at that time. A company spokesman said that the design changes were "not in response to some specific problem with Explorers or even the (recalled tires)" [26]. The details of the tests were not reported; hence, it is not possible to determine if the tests reproduced conditions that the tires would actually encounter.

What is your opinion regarding the appropriate level of government involvement in determining whether consumer products are safe?

An Unfortunate Precedent

In May and July of 1978, executives from Firestone Tire and Rubber Co. (before it was purchased by Bridgestone), testified before Congress that another tire (the Firestone 500 radial) was not defective, and the reason for excessive tread separation incidents was incorrect use (under-inflation) by the consumer. Sound familiar? At the time, the NHTSA did not agree and Firestone was ordered to recall 14.5 million tires, and to pay a $500,000 civil penalty. The poor publicity and loss of sales nearly resulted in bankruptcy for Firestone. At that time, the government's testing standards were declared inadequate for protecting public safety [27]. Yet, the tests were not updated, and Firestone again is facing allegations of knowingly selling tires that it knew to be defective.

Independent Analysis

An independent tire expert, hired by Bridgestone/Firestone, reported that the tread separation resulted from growth of a crack between two of the belts in the tire. The tire expert, Dr. Sanjay Govindjee, an associate professor of civil engineering at University of California, Berkeley indicated that the reasons for the crack initiation, and its subsequent growth to a critical size were due to several different factors. He cited climate (high temperature), low tire inflation, tire design and manufacturing differences between Firestone plants. At a press conference, he said that the most important factor in the growth of the cracks was "the weight placed on the tire by the vehicle." [28].

The Impact of Product Related Liability

For the case of the Challenger accident, the question of responsibility was answered after several inquires and many investigations. The question of responsibility still has not completely been resolved in this case. On May 31, 2001, Bridgestone/Firestone, Inc. (Firestone) submitted a "request" that NHTSA open a safety defect investigation regarding the handling and control characteristics of Ford Explorer sport utility vehicles (SUV) following a tread separation on a rear tire. The request was filed shortly after Ford Motor Company (Ford) announced on May 22, 2001 that it would provide free replacements for all Firestone Wilderness AT tires on Ford vehicles. It was part of Firestone's effort to argue that the crashes (many of which involved rollovers) that occurred in Explorers following tread separations of Firestone ATX and Wilderness AT tires were to a large extent due to the design of the Explorer rather than a defect in the tires. The fallout has impacted Bridgestone/Firestone and its parent company Bridgestone greatly, with profits for Bridgestone dropping 80% for the 2000 annual year [29].

It appears that the Ford Motor Company wants to place all of the blame on Bridgestone/Firestone. To quote the CEO of Ford, Jaques Nasser "This is a tire issue, not a vehicle issue" [24]. However, Bridgestone/Firestone's position is that they were making the tires according to the specifications ordered by Ford and that the consumer and Ford were partially to blame because the tires were designed for use at higher tire pressures than recommended by Ford. With all of this blame being shifted back and forth, the issue of communication within each company and between the companies surfaces. The design teams at Bridgestone/Firestone and Ford did not appear to be working very closely together!

18.13 SUMMARY

The degradation of ethical behavior during the past two or three generations has been described. Some of the probable reasons for these behavioral changes were discussed. While there are many controversial issues that can be classified as "dilemma" ethics, there are several "basic" characteristics where the difference between right and wrong is clearly defined. The list of basic virtues was described that include:

- Prudence
- Justice
- Fortitude
- Temperance
- Faith
- Hope
- Charity
- Humility
- Respect

Laws are written to prohibit, by punishment, a well-defined set of vices. These laws control behavior, but only affect the extremely repulsive actions on the part of individuals or corporations.

Individual behavior is important, but as we learned in the Challenger explosion, it is not sufficient. Corporations (business) must operate with the highest ethical standards. The chief executive officer (CEO) establishes the standards and the lower level managers act to establish the corporate character.

Honor codes exist on some college campuses to guide ethical behavior (lying, cheating and stealing) and the students and the faculties respect them. Character development and honor codes go together in producing a meaningful educational experience. Some of the more important attributes in character development are listed below:

1. Excellence
2. Respect
3. Integrity
4. Decisiveness
5. Responsible
6. Temperance
7. Fortitude
8. Understanding

Professional ethics was described by introducing the Code of Ethics of Engineers sponsored by the Accreditation Board for Engineering and Technology (ABET). The code is short with four fundamental principles and seven basic canons.

The fatal accidents of the Challenger and Columbia Space Shuttles, which failed during their missions in 1986 and 2003, were discussed. Many ethical issues dealing with individual and corporate behavior were apparent when the events leading to this fatal accident were reviewed. This example should aid you in developing both your individual and professional character.

A case study that considers the recent failures of the Firestone tires mounted on sports utility vehicles (SUVs) was presented. The tires fail by ply separation resulting in loss of control and roll over of the SUVs with their high center of gravity. The failures were most common on the Ford Explorer and the number of people killed to date exceeds 150. Some background on tire manufacturing was provided together with a discussion of the problem as revealed by media coverage.

REFERENCES

1. Patterson, J. and P. Kim, <u>The Day America Told the Truth: What People Really Believe about Everything that Really Matters</u>, Prentice Hall, New York, NY, 1991.
2. Sommers, C. H., "Teaching the Virtues," *Public Interest,* No. 111, Spring 1993, pp. 3-13.
3. Woodward, K. L. "What is Virtue," *Newsweek*, June 13, 1994.
4. George, R. P., <u>Making Men Moral: Civil Liberties and Public Morality</u>, Oxford University Press, New York, NY, 1994.
5. Baker, D. F. "Ethical Issues and Decision Making in Business," <u>Vital Speeches of the Day</u>, 1993
6. <u>United States Air Force Character Development Manual</u>, USAF Academy, CO, December 1994.
7. Abramson, R., "A Matter of Honor," *Los Angeles Times*, April 3, 1994.
8. Lickona, T. A. <u>Educating for Character</u>, Bantam Book, New York, NY, 1991.
9. Jensen, C., <u>No Down Link:</u> A Dramatic Narrative about the Challenger Accident, Farrar, Straus and Gitoux, New York, NY, 1996.
10. *Time,* February 10, 1986
11. Lewis, R. S., <u>The Voyages of Columbia: The First True Space Ship</u>, Columbia University Press, New York, NY, 1984.
12. *The New York Times* April 23, 1986.
13. Feynman, R. P. <u>What Do You Care What Other People Think?</u> Bantam, New York, NY, 1988.
14. Report to the President by the Presidential Commission on the Space Shuttle Challenger Accident, Ayer Co., Salem, MA, 1986.
15. Anon, Report of Columbia Accident Investigation Board, Volume I, August 26, 2003, See http://www.nasa.gov/columbia/home/CAIB_Vol 1.html.
16. Hall, J. L., "Columbia and Challenger: Organizational Failure at NASA," Space Policy Vol. 19, 2003, pp. 239-247.
17. Vaughan, D., <u>The Challenger Launch Decision: Risky Technology, Culture and Deviance at NASA</u>, The University of Chicago Press, Chicago, 1996.
18. Feynman, R. P., Personal Observations on the Reliability of the Shuttle, http://www.fotuva.org/feynman/challenger-appendix.html.
19. Swoboda, F., "Rubber's Journey to the Road" Washington Post, August 11, 2000.
20. Grimaldi, J. V., "Testimony Indicates Abuses at Firestone", Washington Post, August 13, 2000.
21. Swoboda, F. and Grimaldi, J. V., "Ford Finds Tire Plant Had Many Failures", Washington Post, August 14, 2000.
22. Mayer, C. E. "Ford Advises Inflating SUV Tires to 30 psi", Washington Post, September 23, 2000.
23. Mayer, C. E. and Grimaldi, J. V., "Firestone Narrows Flaw Probe", Washington Post, September 13, 2000.
24. Eisenberg, D., "Anatomy of a Recall", Time, 29-32, September 11, 2000.
25. Skrzycki, C., "U.S. Seeks New Test for Tires", Washington Post, August 12, 2000.
26. Grimaldi, J. V. and Skrzycki, C., "Ford Faces New Probe", Washington Post, September 21, 2000.
27. Skrzycki, C., "The Regulators: 'Firestonewalling Again? Two Decades Later, Echoes of Earlier Testimony", Washington Post, September 12, 2000.
28. Mayer, C. E., "Study Can't Pinpoint Cause of Tire Failure", Washington Post, February 3, 2001.
29. Kashiwagi, A., "Recall Cost Bridgestone Dearly", Washington Post, February 23, 2001.

EXERCISES

18.1 Political leaders are often attacked by the media for lax ethical behavior. Please write a short paper describing three recent lapses of ethical behavior on the part of the leadership in either a State or the Federal government. Did these officials break the law? Is it important that they did or did not break the law?

18.2 List what you consider poor behavior in a college classroom.

18.3 Have you ever cheated on an important exam in high school? What about a college exam? Do you know of someone who cheated? What was your attitude when you observed this cheating?

18.4 Is there an honor code at the University where you are pursuing your studies? Have you read it? Do you abide by the rules?

18.5 Write a paragraph or two explaining your position regarding:
- Prudence
- Justice
- Fortitude
- Temperance

18.6 Write a paragraph or two describing your feelings about:
- Faith
- Hope
- Charity

18.7 If you could write a law that would go on the books tomorrow and be strictly enforced, what behavior would it require or prohibit?

18.8 Why does the CEO of a corporation set the standard for ethical behavior? What are some of the issues that arise on a daily basis that develop corporate character? Is it possible for a corporation with tens of thousands of employees to develop a character like an individual?

18.9 Write a 400-word essay describing your character. Include a discussion of your weaknesses and strengths.

18.10 Examine the code of ethics for engineers given in Fig. 18.1, and describe your opinion of the fundamental principles or canons. Can you live with these expectations if you become a professional engineer? Do you believe ABET's code is adequate, or should it be revised to reflect a more modern viewpoint of what is right and wrong?

18.11 Suppose that you were a lead engineer working for Morton Thiokol on the evening of January 27, 1986, involved in the discussion of the O-rings. What would you have done?
- Early in the teleconference.
- At the critical stage of the teleconference.
- After management had taken over the decision process.

There is no right or wrong answer to these questions. The purpose is to place you in a professional dilemma. Someday you may be placed in a similar situation where there is a trade-off between safety and corporate business interests. Think about your response in advance and be prepared to deal with such a dilemma and its consequences.

18.12 Suppose you were the engineer responsible for maintaining the foam on the external tank of the Space Shuttle. What would you do to prevent the failure of foam during the launch and ascent of a Shuttle?

18.13 Suppose you were the engineer responsible maintaining the RCC composite panels that form the leading edge of the three remaining Orbiters. What changes in your maintenance plan would you make following the Columbia accident?

18.14 Write a review of Chapter 2 "Columbia's Final Flight" from volume I of the Report of the Columbia Accident Investigation Board. Reference 18 provides the URL for the Website.

18.15 Write a review of Chapter 3 "Accident Analysis" from volume I of the Report of the Columbia Accident Investigation Board. Reference 18 provides the URL for the Website.

18.16   Write a review of Chapter 5 "From Challenger to Columbia" from volume I of the Report of the Columbia Accident Investigation Board. Reference 18 provides the URL for the Website.

18.17   Do you believe the federal government should take a more active role in determining if tires and other consumer products are safe?

18.18   Are there government agencies that are concerned with product safety? Name them and describe the way they function.

18.19   Do you believe that Bridgestone/Firestone undervalued public safety in their decision not to recall tires until after the media and the government began to investigate the tire failure incidents?

18.20   Considering the design of a tire, how would you weight the following criteria: performance, cost, safety, ease of assembly, and ease of manufacture?

18.21   Do you believe any of the fundamental principles or canons of the ABET Code of Ethics of Engineers were violated by any of the engineers at either Ford or Bridgestone/Firestone? If yes, which ones?

18.22   Suppose you were in charge of the testing program at Firestone in 1996, and noted that 31 tires out of 229 tires had failed the running high-speed durability tests. What actions would you initiate?

18.23   You are given the responsibility to build a new plant with a manufacturing line for fabricating high performance tires suitable for heavy SUVs. Outline the problems you would attempt to solve based on the information given in Section 18.12. Indicate your approach in solving each problem that you have identified.

CHAPTER 19

SUSTAINABLE ENGINEERING

19.1 ENVIRONMENTAL ISSUES

The global environmental problem threatens our ability to sustain future generations with an improving or even constant standard of living. The environmental problem is difficult to define because it has many facets and many different players. In this country, societal concerns, societal consumption, corporate objectives, corporate behavior, local, state and federal agencies, political agendas, political structure, Congress and the Executive exacerbate the problem. To put it simply— there are many different issues and too many players on all sides of every issue. Let's begin by discussing an environmental assessment formulated by Paul Hawken in his well-recognized book *The Ecology of Commerce* [1].

The fundamental issue is sustainability. Will the Earth provide the resources (food, materials, and energy) to support the population now and in the future? Others have raised this question much earlier in history. Thomas Malthus, in 1798, in *An Essay on the Principle of Population* [2], argued that the population tends to increase faster than the food supply. The population would starve unless war, famine and disease limited population growth. Today, this argument seems preposterous, at least in the U. S., because farmers produce enough food to more than feed the country's population and to export large quantities of meats and grains with less than 2% of our population engaged in agriculture. However, abundant food supplies were not always available. Henry Weaver [3] describes many civilizations and countries that failed because they could not feed their people. The Roman Empire collapsed in famine. French peasants were dying of hunger when Napoleon Bonaparte stormed Europe. As late as 1846, the Irish starved during the potato famine. Currently about 35,000 children die from starvation each day. Two years ago in Afghanistan, the U.N. estimated that more than 1 million Afghans would be at immediate risk of starvation. At that time Afghanistan's rulers, the Islamic Taliban, impose their fundamentalist beliefs on women and gave sanctuary to suspected terrorists. Consequently, few donors were willing to step forward with emergency aid.

Food is in abundant supply today in most places worldwide, even with significant increases in world population, because of marked advances in agricultural machinery and in plant and animal sciences. However, serious food shortages still exist in many localities. About one out of six people in the world live on the brink of starvation surviving on international handouts. These food shortages are usually due to war, natural disasters or political strife.

Finite Resources

Yet even with the current abundance of food and merchandise, many intelligent, concerned and dedicated individuals raise the question of sustainability. The world population is growing exponentially. From 1800 to 1900 the population increased by about 600 million with an average growth rate of 6 million/year. From 1900 to 1950 the world population grew at a rate of 18 million/year. From 1950 to 1975 the growth rate increased to 60 million/year. Today the population growth rate is about 77 million/year and the world population exceeds 6.15 billion. The census bureau projects a world population of 9.1 billion by 2050. However, this projection assumes a marked reduction in the annual growth rate of the population from 1.24% today to only 0.43% in 2050. If this growth rate does not decrease according to projections, the world population will be much larger than the 9 plus billion estimate. When more and more food is produced from less and less land, the population appears to grow at a rate to ensure its consumption.

Another problem associated with an expanding population is urban growth. Current population trends indicate that by 2050, 6 billion people will be living in cities, which is three times more than today. The increase, 4 billion people, represents an average population increase of 80 million per year. To accommodate this urban population growth, will require enlarging existing megacities or developing new megacities. To grasp an idea of the size of this growth, 80 million people per year is equivalent to the formation of 10 new megacities with a population of 8 million each year for 50 years [4].

Consumption is another critical issue. The resources of the world are finite, and the population's understanding of the resource reserves is poor. Oil, gas and coalfields develop on a geological time scale (tens of millions of years per tick mark on a graph). The large corporations, supplying the world's gasoline and fuel oil requirements, will pump all but the very rich fields nearly dry in a few decades. It is clear that the world will run out of fuel even without additional population growth. The only question is when the supplies of crude oil will begin to decrease[1].

The supply of nearly all high-grade ores, from which metals are smelted, is also limited. For example, during World War II, the reserves of high-grade iron ore in Minnesota were seriously depleted as iron and steel production was increased to provide war materials. Today, Minnesota still provides over 70% of the iron ore for the blast furnaces in the U. S., but it is obtained by processing a low-grade ore known as taconite.

A final example of limited resources pertains to water supply. The Ogalala Aquifer, which underlies the Great Plains, is the largest body of fresh water in the U. S. Each year farmers pump 20 billion more gallons of water from this aquifer than is replaced by rainfall. It is predicted that the aquifer will become exhausted in thirty to forty years at the current rate of pumping. What will the farmers in the Great Plain states do when their irrigation wells run dry?

Why, knowing these constraints, do we continue to utilize resources at these high rates?

19.2 CONSUMPTION AND WASTE

Municipal Solid Waste

The people of the world are consumers. This fact is implied by the annual production of $21 trillion of products and commodities. In the U. S., we are the champion consumers as we purchase nearly twice as many goods and services per person as any other people on the face of the Earth. A typical person in the U. S. consumes 136 pounds of resources per week. In addition, about a ton of wastes are generated for each person each week in the process of producing the goods and services consumed. All of this consumption not only depletes the Earth's finite resources, but also leads to significant quantities of wastes that must be disposed of without polluting the air or the ground water.

Most of you are familiar with municipal solid waste. Your waste (garbage) is picked up once or twice a week and trucked away to a landfill—someplace that you never want to see or smell. You are often encouraged by municipal governments to separate the waste into categories—newspapers, metal cans, glass, plastics, etc. You trust that these items are recycled because it is incumbent upon every person to at least believe they are good environmental citizens. But, what really happens to the

[1] There is some evidence that a few oil fields may be regenerating apparently supplied from very deep reservoirs. However, this regeneration capability has not been firmly established for even a single oil field.

glass, plastics or metal trash? Is it recycled or do you have two or three trucks carrying wastes (that you carefully sorted and separated) from the neighborhood to the local landfill? Check with your municipality to determine the effectiveness of its recycling program.

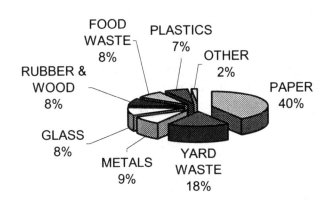

Fig. 19.1 Municipal solid waste generated in the U. S. in 1994. Annual total 180 million tons.

In 1994, about 180 million tons of municipal solid wastes were produced in the U.S. The composition of this solid municipal waste is shown in Fig. 19.1. It is evident that most of these wastes consist of paper (40%) and yard trash (18%). The paper, mostly newsprint and junk mail, can be recycled if the paper mills are prepared to accept it, and if the market will tolerate a lower standard for brightness and a lower strength for paper products produced with a large percentage of recycled newsprint. Many states have passed laws requiring newspaper publishers to employ newsprint that contains a specified percentage of recycled paper. In this instance, the law drives the demand for recycled newsprint that probably would not exist with normal market forces.

Most yard trash can be eliminated from landfills. First, grass clippings can be mulched with the lawn mower and dropped in place as the lawn is cut. Grass clippings should not be accepted in municipally collected wastes. Tree branches can be chipped when collected and converted to mulch. The mulch is of sufficient value for residents of the neighborhood to retrieve from conveniently located distribution stations if they are informed of these locations.

Metals, glass and plastic are also recyclable, but is it economically feasible to collect sort, ship and process these types of wastes? Recycling aluminum cans is profitable because it takes 10% less energy to smelt scrap aluminum than to reduce bauxite to aluminum. Steel is also recyclable and about 25% of scrap steel has been employed for many decades in producing most of the common alloys of steel.

When it is less costly to ship many types of solid wastes to a landfill, the wastes with marginal value are usually discarded. Since tipping fees in the U. S. are still relatively low by world standards[2] at an average of $31 per ton in 1996, valuable resources are often discarded in landfills. Unfortunately, the economic arguments for tipping instead of recycling do not account for costs associated with cleaning up toxic wastes found in landfills and in health hazards related to ground water pollution by fluids leaching from the wide variety of discarded products[3]. In the coming years, everyone must be more creative in producing many different types of products from recycled materials. A few examples will be described later in this chapter indicating successful approaches currently practiced by some companies for recycling materials into new products.

---

[2] Tipping fees at landfills in Japan range from $300 to $400 per ton.
[3] For example, a discarded television contains nearly a pound of lead, thousands of other chemicals, and a vacuum tube that may implode.

National Wastes

While the problems of disposal of municipal solid waste have been well publicized, and recycling efforts have been organized, the problem of national waste has largely been ignored. Consequently, the general population poorly understands the magnitude of the problem. National waste is mostly produced by industry, and the amount is enormous (12,900 million tons per year). National waste, as illustrated in Fig. 19.2, is classified as nonhazardous waste, hazardous waste, mining waste, oil and gas waste, municipal solid waste and other debris from industry and construction. Municipal solid waste makes up only 1.4% of the total national waste. Clearly the public, with its attention on municipal solid waste, is focused only a very small segment of the problem.

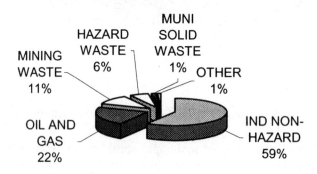

Fig. 19.2 Waste generated in 1994 in the U. S. The annual total is 12,915 million tons.

Industrial wastes are not usually shipped to municipally owned landfills; nevertheless, the refuse goes into what bureaucrats call land disposal units—some located above ground and others out of sight below ground. About 59% (7,600 million tons per year) of this industrial waste is classified as nonhazardous, but even nonhazardous wastes often cause harm. Heavy metals, PCBs, pesticide containers and organic compounds are dumped into these disposal sites. While classified as non-hazardous, these are not environmentally friendly dumping sites.

Hazardous Waste

The hazardous waste sites are much worse than the industrial disposal sites. Unfortunately they are numerous—today there are 90,000 hazardous waste sites in the U. S. Moreover, 1,200 of them have been designated as priority cleanup sites under the Superfund law.

How do you dispose of hazardous waste? Generally, it is stored in a watertight container such as a barrel or tank. This procedure is effective for some period, but eventually the barrel begins to corrode, holes develop and the container begins to leak. At that time, the toxic wastes are released and they may enter a stream or the ground water. When possible, a much more effective disposal solution is to chemically convert the toxic substance to a new chemical compound that is non-hazardous.

The amount of toxic waste produced each year by the chemical industry is huge. For example, in 1986, EPA's inventory of toxins released to the environment by the 50 largest chemical companies in the U. S. was an impressive 270 million tons. Suppose that the chemical industries incinerated[4] this waste instead of dumping it at its disposal sites. At the current rate for incineration of about $100/ton, the cost for burning these hazardous wastes totals $27 billion. This amount is larger by a factor of almost 10 than the entire profits accrued by the 50 large chemical companies in 1986 [1]. **The answer**

---

[4] Incineration as a disposal method also has serious deficiencies that will be discussed later.

is not to dispose of the hazardous wastes by dumping or burning, but to reduce and eventually eliminate their production during the manufacture of chemical products.

Reducing toxic waste is not a trivial matter. Syntex, a pharmaceutical company with a manufacturing faculty in Boulder, Colorado was well recognized as the worst polluter in the city. The toxic release inventory (TRI), issued annually by the Environmental Protection Agency (EPA), identified Syntex as the worst offender in Boulder County and the second worst offender in the entire state of Colorado. The company was under intense public pressure to clean up its chemical processes. They responded with large investments in pollution prevention measures, a vigorous employee-training program and a serious effort to communicate with the environmental organizations in Boulder. They agreed to reduce the total volume of air emissions by 50% over a four-year period. This example shows that progress can be made in reducing toxic wastes, but it is costly and significant reductions may require many years of dedicated efforts by concerned citizens and corporations.

Radioactive Waste

Radioactive waste is a much more significant problem than hazardous waste. Of course, both are hazardous but in different ways. For instance, dioxin is a very toxic chemical, which is stable and does not degrade. It accumulates in algae, insects and fish; eating these fish exposes one to very high levels of a known carcinogenic. Radioactive waste is less prevalent because there are fewer plants generating these wastes than there are chemical facilities producing toxic chemicals. However, while there are fewer sources of this contamination, they are not rare—most of people can identify a storage site for radioactive waste within their home state. Also, the amount of radioactive waste stored at some of these facilities is immense. For instance, at the nuclear weapons research facility in Hanford, Washington, there are enough liquid wastes stored to form a lake 35-feet deep twenty-five square miles in size. Recent news reports indicate that some of the storage tanks containing these radioactive fluids are leaking and polluting the ground water. In time, this seepage may reach the Columbia River and the pollution will spread to large regions in both Washington and Oregon.

There are two significant problems in disposing of radioactive wastes. First, many of the radioactive elements have a half-life of thousands of years or more; thus, producing a long-term storage requirement that exceeds the period of recorded history. Second, a technique for permanent disposal of these wastes has not been developed. The current practice is to place the wastes in containers and maintain them in a retrievable storage facility. Spent reactor fuel rods are maintained in water pools at the reactor sites of nuclear power plants across the country. The current plan is to store the containers housing the radioactive wastes in deep, geologically stable underground tunnels[5]. Unfortunately, no one wants to have this tunnel in his or her locality.

Recently[6] (March 26, 1999) the Department of Energy (DOE) opened the nation's first nuclear waste disposal facility in Carlsbad, NM. The $1.8 billion facility is intended for the permanent storage of radioactive material from 23 sites, which are located in 16 different states. The facility is composed of a series of storage rooms located off of access tunnels that are cut into an ancient salt dome about 2,500 feet underground. The facility will handle only radioactive wastes from DOE laboratories and their contractors. High-level radioactive wastes from utility plants and medical facilities, still located in temporary storage facilities, will be shipped to the Yucca Mountain site for disposal in the near future.

---

[5] The Department of Energy recently received Congressional approval for the use of the Yucca Mountain Site in Nevada for long-term storage of highly radioactive materials. The site is remote—100 miles from the nearest sizable population center (Las Vegas). The climate is dry with less than six inches of rainfall per year and the water table is 800 to 1000 feet below the proposed repository.

[6] The opening of this facility followed nearly a quarter of a century of studies, protests, and lawsuits.

Incineration

Incineration is often employed in many areas of the country where landfills are not readily available. The concept is simple—burn the trash to significantly reduce its volume and truck the residual to a landfill. In some instances the incinerators are an integral part of a power unit with the heat from burning trash used to produce the steam that powers the turbines generating electricity that is consumed by the local municipality. Unfortunately, the trash-to-electric power systems are rare, and even the existing systems generate only a small fraction of the energy required to produce the solid waste consumed in the incinerator.

While incinerators reduce the volume and weight of refuse that eventually is shipped to a landfill by a municipality, it does so by an airborne redistribution process. A modern incinerator burning 1,000 tons of a typical municipal-solid-waste converts this waste to a number of different elements and chemical compounds as indicated in Table 19.1.

Some of the heavy metals are trapped in the fly ash; consequently, this fine sand like material must be encased with a plastic liner when it is initially disposed of in a landfill. Otherwise, the heavy metals will leach out and pollute the ground water. Even more serious are the gases and small particulate matter released to the atmosphere. The area over which these compounds are distributed depends on the height of the smokestack and the direction and velocity of the prevailing wind. With the large quantity of sulfur dioxide, hydrogen chloride and sulfuric acid released to the atmosphere, those unfortunate people residing downwind from the incinerator can anticipate acidic rainfall.

Table 19.1
Emissions from a modern incinerator
1,000 tons of municipal solid waste [1]

Substance	Amount (tons)
Lead	2.22
Mercury	7.55
Cadmium	0.13
Nitrous Oxide	999
Sulfur Dioxide	379
Hydrogen Chloride	345
Sulfuric Acid	38.7
Fluorides	8.0
Small Particulate Matter	43.5
Fly Ash	300

19.3 WASTE GENERATING SYSTEMS

The public is anticipating a technological solution to the environmental problems, especially those dealing with the generation and disposal of wastes. Engineers today and certainly in the future will be called upon to improve ecoefficiency[7]. Strategies for realizing improvements include:

1. Minimizing emissions, increasing yields, reducing the generation of nonproductive material streams, using energy more efficiently and substituting benign materials for hazardous materials.
2. Consider wastes as raw materials for other products.

[7] The World Business Council for Sustainable Development has suggested that ecoefficiency is the delivery of competitively priced goods and services that satisfy human needs and bring quality of life and the progressive reduction of ecological impacts and resource intensity throughout the life cycle to a level commensurate with the Earth's estimated carrying capacity.

Engineers are interested in examining processes as systems. With a systems approach, one examines the complete process—not a small portion of it. The advantage of systems analysis is that techniques for reducing or eliminating waste become more apparent. Let's begin a systems analysis by considering a linear waste model that is typical of a poorly designed process generating a significant amount of waste.

A Linear Waste Model

A linear waste model, that depicts the generation of the maximum amount of waste and the consumption of the maximum amount of power per unit of production, is illustrated with a block and arrow drawing in Fig. 19.3.

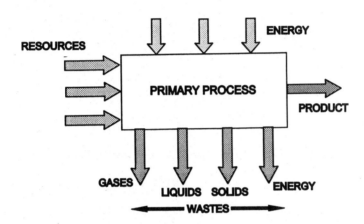

Fig. 19.3 A linear waste model that generates maximum waste without energy recovery.

The linear waste model is for a process that utilizes 100 percent virgin resources as input material. Energy in the form of electricity, steam and hot water is also provided as input to the primary production process. The resources and energy are consumed in the process as the product is produced. The product might be a lawn mower, plastic for soda bottles or carpeting for floors.

The wastes created in the process are rejected and transported from the facility as gases, liquids or solids. The gases are usually emitted from stacks (often tall) into the atmosphere. The small particulate matter and chemical compounds carried by the gas are distributed over a large area, which usually exceeds several hundred square miles. The solid waste generated in a linear waste process is either incinerated or shipped directly to a disposal site. The liquid waste is usually carried in a pipeline that eventually dumps into a municipal sewerage line or perhaps directly into a stream or river. Have you ever noticed the large number of factories located on expensive water front property? Have you wondered why?

Energy losses are also a form of waste. Steam exhausted to the atmosphere is not a pollutant since it is pure water. However, there is a significant amount of available energy in that steam that is wasted when it is released to the atmosphere. It could be employed to heat the water used throughout the facility. If hot water is released into a stream or river, energy is again wasted because that hot water could be used to heat the building in the winter or for some other useful purpose.

An improved linear waste model is presented in Fig. 19.4. This model is similar to that shown in Fig. 19.3 except that some of the waste is utilized in a secondary process to produce a by-product (a second product). There are five benefits that result from improvement of waste management associated with a manufacturing process.

1. The resources input to the secondary process are obtained at essentially no cost.

2. The by-product has value and its distribution increases the company's profits.
3. Society benefits from a more effective use of resources and energy.
4. Less waste is generated and the impact to the environment is less severe.
5. The cost to the company for disposing of the waste is reduced.

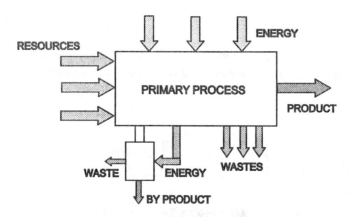

Fig. 19.4 An improved linear model with both product and by-product production.

A Waste Model with Limited Recycling

A waste model with limited recycling of resources consumed in a primary process is shown in Fig. 19.5.

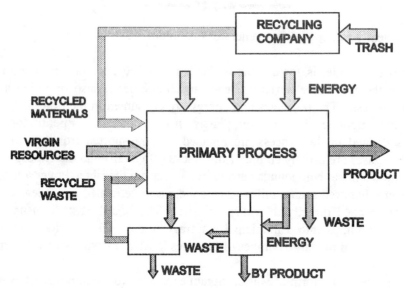

Fig. 19.5 A waste model with limited recycling.

The model demonstrates three fundamental methods that may be employed to reduce wastes and to reduce the consumption of virgin resources.

- Utilize waste as a resource to create by-products of value to society.
- Recycle waste internally, processing it as required to convert it to the quality of a virgin resource.
- Employ recycled materials obtained from external companies that recycle industrial and municipal wastes.

Let's consider some examples that demonstrate the savings accrued by employing each of these three waste reduction methods. Dupont, one of the largest chemical companies in the world, produces an enormous amount of waste. Responding to growing social pressures, Edgar S. Woolard, Dupont's CEO, initiated a broad and aggressive program in 1989 to reduce the quantity of waste generated in their production processes. One of their most successful approaches was to view chemical wastes as a production product looking for a market. They found two waste products generated in the primary production process for nylon—methylglutaronitrite and di-basic-acid. With some additional processing, methylglutaronitrite is a suitable resource for the manufacture of epoxy coatings for concrete and for general-purpose epoxy adhesives. Di-basic-acid is converted into a strong solvent that is an excellent substitute for acetone and methylene chloride that are hazardous solvents. In this case, wastes were converted into valuable by-products with ready markets, and the costs and environmental impact of disposal were eliminated.

Gillette, the razor blade company, saves money and reduces the negative effects of wastewater disposal on the local municipality and its environment by recycling internally generated wastewater. They installed a water recirculation system permitting them to recycle water used in the manufacturing process. They reduce the consumption of water from 730 million to 156 million gallons per year. The result was an annual reduction of $1.5 million in water and sewerage costs and a decrease of 574 million gallons of wastewater imposed on the municipal sewerage system. Gillette found that working to eliminate or reduce waste was not only environmentally friendly, but also provided increased profits.

Resources (materials) that feed primary or secondary processes need not be virgin materials. Materials that contain significant recycled content may often be employed. Most metals, such as aluminum, steel, lead, zinc and copper, contain a significant percentage of "scrap" without loss of performance. Both Ford Motor Company and Chrysler have changed the composition of several components installed in some of their automobiles to 100% recycled plastic and have made significant reductions in costs. The position taken by the automakers and most other manufacturers is to use recycled materials if they cost less and perform as well as virgin materials.

3M (Minnesota Mining and Manufacturing) was one of the first large corporations to address waste management by initiating their "Pollution Prevention Pays (3P)" program in 1975. Since then they have implemented hundreds of employee suggestions for reducing wastes and cutting emissions. From 1975 to 1990, they have documented saving in excess of $500 million and have reduced air emissions by 120,000 tons, solid wastes by 410,000 tons and waste water by one billion gallons [1]. The 3P program at 3M probably offers the best guide to other corporations for improving their manufacturing processes to reduce wastes while enhancing corporate profits.

It is clear to most high-level corporate managers that generating wastes is an inefficient utilization of resources that increases costs, decreases profits and pollutes the environment. Environmental regulations, discussed in a later section of this chapter, will force changes to improve both primary and secondary manufacturing processes. However, the recognition of the costs associated with the generation and disposal of wastes by corporate managers and employees will produce even more changes. Most corporate executives are moving from a compliance phase, where regulations were followed and met, to a leadership and persuasion phase, where managers and employees are encouraged to make changes to improve ecoefficiency.

Lifelong Ownership

Lifelong ownership is a concept introduced by the town of Sanger, California and the court [1]. The town sued three large chemical companies for contaminating the town's water supply with the pesticide dibromochloropropane (DBCP). The court ruled in favor of the town and established the precedent that the corporations were responsible for their product even after the time of the sale. The town was awarded $15 million to clean up its water supply.

Corporate responsibility for a product was established much earlier in cases where products with inherent dangers were sold to the public. Product liability lawsuits are an every day occurrence that has alerted corporations to the costs associated with releasing unsafe products to the market. Since the class action lawsuit addressing the gasoline fires with the Ford Pinto, both the public and the corporations clearly recognize the corporation's responsibility to produce a product that is safe to operate for the entire life of the product. In the near future, it appears that the corporation's responsibility for producing products that are safe for the environment will be more firmly established by the courts.

The concept of life-long ownership addresses the basic environmental problem— responsibility for toxic wastes and pollution belong to the manufacturer and not the customer. The advantage of this approach is to place the responsibility and the cost of the disposal with the corporation making the product. With the cost of disposal added to the original cost of producing the product, there is a strong economic incentive for redesign of both the processes and the products to reduce life cycle costs[8].

While the concept of lifelong ownership may appear to be new and radical, some countries have already enacted laws requiring corporations to assume responsibility for disposal. In 1992 Japan passed legislation requiring durable goods manufacturers to establish resource recovery centers. In Belgium a new law requires the last owners to hand in end-of-life vehicles to licensed recovery centers or to new car dealers if they purchase a new car. The percentage of reused or recycled content in the vehicles is also mandated [5]. In Germany manufacturers and distributors are required to take back and to recycle product packaging. Customers are permitted to leave all packaging materials for the products they purchase in the retail outlet for recycling.

A few companies already recognize their environmental responsibilities and the potential profits in recycling or remanufacturing their own products. Again in Germany, the automaker BMW has established a disassembly plant to recycle most of the components in its scrapped autos. Also, the newer model automobiles are being designed to permit rapid disassembly. In Belgium, an agreement was reached for the auto dealers to take back old models for recycling [6]. In the U. S., Ford is opening a chain of used auto parts centers [7]. Ford is interested in being perceived as the world's most environmentally sensitive automaker. The company has set a goal of increasing the amount of a junked car that is recycled from the current 80% to 95%.

In the U. S., Xerox—the copy machine manufacturer—has established an "Asset Recycle Management" program. Early in the development of the copy machine, Xerox recognized the potential health hazard associated with disposal of photoreceptor drums with selenium and arsenic coating. Xerox accepted ownership of the discarded photoreceptor drums, and discovered that they could be remanufactured at a fraction of the cost of making new drums. Xerox has expanded their remanufacturing program and today they collect older models of copy machines that are traded in when a customer purchases a newer more sophisticated model. The old machines are remanufactured by checking all of the components and replacing only those that are worn or those with a limited life expectancy. These remanufactured, but older models are sold to customers in Central and South America or Eastern Europe where cost is more important than the latest new feature. Xerox believes in reusing serviceable components to reduce costs and their demand for virgin resources. The customer benefits from lower prices for new/remanufactured copiers and society benefits from the conservation of virgin resources and reduction of wastes imposed on the environment.

In some instances, two companies cooperate to reduce waste and produce a new product using waste as a raw material from one company in a product produced by another. Chaparral Steel produces steel and the process generates slag as a waste product. Texas Industries is a company producing Portland cement. In a joint research program, Chaparral Steel and Texas Industries developed a patented process that uses steel slag in a cement kiln to create high quality Portland

---

[8] Life cycle costs include the costs of production, operation, maintenance and disposal.

cement [8]. The joint venture has increased profits for both companies, reduced waste disposal impact on the environment, cut energy consumption in producing cement and reduced CO_2 emissions.

Waste Metrics

Corporate America understands that environmental issues have a direct impact on the bottom line; therefore, these issues affect the way products are designed. When any factor affects the bottom line, corporate America seeks to devise a set of measurements for assessing the effectiveness of the corporation in managing this factor. Robert J. Eaton, the former CEO of Daimler Chrysler, has described the characteristics that any metric for measuring ecoefficiency must exhibit [9].

- The metrics must meet the needs of all of the stakeholders—community, government, and business.
- They must facilitate innovation and growth with continuous improvement as a cornerstone.
- They must be harmonized at the local, state, national and international levels.
- They must be fully compatible with existing business systems. If they do not add value, they will not support continuous improvement and will not be used.
- They must measure the right things. Only the quantities measured receive management's attention.

Let's suppose that you have been assigned the task of reducing waste generated by a production facility by 45%. Management (your boss) asks you to prepare a quarterly report describing your activities, the costs of implementing your program and periodic measurements of progress. How do you propose measuring the progress made in reducing the waste generated in a production process? There are two approaches—one based on the weight of the waste and the other on costs. The latter is more important than the former, but it is often more difficult to measure.

Let's consider a waste management index based on weight. This approach is relatively easy to implement because production control staff usually maintains careful records of unit production of primary products, by-products and the weight of all materials employed. Solid waste generated is collected and weighed. Wastewater and other liquid emissions are measured with a flow meter. Measurements of gas emissions from vents and/or stacks yield the weight of various chemical compounds released to the atmosphere. The weight of all of these wastes is summed to give:

$$W_{waste} = W_{solids} + W_{liquids} + W_{gases} \qquad (19.1)$$

The weight of the total output from the production process is determined from:

$$W_{total} = W_{product} + W_{by\text{-}product} + W_{waste} \qquad (19.2)$$

The waste measurement index (WMI) is a ratio determined from:

$$WMI = (W_{waste}/W_{total}) \times 100 \qquad (19.3)$$

Both Eqs. (19.1) and (19.3) are useful in measuring the effectiveness of a waste reduction effort. The weight of the waste released to the environment given by Eq. (19.1) is of interest to the surrounding communities and to management. However, the weight of the waste is affected by the production rate. If the production process is expanded from eight hours per day five days a week to twelve hours per day six days a week, the weight of waste generated is probably going to increase even if your waste reduction efforts have been markedly successful. However, the waste reduction index, given in Eq. (19.3) is not affected by the production rate because it is a ratio. Using a ratio gives a much better indicator of progress in a waste reduction program when the production rate varies with time.

Cost Analysis of Waste

In a waste reduction program one of the first tasks is to identify all of the sources of waste. While identifying sources seems obvious, it is easy to overlook some sources such as cleaning solvents, packaging materials, lubricating oils and greases, scrapped equipment and tools and other debris. The weight of these less obvious items of waste is measured and added to the weight of the waste as determined in Eq. (19.1). Finally, an estimate of the cost of the waste generated in a production process can be determined from the following factors.

1. Cost of raw materials to replace the wasted resources.
2. Labor costs to remove and treat wastes.
3. Storage costs to hold wastes at the production facility prior to final disposal.
4. Shipping costs to load and transport wastes to a reclamation center or a disposal site.
5. Disposal fees at the disposal site or at an incinerator.
6. Intangible costs such as insurance fees, regulatory costs, or the cost of EPA and other local regulatory agency issued pollution permits.

A cost analysis is essential for a waste reduction program because the expense of dealing with waste is much larger than most managers appreciate. Also, the EPA imposed regulations require extensive reports describing the quantities of waste generated and the methods used in their disposal. Preparing these reports requires time and effort. Regulatory costs, which exceeded $115 billion or 2.1 % of the gross domestic product (GPD) in 1990, are estimated to increase to 2.8% of the GPD by the year 2,000 [10].

A second benefit of the cost analysis of waste is in the identification of high cost waste streams. Maximum cost savings are realized when investments are made in equipment to reduce high cost waste. Employee training can be focused on elimination of these specific wastes. Research expenditures may be justified in developing by-products that consume these wastes.

19.4 AIR QUALITY

Perhaps the environmental issue that is most widely discussed in the news media is air quality. Stories of the effect of acid rain on the foliage of the North Eastern states are common. The pollution of air over many major cities such as Los Angeles, San Diego, Houston, Denver and Washington is a major concern to large segments of our population. The air quality index is reported daily in the weather section of most newspapers. These concerns lead Congress to pass the 1990 Clean Air Act that enables the U. S. Environmental Protection Agency (EPA) to establish limits on the amount of various pollutants allowed in the air anywhere in the U. S.

The 1990 Clean Air Act

The EPA has established two sets of limits on the amount of specified pollutants in the atmosphere—one set, called the primary standard, is to protect health. The other set, called the secondary standard, is to protect the environment. If one of the several EPA designated geographic regions in the U. S. exceeds either of these limits, it is classified as a non-attainment region; today about 90 million Americans live in non-attainment regions [11] indicating that much remains to be accomplished in reducing air pollution.

The 1990 Clean Air Act (CAA) vests the EPA with significant regulatory and enforcement power. The EPA requires industry, state and local governments to periodically report on emissions to the atmosphere. The CAA gives the EPA teeth and claws; they can fine violators about as easily as a police officer can write a traffic ticket. The fines are large and prison sentences[9] even for first offenders are common. The CAA defines seven categories of pollution offenders—one misdemeanor and six felonies. Moreover, the EPA and the Federal Bureau of investigation (FBI) have a memorandum of understanding. The EPA can call on the services of nearly 10,000 FBI field agents in 56 field offices to investigate potentially responsible parties. Clearly, cleaning up the air is intended to be a very serious business.

The EPA has established deadlines for the states, local government and industry to reduce air pollution. States are required to develop implementation plans outlining procedures for reducing specified pollutants such as carbon monoxide, sulfur dioxide, nitrogen oxides, particulate matter, etc. If in the opinion of the EPA, the state's implementation plan is deficient, the EPA has the power to enforce its mandates in that state.

The 1990 Clean Air Act is long and complicated. In this section, select aspects of the Act will be briefly described to demonstrate the magnitude of the effort currently underway to improve air quality and the environment in general. Today, it is estimated that industry spends more than $150 billion each year complying with regulations imposed on them by the EPA [10]. The costs of reducing air pollution are significant and as the EPA reduces the allowable limits in the near future[10], these costs will increase dramatically. The regulations and the burden of reports required by the CAA are costly and confining. In addition, industry must also deal with several other environmental laws including:

1. The Clean Water Act.
2. The Comprehensive Environmental Response, Compensation and Liability Act.
3. The Hazardous Materials Transportation Act.
4. The Resource Conservation and Recovery Act.
5. The Safe Drinking Water Act.
6. The Superfund Amendments and Reauthorization Act.
7. The Toxic Substances Control Act.
8. The Endangered Species Act.

Since each Act is the basis for a long stream of Federal and State regulations, the number of regulations limiting corporate activities has increased markedly in recent years. Regulations clearly affect business—compliance takes time, effort and significant capital investment. Indeed, in 1991 the Pharmaceutical Manufacturers Association stated, "If the current trend continues, environmental regulation, not FDA[11] approval, may cause the greatest delay in new drug introduction. The time to deal with regulatory agencies is also a major factor in the design of automobiles. Eaton [9] has stated, "Easily half of the three-year product introduction cycle can be consumed by acquiring air emission permits." Construction or modification of facilities needed for producing the new design may not be initiated until a permit is issued. Clearly, dealing with the bureaucrats not well trained in technology, lacking an understanding of the benefits of first to market or familiar with the principles of continuous improvement is a serious issue.

---

[9] Of the 65 criminal cases the EPA referred to the Justice Department in 1991, 55 convictions were obtained in the courts. The sentences resulted in a total of 62 years of imprisonment or an average of 1.8 years for each individual found guilty. This sentence is equivalent to the time given to those convicted of first time possession of heroin [11].

[10] In July of 1997, the EPA issued new regulations requiring states to dramatically reduce the amount of ozone and microscopic soot that they may allow in the air. In a recent lawsuit by a number of industrial groups, a three-judge panel of the U. S. Court of Appeals ruled that the EPA failed to justify the pollution levels selected as minimum requirements to protect public health. The ruling is being appealed by the Justice Department.

[11] The very long time required for approval for introducing new drugs by the Federal Drug Administration (FDA) is well recognized by the population in general and in particular by those anxiously waiting for a drug known to be effective that has been available in Europe for several years.

Auto and Truck Emissions

Much of the air pollution comes from three primary sources—auto and truck emissions, coal burning power plants and the exhaust gasses from large and small businesses. Let's consider the pollution generated by the internal combustion engines used to power autos and trucks.

When you burn gasoline or diesel fuel in an internal combustion engine, the chemical reaction produces hydrocarbon emissions that pollute the atmosphere. In some cities with high-density traffic and confining hills or mountains, smog forms which is always irritating and often dangerous. Reduction of hydrocarbon emissions has been addressed with considerable success using two different approaches. The first was to install catalytic converters on all new autos manufactured after 1971. These converters were very effective and reduced emissions by 95%. Unfortunately, the smog did not disappear; 50 million more autos were added to the national fleet since introducing the catalytic converters; each vehicle is driven more now than in 1971; and many of these new vehicles are pick-up trucks and sport utility vehicles[12] (SUVs) which consume copious quantities of gasoline. The pollution problem was mitigated with the converters, but it was not eliminated. Air pollution continues to be a serious concern as the population increases and becomes more affluent.

A second more costly approach is to reformulate the gasoline so that it releases a lower concentration of hydrocarbons in the combustion process. Makower [12] has estimated that the cost of reformulating gasoline to meet the emissions standards imposed in California will be in excess of $10,000 per ton of hydrocarbon emissions. This cost will escalate in a few years when more stringent standards are adopted. Catalytic converters, with an estimated cost of $1,000 per ton, represent a much more cost-effective method for reducing hydrocarbon emissions than gasoline reformulation.

There are other novel approaches to reducing emissions from the internal combustion engine. In 1990 Unocal Corporation, in an effort to reduce pollution in Southern California, purchased many old clunkers (pre 1971 automobiles without catalytic converters) from their owners for a flat price of $700. For less than $6 million, they removed 8,376 high-polluting automobiles from the roads and placed them in recycling centers (junkyards). Unocal environmental engineers estimate that this action was equivalent to removing 150,000 new automobiles from the road. The costs were relatively low at about $940 per ton of hydrocarbon emissions[13] (12.8 million pounds) removed from the atmosphere.

A very simple approach to reducing hydrocarbon emissions from trucks and autos is to reduce the number of miles driven and/or the amount of gasoline and diesel fuel consumed. This can be accomplished with one of the fundamental laws in economics—the law of supply and demand. Gasoline and diesel fuel is incredibly cheap in the U. S. when compared to the costs in Europe. The difference is about 300% (roughly $1.50 in the U. S. and $4 to $5 in Europe. The tax policy in Europe is to limit consumption of gasoline and the number of miles driven. In the U. S., the relatively low taxes imposed on gasoline and diesel fuel encourage consumption, unnecessary driving and the purchase of large gas guzzling vehicles. The effect of government policies on environmental issues will be discussed later in Section 19.6.

Power Plant Emissions

Many of the utility companies in the U. S. burn coal in the production of electricity that is distributed to homes and businesses. They consume an enormous quantity of coal; unfortunately, this coal is not a pure hydrocarbon. It contains small quantities of many different elements. Perhaps the most troublesome of these is sulfur. When sulfur is involved in the combustion process it reacts with

---

[12] It is interesting to observe the high percentage of SUVs driven by individuals who claim to strongly support environmental issues.

[13] At the same time Unocal was working to reduce hydrocarbon emissions from its refinery. Their engineers estimated that removing the equivalent weight of emissions from the refinery would have cost $160 million.

oxygen to create sulfur dioxide (SO_2). Scrubbers installed in the smokestacks of the utilities to remove the sulfur from the flue gases are effective in reducing, but not eliminating the release of SO_2 to the atmosphere.

The emissions from the smokestacks of utilities vary from one plant to another. Some utilities burn natural gas or oil with much lower sulfur content; consequently, they do not contribute significantly to the quantity of SO_2 in the atmosphere. Other utilities burn coal with a high sulfur content that overburdens the pollution control equipment installed in their smokestacks. Clearly in 1990, compliance with the provisions of the CAA was not uniform.

To address the problem of non-uniform compliance, the CAA enabled the EPA to issue permits permitting the coal burning utilities to release 2.5 pounds of SO_2 for each million of BTU[14] of heat generated. The idea in issuing permits is to have the market drive the efforts to reduce SO_2 emissions. If a utility releases less than the allotted amount of SO_2, it may sell its permits to a company that has difficulty in meeting the limits established. This practice rewards those utilities that have achieved early compliance and penalizes those not yet in compliance.

The Tennessee Valley Authority (TVA) [15] in 1995 made one of the first purchases of a pollution permit. This agency purchased from Wisconsin Power and Light the right to emit 10,000 tons of SO_2. The cost was nearly $3 million or about $300 per ton. The TVA basically purchased the time needed to procure and install more efficient scrubbers, to convert to a coal with lower sulfur content or to convert their boilers to burn oil or natural gas.

The CAA established standards to reduce SO_2 emissions by 50% by the year 2001. The costs to the utilities for new pollution control equipment to achieve this level of compliance have been estimated at $4 to $7 billion [12]. The pollution permit concept establishes a market that places a price on pollution. It provides an economic incentive for the utilities to purchase pollution control equipment to reduce their emissions of damaging and obnoxious gases.

19.5 THE KYOTO PROTOCOL CARBON EMISSIONS GOALS

The Kyoto Protocol was developed at the International Kyoto Climate Change Conference held in Japan in early December 1997. The objective of the conference was to accelerate the pace of international efforts under the United Nations Framework Convention on Climate Change. After intense negotiations, officials from 160 countries reached agreement on a legally binding Protocol under which industrialized countries would reduce their collective emissions of six greenhouse gases 5.2% below their 1990 levels by 2008-2012. For the U. S. these reductions are even more severe—a 7% reduction. The American Society for Mechanical Engineers (ASME)[16] has prepared a position paper on the technical implications of the Kyoto Protocol [13]. While the ASME does not take a position on the need to achieve the reductions specified in the Protocol, it has evaluated the feasibility of achieving these reductions in the U. S.

In 1990 the carbon emissions in the U. S. were about 1,346 million metric tons[17] (MMT); however, under a business-as-usual scenario carbon emissions are expected to increase by about 34% to 1,803 MMT by 2010. To meet the goals of the Protocol, these anticipated carbon emissions must be reduced by 551 MMT to 1,252 MMT by 2010. Is this a realistic goal knowing the propensity of

---

[14] The British Thermal Unit (BTU) is a measure of heat.

[15] The TVA is an agency of the Federal Government producing electrical power from hydroelectric dams, nuclear power plants, and coal fired plants.

[16] The 125,000-member ASME is a worldwide engineering society that deals with technical, educational, and research issues. It operates an extensive publishing operation, sponsors many technical conferences and professional development courses each year, and establishes many industrial and manufacturing standards.

[17] A metric ton is a unit of mass equal to 1000 kilograms. When converted to a force on Earth, it is 2205 pounds.

Americans to consume fuel? Does technology exist to reduce the amount of carbon employed in combustion processes?

The major sources of greenhouse gases are transportation, industry, buildings and electric power generation. Let's consider these sources individually, and summarize the findings of the ASME technical committee regarding each of them.

Transportation Sector

In a life-goes-on-as-usual scenario, energy consumption for transportation is expected to increase by about 2.4% per year until 2010. This estimate is based on expectations of continued low prices for fuel and for higher disposable personal income. With more money and cheap gas, people will buy more powerful and larger vehicles in increasing numbers and vehicle-miles driven will increase by an estimated 1.5% per year. At the same time, engine improvements will increase efficiency—but only by about 0.2% per year. The business expansion is anticipated to continue and the growth of the economy will result in an increase in freight transport with an attendant increase in the use of diesel fuel by 1.4% per year. Also, air travel was expected to grow during this period causing jet fuel usage to increase by 2.8% per year. However, since the tragedy at the World trade Center on 9/11, the airline industry has contracted and air travel has decreased in the U. S. and in other countries.

All segments of the transportation sector indicate a relatively large increase in the consumption of carbon. By 2010 the transportation sector is expected to generate 628 MMT of carbon emissions, which is an increase of 196 MMT (45%) above the 1990 level. These estimates consider technology, economic growth and human behavior. It is apparent that the carbon consumption for transportation is expected to increase at a high rate—not decrease in accordance with the Kyoto Protocol.

The transportation sector presents significant opportunities for advanced technology to reduce carbon emissions. Some possible areas for improved efficiencies or lower emissions are listed below:

1. Engine improvements—the direct-injection, stratified charge gasoline engine and turbocharged direct-injection diesel engine should provide efficiency improvements of 15 to 30%.
2. Hybrid electric vehicles—with a combined electric drive, an auxiliary power unit and an energy storage device offer opportunities for reductions of carbon emissions of up to 50%.
3. Advanced fuel cells—proton-exchange-membrane fuel cells that operate at relatively low temperature (180-300 °F) would offer significant reductions in emissions. Unfortunately, fuel cell development is in the research stage and may not become commercially feasible for several more years.
4. Alternative fuels—fuels such as compressed natural gas or propane would reduce carbon emissions by 10 to 20%. More effective would be a bio-fuel such as ethanol or methanol produced from feedstock. The uncertainties with this approach are the increased cost of alcohol relative to gasoline and the availability of sufficient quantities of feedstock materials to produce the large quantities needed.
5. A high-speed integrated light rail system—operating at 200 to 300 MPH— magnetically levitated or with steel wheel rail cars. This system would substitute electricity for gasoline or jet fuel while relieving both air traffic and highway congestion.
6. An effective information system—providing up-to-date information on traffic congestion and alternative routes would save time and reduce fuel consumption.

Industrial Sector

Small increases in carbon emissions are expected in the industrial sector in the coming decade. The trend of energy usage in this sector is toward electrification and a shift to energy produced by electric utilities. However, some larger industrial installations will continue to use coal to generate electrical power and to provide heat for manufacturing processes. Increasing use of cogeneration, where the steam from turbine driven generators is utilized as heat for production processes, will reduce carbon consumption. The challenge in cogeneration is not technology, but the necessity for coordination between industry, business and municipal governments. Another major area where reductions of carbon emissions are possible is by improving production processes and eliminating wastes as discussed previously in this chapter.

Building Sector

The carbon consumption in the building sector is for heating, air conditioning, lighting and other building services. The building sector accounts for about 1/3 of the total carbon emissions. With the business-as-usual scenario, the energy usage is expected to continue to grow with carbon emissions increasing to 612 MMT by 2010. This growth in consumption represents an increase of 33% from the 1990 levels—not a 7% reduction. Some of the methods to reduce carbon consumption in the building sector are listed below:

1. Advanced insulation technology—utilizing gas filled and evacuated panels with much higher R-values for construction.
2. Smart climate control systems—capable of adjusting temperatures automatically to accommodate actual usage of each room in a building 24 hours a day seven days a week.
3. Smart occupancy based light systems—capable of adjusting light intensity with voice-activated controls.
4. Advanced lighting systems—include more efficient light sources, efficient dimmable power supplies and highly reflective and efficient optical designs.
5. Cogeneration—small natural gas fired turbines to provide both the hot water and electricity required for individual facilities with improved efficiencies.

Electric Power Sector

The electric power sector emitted 477 MMT of carbon in 1990 and is expected to emit 663 MMT by the year 2010, which represents growth in demand for electricity of 1.25% per year. The fuels used to generate the power in 1995 were coal (54.3%), nuclear (21.8%), hydro (10.4%), natural gas (10.4%), and oil (2.1%). The prospect for reducing carbon emissions in the electric power sector is not promising. A significant fraction of the power produced is from nuclear plants, and many of these plants are nearing the end of their licensed life. If these plants are not relicensed, it is unlikely that electrical supply will satisfy the growing demand[18]. However, when a nuclear plant is not relicensed, it will be necessary to replace it with a new plant fueled with either coal or natural gas. The consequence is even higher carbon emissions than predicted.

More than half of the electricity generated in the U. S. is from burning coal. Power generation companies could convert some of these coal-burning units to natural gas; however, the demand for natural gas will increase its price significantly and seriously deplete the limited reserves. A more rational approach includes inspection and relicensing all of the nuclear plants deemed safe to operate, development of advanced gas and coal cycles and development of advanced nuclear plants. It appears that it will be necessary for society to decide among the following alternatives.

---

[18] It appears that the operating licenses of many of the nuclear power plants will be extended.

- Agree to accept severe reductions in consumption of both gasoline and electricity.
- Agree to the construction and operation of additional nuclear power plants[19].
- Agree not to conform to the limits set by the Kyoto Protocol.

Much has been written about the use of renewable energy technologies, but the hope of generating significant amounts of power from renewable sources is dim. Hydropower represents the best potential but the environmental permits required for adding turbines to existing dams will be difficult to obtain. Indeed, consideration is being given to removal of existing hydroelectric dams because they are harmful to fish. The application of solar energy suffers from its limited availability and the relatively low temperatures that can be generated from the sun without focusing devices. The potential for biomass in generation of electricity is limited by the availability of farmland needed for energy crop production. Wind energy is limited by available sites and by availability of cost-effective, energy-storage methods. The fact that new wind turbines for generating electricity are currently being installed in many parts of the U. S. is due to generous government subsidies based on the generator's electrical output.

Discussion

The limits imposed on carbon consumption by the Kyoto Protocol will be difficult if not impossible to achieve unless significant changes are made in government policy and the attitudes of the country's population relative to conservation. Achievement of the goals will require several strategies to be implemented in an extremely short time[20]. Because actions by the country's population and Federal, State and Local governments are not swift, there is little reason for optimism.

The transportation and electric power sectors afford the best opportunities for improvement. Some of the strategies for reduction of carbon emissions include:

- A carbon tax to reduce consumption.
- Relicensing nuclear plants to extend their safe life.
- Replace coal-burning plants with state-of-the-art natural gas combined cycle plants.
- Expansion of hydropower-generating capacity.
- In the long term, construction of new safe nuclear power plants.
- Development of hybrid vehicles with markedly improved efficiency.
- Development of electric vehicles with fuel cell power sources.
- Development of an effective information system that significantly reduces traffic congestion.
- Development of an integrated air transport and high-speed rail system.

To achieve sustainability, environmental impacts will have to be reduced and a balance between growth and the environment established. There are three approaches to accomplish this balance [14]:

1. Stabilize population.
2. Decrease wealth and affluence.
3. Apply existing technology to the fullest extent and develop new methods for producing goods and services with minimal impact on the environment and the resource reserves.

[19] Implicit in this agreement is the acceptance of one or more disposal sites for nuclear wastes.
[20] It should be recognized that the time to construct a new power plant today is about ten years with much of that time consumed in obtaining permits and dealing with environmental constraints.

Population stabilization does not appear to be feasible because of different political and cultural values among the populations of different countries. Decreasing wealth and hence consumption by design is also an unlikely event. People will not voluntarily reduce their standard of living. Only a crisis of major magnitude will result in reduced wealth and consumption without political upheaval. Indeed, the wisdom of this approach for achieving environmental objectives should be questioned. It is a well-known fact that birth rates decrease when standards of living and educational levels are raised.

By a process of elimination, it is clear that technology is the most suitable approach for achieving the balance between economic growth and the environment. Engineers must develop sustainable processes and systems that create, deliver and manage high quality goods and services.

19.6 GOVERNMENT POLICIES

The policies of the governments—federal, state and local—markedly affect the way we live and act. During World War II, in a state of emergency, the Federal government suspended production of almost all durable equipment including automobiles, rationed food, severely restricted gasoline consumption and drafted into the armed services nearly every physically fit man between the ages of 18 and 35. Clearly, the government has the power to markedly change the way we live and the amount of resources we consume. On the other hand, we have the power to vote them out of office.

In times of peace, the government affects the way we live using much more subtle methods. Tax policy is one of the most important and effective governmental tools. While taxes are levied to raise money to conduct the affairs of governments, these taxes also encourage or discourage certain activities by the population. For example, deducting interest cost for home mortgages from income on a federal tax return encourages home ownership. Similarly taxing income derived from interest on bonds, certificates of deposit, etc. discourages savings and encourages consumption.

The tax system at all levels of government is difficult to understand. By taxing—income, business profits, capital gains and interest from savings—governmental tax policy discourages many good activities, such as new investment and new business creation, which lead to a robust economy and prosperity. Moreover, the annual filing of the income tax forms for the Internal Revenue Service (IRS) has become a costly headache of major proportions. The Tax Foundation, a Washington non-profit research group, estimates businesses and individuals spend more than $125 billion annually on compliance [15]. Perhaps the federal government should lead the way in eliminating waste by markedly reducing the costly and wasteful process employed in the collection of taxes.

A New Approach to Taxation

Paul Hawken [1] has outlined a new approach for governments to follow in writing tax laws. Instead of taxing the productive activities such as producing income, capital gains, interest on savings, etc, the government should tax the harmful activities such as utilization of virgin resources, emissions, disposal of wastes and consumption of products and services. Of course, people and the governments need time to adjust to a new system; hence, Hawken proposes that the shift from one system to the other take place over a twenty-five year period. Also the shift would be revenue neutral. For every dollar collected with a green fee or green tax, taxes on the productive activities like earning income would be reduced by a dollar.

Expenditures by the Federal government for the fiscal year 2003 are estimated at $2.14 trillion. While this is a huge amount of money, it is only about 20% of the Gross Domestic Product (GDP) that is estimated at nearly $10.7 trillion. Transforming the tax system over a twenty-five year period would imply changes in tax collection rules that would amount to less than one percent of the GDP per year. This rate of change is very slow particularly when compared to drastic actions taken by the federal government during the crisis of World War II.

The advantages of taxing actions that deplete resources and degrade the environment are numerous. Green taxes would be included in the price a customer pays for consuming products and services. The taxes provide motivation for companies to improve their processes and the design of their product lines to reduce these costs. The consumer will recognize the added cost of the green tax and have the opportunity to modify their buying habits to avoid these charges. This approach gives the customer and corporation strong incentives for minimizing green taxes. In doing so, society will save virgin resources, reduce pollution and lower carbon consumption.

Gasoline Tax

On May 1, 2003 a barrel (42 gallons) of light sweet crude oil was priced on the New York futures market at about $25.24[21] per barrel or 60 cents per gallon. Shipping, refining and distribution costs including profit add about another 30 to 50 cents. Current state and federal taxes combined average about 34 cents per gallon taxes; thus, unleaded regular gasoline is sold at about $1.30 to $1.70 per gallon in most regions of the U. S. In other countries, the tax on gasoline is much higher—$2 to $3 per gallon. Americans love cheap gasoline; they buy large and heavy vehicles (SUVs and pickup trucks) that consume large quantities of fuel and they drive to their exercise facility. The problem is the high rate of consumption of finite virgin resources of petroleum and the emissions of smog producing hydrocarbons that result from this tax policy.

Currently, about 125 billion gallons of gasoline is consumed each year in the U. S. Suppose taxes on this gasoline were increased from the average of 34 cents to $2.00. The increased tax collections would amount to $207 billion per year if driving habits did not change. Increased gasoline taxes should markedly decrease consumption, conserving valuable reserves of petroleum and significantly decreasing pollution and carbon emissions.

Howls of protest were heard when Ross Perot, a presidential candidate in 1992, suggested increasing the gasoline tax ten cents a gallon each year for five years to raise the revenue needed to eliminate the deficit in the federal budget that existed at that time. People complained that it was not fair to those living in rural areas because they had to drive longer distances than those living in urban regions do. This may be true, but those living in rural areas have the opportunity to increase the prices of the products from their farms and ranches to better reflect the energy consumed in their production. Trucking firms complained that they would have to increase freight rates. Increasing freight rates is appropriate because the cost of a virgin resource (petroleum) would be more accurately assessed. Indeed, higher freight rates by the trucking industry would lead to increasing use of both rail and water transportation which are much more energy efficient methods of transport. It would also reduce the number of trucks on the interstate highways and reduce highway fatalities.

Paul Hawken [1] argues persuasively that "by imposition of incremental and eventually large green fees, businesses are positively encouraged not merely to meet regulations, but to embrace and exceed them, because the lower the green fees, the lower their costs".

Carbon Tax

The Kyoto Protocol, which constituted a worldwide effort to address the environmental problem of excess emissions of CO_2 has already been discussed. The U. S. is responsible for a large share of this problem. Today about 25 billion tons of CO_2 is released worldwide to the atmosphere each year. The U. S. is responsible for almost 25% of this total while supporting only 4% of the world's population.

---

[21] The price of crude oil on the world market varies from day to day. In the past five years, the price of crude oil varied from about $10 in 1998 to over $37 per barrel in 2003 before dropping to about $25 per barrel following the end of the conflict in Iraqi. Retail prices in some regions of the country do not always reflect true costs, and prices above the national average are common.

A careful study of uses of carbon in the U. S. by the ASME cast serious doubts on the country's ability to meet the 7% reduction on carbon emissions proposed by the protocol. Perhaps a better approach to reducing CO_2 emissions would be to implement a tax on carbon consumption. In the U. S., significant amounts of energy are wasted. Industries in Japan and Europe operate with much more efficiency using only about half the energy per unit output than comparable industries in the U. S. With significantly higher costs for energy in Japan and Europe there is much more incentive to strive for more efficient processes and practices.

Technologies for significant reductions in energy consumption exist, but often they are not fully implemented. Lighting is an excellent example of a newly developed technology[22] waiting for full implementation. The Electric Power Research Institute (EPRI) estimates that if highly efficient lighting were fully implemented nationwide, electricity required for lighting would be reduced by 50% [10]. Aggregate demand for electricity would be reduced by 10%, emissions of both CO_2 and SO_2 would be reduced and fuel would be conserved. The problem is that only a small portion of the population has taken advantage of these new lamps. Green taxes imposed on consumption of electricity would motivate both industry and ordinary households to convert to more efficient lighting.

Reducing energy consumption often requires a capital investment. With the very low cost of energy, which does not adequately assess external environmental costs, management and homeowners often decide to continue wasteful energy practices. With green fees or taxes, the external environmental costs would be reflected in higher costs of energy and individuals and companies would be motivated to invest in capital improvements to reduce energy consumption.

As green taxes and fees are increased, it is important to reduce taxes on income, profits and capital gains. The changes to the tax system should be revenue neutral. Legal tax avoidance opportunities provide incentives promoting investment in capital equipment to reduce energy consumption.

19.7 ENGINEERING RESPONSIBILITIES

Environmental issues are currently a serious concern to the engineering profession and will become even more significant in the coming decades. As the finite limits of resources becomes better understood by all concerned and the effects of carbon emissions on global warming become even more apparent, engineering efforts to improve efficiencies and to design more environmentally friendly products and processes will intensify.

In their efforts to design environmentally better systems, engineers serve three different constituencies—industry, governments and society as illustrated in Fig. 19.6. Unfortunately the signals received from each of the constituencies are mixed. Society clearly wants clean air and water, yet the population shows no desire to conserve. People buy bigger and bigger autos and drive more and more miles. Most folks spend every dime they earn. The government establishes the EPA and passes laws placing limits on pollutants released to the atmosphere. Yet the government pursues tax policies that encourage consumption and discourage conservation. It is like the Surgeon General warning of the dangers of smoking tobacco while the Agriculture Department continues to subsidize farmers growing tobacco.

The attitude and policies followed by industry is not uniform. Some companies like 3M have vigorously pursued environmentally friendly ways of doing business for the past thirty years. Other companies are still in the denial or avoidance stage; they do the minimum required by government-imposed regulations to avoid penalties.

---

[22] A 20-watt electronic compact fluorescent bulb provides light comparable to a 75-watt incandescent bulb and has a life of 10,000 hours.

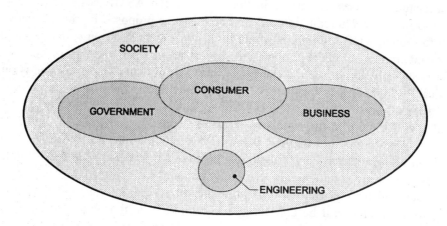

Fig. 19.6 Three constituencies served by engineers.

As a new engineer beginning the first assignment in your career it is important to carefully assess the position of your company relative to environmental issues. Perhaps they recognize the importance of eliminating wastes, making the production processes more efficient and designing lightweight energy efficient products with a long life. On the other hand, they may largely ignore environmental issues and focus on making profits without much concern for the environment. It is also important to assess the attitude of your manager. Without the boss on your side, giving his or her approval, you will not make significant progress on any ideas you have for changing the company's practices.

The actions you take in the first month or two on the job are very important because you establish your reputation. Do not try to become perfectly green. If you become known as a tree hugger, any attempts to initiate programs for eliminating wastes or making processes more efficient will not be given the attention they deserve. Observation is in order for the first few months to learn about the processes used and the company's attitudes and practices. When you understand the implications of waste generated in the processes and apparent inefficiencies, you can make suggestions for reducing wastes or improving the efficiency of a process. These suggestions should be accompanied with an analysis of the costs to indicate the saving accomplished by reducing wastes. If the suggestion involves investment in capital equipment, the cost analysis should incorporate the interest expenses associated with this equipment. There are significant economic returns associated with the elimination of wastes. The costs of disposal of wastes have escalated in recent years and new assessments of old waste problems may indicate more profitable solutions that are environmentally friendly.

Perhaps your first opportunity as an engineer will be in designing a new product or process or the modification of one or the other. In either case, the design process should include careful consideration of the environment. While you design for easy manufacturing, assembly and maintenance, it is important to add the environment to this list. Good environmental design is inherently good for business. If the use of virgin resources is minimized, usually the costs of producing a product are reduced. If the wastes are recycled there will be additional cost benefits. If the production process is made more energy efficient, costs are again reduced. If paper or plastics used in packaging the product are minimized, it costs less. There is a consistency between good environmental design, lower cost and increased profits. Well-managed corporations recognize this consistency.

19.8 SUMMARY

The fundamental environmental issue—sustainability—has been described. It is not possible to predict when shortages in virgin resources will develop, but it is a certainty that many of the world's resources are finite. Today the generation and disposal of wastes is one of the country's most severe environmental problems. It is serious for two reasons—the waste of valuable virgin resources and pollution of the environment.

Waste has been categorized into several different types and current methods for either storing or disposing of these wastes have been briefly described. The particularly acute problems associated with handling both hazardous and radioactive wastes were discussed. Many methods for handling wastes are known, but the best solution is to eliminate waste. Several waste models have been introduced beginning with the most wasteful (the linear model) and evolving to the concept of lifelong ownership. Laws, court actions, public attitude and the recognition of corporate responsibility will eventually force significant reductions in waste generation and the utilization of virgin resources. Methods for measuring progress of waste reduction programs based on both cost and weight of the waste generated have been outlined.

Congress has, in the past two decades, enacted several different environmental laws that constrain the activities of corporations operating in the U. S. Some aspects of the 1990 Clean Air Act (CAA) have been described because it is the most widely discussed of the many different environmental acts. The EPA is empowered to enforce the CAA. Offenders are usually charged with felonies, and large fines and significant prison sentences may result after a conviction. Corporations often consider the environmental acts a severe constraint; however, there is a growing tendency for the more progressive companies to recognize public attitudes, the intent of the law and the opportunities for cost reductions by eliminating wastes. Methods used in the auto and truck industries for reducing hydrocarbon emissions have been described. Also, the concept of issuing permits for SO_2 emissions for electric utilities, using market forces to promote the installation of pollution control equipment, has been discussed.

The Kyoto Protocol is an international effort aimed at reducing the amount of CO_2 released in the atmosphere. The concept is to relieve the greenhouse effect that is believed to produce global warming. The Protocol limits the carbon emissions from each country in the world. This basically amounts to restrictions placed on the industrialized countries since they are responsible for most of the carbon emissions. The results of an ASME position paper describing the impact of the Protocol on energy utilization for transportation, industry, buildings and electrical power have been presented. Strategies for reducing carbon emissions in the future were listed.

Paul Hawken's ideas about tax policies by the federal government have been described. Conversion of a tax structure from one that imposes penalties on the productive activities to one that taxes the harmful activities is presented. A green tax, imposed on the consumption of gasoline and electrical energy, was discussed. The economic incentives to reduce wastes and pollution and to conserve virgin resources with a new green tax policy have been described.

Finally, the dilemma that you may face upon entering the engineering profession has been introduced. You will serve three different constituencies and the message each provides is not consistent. It is recommended that you proceed carefully and avoid being classified by management and peers as an environmental extremist. With time and a favorable corporate policy regarding the environment, you will recognize opportunities for reducing wastes, designing lightweight efficient products, and conserving energy. Suggestions are presented for pursuing engineering activities associated with improving the environment and conserving resources.

REFERENCES

1. Hawken, P., <u>The Ecology of Commerce: A Declaration of Sustainability</u>, HarperBusiness, 1993.
2. Malthus, T. R., <u>An Essay on the Principal of Population</u>, 1798.
3. Weaver, G. R., <u>The Mainspring of Human Progress</u>, 13th Edition, Foundation for Economic Education, Irvington-on-Hudson, New York, NY, 1984
4. Frosch, R. A., "Editorial," *The Bridge*, Vol. 29, No. 1, Spring, 1999.
5. Makower, J., <u>Beyond the Bottom Line: Putting Social Responsibility to Work for Your Business and the World</u>, Simon & Schuster, New York, NY, 1994.
6. Anon, *Business and the Environment*, Vol. IX, No. 11, November 1998.
7. Akre, B. S., *Ford Opening Auto Parts Recycling Centers*, Knoxville, News-Sentinel, May 1, 1999.
8. Forward, G. and A. Mangan, "By-Product Synergy," *The Bridge*, Vol. 29, No. 1, Spring, 1999.
9. Eaton, R. J. "Getting the Most Out of Environmental Ethics," *The Bridge*, Vol. 29, No. 1, Spring, 1999.
10. Denton, D. K., <u>Enviro-Management: How Smart Companies Turn Environmental Costs into Profits</u>, Prentice Hall, Englewood Cliffs, NJ, 1994.
11. Anon, Plain English Guide to the Clean Air Act, EPA-400-K-93-001, April, 1993, EPA web site, http://www.epa.gov/oar/oaps/peg_cac/pegcaain.html.
12. Makower, J., <u>The e Factor: The Bottom Line Approach to Environmental Responsible Business</u>, Tilden Press, Times Books, Random House, New York, NY, 1993.
13. Anon, "Technology Implications for the U. S. of the Kyoto Protocol Emission Goals," Global Climate Change Task Force of the Council on Engineering and Council on Public Affairs, ASME, New York, NY, December, 1998.
14. Richards, D. J., "Harnessing Ingenuity for Sustainable Outcomes," *The Bridge*, Vol. 29, No. 1, Spring, 1999.
15. Tax Report, Wall Street Journal, November 22, 2000, p A-1.
16. Committee on Industrial Environmental Performance Metrics, <u>Industrial Environmental Performance Metrics</u>, National Academy Press, 1999.
17. Richards, D. J. and G. Pearson, Editors, <u>The Ecology of Industry</u>, National Academy Press, 1998.

EXERCISES

19.1 Prepare a graph showing the world population from 1,750 to 2,000. Fit an exponential equation ($P = Ae^{Bt}$) to this curve of population versus time. Finally, predict the world population in the years 2020, 2050 and 2100. Comment on sustainability of the population in the future. Note that P is the population, t time, and A and B constants in the equation shown in parenthesis.

19.2 Compare your prediction of the world population with the one provided by the government on the web site at http://www.census.gov/ipc/www/worldpop.html.

19.3 Prepare a list of items you consume per week and estimate the weight of these items. Compare the total weight with the average of 136 pounds per week consumed by a typical American. Did you take into account the gasoline used?

19.4 Where is the landfill that is used by the municipality in which you live? How many miles is the municipal solid waste shipped before being disposed of in this landfill? Is it possible to see or smell the landfill? What are the tipping fees at this landfill? What are the charges per month for the disposal service that takes care of the waste you and your family generates?

19.5 Does the municipal government where you reside collect yard waste? Do they chip tree limbs and branches to make mulch that is available to the citizens of the municipality?

19.6 Does the municipal government in your hometown require that you sort and separate the wastes from your home? If so, determine the final disposition of the newsprint, glass containers, steel and aluminum cans and plastics.

19.7 Identify the three corporations in your home state that are the major sources of toxic wastes. Hint, the EPA issues a Toxic Release Inventory (TRI) annually that may be accessed on their web site—http://www.epa.gov.

19.8 Where is the location of the closest site for the temporary storage of high-level radioactive wastes relative to the university that you are attending?

19.9 Where is the location of the closest site for permanent storage of high-level radioactive waste relative to your hometown?

19.10 Where is the location of the closest incinerator that burns:
- Municipal wastes;
- Hazardous wastes;
- Low-level radioactive wastes?

19.11 The data in Table 19.1 indicates that incineration of 1,000 tons of municipal solid waste results in an output in excess of 2,000 tons. Explain the doubling of the total weight of the waste product in the incineration process.

19.12 Prepare an engineering brief describing a linear waste model. Include in the brief an outline of a plan to reduce the wastes generated in the manufacturing process.

19.13 Cite the benefits of a waste model of a process that utilizes wastes as the primary resource for a by-product.

19.14 If you were the vice president of engineering, what policy would you adopt for the use of recycled materials in your product?

19.15 Do you believe engineers should redesign products so that recycled materials with lower performance characteristics than virgin materials would be suitable?

19.16 Search the literature and write a 200-word paper describing a successful program to reduce waste that was implemented by a corporation.

19.17 Write an engineering brief describing the implications of lifelong ownership of a product by the corporation producing it.

19.18 Write a paper describing the implications of lifelong ownership of a product and the possible benefits of lifelong ownership to society and the environment.

19.19 Write an engineering brief describing a new automobile that would be manufactured by the Forever Durable Corporation. Understand that this corporation has a cradle to grave philosophy and assumes ownership of the automobile over its entire life.

19.20 You have been in charge of a waste reduction program for two years. Data collected from the process control staff over this period are summarized in the table shown below.

PERIOD	W_{solids} (Tons)	$W_{liquids}$ (Tons)	W_{gases} (Tons)	$W_{product}$ (Tons)	$W_{by-product}$ (Tons)
1ST QTR	500	2,000	10	1,000	0
2ND QTR	600	2,400	12	1,200	0
3RD QTR	550	2,400	12	1,200	50
4TH QTR	600	2,800	11	1,400	100
5TH QTR	600	1,700	13	1,700	150
6TH QTR	800	2,000	15	2,000	200
7TH QTR	600	1,500	11	1,500	150
8TH QTR	725	1,800	13	1,800	175

Determine the waste management index for each of the eight quarters. Also determine the total waste generated each quarter. Write an engineering brief assessing the success or lack thereof of your waste reduction program.

19.21 Write a short paper describing the effectiveness of the Clean Air Act of 1990. In the paper give reasons for your opinions.

19.22 In your hometown or city, do you have any problems with air quality? What is the air quality index on a good day and on a bad day? Where do you find the air quality index?

19.23 Is the formulation of gasoline sold in your region of the country changed from time to time to improve air quality? If so does the price increase or decrease when the formulation is changed? Does the performance of your auto as measured by the miles per gallon improve or degrade with the new formulation?

19.24 Consider one of the recent environmental laws enacted by Congress, and write a paper describing its effect on industry located in your hometown. A list of these laws is given in Section 19.4.

19.25 How many gallons of gasoline do you consume per year when driving your automobile? Would you be willing to reduce this consumption? If so by what percentage?

19.26 Would you be in favor of gasoline rationing to improve the air quality in select cities?

19.27 Would you be in favor of gasoline rationing to save a valuable finite resource?

19.28 Review the usage of electricity at the College you are attending. Prepare a list of suggestions for reducing the consumption of electricity. The new practices that you suggest should not interfere with the educational process.

19.29 What is the name of the utility company serving your hometown? What fuel is used to fire the boilers in this utility? Are they considered as a significant polluter in your locality? Is it necessary for them to purchase SO_2 permits?

19.30 Write a paper describing both the advantages and the disadvantages of the law, which enables the EPA to issue pollution permits.

19.31 Describe the Kyoto carbon emission goals. State your assessment of the ability of the U. S. to meet these goals.

19.32 Prepare a list of improvements that will result in a reduction of carbon emissions:
  • in the transportation sector
  • in the building sector
  • in the electric power sector

19.33 Do you concur with Paul Hawken's philosophy that tax policies of the state and federal governments are misdirected? Please state the arguments for your conclusion.

19.34 Do you believe green taxes and fees in place of income taxes will result in the conservation of resources? Please state the arguments for your conclusion.

19.35 Upon graduation with a B. S. in engineering you begin to work for a typical corporation at a salary of $46,000 per year. Determine your federal and state income taxes assuming that you are single.

19.36 Upon graduation with a B. S. in engineering you begin to work for a typical corporation at a salary of $46,000 per year. Determine your federal and state income taxes—assume that you are married with no children and your spouse's income is equal to yours. Is the tax twice the amount you and your spouse would have paid as individual taxpayers?

19.37 Would you prefer to pay income taxes or green taxes? Do you believe the politicians will give you a choice?

19.38 Write a paper describing your position relative to environmental issues when you begin your career in industry.

19.39 In a recent article[23] Arthur Robinson and Noah Robinson, chemists at the Oregon Institute of Science and Medicine, wrote that global warming, which is the primary reason for the Kyoto treaty, is a long-term trend that began about 300 years ago. In the article they show a graph of temperatures over the past 3,000 years that indicates the current temperature is about one degree Fahrenheit below the average temperature over this period. They also show that in medieval times (1,000 AD), the temperature was about 3 degrees higher than the current temperatures. These conditions produced what is considered an optimum climate. Prepare a paper stating your position with respect to the Kyoto treaty that considers the information in this article.

---

[23] See article titled "Global Warming is 300-Year-Old News," The Wall Street Journal, January 18, 2000.

APPENDICES

APPENDIX A

DEVELOPMENT TEAMS

A.1 INTRODUCTION

The development of quality, leading edge, and high-performance products requires the coordinated efforts of many skillful individuals from different disciplines over an extended period of time. This group of individuals is organized within the corporation to act in an integrated manner to successfully complete the development. The organizational structure employed varies from one corporation to another; it also depends on the size of the company engaged in the development. For relatively new firms in the entrepreneurial stage, team structure is usually not an issue. These firms are small with a single product and only ten to twenty individuals are involved in the development process. Everyone on the payroll is deeply committed to this product, and communication is usually accomplished around the lunch table. On the other hand for very large corporations, where the number of employees may exceed 100,000, the company is often organized into divisions or operating groups. These large companies have many products or services that are offered by each division, and in some instances one division competes to some degree with another. For example, General Motors with its five automotive divisions produces many different models of each of its lines that compete for the same market segment.

In a large corporation, the total number of new products that are introduced each year may exceed 100. Communication is often very difficult because hundreds or even thousands of miles sometimes separate personnel involved in the development of a given product. Sometimes different divisions are in different countries, and business is conducted in different languages. Clearly, in a large firm, organization of the product development team and the physical location of its team members becomes extremely important.

In the past decade, there have been many examples [1] demonstrating that a cross disciplinary team provides a very effective organizational structure for product development. The idea of forming a cross disciplinary team to design a new product appears simple until one examines the existing organizational structure of all but the smaller of the product oriented corporations. Larger corporations are usually organized along functional lines. The organization chart presented in Fig. A.1 shows many of the functional departments that provide personnel needed on a typical product development team. Each department has a functional manager. Also, several departments are often grouped to form a division, and of course, each division has a director. Titles for the division leaders vary from firm to firm—but Director or Vice President are commonly employed.

The functional organization with its leadership and its ongoing operations presents a management problem when forming product development teams. As a team member, do you report to the team leader (product manager) or to the functional manager? When important decisions must be made, are they made by the product manager, by one or more of the functional managers or by the division directors? Clearly, the establishment of a product development team within a company that is organized along functional lines creates several operational problems for management. Functional organizations may also create problems for working engineers attempting to serve several different managers. In the next section, these problems are addressed, and a popular team structure employed in many corporations that are organized along functional lines is described.

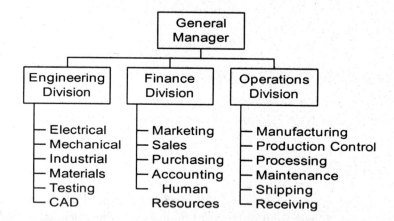

Fig. A.1 Typical organization chart showing the disciplinary functions. The organization includes three divisions and three levels of management.

A.2 A BALANCED TEAM STRUCTURE IN A FUNCTIONAL ORGANIZATION

There are several organizational arrangements frequently used in forming teams within companies organized along functional lines. They include:

- Functional teaming.
- Modified functional teaming.
- Balanced team structure.
- Independent team organization.

Each of these techniques for forming product development teams in a functional organization has advantages and disadvantages. A complete discussion of these four different team structures is given in reference [2]. To introduce you to the concept of forming teaming within a functional organization, the balanced team structure the will be described.

The balanced team structure is represented in the organization chart presented in Fig. A.2. The team is formed with members drawn from the functional departments; however, in most situations, the team members are located in close proximity in a separate room or a building wing reserved for the team. The team members are usually dedicated to a single project, and their responsibilities are coordinated and focused. The team members retain their reporting relationship to the functional manager, but the contact is less frequent than in functional teaming arrangements, and the relation is less intense. Members of a balanced team often consider the product manager to be more important than their functional manager. The product manager is frequently a senior technical administrator with more experience, status and rank than the typical functional manager. The status of the product manager is dependent on the product development costs and the size of the development team. The product manager reports to the general manager, and usually is equal in status to the division directors in the organization. The senior program manager with significant status and clout leads the team, and insures that resources will be available as required to meet schedules. The team members recognize the freedom from functional constraints that the balanced team structure implies. They assume ownership of the product specifications and commit fully to the success of the development activities. Communication is accomplished though the use of division and department coordinators and daily meetings. The team makes most of the routine decisions and only the higher-level decisions require the approval of the appropriate division directors.

The primary disadvantage of the balanced team structure is the dedication of the team members to a single project. Many talented team members are needed by the functional managers to support design activities across several product lines. With some of the functional resources dedicated to a single product development, other products may not receive adequate attention. The balanced team structure enhances the development capability of a given product at the cost of reducing the technical capability that can be applied across the company's entire product line. A second disadvantage is that some team members from a given department or division may not have sufficient expertise to perform adequately. With loose and distant functional supervision and review, this fact may not be apparent. The new design may not be at the leading edge of technology.

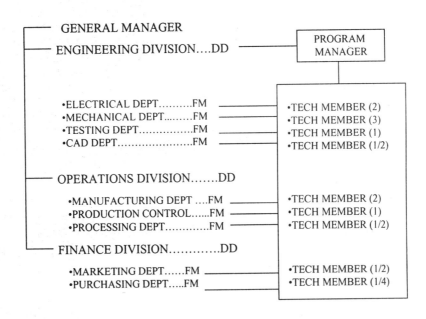

DD ....DIVISION DIRECTOR
FM.....FUNCTIONAL MANAGER
1/2......PART TIME (50%)

Fig. A.2 Organization with a balanced team structure.

A.3 DEVELOPMENT PLANNING

As the development team becomes more independent and moves away from the functional departments, planning becomes more important. Team planning documents are often prepared by a small group of senior team members with talents in business, marketing, finance, engineering design and manufacturing. These planning documents incorporate detailed descriptions of the product, market analysis, business strategy for capturing market share, product performance specifications, development schedules and budgets, team staffing requirements, material selections, design strategies, technology prerequisites, manufacturing processes, tooling requirements, inspection procedures, test specifications, facility availability and distribution capabilities. An element of the plan covers techniques and personnel that are to be employed to insure quality and product reliability. Finally, the plan includes a section specifying the deliverables and another section on methods to measure performance of the product and the productivity of the team.

The development planning documents are extremely important since they outline strategies for achieving the company's goals and objectives in developing a new product. Some of the documents may be used by management to monitor progress and to judge the team performance during periodic design reviews. The plan is a well-defined contract between the team and the division management involved in the product development. The team members are expected to sign-on; to commit to the level of effort required to achieve the goals, meet the schedules and to produce the deliverables. Management is expected to fund the program, to provide adequate staffing and to provide the necessary facilities, tooling and equipment in a timely manner.

A.4 STAFFING

Cross-disciplinary teams are necessary in staffing a product development team regardless of the structure employed by the organization. Usually, at least five disciplines or functions should be represented on the team including: marketing, finance, design engineering, operations or manufacturing, and quality control. The level of participation of each team member depends on the product being developed and the stage of the development. In some cases, team members work on more than one product and have responsibilities on two or more different projects. In other instances, several engineers from the same discipline are needed to complete long and complex tasks to meet the schedule for a single product. These members are usually dedicated to a single project.

Staffing on a product development team usually changes with time over the duration of the development period. Initially the team is small and staffed with senior personnel with demonstrated skills and talent. This small cadre may stay with the project for its duration. As the development proceeds, additional design engineers are required as technical strategies are converted into design concepts, then to design proposals and finally into detailed engineering drawings. Manufacturing engineers are needed in larger numbers when the early prototypes are being produced. After release of the design by engineering, the number of designers is reduced, and the emphasis is shifted to production where staffing from operations (production control, quality control, purchasing and plant maintenance) increases sharply. Clearly, the staffing of the development team is dynamic. Usually only a relatively small proportion of the total team is committed to the project on a full time basis from its initiation to its completion. In some companies, individuals are committed to the development team for the duration of the project, but at varying levels of participation. This approach enhances ownership of the project by the individual team members.

A.5 TEAM LEADERS

The experience and leadership skills required of the product manager are strongly dependent on the team structure employed as well as the size and cost of the product development. Some products can be developed with a relatively small commitment of staff (say less than 10 members). In these circumstances, the team established is often structured along functional lines, and the product manager is usually one of the functional managers. The functional managers are experts in their respective disciplines, skilled in managing personnel and knowledgeable (but not expert) in the other disciplines. They are also familiar with the company's entire product line.

Other product developments require a larger commitment of personnel (say a team numbering 10 to 50), and the balanced team structure shown in Fig. A.2 is probably the most suitable organization. The product manager is a seasoned staff member well-known for his or her expertise and respected for performing at a level that exceeds expectations. The appointment as a product manager (equal in rank and status to the functional department managers) often represents a promotion for the individual involved and a first experience in managing a technical group. The functional managers and the divisional directors provide close support through periodic reviews that monitor progress and reveal problems early in the development process.

As the size of development team becomes even larger, 100 or more members, team leadership assumes major importance. The product manager in this situation is a senior member of the company's executive management with a well-established record of successful accomplishments. He or she is senior to the managers of the functional departments, and is at least equal in status to the division directors. The authority and responsibility given the product manager is significant, and his or her career depends strongly on the success of the product in the market and the productivity of the development team.

A.6 TEAM MEMBER RESPONSIBILITIES

Members of a product development team have two different sets of responsibilities—one to his or her disciplinary function and the other to the team. A team member, from say the electrical engineering department, provides the technical expertise in his or her specialty and ensures that the product is correctly designed with leading edge, state-of-the-art technology. The team member acts to bring all of the important functional issues that affect product performance to the attention of the team. The team member also represents the functional department to ensure that the product development team maintains the principles, goals, and objectives of the functional discipline.

The most important team activity for the individual member is to share responsibility. The team member shares responsibilities with the others to increase the team's effectiveness and to insure its success. Next, the team member must recognize and understand all of the product features, and fully participate in the techniques employed to meet the design objectives. Individual team members must assess team progress and participate in improving team performance. The team member must cooperate in establishing all of the reporting relationships required to maintain the communication among the team and functional departments. In many of the team structures, an individual may report to three different managers—the Division Director, the Department Manager and the Program Manager. If a Functional Coordinator is assigned to the team representing a certain discipline, it is necessary to keep him or her informed of your progress and plans. The matrix organization structure, inherent in forming development teams, creates complex reporting relations. Flexibility and cooperation on the part of the individual team members are required for the team structure to be effective.

Finally, the team members must be able to communicate in a clear concise manner in all three modes: **writing, speaking and graphics**.

A.7 TEAM MEMBER TRAITS

When a group of individuals work together on a team to achieve a common goal, they can be extremely effective [3]. The team interaction promotes productivity for several different reasons. First, meeting together is synergistic in that one's ideas freely expressed stimulate additional ideas by other team members. The net result is many more original ideas than would have been possible by the same group of individuals working independently. Another advantage is in the breadth of knowledge available in the team. The team is cross-disciplinary and the very wide range of skills necessary for the product development process is included within the team. Each member of the team is different with some combination of strengths and weaknesses. Acting together the team can build on the strengths of each member and compensate for any weaknesses. The grouping of individuals provides social benefits that are also very important to the individuals and to the corporation. There is a bonding that occurs over time, and a support system develops which team members appreciate. Feelings of trust and mutual understanding, important to everyone involved, develop over the duration of the project. The team develops a sense of autonomy valued by each of its members. The team develops solutions, implements them and then monitors the results with little or no direction from

428 — Appendix A
Development Teams

management. Empowered teams assume ownership of the project. They commit to the product development process and consistently perform beyond expectations.

While cross disciplinary teams have been employed in many different companies for a decade or more, little has been done by the educational system to develop team skills. On the contrary, both the secondary school and many college systems tend to encourage competitive attitudes and independence. Students compete for the few "A" grades that might be given in a course, and are discouraged from cooperating on assignments. This system prepares students to work independently, but team-building skills are not addressed.

The industrial workplace is much different. Team members compete, but not within the team. They compete with a similar development team from a rival corporation. Team members bond, cooperate and consistently help each other by sharing assignments. Competition between team members is discouraged. Recognition and rewards go to the team as a whole much more often than to select individuals.

There is a set of characteristics that describe a good team member, and another set that depicts an individual who can destroy the efforts of a team.

The characteristics of a good team member are:

1. Treat every team member with respect, trust their judgment and value their friendship.
2. Maintain an inquiring attitude free of predetermined bias so that team members will all participate and share knowledge and opinions in an open and creative manner. Listen carefully to the other team members.
3. Pose questions to those hesitant team members to encourage them to share their knowledge and experience more fully. Share your experiences and opinions in a casual, easy-going manner. Help other members to relax and enjoy the interactive process involved in team cooperation. Participate but do not dominate.
4. Observe the body language of the other team members because it may indicate lack of interest, defensive-attitudes, hostility, etc. Act with the team leader to defuse hostility and to stimulate interest. Disagree if it is important, but with good reason and in good taste.
5. Emotional responses will occur when issues are elevated to a personal level. It is important to accept an emotional response even when you are opposed. It is also important to control your emotions and to think and speak objectively. You should be self confident in your discussions, but not dogmatic.
6. Seating arrangements once established tend to become permanent. It is productive to break these patterns, change the seating pattern and to mix team members. The result, over time, is to enhance bonding and to stimulate new ideas generated by different combinations of members.
7. Allocate time for self-assessment so the team can determine if it is performing up to expectations. Make suggestions during these assessment periods to build team skills.
8. Be comfortable with your disciplinary skills but continue to study to improve your capabilities. Communicate effectively by speaking clearly in good English, writing clearly and concisely and illustrating with modern graphics.

The characteristics of a destructive team member are:

1. No member of the team should ever participate in a conversation that is derogatory about a person on the team or in the corporation. If you cannot make complimentary remarks about a person, keep the negative thoughts to yourself. Respect, trust and friendship are vital elements in the founding of a successful product development team. Derogatory remarks destroy the foundation necessary for respect, trust and friendship.

2. Arguments among team members are to be avoided. You are encouraged to introduce a different opinion and participate in a discussion with a different viewpoint, but the discussion should never degenerate into an argument. Every member on the team is responsible for quickly resolving arguments that arise among the team members.
3. All members of the team are responsible for on-time attendance at their meeting. If a member is absent, the leader should know the reason beforehand and explain it to the team. You must be certain that all members appreciate and respect the reason for everyone's role on the team and the importance of every member attending every meeting.
4. Team progress is hindered when one member dominates the meeting. These people are often overly critical, intimidating and stimulate confrontations. If you have these characteristics, work hard to suppress them.
5. The team leader must be extremely careful about intervention. If the team is moving and working effectively, the leader should be quiet because intervention in this instance is counter productive. When the team is having difficulty, intervention may be necessary depending on the problem. In the event a team member becomes hostile, intervention should be quick, and the disagreement producing the hostility should be dealt with immediately. When a team is seeking consensus, the process may require time and the leader should wait 5 to 10 minutes before intervening.
6. No member of the team can be opinionated including the leader. He or she is not a judge determining the correct solution. The leader seeks to facilitate so that the team can reach a consensus on the correct solution. It is only when the team accepts the solution that implementation can begin.

Compare your traits with those listed above. You will probably exhibit both good and bad characteristics. To be effective team members, you must work to enhance the favorable traits and to suppress those that are destructive to the efforts of the team.

A.8 EVOLUTION OF A DEVELOPMENT TEAM

If a number of workers or students are assigned to a new team, it is possible to observe several behavioral phases as the team bonds, melds and matures [4]. In the very beginning (Phase I), the team members usually exhibit both excitement and concern. They are excited about working on a high-profile project with a new group of talented people. It should be an opportunity to learn, make new friends, advance in position and/or stature and have fun in the process. At the same time, the team members exhibit signs of concern. They are worried about meeting and understanding other team members. The tasks assigned to the team are probably extremely vague at this stage of the development, and vagueness leads to uncertainty. The skill levels and areas of expertise of fellow team members have yet to be determined. The personality of the team leader is unclear. There is worry over the new reporting relations that the new team membership implies. In this orientation phase, the team members are searching for their role in the team and evaluating their possibilities for success or failure. It is a tense time.

Dissatisfaction is the second phase in the evolution of the team. You have met your fellow team members, and the news is not good. You recognize the differences in personalities and in work habits. A few of the team members do not understand how to tell time and never arrive on time when attending team meetings. Member schedules make it very difficult for the entire team to meet. Some team members lack minimal social graces. Some members are inexperienced and unable to cope with their assignments. You have met the team leader and he or she is very difficult (stressed, demanding, exacting, impatient and abrupt). The schedule calls for the team to leap over mountains on a daily basis. The budget for the project is totally inadequate. Surely you are in a lose—lose situation. Your thoughts are dominated by schemes to transfer to a better team. During this phase, the progress of the team toward scheduled milestones is exceedingly slow.

With time progress is made, and the team members start learning to work together. This is the resolution phase (Phase III). Everyone agrees to attend meetings and arrive on time. Personality conflicts are smoothed over. Team members may never be true friends, but everyone manages to get along and show mutual respect. Team members have come to terms with the schedule, and have committed to the extra effort that it requires. The manager has mellowed, or team members have learned how to handle his or her impatient demands. Solutions for reducing development costs have been found. There appears to be a good possibility of meeting the product development plan. The team is beginning to perform well, and more members are committed.

The team has melded into an efficient unit. Conflicts rarely arise. Lasting friendships begin to be formed. The strengths of each team member are fully utilized. The manager recognizes the talents of the team. Tasks are completed ahead of schedule and under budget. Upper level managers recognize the accomplishments of the team. You now find yourself in a win—win situation. This is the production phase of the development team.

The project has been completed and the ribbons have been cut. The team prepares to disband. In this termination phase (Phase V), you reflect on your experiences, good and bad, over the duration of the project. You examine your individual performance and evaluate various factors that improved the effectiveness of the team. You meet with the manager for a performance evaluation. The team meets to share a sense of accomplishment in achieving the goals and objectives of the development.

A.9 A TEAM CONTRACT

Teams are effective because they meet together to focus their wide range of disciplinary skills and natural talents toward the solution of a well-formulated problem. The team meeting is the format for synergistic efforts of the team. Unfortunately, not all teams are effective because some members are so disruptive they destroy the cohesiveness of the team. Bonding of team members is vital if the team is to successfully solve the multitude of problems that arise in the product development process.

Teams **fail** because of three main reasons. First, they deviate from the goals and objectives of the development, and cannot meet the milestones on the development schedule. Second, the team members become alienated and bonding, trust and understanding, critical to the success of a team, never develop. Third, when things begin to go wrong and adversity occurs, "finger pointing" begins and an attempt to fix blame on someone replaces creative actions.

Clearly, many problems will arise to impair the progress of a development team. It is recommended that the team establish a set of simple rules for working together. Suggestions for your team to consider are provided below. The members should meet and establish a set of rules it considers necessary to govern the team.

- The information discussed by team members will remain confidential.
- Members will acknowledge problems and deal with them.
- Members will be supportive rather than judgmental.
- Members will respect differences.
- Members will provide responses directly and openly in a timely fashion.
- Members will provide information that is specific and focused on the task and not on personalities.
- Members will be open, but will respect the right of privacy.
- Members will not discount the ideas of others.
- Members are each responsible for the success of the team experience.
- The team will have the resources needed to solve the problems that arise.
- Every member of the team will contribute to its success.

- Members will try to know and understand fellow team members and identify ways for enhancing everyone's professional development.
- Members recognize the importance of team meetings to the success of the group, and accordingly agree to:

1. Always attend the scheduled team meetings.
2. When attendance is impossible, notify the team leader and as many other members as possible in advance.
3. Use meeting time wisely.
4. Start on time.
5. Limit breaks and return from them on time.
6. Keep focused on the goals.
7. Avoid side issues, personality conflicts and hidden agenda.
8. Share responsibility for briefing members when they miss a meeting.
9. Avoid making phone calls or engaging in side conversations that interrupt the team.

Writing a team contract has been included in this chapter in Exercise A.19. It is recommended that each team prepare a contract and each team member sign and date it.

A.10 EFFECTIVE TEAM MEETINGS

Leadership is important in ensuring that the team remains focused on the overall project goals and objectives and on the agenda at each meeting. Creating the correct team environment or atmosphere is essential so that the team members cooperate, share ideas and support each other to achieve solutions [5]. It is important for the team leader to assess the performance of the entire team at each meeting. When the team fails to make progress, it is critical for the team leader or possibly an individual member to make changes as necessary to enhance the team's effectiveness. There are several measures of team effectiveness that may be exercised by anyone on the team during a typical team meeting. Corrective actions to improve team performance are not difficult to implement. Some typical difficulties encountered during team meetings are identified in the following paragraphs. Also actions by members for enhancing the team's productivity are suggested.

One common problem impairing team productivity occurs when the goals and objectives of the meeting are not clear or they appear to be changing. While in many cases team meetings are planned well in advance and are supposed to follow an agenda, new topics are introduced and the meeting drifts from one item to another. This team behavior indicates that either some of the team members do not understand the goals, or they do not accept them. Instead, they are marching to their own drummer and trying to take the team with them. Unless corrective actions are taken to focus the team on the original goals, this team meeting will fail and time and effort will be lost. A focused team stays on the agenda, and its team members accept the meeting objectives. Team discussions deal with the issues important in generating a wide set of solutions to the problem being considered. The discussions continue until team members reach a consensus. This meeting is successful because an issue has been resolved and all of the team members accept the solution. The team can move forward and organize its talents to address the next problem.

Team leadership is an essential element for its effectiveness. The most effective teams are usually democratic with a large amount of shared leadership. While there is a recognized leader, with a considerable degree of responsibility and authority, the team leader is supported by each team member. The team member with the appropriate expertise (relative to the problem being considered) will usually lead the discussion and effectively act as leader of the team during this period. However, in some instances, the team leader maintains strict control of the meeting and never shares the leadership role with any team member. Some team members resent this style of leadership and may not participate as fully as possible. The result is that complete utilization of the resources of the team does not occur. The "tightly controlled" team leader will almost always insist on sitting at the head of the table.

Attitude of the team members is an important element in the success of the team. Are the team members committed? This commitment is evidenced in their attitude. If they come to the meeting table exhibiting interest and willingness to participate, they are committed and will make a positive contribution. However, if they are bored and indifferent, they will not become engaged in a meaningful way. In fact, if they are sufficiently disinterested, they may initiate side conversations or arguments and destroy the atmosphere of trust necessary for an effective meeting. It is important to evaluate each team member, and to secure their commitment to the goals and objectives of the project and to the agenda for each meeting.

As the meeting progresses, it is important to assess the discussions that are occurring. For a meeting to be successful, the discussion should include everyone on the team. A discussion should stay focused on an agenda item with few if any deviations to unrelated topics. The team members must listen carefully to one another. All of the ideas presented are given a serious hearing. No one is intimidated or is made to appear foolish for making suggestions that may appear to be too radical. There is an informal and relaxed attitude exhibited by the participants. Teams tend to accomplish less when the discussion is dominated by only a few of the participants. They may talk well, but they usually do not bother to listen to others. They may introduce a topic that is not on the agenda, and ignore suggestions for getting back on track. These members tend to intimidate others in their efforts to control and dominate the team. The result is disastrous because they essentially eliminate the contributions of other team members.

Teams do not operate with total agreement on every issue. There must be accommodation for disagreements and for criticism. All of the ideas or suggestions made by every team member at any meeting will not be outstanding. Indeed, some of them may be ridiculous; criticism of these ideas must occur. However, the criticism should be frank and without hostility. Personal attacks must not be a part of the critique of an idea. If the criticism is to improve an idea or to eliminate a false concept, then it is of benefit to the progress of the team. The criticism should be phrased so that the team member advancing the flawed idea is not embarrassed. When disagreements occur, they should not be suppressed. Suppressed disagreements breed hostility and distrust. It is much better for the team to deal with disagreements when they arise. The root causes of a disagreement should be ascertained, and the team should take the actions necessary to resolve them. On some occasions voting is a mechanism used to resolve conflict. This procedure must be used with care particularly if the vote indicates the team is split almost equally in their opinion. A better practice is to discuss the issues involved and attempt to reach a consensus. The time for a team to reach consensus, may require more time than a simple vote, but the results are worth the effort. Consensus implies that all members of the team are in general agreement and willing to accept the decision of the team. When the team votes, a simple majority is sufficient to resolve an issue. However, the minority members may become resentful if they are always on the short end of the vote. In a very short time, they will not accept the outcome, and they will not commit to the actions necessary to implement the team's decision.

The team meetings must be open. The agenda should be available in advance of the meetings and subject to change with added topics introduced as "new business". A sample agenda is shown in Table A.1. Team members should believe that they have the authority to bring new agenda topics before the team. They should feel free to discuss procedures used in the team's operation. Hidden agenda items detract from the harmony and trust developed through open and fair operation of the team. Secret meetings of an inside group should be carefully avoided. When the fact leaks that secret meetings are being held and that issues are prejudged by a select few, the effectiveness of the team is destroyed.

Table A.1
Sample agenda
Weekly meeting of the Tiger Team
September 10, 2003 7:59 am

1. Weekly status report — Team Leader
2. Review of outstanding action items
   - Action item #14 — Member responsible
   - Action item #15 — Member responsible
   - Action item #N — Member responsible
3. Report on progress
   - Subsystem –Power — Member responsible
   - Subsystem –Electrical — Member responsible
   - Subsystem – Mechanical — Member responsible
   - Subsystem –Control — Member responsible
   - Market study results — Marketing representative
4. Identify new problems — All members participate
5. Assignment of action items — Team leader and volunteers
6. New business — All members participate
7. Summary — Team leader
8. Adjourn at 8:59 am

At least once during a meeting it is useful to evaluate the progress of the team. Is the team following the agenda? Is someone dominating the meeting? Is the discussion to the point and free of hostility? Is everyone properly prepared to address his or her agenda items? Are the team members attentive, or are they bored and indifferent? Is the team leader leading or is he or she pushing? If a problem in the operation of the team is identified during the pause for self-appraisal, it should be resolved immediately through open discussion.

Table A.2
Action items
Tiger Team—Meeting of September 10, 2003

Decisions made	Follow-up required	Responsible member	Date complete
1.	1.	1.	1.
2.	2.	2.	2.
3.	3.	3.	3.
4.	4.	4.	4.
Action at next meeting	**Preparation required**	**Responsible member**	**Date complete**
1.	1.	1.	1.
2.	2.	2.	2.
3.	3.	3.	3.
4.	4.	4.	4.

Finally, the team must act on the issues that are resolved and the problems that are solved. To discuss an issue and to reach a consensus is part of the process, but not closure. Implementation is required for closure, and implementation requires action. When decisions are made, team members are assigned action items. These action items are tasks to be performed by the responsible individual;

only when the tasks are completed is an issue considered closed. It is important that the action items are clearly defined, and the role of each team member in completing their respective tasks is understood. A realistic date should be set for the completion of each action item. The individual responsible should be clearly identified; he or she must accept the assignment without objection or qualification. A checking system must be employed to follow up on each action item to insure timely completion. If there is a delay, the schedule for the entire product development cycle may be at risk. It is important to deal with delays immediately and for the team to participate in the development of plans to eliminate the cause of the delay. A form in Table A.2 has been provided to facilitate the assignment of action items and the follow-up on each item.

A.11 PREPARING FOR MEETINGS

Effective team meetings do not just happen. It is necessary to prepare for the meeting and to execute post-meeting activities to insure success. The preparation usually involves selecting, arranging and equipping the meeting room, scheduling the meeting so that the necessary personnel are in attendance, preparing and distributing an agenda in advance of the meeting and conducting the meeting following a set of rules accepted by the team. Equally important are actions taken by the team members following the meeting to implement the decisions that have been reached.

The meeting room and its equipment also affect the team's progress. The room should be sized to accommodate the team and any visitors that have been invited. Rooms that are too large permit the members to scatter and the distance between some members becomes too long for effective and easy communication. The chairs should be comfortable, but not so soft as to promote napping. The seating arrangement should be around a table so that everyone can observe each other's face and body language. It is very difficult to engage anyone in a meaningful conversation if they have their back to you. Water, coffee, tea or soft drinks should be available if the meeting duration exceeds an hour. Smoking is strictly prohibited. The equipment that will be needed for presentations should be available, and its operation should be checked prior to the start of the meeting. Flip charts and white boards that are useful for recording the key ideas generated during the meeting are essential. Tape, pens, markers and other supplies necessary for preparing and displaying charts should also be available. A computer with projection capability is of growing importance in a well-equipped meeting room. The availability of a computer during the meeting permits one to draw from a large database and to modify the presentation in real time. Also results of analyses or experiments can be displayed in graphical form for the entire team to review. Finally, adjust the room temperature until it is comfortable. Check to ascertain if the light intensity can be controlled from low for projecting overheads to high for round table discussions.

Prior to the scheduled meeting, it is important to make careful preparations. Minutes from the previous meeting should be distributed with sufficient time for review before the next scheduled meeting. A detailed agenda, similar to the one shown in Table A.1, is distributed to inform team members of the topics and/or problems that will be addressed in the next meeting. Individual team members responsible for specific agenda items are identified. The details are clear to all and responsibility shared by individual members is defined. In some instances, information will be needed from corporate employees, suppliers or visitors that are not members of the team. In these cases, arrangements must be made to invite these people to the meeting so that they can provide the necessary expertise and respond to questions from all of the disciplines represented on the team.

In conducting the meeting, it is important to start on time. It is very annoying to the majority of the members to wait five or ten minutes for a straggler or two. The team leader must make certain that everyone involved understands that 8:00 am means 8:00 am and not 8:08 or 8:12 am. The objectives of the meeting should be clearly understood and the time scheduled for each agenda item should be estimated. A team member should be assigned as the timekeeper, and another should act as a secretary to record notes and to prepare the minutes of the meeting. If team meetings are frequent and held over a long time (several months), the duties of the timekeeper and the secretary should be shared with others on the team. The meeting should follow a set of rules that govern the behavior of individual members. An assignment involving the preparation of a list of rules that should be followed in conducting an effective meeting is given in Exercise A.17. As the meeting draws to a conclusion, the team leader should take a few minutes to summarize the outcome of the discussions. This summary gives an ideal opportunity to insure that assignments are understood, responsibility accepted and completion dates established. The meeting must be completed on time and everyone's schedule should be respected. If the meeting is not periodic (e.g. every Tuesday at 8:00 am), then it is important that the time and place for the next meeting be scheduled.

After the meeting, it is productive for the team leader to make certain that the complete minutes have been prepared and distributed. If any member was not able to attend the meeting, the team leader should brief that person on the team's progress. Finally, the team leader should follow up on each action item to ensure that progress is being made, and that new or unanticipated problems have not developed. It is clear that effective meetings do not happen by accident. Many members of the team work diligently before, during and after the meeting to make certain that the goals and objectives are clearly defined, that the issues and problems are thoroughly discussed and that the solutions developed result in assignments that are executed with dispatch.

A.12 POSITIVE AND NEGATIVE TEAM BEHAVIOR

All the members on a team and particularly the team leader must act in a positive manner for the team to function effectively. Frequent negative behavior by a team member is detrimental and inhibits progress and delays the product development cycle. Group behavior is often divided into three categories including [6]:

- Leading
- Supporting
- Hindering

The leading and supporting roles are consistent with team progress and positive contributions to the development project. However, the hindering role is destructive as it impedes progress and causes conflict and concern to many team members.

Leading Roles

Any team member can play a leading role during a team meeting. Of course, you expect the team leader to provide direction, but leadership is shared in a high-performance development team. Leading tasks include:

1. **Initiate:** Suggest ideas or concepts for the design. Propose objectives or goals. Define a problem and give an approach for its solution. Raise issues that are important to the progress of the team. Recommend procedures for completing tasks.
2. **Provide:** Give relevant information pertaining to issues before the team. Offer suggestions, ideas and theories. Furnish facts and background knowledge. Present opinions with supporting material.
3. **Seek:** Ask for suggestions and ideas. Solicit options. Request facts relative to a team problem. Look for information leading to resolution of team concerns.
4. **Clarify:** Indicate alternatives and issues for the team. Provide examples to support another's statement. Interpret ideas and suggestions. Define issues and terms. Offer illustrative conclusions.

5. **Test:** Check to ascertain the degree of agreement within the team. Inquire to determine if the team is prepared to reach a consensus.
6. **Summarize:** Offer conclusions or decisions at the close of a team discussion. Gather together related ideas offered by several team members. Restate suggestions after team discussions.

Supporting Roles

Many team members are not always comfortable with the leading roles. However, they may make significant contributions in a supporting role. Team members providing support engage in the following activities:

1. **Monitor:** Assist in maintaining open communication among team members. Recommend procedures for sharing ideas. Promote the participation of less active team members.
2. **Encourage:** Respect other team members by providing opportunities for their recognition. Appear friendly, warm and responsive to all team members. Accept others and recognize their contributions.
3. **Compromise:** When team conflicts arise, offer a compromise approach. Admit errors readily. Modify ideas or concepts in the interest of team cohesion and growth.
4. **Sense:** Share your feelings with the team. Identify feelings, moods and relationships within the team.
5. **Harmonize:** Act to defuse conflict. Reconcile disagreements. Reduce tension and stress. Encourage team members to explore their differences.
6. **Testing:** Ascertain if the team is satisfied with the procedures in place. Suggest new procedures or guidelines to improve the operation of the team.

Hindering Roles

Unfortunately, sometimes you will find "lemons" in a group. These are team members who do more harm than good. Behaviors indicative a team member impeding progress include:

1. **Uncooperative:** Stubborn resistance of ideas of others for personal motives. Disagrees for no apparent reason. Employs a hidden agenda to deliberately impede the advancement of another's idea.
2. **Degrading:** Jokes in a sarcastic manner. Hostile criticism of ideas and suggestions of others. Makes comments harmful to the esteem of other team members.
3. **Dominating:** Interrupts other team members when they attempt to contribute. Asserts authority to manipulate one or more team members. Controls some team members by manipulation.
4. **Avoiding:** Absent or late to team meetings on many occasions without valid justification. Frequent changing of the topic under discussion. Uncomfortable with conflict.
5. **Withdrawing:** Often avoids participation in team activities. Offers few if any suggestions, ideas or recommendations. Responds to questions with abrupt and brief answers.
6. **Distracting:** Ignores the team meeting and begins side conversations with other team members.

Most of team members are capable of playing all three roles at any time. You may help or hinder the progress of the team. It is important that you pause occasionally during a team meeting to assess the roles played by each team member. An assessment form has been provided in Exercise A.20 to facilitate recording data from your assessment.

A.13 SUMMARY

Arguments for the importance of development teams in industry have been presented. Experience has shown that effective development teams are vital if a corporation is to develop highly successful products and introduce them in time to win a major share of the market.

Functional organizations that exist within most corporations have been described. Reasons why a functional organizational structure often interferes with rapid low-cost product development have been given. The balanced team structure commonly employed in establishing development teams within large functionally organized corporations has been described.

The importance of location of team members in either enhancing or inhibiting communication has not been considered. However, you should be aware of the important role distance plays in effective and timely communication. The rule is simple. To enhance communication, minimize the distance between those needing to communicate with each other. The ideal situation is to place the entire development team in the same room.

Development planning and staffing of the development team has been covered. The success of the team depends strongly on generating a comprehensive plan. The plan is used initially to justify the development budget and its schedule. During the course of the development, the plan provides guidance for monitoring the progress and for judging both the team and product performances. Team staffing is cross-disciplinary with adequate representation from both the engineering and finance divisions. Engineers interact with the business personnel from marketing, sales and purchasing. Responsibility is shared between disciplines and functions.

The team's organizational structure and the team leader's management status have been discussed. This relationship is important because of the executive authority that corresponds with a higher level of management. Higher-level managers carry more authority, many decisions can be made more rapidly, communication is more effective and team progress is often enhanced if the leader has status and clout associated with executive management.

Next, team member responsibilities and team member traits were discussed in considerable detail. Team members work in a matrix organization and often report to two or more managers. This arrangement is sometimes difficult particularly when you get mixed signals from different managers. It is important to be flexible and remain cooperative with the program manager and the functional managers. Team member traits are very important to your career. Evaluate your traits in an honest self-assessment. This is a critical first step in building team skills.

Most design teams evolve over the duration of a development project. Five phases usually experienced by the team in this evolution have been described. Not all of the phases provide pleasant experiences. Early in the evolution, a team encounters the dissatisfaction phase with very slow progress, low productivity and member gloom. To minimize the time in this phase, it is suggested that the team prepare and execute a contract that provides guidelines for member behavior.

Finally, guides for conducting successful team meetings have been provided. The progress of the project is dependent on the team's ability to control its meetings. You will spend more time than you can imagine in team meetings. From the outset, learn the techniques for a successful meeting. Reasons for the failure of some meetings and the success of others have been presented. The role of the team leader and the behavior of the team members are described. The meeting room is more important that most engineers imagine. Features and furniture arrangements in a room affect the ability of a team to function effectively. Recommendations are made for meeting room size and for arrangements to enhance team productivity during meetings. Positive and negative behaviors of individual team members are illustrated. Explicit examples of both types of behaviors are discussed.

REFERENCES

1. Smith, P. G., D. G. Reinertsen, Developing Products in Half the Time, Van Nostrand Reinhold, New York, NY, 1991.
2. Cunniff, P. F., Dally, J. W., Herrmann, J. W., Schmidt, L. C. and Zhang, G., Product Engineering and Manufacturing, College House Enterprises, Knoxville, TN, 1998.
3. Barczak, G. and Wilemon, D., "Leadership Differences in New Product Development Teams," Journal of Product Innovation Management, Vol. 6, 1989, pp. 259-267.
4. Lacoursiere, R. B. The Life Cycle of Groups: Group Development State Theory, Human Service Press, New York, NY, 1980.
5. Barra, R. Putting Quality Circles to Work, McGraw Hill, New York, NY, 1983.
6. Foxworth, V., "How to Work Effectively in Teams," Class Notes ENME 371, University of Maryland, College Park, September 5, 1997.
7. Clark, K. and Fujimoto, T., Product Development Performance, Harvard Business School Press, Boston, MA, 1991.
8. Wheelwright, S. C. and Clark, K. B., Revolutionizing Product Development, Free Press, New York, NY, 1992.

EXERCISES

A.1 Write an engineering brief describing why team structure in a large corporation is so important in the product realization process. Add a second paragraph indicating why team structure is much less important in a very small company.

A.2 Prepare an organization chart, like the one shown in Fig. A.1, for the University or College that you are attending. Describe the logic, as you see it, for this organizational structure.

A.3 List the advantages and disadvantages of the balanced team structure.

A.4 Outline a development plan that you would prepare if you were a senior team member representing an engineering function (discipline) on a newly formed development team. The product to be developed is a new scooter.

A.5 Write an engineering brief describing staffing required for the development team for the scooter. Give the reasons for changing the staff as the scooter evolves during the development process.

A.6 Describe the experience and status of the product manager for the balanced team structure.

A.7 Describe a matrix organization structure. Explain why this organization impacts an engineer assigned to a product development team with a balanced structure. If you were this engineer, how would you respond to inquires from your functional manager?

A.8 Write a letter to a prospective employer describing your teamwork skills.

A.9 Write a letter, with the best intentions, to a friend explaining why you think he or she is not an effective team member.

A.10 You have been a member of a functional team for four months. The team is not doing well and is beginning to miss milestones in the schedule for the development plan. You are to meet with your functional supervisor for your formal six month performance review. Write a script for a two-person play, with your supervisor's questions and comments, and your answers and comments covering what you expect to occur during your review.

A.11 You (a male/female figure) are attending weekly team meetings and are always seated next to Sally/Bill who is young and very attractive. Sally/Bill is apparently interested in you because she/he frequently involves you in side conversations that are not related to the ongoing team discussions. Write a plan describing all of the actions you will take to handle this situation.

A.12 Brad, the team leader, is a great guy who believes in leading his team in a very democratic manner. He encourages open discussion of the issue under consideration for a defined time period, usually 5 or 10 minutes. At the end of the time period, he intervenes and calls for a team vote to resolve the issue. Write a critique of this style of leadership.

A.13 You together with Zack and Mary are senior members on a development team with 14 other members. The three of you are really very knowledgeable and experienced. You get along famously, and have developed a habit of gathering together at the local watering hole the evening before the scheduled team meeting. During the evening you discuss the issues on the agenda and arrive at some design decisions. Write a brief describing the consequences of continuing this behavior.

A.14 Sue has a wonderful personality, and as a team leader is skillful in promoting free and open discussion by all of the team members. After the team reaches a consensus, she calls for volunteers to implement the required action. When two or more people volunteer, she assigns them as a group to handle the action item. When no one volunteers, she assigns the least active person on the team to the action item and moves promptly to the next item on the agenda. Write a critique of this leadership style.

A.15 Efficient Ed prepares a meeting agenda that includes 16 major items. He schedules each item on the agenda with a 15-minute discussion period without indicating which team member will lead the discussion on each topic. The meeting is scheduled to begin at 8:00 am and to adjourn at 11:59 am. Write a description of what you imagine would happen during the meeting. If you were the team leader, how would you handle the 16 issues that had to be resolved in a short time frame?

A.16 You are the leader of an eight-member team meeting for the first time to develop a new hair dryer. During this initial meeting the following events occur.

- The temperature in the meeting room is 88 °F.
- The room is set up for a seminar speaker.
- The meeting is scheduled to begin at 9:30 am and at 9:35 am only five of the eight members have arrived.
- During the meeting Horrible Harry begins to verbally abuse Shy Sue.
- At 9:50 am Talkative Tom begins to describe his detailed positions on the world situation and is still going at 10:10 am.
- The team breaks for coffee at 10:15 am and has not returned at 10:30 am.
- Sleepy Steve begins to snore.
- Procrastinating Peter refuses to accept an action item after leading the discussion on the issue for 15 minutes.

Describe the approach that you, acting as the team leader, would follow in handling each of these situations.

A.17 Prepare a list of ground rules that should be followed by all the members of a team in conducting an effective meeting.

A.18 You are a member of a team with an ineffective leader. What can you do to improve the productivity of the team? In responding to this question consider the irresponsible actions on the part of certain team members that are listed in Exercise A.16. Remember in your response that you are not the leader and the most that you can do is to share the leadership role from time to time.

A.19 Meet with your team and discuss a list of rules that will be used in guiding the team's conduct during the semester. Incorporate the team's list of rules in the form of a contract. As a final step, each member of the team is to sign the contract indicating his or her commitment to the rules of conduct.

A.20   Record your assessment of the leading, supporting and hindering roles that are played by each member of your team. Use the form shown below to facilitate this process. Assign a score from 1 to 10 in each cell with 1 indicating very poor team skills and 10 superb team skills. Place the name of each team member in a column heading.

TEAM MEMBER NAMES						
SCORES	1 to 10	1 to 10	1 to 10	1 to 10	1 to 10	1 to 10
LEADING ROLES						
1. Initiate						
2. Provide						
3. Seek						
4. Clarify						
5. Test						
6. Summarize						
SUPPORTING ROLES						
1. Monitor						
2. Encourage						
3. Compromise						
4. Sense						
5. Harmonize						
6. Test						
HINDERING ROLES						
1. Uncooperative						
2. Degrading						
3. Dominating						
4. Avoiding						
5. Withdrawing						
6. Distracting						
TOTAL SCORE						

APPENDIX B

STATISTICAL METHODS

B.1 INTRODUCTION

Experimental measurements of quantities such as pressure, temperature, length, force, stress, or strain will always exhibit some variation if the measurements are repeated a number of times with precise instruments. This variability, which is fundamental to all measuring systems, is due to two different causes. First, the quantity being measured may exhibit significant variation. For example, in a materials study to determine fatigue life at a specified stress level, large differences in the number of cycles to failure are noted when a number of specimens are tested. This variation is inherent in the fatigue process and is observed in all fatigue life measurements. Second, the measuring system, which includes the sensor, signal conditioning equipment, analog-to digital converter, recording instrument, and the operator, may introduce error in the measurement. This error may be systematic or random, depending upon its source. An instrument operated out of calibration produces a systematic error, whereas, reading errors due to interpolation on a chart are random. The accumulation of random errors in a measuring system produces a variation that must be examined in relation to the magnitude of the quantity being measured.

The data obtained from repeated measurements represent an array of readings, not an exact result. Maximum information can be extracted from such an array of readings by employing statistical methods. The first step in the statistical treatment of data is to establish the distribution. A graphical representation of the distribution is usually the most useful form for initial evaluation. Next, the statistical distribution is characterized with a measure of its central value, such as the mean, the median, or the mode. Finally, the spread or dispersion of the distribution is determined in terms of the variance or the standard deviation.

With elementary statistical methods, the experimentalist can reduce a large amount of data to a very compact and useful form by defining the type of distribution, establishing the single value that best represents the central value of the distribution (mean), and determining the variation from the mean value (standard deviation). Summarizing data in this manner is the most meaningful form of presentation for application to design problems or for communication to others who need the results of the experiments.

The treatment of statistical methods presented in this chapter is relatively brief; therefore, only the most commonly employed techniques for representing and interpreting data are presented. A formal course in statistics, which covers these techniques in much greater detail as well as many other useful techniques, should be included in the program of study of all engineering students.

B.2 CHARACTERIZING STATISTICAL DISTRIBUTIONS

For purposes of this discussion, consider that an experiment has been conducted n times to determine the yield strength of a particular type of cold-drawn mild steel. The data obtained represent a sample of size n from an infinite population of all possible measurements that could have been made. The simplest way to present these data is to list the strength measurements in order of increasing magnitude, as shown in Table B.1.

Table B.1
Listing of data in order of increasing magnitude of the yield strength of cold-drawn mild steel

Sample No.	Strength		Sample No.	Strength	
	ksi	MPa		ksi	MPa
1	65.0	448	11	79.0	545
2	68.3	471	12	79.2	546
3	72.2	498	13	79.9	551
4	73.5	507	14	80.3	554
5	74.0	510	15	81.1	559
6	75.2	519	16	82.6	570
7	76.8	530	17	84.0	579
8	77.7	536	18	85.5	590
9	78.1	539	19	87.0	600
10	78.8	543	20	89.8	619

These data can be arranged into five groups to give a frequency distribution as shown in Table B.2.

Table B.2
Frequency distribution of yield strength

Group Intervals		Observations	Relative	Cumulative
ksi	MPa	In the Group	Frequency	Frequency
65.0–69.9	448–482	2	0.10	0.10
70.0–74.9	483–516	3	0.15	0.25
75.0–79.9	517–551	8	0.40	0.65
80.0–84.9	552–585	4	0.20	0.85
85.0–89.9	586–620	3	0.15	1.00
Total		20		

The advantage of representing data in a frequency distribution is that the central tendency is more clearly illustrated.

Graphical Representations of the Distribution

The shape of the distribution function representing the yield strength of the cold-drawn, mild-steel is indicated by the data groupings of Table B.2. A graphical presentation of this group data, known as a histogram, is shown in Fig. B.1. The histogram method of presentation shows the central tendency and variability of the distribution much more clearly than the tabular method of presentation of Table B.2. Superimposed on the histogram is a curve showing the relative frequency of the occurrence of a group of measurements. Note that the points for the relative frequency are plotted at the midpoint of the group interval.

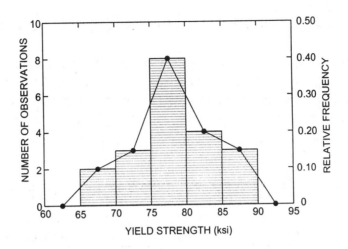

Fig. B.1 A histogram with a superimposed relative frequency diagram.

A cumulative frequency-diagram, shown in Fig. B.2, is another way of representing the yield-strength data from the experiments. The cumulative frequency is the number of readings having a value less than a specified value of the quantity being measured (yield strength) divided by the total number of measurements. As indicated in Table B.2, the cumulative frequency is the running sum of the relative frequencies. When the graph of cumulative frequency versus the quantity being measured is prepared, the end value for the group intervals is used to position the point along the abscissa.

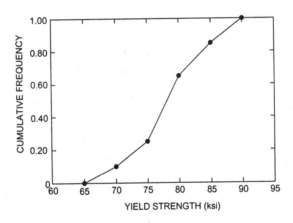

Fig. B.2 Cumulative frequency diagram.

Measures of Central Tendency

While histograms or frequency distributions are used to provide a visual representation of a distribution, numerical measures are used to define the characteristics of the distribution. One basic characteristic is the central tendency of the data. The most commonly employed measure of the central tendency of a distribution of data is the sample mean x, which is defined as:

$$\bar{x} = \sum_{i=1}^{n} \frac{x_i}{n} \tag{B.1}$$

where x_i is the ith value of the quantity being measured.
   n is the total number of measurements.

Because of time and costs involved in conducting tests, the number of measurements is usually limited; therefore, the sample mean $\bar{x}$ is only an estimate of the true arithmetic mean μ of the population. It is shown later that $\bar{x}$ approaches μ as the number of measurements increases. The mean value of the yield-strength data presented in Table B.1 is $\bar{x}$ = 78.4 ksi (541 MPa).

The median and mode are also measures of central tendency. The median is the central value in a group of ordered data. For example, in an ordered set of 21 readings, the 11th reading represents the median value with 10 readings lower than the median and 10 readings higher than the median. In instances when an even number of readings are taken, the median is obtained by averaging the two middle values. For example, in an ordered set of 20 readings, the median is the average of the 10th and 11th readings. Thus, for
the yield-strength data presented in Table B.1, the median is ½ (78.8 + 79.0) = 78.9 ksi (544 MPa).

The mode is the most frequent value of the data; therefore, it is the peak value on the relative-frequency curve. In Fig. B.1, the peak of the relative probability curve occurs at a yield strength S_y = 77.5 ksi (535 MPa); therefore, this value is the mode of the data set presented in Table B.1.

It is evident that a typical set of data gives different values for the three measures of central tendency. There are two reasons for this difference. First, the population from which the samples were drawn may not be Gaussian where the three measures are expected to coincide. Second, even if the population is Gaussian, the number of measurements n is usually small and deviations due to a small sample size are to be expected.

Measures of Dispersion

It is possible for two different distributions of data to have the same mean but different dispersions, as shown in the relative-frequency diagrams of Fig. B.3. Different measures of dispersion are the range, the mean deviation, the variance, and the standard deviation. The standard deviation S_x is the most popular and is defined as:

$$S_x = \left[\sum_{i=1}^{n} \frac{(x_i - \bar{x})^2}{n-1} \right]^{1/2} \tag{B.2}$$

Because the sample size n is small, the standard deviation S_x of the sample represents an estimate of the true standard deviation σ of the population. Computation of S_x and $\bar{x}$ from a data sample is easily performed with a scientific calculator.

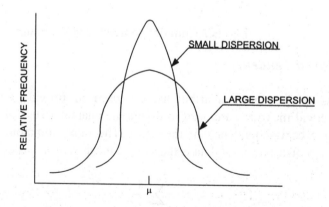

Fig. B.3 Relative frequency diagrams with large and small dispersions.

Expressions for the other measures of dispersion, namely, range R, mean deviation d_x and variance S_x^2 are as follows:

$$R = x_L - x_S \qquad \text{(B.3)}$$

where x_L is the largest value of the quantity in the distribution.
        x_S is the smallest value of the quantity in the distribution.

$$d_x = \frac{\sum\limits_{i=1}^{n}(x_i - \bar{x})}{n} \qquad \text{(B.4)}$$

Equation (B.4) indicates that the deviation of each reading from the mean is determined and summed. The average of the n deviations is the mean deviation. The absolute value of the difference $(x_i - \bar{x})$ must be used in the summing process to avoid cancellation of positive and negative deviations.

$$S_x^2 = \sum\limits_{i=1}^{n}\frac{(x_i - \bar{x})^2}{n-1} \qquad \text{(B.5)}$$

The variance of the population σ^2 is estimated by S_x^2 where the denominator $(n - 1)$ in Eqs. (B.2) and (B.5) serves to reduce error introduced by approximating the true mean σ with the estimate $\bar{x}$. As the sample size n is increased the estimates of $\bar{x}$, S_x, and S_x^2 improve as shown in the discussion of Section 4. Variance is an important measure of dispersion because it is used in defining the normal distribution function.

Finally, a measure known as the coefficient of variation C_v is used to express the standard deviation S_x as a percentage of the mean $\bar{x}$. Thus:

$$C_v = \frac{S_x}{\bar{x}}(100) \qquad \text{(B.6)}$$

The coefficient of variation represents a normalized parameter that indicates the variability of the data in relation to its mean.

B.3 THE GAUSSIAN OR NORMAL DISTRIBUTION FUNCTION

As the sample size is increased, it is possible in tabulating the data to increase the number of group intervals and to decrease their width. The corresponding relative-frequency diagram, similar to the one illustrated in Fig. B.1, will approach a smooth curve (a theoretical distribution curve) known as a distribution function.

A number of different distribution functions are used in statistical analyses. The best-known and most widely used distribution in engineering is the Gaussian or normal distribution. This distribution is important because it describes random errors in measurements and variations observed in strength determinations. Other useful distributions include binomial, exponential, hypergeometric, chi-square (χ^2), F, Gumbel, Poisson, Student's t, and Weibull distributions. The reader is referred to references [1–3] for a complete description of these distributions. Emphasis here will be on the Gaussian or normal distribution function because of its wide range of application in engineering.

Gaussian Distribution

The Gaussian or normal distribution function, as represented by a normalized relative-frequency diagram, is shown in Fig. B.4. The Gaussian distribution is completely defined by two parameters; the mean μ and the standard deviation σ. The equation for the relative frequency f in terms of these two parameters is:

$$f(z) = \frac{1}{\sqrt{2\pi}}\, e^{-(z^2/2)} \qquad\qquad (B.7)$$

where

$$z = (x - \mu)/\sigma \qquad\qquad (B.8)$$

Experimental data (with finite sample sizes) can be analyzed to obtain $\overline{x}$ as an estimate of μ and S_x as an estimate of σ. This procedure permits the experimentalist to use data drawn from small samples to represent the entire population.

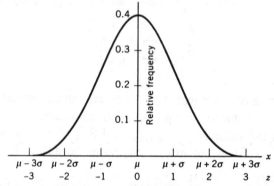

Fig. B.4 The Gaussian or normal distribution function.

The method for predicting population properties from a Gaussian (normal) distribution function utilizes the normalized relative-frequency diagram shown in Fig. B.4. The area A under the entire curve is given by Eq. (B.7) as:

$$A = \frac{1}{\sqrt{2\pi}} \int_{-\infty}^{\infty} e^{-(z^2/2)}dz = 1 \qquad\qquad (9)$$

Equation (B.9) implies that the population has a value z between $-\infty$ and $+\infty$ and that the probability of making a single observation from the population with a value $-\infty < z < +\infty$ is 100 percent. While the previous statement may appear trivial and obvious, it serves to illustrate the concept of using the area under the normalized relative-frequency curve to determine the probability p of observing a measurement within a specific interval. Figure B.5 shows graphically (shaded area under the curve) the probability that a measurement will occur within the interval between z_1 and z_2. Thus, from Eq. (B.7) it is evident that:

$$p(z_1, z_2) = \int_{z_1}^{z_2} f(z)dz = \frac{1}{\sqrt{2\pi}} \int_{z_1}^{z_2} e^{-(z^2/2)}dz \qquad\qquad (10)$$

Evaluation of Eq. (B.10) is most easily made by using tables that list the areas under the normalized relative-frequency curve as a function of z. Table B.3 lists one-side areas between limits of $z_1 = 0$ and z_2 for the normal distribution function.

Because the distribution function is symmetric about $z = 0$, this one-sided table is sufficient for all evaluations. For example:

$$A(-1, 0) = A(0, +1)$$

Therefore;

$$A(-1, +1) = p(-1, +1) = 0.3413 + 0.3413 = 0.6826$$
$$A(-2, +2) = p(-2, +2) = 0.4772 + 0.4772 = 0.9544$$
$$A(-3, +3) = p(-3, +3) = 0.49865 + 0.49865 = 0.9973$$
$$A(-1, +2) = p(-1, +2) = 0.3413 + 0.4772 = 0.8185$$

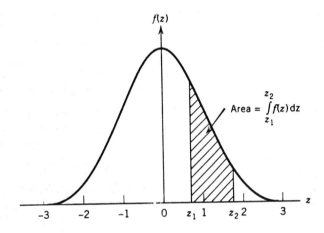

Fig. B.5 Probability of measuring x between the limits of z_1 and z_2. Total area under the curve f(z) is 1.

Because the normal distribution function has been well characterized, predictions can be made regarding the probability of a specific strength value or measurement error. For example, one may anticipate that 68.3 percent of the data will fall between limits of $\overline{x} \pm S_x$: 95.4 percent will fall between the limits of $\overline{x} \pm 2S_x$ and 99.7 percent between limits of $\overline{x} \pm 3S_x$. Also, 81.9 percent of the data should fall between limits of $\overline{x} - S_x$: and $\overline{x} + 2S_x$.

In many problems, the probability of a single sample exceeding a specified value z_2 must be determined. It is possible to determine this probability by using Table B.3 together with the fact that the area under the entire curve is unity (A = 1); however, Table B.4, which lists one-sided areas between limits of $z_1 = z$ and $z_2 \Rightarrow \infty$, yields the results more directly.

The use of Tables B.3 and B.4 can be illustrated by considering the yield strength data presented in Table B.1. By using Eqs. (B.1) and (B.2), it is easy to establish estimates for the mean and standard deviation as $\overline{x} = 78.41$ksi (541 MPa) and $S_x = 6.04$ ksi (41.7 MPa). These values of $\overline{x}$ and S_x characterize the population from which the data of Table B.1 were drawn. It is possible to establish the probability that the yield strength of a single specimen drawn randomly from the population will be between specified limits (by using Table B.3), or that the yield strength of a single sample will not be above or below a specified value (by using Table B.4). For example, one determines the probability that a single sample will exhibit a yield strength between 66 and 84 ksi by computing z_1 and z_2 and using Table B.3. Thus:

$$z_1 = (66 - 78.4)/6.04 = -2.05 \qquad z_2 = (84 - 78.4)/6.04 = 0.93$$

$$p(-2.05, 0.93) = A(-2.05, 0) + A(0, 0.93) = 0.4798 + 0.3238 = 0.8036$$

This simple calculation shows that the probability of obtaining a yield strength between 66 and 84 ksi from a single specimen is 80.4 percent. The probability of the yield strength of a single specimen being less than 65 ksi is determined by computing z_1 and using Table B.4. Thus, $z_1 = (65 - 78.4)/6.04 = -2.22$ and $p(-\infty, -2.22) = A(-\infty, -2.22) = A(2.22, \infty) = 0.0132$. Clearly, the probability of drawing a single sample with a yield strength less than 65 ksi is 1.3 percent.

Table B.3
Areas under the normal-distribution curve from $z_1 = 0$ to z_2 (One Side)

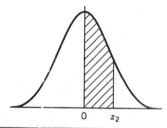

$z_2 = \frac{x-\bar{x}}{S_x}$	0.00	0.01	0.02	0.03	0.04	0.05	0.06	0.07	0.08	0.09
0.0	.0000	.0040	.0080	.0120	.0160	.0199	.0239	.0279	.0319	.0359
0.1	.0398	.0438	.0478	.0517	.0557	.0596	.0636	.0675	.0714	.0753
0.2	.0793	.0832	.0871	.0910	.0948	.0987	.1026	.1064	.1103	.1141
0.3	.1179	.1217	.1255	.1293	.1331	.1368	.1406	.1443	.1480	.1517
0.4	.1554	.1591	.1628	.1664	.1700	.1736	.1772	.1808	.1844	.1879
0.5	.1915	.1950	.1985	.2019	.2054	.2088	.2123	.2157	.2190	.2224
0.6	.2257	.2291	.2324	.2357	.2389	.2422	.2454	.2486	.2517	.2549
0.7	.2580	.2611	.2642	.2673	.2704	.2734	.2764	.2794	.2823	.2852
0.8	.2881	.2910	.2939	.2967	.2995	.3023	.3051	.3078	.3106	.3233
0.9	.3159	.3186	.3212	.3238	.3264	.3289	.3315	.3340	.3365	.3389
1.0	.3413	.3438	.3461	.3485	.3508	.3531	.3554	.3577	.3599	.3621
1.1	.3643	.3655	.3686	.3708	.3729	.3749	.3770	.3790	.3810	.3830
1.2	.3849	.3869	.3888	.3907	.3925	.3944	.3962	.3980	.3997	.4015
1.3	.4032	.4049	.4066	.4082	.4099	.4115	.4131	.4147	.4162	.4177
1.4	.4192	.4207	.4222	.4236	.4251	.4265	.4279	.4292	.4306	.4319
1.5	.4332	.4345	.4357	.4370	.4382	.4394	.4406	.4418	.4429	.4441
1.6	.4452	.4463	.4474	.4484	.4495	.4505	.4515	.4525	.4535	.4545
1.7	.4554	.4564	.4573	.4582	.4591	.4599	.4608	.4616	.4625	.4633
1.8	.4641	.4649	.4656	.4664	.4671	.4678	.4686	.4693	.4699	.4706
1.9	.4713	.4719	.4726	.4732	.4738	.4744	.4750	.4758	.4761	.4767
2.0	.4772	.4778	.4783	.4788	.4793	.4799	.4803	.4808	.4812	.4817
2.1	.4821	.4826	.4830	.4834	.4838	.4842	.4846	.4850	.4854	.4857
2.2	.4861	.4864	.4868	.4871	.4875	.4878	.4881	.4884	.4887	.4890
2.3	.4893	.4896	.4898	.4901	.4904	.4906	.4909	.4911	.4913	.4916
2.4	.4918	.4920	.4922	.4925	.4927	.4929	.4931	.4932	.4934	.4936
2.5	.4938	.4940	.4941	.4943	.4945	.4946	.4948	.4949	.4951	.4952
2.6	.4953	.4955	.4956	.4957	.4959	.4960	.4961	.4962	.4963	.4964
2.7	.4965	.4966	.4967	.4968	.4969	.4970	.4971	.4972	.4973	.4974
2.8	.4974	.4975	.4976	.4977	.4977	.4978	.4979	.4979	.4980	.4981
2.9	.4981	.4982	.4982	.4983	.4984	.4984	.4985	.4985	.4986	.4986
3.0	.49865	.4987	.4987	.4988	.4988	.4988	.4989	.4989	.4989	.4990

Table B.4
Areas under the normal-distribution curve from z_1 to $z_2 \Rightarrow \infty$ (One Side)

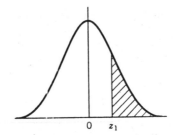

$z_1 = \frac{x-\bar{x}}{S_x}$	0.00	0.01	0.02	0.03	0.04	0.05	0.06	0.07	0.08	0.09
0.0	.5000	.4960	.4920	.4880	.4840	.4801	.4761	.4721	.4681	.4641
0.1	.4602	.4562	.4522	.4483	.4443	.4404	.4364	.4325	.4286	.4247
0.2	.4207	.4168	.4129	.4090	.4052	.4013	.3974	.3936	.3897	.3859
0.3	.3821	.3783	.3745	.3707	.3669	.3632	.3594	.3557	.3520	.3483
0.4	.3446	.3409	.3372	.3336	.3300	.3264	.3228	.3192	.3156	.3121
0.5	.3085	.3050	.3015	.2981	.2946	.2912	.2877	.2843	.2810	.2776
0.6	.2743	.2709	.2676	.2643	.2611	.2578	.2546	.2514	.2483	.2451
0.7	.2420	.2389	.2358	.2327	.2296	.2266	.2236	.2206	.2177	.2148
0.8	.2119	.2090	.2061	.2033	.2005	.1977	.1949	.1922	.1984	.1867
0.9	.1841	.1814	.1788	.1762	.1736	.1711	.1685	.1660	.1635	.1611
1.0	.1587	.1562	.1539	.1515	.1492	.1469	.1446	.1423	.1401	.1379
1.1	.1357	.1335	.1314	.1292	.1271	.1251	.1230	.1210	.1190	.1170
1.2	.1151	.1131	.1112	.1093	.1075	.1056	.1038	.1020	.1003	.0985
1.3	.0968	.0951	.0934	.0918	.0901	.0885	.0869	.0853	.0838	.0823
1.4	.0808	.0793	.0778	.0764	.0749	.0735	.0721	.0708	.0694	.0681
1.5	.0668	.0655	.0643	.0630	.0618	.0606	.0594	.0582	.0571	.0559
1.6	.0548	.0537	.0526	.0516	.0505	.0495	.0485	.0475	.0465	.0455
1.7	.0446	.0436	.0427	.0418	.0409	.0401	.0392	.0384	.0375	.0367
1.8	.0359	.0351	.0344	.0336	.0329	.0322	.0314	.0307	.0301	.0294
1.9	.0287	.0281	.0274	.0268	.0262	.0256	.0250	.0244	.0239	.0233
2.0	.0228	.0222	.0217	.0212	.0207	.0202	.0197	.0192	.0188	.0183
2.1	.0179	.0174	.0170	.0165	.0162	.0158	.0154	.0150	.0146	.0143
2.2	.0139	.0136	.0132	.0129	.0125	.0122	.0119	.0116	.0113	.0110
2.3	.0107	.0104	.0102	.00990	.00964	.00939	.00914	.00889	.00866	.0084
2.4	.00820	.00798	.00776	.00755	.00734	.00714	.00695	.00676	.00657	.00639
2.5	.00621	.00604	.00587	.00570	.00554	.00539	.00523	.00508	.00494	.00480
2.6	.00466	.00453	.00440	.00427	.00415	.00402	.00391	.00379	.00368	.00357
2.7	.00347	.00336	.00326	.00317	.00307	.00298	.00288	.00280	.00272	.00264
2.8	.00256	.00248	.00240	.00233	.00226	.00219	.00212	.00205	.00199	.00193
2.9	.00187	.00181	.00175	.00169	.00164	.00159	.00154	.00149	.00144	.00139

B.4 REGRESSION ANALYSIS

Many experiments involve the measurement of one dependent variable, say y, which may depend upon one or more independent variables, x_1, x_2, ... x_k. Regression analysis provides a statistical approach for conditioning the data obtained from experiments where two or more related quantities are measured.

Linear Regression Analysis

Suppose measurements are made of two quantities that describe the behavior of a process exhibiting variation. Let y be the dependent variable and x the independent variable. Since the process exhibits variation, there is not a unique relationship between x and y and the data, when plotted, exhibit scatter, as illustrated in Fig. B.6. Frequently, the relation between x and y that most closely represents the data, even with the scatter, is a linear function. Thus,

$$Y_i = mx_i + b \qquad \text{(B.11)}$$

where Y_i is the predicted value of the dependent variable y_i for a given value of the independent variable x_i.

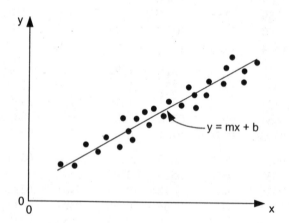

Fig. B.6 Linear regression analysis is used to fit a least-squares line through scattered data points.

A statistical procedure used to fit a straight line through scattered data points is called the least-squares method. With the **least-squares method**, the slope m and the intercept b in Eq. (B.11) are selected to minimize the sum of the squared deviations of the data points from the straight line shown in Fig. B.6. In utilizing the least-squares method, it is assumed that the independent variable x is free of measurement error and the quantity

$$\Delta^2 = \sum (y_i - Y_i)^2 \qquad \text{(B.12)}$$

is minimized at fixed values of x. After substituting Eq. (B.11) into Eq. (B.12) this implies that:

$$\frac{\partial \Delta^2}{\partial b} = \frac{\partial}{\partial b} \sum (y_i - mx - b)^2 = 0$$

$$\frac{\partial \Delta^2}{\partial m} = \frac{\partial}{\partial m} \sum (y_i - mx - b)^2 = 0$$

(a)

Differentiating yields:

$$2\sum (y_i - mx - b)(-x) = 0$$

$$2\sum (y_i - mx - b)(-1) = 0$$

(b)

Solving Eqs. (b) for m and b yields:

$$m = \frac{\sum x \sum y - n \sum xy}{\left(\sum x\right)^2 - n \sum x^2}$$

$$b = \frac{\sum y - m \sum x}{n}$$

(B.13)

where n is the number of data points. The slope m and intercept b define a straight line through the scattered data points such that Eq. (B.12) is minimized.

In any regression analysis it is important to establish the correlation between x and y. Equation (B.11) does not predict the values that were measured, exactly, because of the variation in the process. To illustrate, assume that the independent quantity x is fixed at a value x_1 and that a sequence of measurements is made of the dependent quantity y. The data obtained would give a distribution of y, as illustrated in Fig. B.7. The dispersion of the distribution of y is a measure of the correlation. When the dispersion is small, the correlation is good and the regression analysis is effective in describing the variation in y. If the dispersion is large, the correlation is poor and the regression analysis may not be adequate to describe the variation in y.

The adequacy of regression analysis can be evaluated by determining a correlation coefficient R^2 that is given by the following expression:

$$R^2 = 1 - \frac{n-1}{n-2}\left[\frac{\{y^2\} - m\{xy\}}{\{y^2\}} \right]$$

(B.14)

where

$$\{y^2\} = \sum y^2 - \left(\sum y\right)^2 / n$$

$$\{xy\} = \sum xy - \left(\sum x\right)\left(\sum y\right) / n$$

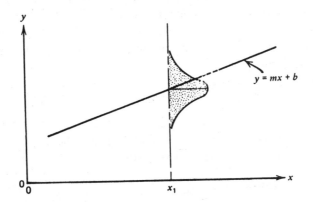

Fig. B.7 Distribution of y at a fixed value of x superimposed on the linear regression line.

When the value of the correlation coefficient $R^2 = 1$, perfect correlation exists between y and x. If $R^2 = 0$ no correlation exists and the variations observed in y are due to random fluctuations and not changes in x. Because random variations in y exist, a value of $R^2 = 1$ is not obtained even if y(x) is linear. To interpret correlation coefficients $0 < R^2 < 1$, the data in Table B.5 is used to establish the probability of obtaining a given R^2 due to random variations in y.

Table B.5
Probability of obtaining a correlation coefficient R^2 due to random variations in y.

n	\multicolumn{4}{c}{Probability}
	0.10	0.05	0.02.	0.01
5	0.805	0.878	0.934	0.959
6	0.729	0.811	0.882	0.917
7	0.669	0.754	0.833	0.874
8	0.621	0.707	0.789	0.834
10	0.549	0.632	0.716	0.765
15	0.441	0.514	0.592	0.641
20	0.378	0.444	0.516	0.561
30	0.307	0.362	0.423	0.464
40	0.264	0.312	0.367	0.403
60	0.219	0.259	0.306	0.337
80	0.188	0.223	0.263	0.291
100	0.168	0.199	0.235	0.259

As an example, consider a regression analysis with n = 15 which gives $R^2 = 0.65$ as determined by Eq. (B.14). Reference to Table B.5 indicates that the probability of obtaining $R^2 = 0.65$ due to random variations is slightly less than one percent. Thus one can be 99 percent certain that the regression analysis represents a true correlation between y and x.

Multivariate Regression

Many experiments involve measurements of a dependent variable y, which depends upon several independent variables $x_1, x_2, x_3, ...,$ etc. It is possible to represent y as a function of $x_1, x_2, x_3, ...,$ etc. by employing the multivariate regression equation

$$Y_i = a + b_1 x_1 + b_2 x_2 + ... + b_k x_k \qquad (B.15)$$

where $a, b_1, b_2, ... b_k$ are regression coefficients.

The regression coefficients $a, b_1, b_2, ... b_k$ are determined by using the method of least squares in a manner similar to that employed for linear regression analysis where the quantity $\Delta^2 = \Sigma(y_i - Y_i)^2$ is minimized. Substituting Eq. (B.12) into Eq. (B.15) yields:

$$\Delta^2 = \Sigma(y_i - a - b_1 x_1 - b_2 x_2 - ... - b_k x_k)^2 \qquad (B.16)$$

Differentiating yields:

$$\frac{\partial \Delta^2}{\partial a} = 2\left[\sum (y_i - a - b_1 x_1 - b_2 x_2 - \cdots - b_k x_k)(-1)\right] = 0$$

$$\frac{\partial \Delta^2}{\partial b_1} = 2\left[\sum (y_i - a - b_1 x_1 - b_2 x_2 - \cdots - b_k x_k)(-x_1)\right] = 0$$

$$\frac{\partial \Delta^2}{\partial b_2} = 2\left[\sum (y_i - a - b_1 x_1 - b_2 x_2 - \cdots - b_k x_k)(-x_2)\right] = 0 \qquad \text{(B.17)}$$

$$\cdots = \qquad\qquad\qquad\qquad\qquad\qquad \cdots = 0$$

$$\frac{\partial \Delta^2}{\partial b_k} = 2\left[\sum (y_i - a - b_1 x_1 - b_2 x_2 - \cdots - b_k x_k)(-x_k)\right] = 0$$

Equations (32) lead to the following set of $k + 1$ equations, which can be solved for the unknown regression coefficients $a, b_1, b_2, \ldots b_k$.

$$a\,n + b_1\,\Sigma x_1 + b_2\,\Sigma x_2 + \ldots + b_k\,\Sigma x_k = \Sigma y_i$$

$$a\,\Sigma x_1 + b_1\,\Sigma x_1^2 + b_2\,\Sigma x_1 x_2 + \ldots + b_k\,\Sigma x_1 x_k = \Sigma y_i x_1$$

$$a\,\Sigma x_2 + b_1\,\Sigma x_1 x_2 + b_2\,\Sigma x_2^2 + \ldots + b_k\,\Sigma x_1 x_k = \Sigma y_i x_2 \qquad \text{(B.18)}$$

$$\ldots\ldots = \ldots\ldots$$

$$a\,\Sigma x_k + b_1\,\Sigma x_1 x_k + b_2\,\Sigma x_2 x_k + \ldots + b_k\,\Sigma x_k^2 = \Sigma y_i x_k$$

The correlation coefficient R^2 is again used to determine the degree of association between the dependent and independent variables. For multiple regression equations, the correlation coefficient R^2 is given as:

$$R^2 = 1 - \frac{n-1}{n-k}\left[\frac{\{y^2\} - b_1\{yx_1\} - b_2\{yx_2\} - \cdots b_k\{yx_k\}}{\{y^2\}}\right] \qquad \text{(B.19)}$$

where

$$\{yx_k\} = \sum yx_k - \left(\sum y\right)\left(\sum x_k\right)/n$$

$$\{y^2\} = \sum y^2 - \left(\sum y\right)^2/n$$

This analysis is for linear, noninteracting, independent variables; however, the analysis can be extended to include cases where the regression equations would have higher-order and cross-product terms. The nonlinear terms can enter the regression equation in an additive manner and are treated as extra variables. With well-established computer routines for regression analysis, the set of $(k + 1)$ simultaneous equations given by Eqs. (B.18) can be solved quickly and inexpensively and no difficulties are encountered in adding extra terms to account for nonlinearities and interactions.

B.5 SUMMARY

Statistical methods are extremely important in engineering, since they provide a means for representing large amounts of data in a concise form that is easily interpreted and understood. Usually, the data are represented with a statistical distribution function that can be characterized by a measure of central tendency (the mean $\bar{x}$ and a measure of dispersion (the standard deviation S_x). A normal or Gaussian probability distribution is by far the most commonly employed; however, in some cases, other distribution functions may have to be employed to adequately represent the data. The most significant advantage resulting from use of this probability distribution function in engineering applications is the ability to predict the occurrence of an event based on a relatively small sample.

Regression analysis can be used effectively to interpret data when the behavior of one quantity (say, y) depends upon variations in one or more independent quantities (say, x_1, x_2, ... x_n). Even though the functional relationship between quantities exhibiting variation remains unknown, it can be characterized statistically. Regression analysis provides a method to fit a straight line or a curve through a series of scattered data points on a graph. The adequacy of the regression analysis can be evaluated by determining a correlation coefficient. Methods for extending regression analysis to multivariate functions exist. In principle, these methods are identical to linear regression analysis; however, the analysis becomes much more complex. The increase in complexity is not a concern, however, because computer subroutines are available that solve the tedious equations and provide the results in a convenient format.

REFERENCES

1.   Bethea, R. M. and R. R. Rhinehart: Applied Engineering Statistics, Dekker, New York, 1991.
2.   Bethea, R. M., B. S. Duran, and T. L. Boullion: Statistical Methods for Engineers and Scientists, 2nd ed., Dekker, New York, 1985.
3.   Blackwell, D.: Basic Statistics, McGraw-Hill, New York, 1969.
4.   Bragg, G. M.: Principles of Experimentation and Measurement, Prentice-Hall, Englewood Cliffs, NJ, 1974.
5.   Chou, Y.: Probability and Statistics for Decision Making, Holt, Rinehart & Winston, New York, 1972.
6.   Snedecor, G. W. and W. G. Cochran: Statistical Methods, 8th ed., Iowa State University Press, Ames, IA, 1989.

EXERCISES

B.1   The air pressure (in psi) at a point near the end of an air-supply line is monitored at 15-min intervals over an 8-h period. The readings are listed in the four columns below:

Column 1	Column 2	Column 3	Column 4	Column 5	Column 6	Column 7	Column 8
113	95	105	118	100	81	97	110
107	92	101	112	96	85	100	113
109	103	98	103	93	91	103	120
97	117	123	101	87	94	105	128

List the pressure readings in order of increasing magnitude. Rearrange the data in five group intervals to obtain a frequency distribution. Show both the relative frequency and the cumulative frequency in this tabular rearrangement. Select the median pressure reading from the data.

B.2 Construct a histogram for the data listed in Exercise B.1. Super-impose a plot of the relative frequency on the histogram.

B.3 Prepare a cumulative frequency curve for the data of Exercise B.1.

B.4 For the air-pressure data listed in Exercise B.1:
(a) Compute the sample mean $\bar{x}$ of the individual columns.
(b) Compute the sample means of columns 1 + 2 and columns 3 + 4.
(c) Compute the sample mean of the complete set of data.
(d) Comment on the results of (a), (b), and (c).

B.5 Determine the mode for the air-pressure data of Exercise B.1 and compare it with the median and the mean of the data.

B.6 A quality control laboratory monitors the tensile strength of paper by testing a small sample every 10 min. The data shown below were reported over an 8-h shift.

Column 1	Column 2	Column 3	Column 4	Column 5	Column 6
980	887	1043	968	920	953
992	913	1031	963	897	944
969	972	999	950	871	981
929	987	990	942	847	1013
892	1021	983	934	843	1043
860	1066	982	936	868	1072
820	1079	977	939	884	1091
862	1080	971	922	907	1115

List the strength readings in order of increasing magnitude. Rearrange the data in seven group intervals to obtain a frequency distribution. Show both the relative frequency and the cumulative frequency in this tabular rearrangement. Select the median strength reading from the data.

B.7 Construct a histogram for the data listed in Exercise B.6. Super-impose a plot of relative frequency on the histogram.

B.8 Prepare a cumulative frequency curve for the data of Exercise B.6.

B.9 For the tensile-strength data listed in Exercise B.6,
(a) Compute the sample mean x of the individual columns.
(b) Compute the sample means of columns 1 + 2, 3 + 4, and 5 + 6.
(c) Compute the sample mean of the complete set of data.
(d) Comment on the results of (a), (b), and (c).

B.10 Determine the mode for the tensile-strength data of Exercise B.6 and compare it with the median and the mean of the data.

B.11 For the data presented in Exercise B.1, determine:
(a) The standard deviation S_x.
(b) The range R.
(c) The mean deviation d_x.
(d) The variance S_x^2.
(e) The coefficient of variation C_v.

B.12 For the data presented in Exercise B.6, determine:
(a) The standard deviation S_x.
(b) The range R.
(c) The mean deviation d_x.
(d) The variance S_x^2.

(e) The coefficient of variation C_v.

B.13 A quality control laboratory associated with a manufacturing operation periodically makes a measurement that has been characterized as normal or Gaussian with a mean of 80 and a standard deviation of 3. Determine the probability of a single measurement:

    (a) Falling between 73 and 83.
    (b) Falling between 70 and 86.
    (c) Falling between 67 and 89.
    (d) Falling between 80 and 82.
    (e) Falling between 79 and 80.
    (f) Being less than 77.
    (g) Being greater than 86.

B.14 Careful measurement of diameter was made after a large shipment of electric motor shafts was received. Measurements of the bearing journal indicate that the mean is 7 mm, with a standard deviation of 0.015 mm. Determine the probability of a journal measurement being:

    (a) Greater than 7.04 mm.
    (b) Greater than 7.03 mm.
    (c) Less than 7.00 mm.
    (d) Less than 6.94 mm.
    (e) Between 7.01 and 7.03 mm.
    (f) Between 6.96 and 6.99.mm.
    (g) Between 6.97 and 7.03 mm.

B.15 Determine the slope m and the intercept b for a linear regression equation $y = mx + b$ representing the following set of data.

x	y	x	y
0.5	1.4	2.8	8.4
0.9	2.9	3.2	9.8
1.4	4.4	3.5	10.7
2.0	6.2	3.9	11.4
2.3	7.1	4.2	12.4

B.16 Determine the slope m and the intercept b for a linear regression equation $y = mx + b$ representing the following set of data.

x	y	x	y	x	y
11.7	23.0	17.2	34.8	21.8	43.7
13.0	26.4	18.0	36.2	23.2	46.0
13.9	28.1	18.9	38.0	24.5	49.6
14.7	29.1	19.7	39.3	25.1	52.5
16.1	32.4	20.9	42.1	27.4	54.4

B.17 Determine the correlation coefficient R^2 for the regression analysis of Exercise B.15.

B.18 Determine the correlation coefficient R^2 for the regression analysis of Exercise B.16.

B.19 The solvent concentration in a coating as a function of time is governed by an equation of the form:

$$C = ae^{-mt}$$

t (s)	C (%)	t (s)	C (%)	t (s)	C (%)	t (s)	C (%)
10	2.05	30	1.96	50	1.89	200	1.68
20	2.00	40	1.92	100	1.79	500	1.45

where C is the concentration, t is the time and a and m are constants that depend upon the diffusion process.

For the given data, determine the constants a and m that provide the best fit.

B.20 Determine the regression coefficients a, b_1, b_2, and b_3 for the following data set.

y	x_1	x_2	x_3
6.9	1.0	2.0	1.0
7.9	1.5	2.0	1.5
8.0	2.0	2.0	2.0
8.1	2.5	3.0	1.0
8.4	3.0	3.0	1.5
8.5	3.5	3.0	2.0
8.6	4.0	4.0	1.0
8.7	4.5	4.0	1.5
9.0	5.0	4.0	2.0
9.2	5.5	5.0	1.0
9.4	6.0	5.0	1.5
9.7	6.5	5.0	2.0

B.21 Determine the correlation coefficient R^2 for the solution of Exercise B.20.

APPENDIX C

GUIDES AND FORMS FOR DESIGN TEAMS

Team Assessment Form

TEAM MEMBERS' NAMES						
SCORES	1 -10	1 -10	1 -10	1 -10	1 -10	1 -10
LEADING ROLES						
1. Initiate						
2. Provide						
3. Seek						
4. Clarify						
5. Test						
6. Summarize						
SUPPORTING ROLES						
1. Monitor						
2. Encourage						
3. Compromise						
4. Sense						
5. Harmonize						
6. Test						
HINDERING ROLES						
1. Uncooperative						
2. Degrading						
3. Dominating						
4. Avoiding						
5. Withdrawing						
6. Distracting						
TOTAL SCORE						

TEAM MEMBERS' AGREEMENT

- The information discussed by our team members will remain confidential.
- We will acknowledge problems and deal with them.
- We will be supportive rather than judgmental.
- We will respect differences.
- We will provide responses directly and openly in a timely fashion.
- We will provide information that is specific and focused on the task and not on personalities.
- We will be open, but will respect the right of privacy.
- We will not discount the ideas of others.
- We are each responsible for the success of the team experience.
- The team has all of the resources needed to solve the problems that arise during the development process.
- Every member of the team will contribute to its success.
- We will try to know and understand our fellow team members so as to identify ways for enhancing everyone's professional development.
- We recognize the importance of the team meeting to the success of the group, and accordingly we agree to:
- Always attend the scheduled team meetings.
- When attendance is impossible, notify the team leader and as many other members as possible in advance.
- Use meeting time wisely
- Start on time.
- Limit our breaks and return from them on time.
- Keep focused on our goals.
- Avoid side issues, personality conflicts and hidden agenda.
- Share responsibility for briefing members when they miss a meeting.
- Avoid making phone calls or engaging in side conversations that interrupt the team.
- _____.
- _____.
- _____.
- _____.
- _____.
-

SIGNATURES:

_____   _____

_____   _____

_____   _____

Date_____

Agenda
Weekly Meeting
Date _____ Time _____

Team Member Responsible

1. Weekly status report   _____

2. Review of outstanding action items

    - Action item #_____   _____
    - Action item #_____   _____
    - Action item #_____   _____
    - Action item #_____   _____
    - Action item #_____   _____

3. Report on progress

    - Subsystem _____   _____
    - Subsystem _____   _____
    - Subsystem _____   _____
    - Subsystem _____   _____
    - Subsystem _____   _____

4. Identify new problems

    - Problem #_____   _____
    - Problem #_____   _____
    - Problem #_____   _____
    - Problem #_____   _____
    - Problem #_____   _____

5. Assignment of action items

    - Action item #_____   _____
    - Action item #_____   _____
    - Action item #_____   _____
    - Action item #_____   _____
    - Action item #_____   _____

6. New business
7. Summary
8. Adjourn

Agenda
Weekly Meeting
Date _____ Time _____

Team Member Responsible

9. Weekly status report   _____

10. Review of outstanding action items

- Action item #_____   _____
- Action item #_____   _____
- Action item #_____   _____
- Action item #_____   _____
- Action item #_____   _____

11. Report on progress

- Subsystem _____   _____
- Subsystem _____   _____
- Subsystem _____   _____
- Subsystem _____   _____
- Subsystem _____   _____

12. Identify new problems

- Problem #_____   _____
- Problem #_____   _____
- Problem #_____   _____
- Problem #_____   _____
- Problem #_____   _____

13. Assignment of action items

- Action item #_____   _____
- Action item #_____   _____
- Action item #_____   _____
- Action item #_____   _____
- Action item #_____   _____

14. New business
15. Summary
16. Adjourn

Action Item Record

Team Meeting of _____

Decisions made	Follow-up required	Responsible member	Date complete
1.	1.	1.	1.
2.	2.	2.	2.
3.	3.	3.	3.
4.	4.	4.	4.
Action at next meeting	**Preparation required**	**Responsible member**	**Date complete**
1.	1.	1.	1.
2.	2.	2.	2.
3.	3.	3.	3.
4.	4.	4.	4.

Team Meeting of _____

Decisions made	Follow-up required	Responsible member	Date complete
1.	1.	1.	1.
2.	2.	2.	2.
3.	3.	3.	3.
4.	4.	4.	4.
Action at next meeting	**Preparation required**	**Responsible member**	**Date complete**
1.	1.	1.	1.
2.	2.	2.	2.
3.	3.	3.	3.
4.	4.	4.	4.

Team Meeting of _____

Decisions made	Follow-up required	Responsible member	Date complete
1.	1.	1.	1.
2.	2.	2.	2.
3.	3.	3.	3.
4.	4.	4.	4.
Action at next meeting	**Preparation required**	**Responsible member**	**Date complete**
1.	1.	1.	1.
2.	2.	2.	2.
3.	3.	3.	3.
4.	4.	4.	4.

CONCEPT SELECTION
PUGH CHART

CRITERIA	CONCEPTS			
	B	C	D	A
				DATUM

S - CONCEPT EQUAL TO DATUM

+ CONCEPT SUPERIOR TO DATUM

− CONCEPT INFERIOR TO DATUM

WEEKLY SCHEDULE FORM

WEEKLY SCHEDULE							
FALL SEMESTER 2001							
NAME		ADDRESS			PHONE NUMBER		
TIME	SUN	MON	TUES	WED	THUR	FRI	SAT
MORNING							
6:00							
7:00							
8:00							
9:00							
10:00							
11:00							
12:00							
AFTERNOON							
1:00							
2:00							
3:00							
4:00							
5:00							
EVENING							
6:00							
7:00							
8:00							
9:00							
10:00							
NIGHT							
11:00-6:00							

DAILY SCHEDULE FORMS

DAILY SCHEDULE					
DAY					
DATE					
NAME					
TIME MORNING	TASK	AFTERNOON	TASK	EVENING	TASK
6:00		1:00		6:00	DINNER
		2:00		7:00	
7:00	BREAKFAST	3:00		8:00	
8:00		4:00		9:00	
9:00		5:00		10:00	
10:00				NIGHT	
11:00				11:00-6:00	
12:00	LUNCH				

DAILY SCHEDULE					
DAY					
DATE					
NAME					
TIME MORNING	TASK	AFTERNOON	TASK	EVENING	TASK
6:00		1:00		6:00	DINNER
		2:00		7:00	
7:00	BREAKFAST	3:00		8:00	
8:00		4:00		9:00	
9:00		5:00		10:00	
10:00				NIGHT	
11:00				11:00-6:00	
12:00	LUNCH				

DAILY SCHEDULE					
DAY					
DATE					
NAME					
TIME MORNING	TASK	AFTERNOON	TASK	EVENING	TASK
6:00		1:00		6:00	DINNER
		2:00		7:00	
7:00	BREAKFAST	3:00		8:00	
8:00		4:00		9:00	
9:00		5:00		10:00	
10:00				NIGHT	
11:00				11:00-6:00	
12:00	LUNCH				

DAILY SCHEDULE FORMS

DAILY SCHEDULE					
DAY					
DATE					
NAME					
TIME					
MORNING	TASK	**AFTERNOON**	TASK	**EVENING**	TASK
6:00		1:00		6:00	DINNER
		2:00		7:00	
7:00	BREAKFAST	3:00		8:00	
8:00		4:00		9:00	
9:00		5:00		10:00	
10:00				**NIGHT**	
11:00				11:00-6:00	
12:00	LUNCH				

DAILY SCHEDULE					
DAY					
DATE					
NAME					
TIME					
MORNING	TASK	**AFTERNOON**	TASK	**EVENING**	TASK
6:00		1:00		6:00	DINNER
		2:00		7:00	
7:00	BREAKFAST	3:00		8:00	
8:00		4:00		9:00	
9:00		5:00		10:00	
10:00				**NIGHT**	
11:00				11:00-6:00	
12:00	LUNCH				

DAILY SCHEDULE					
DAY					
DATE					
NAME					
TIME					
MORNING	TASK	**AFTERNOON**	TASK	**EVENING**	TASK
6:00		1:00		6:00	DINNER
		2:00		7:00	
7:00	BREAKFAST	3:00		8:00	
8:00		4:00		9:00	
9:00		5:00		10:00	
10:00				**NIGHT**	
11:00				11:00-6:00	
12:00	LUNCH